The Marine Officer's Guide

The Marine Officer's Guide

Colonel Robert D. Heinl, Jr.,
U. S. Marine Corps (Ret.)

Foreword by
General Louis H. Wilson, Jr., Commandant,
U. S. Marine Corps

Fourth Edition

NAVAL INSTITUTE PRESS
Annapolis, Maryland
1977

Copyright © 1956, 1964, 1967, 1977
by United States Naval Institute
Annapolis, Maryland

Fourth Edition
Third printing, 1983

Library of Congress
Catalogue Card No. 77-16730
ISBN: 0-87021-366-0
Printed in the United States of America

First Edition by

General Gerald C. Thomas, U. S. Marine Corps (Retired)
Colonel Robert D. Heinl, Jr., U. S. Marine Corps (Retired)
Rear Admiral Arthur A. Ageton, U. S. Navy (Retired)

Second and Third Editions revised by

Colonel Robert D. Heinl, Jr., U. S. Marine Corps (Retired)
Rear Admiral Arthur A. Ageton, U. S. Navy (Retired)

*To All Young Marine Officers,
Into Whose Keeping The Corps
Is Year By Year Entrusted*

"Once a Marine, always a Marine . . ."

Foreword

It is now over twenty years since *The Marine Officer's Guide* first appeared. In that time many things have happened to the Nation and to the Marine Corps including the fighting of a major war. Thousands of American men and hundreds of American women have passed through our commissioned ranks. The lieutenants and captains of 1956 are the lieutenant colonels and colonels of 1977. Undoubtedly many still have their copies of the *Guide*, well-thumbed, I am sure, with use.

I can say this with confidence because my own copy of the first edition has been much used. This is the edition which was in print while I was the Commanding Officer of The Basic School in 1960 and 1961. In those days, as now, the *Guide* was regarded as an invaluable, if unofficial, text book for second lieutenants and other newly-commissioned officers. As I compare the first edition with the new fourth edition, I am struck not so much by things that have changed in the Corps as by the things that have stayed the same. I commend to the reader's particular attention those portions of the *Guide* that deal with the traditions, customs, and history of our Service. This is what binds us together and does much to ensure the continuity of the Corps.

A remarkable amount of advice and information has been concentrated in this volume. It is easy to get at. The arrangement is logical and the indexing is good. One of the further values of the book is that it not only advises and informs, but also points the reader in the direction of more detailed, possibly more timely, information. The purpose and the substance of the book fit very nicely the title: *The Marine Officer's Guide*.

LOUIS H. WILSON
General, U. S. Marine Corps
Commandant of the Marine Corps

BY THE SAME AUTHOR:
　The Defense of Wake
　Marines at Midway
　Soldiers of the Sea
　A Dictionary of Military and Naval Quotations
　Victory at High Tide
　Handbook for Marine NCOs

CO-AUTHOR:
　The Marshalls: Increasing the Tempo
　Written in Blood

Preface to the Fourth Edition

"Here's health to you, and to our Corps, which we are proud to serve..."

This fourth edition of *The Marine Officer's Guide* marks the end of two decades since this work first appeared. For the Marine Corps, as for the Department of Defense, these past 21 years have been years of change and upheaval and professional activity.

To assist new officers to learn the ropes as quickly as they can, to digest for all readers the continuing changes which have beset the Defense Establishment, and to help the entire officer corps keep professionally up to date, are the main aims of this new edition.

As in earlier years, however, the *Guide* holds to its standing objectives:

First, to introduce the Marine Corps to potential officers— that is, to describe the Corps, to give a general picture of the life and duties of a Marine officer, and to explain how one becomes an officer.

Second, to guide and advise inexperienced officers, and to provide a stockpile of information about their new career—as one older officer remarked on seeing the first edition, "To include in a book the things you spend your first four or five years finding out."

Third, to provide a source of Marine Corps reference material, at officer level, for everyone—regular or reserve, soldier or civilian.

To achieve these objectives I have tried to tell what the Marine Corps is and does, and to describe the Defense and Navy Department framework within which the Corps functions. I have outlined the history and tradition of the Marine Corps, as well as the most important social customs and usages among Marine officers and their families. Finally—and perhaps most important—I have endeavored to provide essential information and sound guidance on professional matters which affect the individual Marine officer, his career, and his performance of duty.

As always, I ask all readers to remember that this work is a *guide*. It does not and cannot present complete texts, commas, provisos, and whereases of regulations and instructions. Nor does it confine itself only to matters dealt with in official publications. Indeed, I have done everything possible to describe service customs and usage in matters not covered by "the book."

Although the opinions and assertions contained in this *Guide* are those of the author and are not to be construed as official or as reflecting the views of the Navy Department or of the Naval Services at large, no effort has been spared to make this edition as correct and authoritative as such a work can be.

A word is in order here regarding illustrations. Readers occasionally complain that some photograph or other is "out of date"—old-style chevrons, M-1 rifles, blue cap-covers, whatever. Optimally, every picture in the book should be fully current, but this is not always possible. Moreover, the function of a picture is to make or illustrate a point, not to display uniforms or equipment, which are mostly incidental to the intended purpose of the artwork. Since this *Guide* first appeared, for example, the Corps has gone through three families of rifles, and the Navy has abolished traditional sailor-suits (thereby rendering obsolete nine tenths of scenes showing Navy enlisted men). A further example, closer home, is the Marine Corps shift to camouflaged utilities, which dates the bulk of existing pictures of utility-clad Marines. Thus, many useful pictures are in one detail or another out of date. If a more current substitute cannot be found, this and future editions will use an old picture while looking constantly for a newer one.

While the proven organization of earlier editions has been adhered to, Chapters 2, 3, and 5 have again been largely rewritten to reflect continuing changes in the defense organization at all levels. Chapter 23 contains much new material in the ever-changing field of personal affairs. Appendix V, dealing with foreign Marines, has been greatly expanded. All other chapters and appendices have been thoroughly edited and updated as of this printing. However, the Marine Corps is a dynamic organization, and changes are constantly being made, outdating some information between editions.

A work of this kind depends particularly on the willingness to help and advise on the part of many. The staffs at Headquarters Marine Corps and Marine Corps Development and Education Command have rendered invaluable assistance, as have the respective information directorates of all the Armed Services.

At the top of the list of individuals who have contributed to this edition, by review, criticism, cooperation, and in a myriad of other ways, stands General Louis H. Wilson, Jr., Commandant of the Marine Corps, followed closely by Lieutenant Generals L. E. Brown, J. S. Fegan, Jr., J. N. McLaughlin, and R. L. Nichols; Major Generals C. W. Hoffman and A. J. Poillon; Brigadier Generals J. K. Davis and E. H. Simmons; Colonels M. A. Brewer, A. J. Croft, J. R. Dopler, M. C. Jones, Jr., M. D. Julian, J. P. King, R. D. Miller, and R. D. Revie; Lieutenant Colonels H. V. Bucknam, H. C. Campbell, P. J. Canzano, C. C. Seabrook, and W. R. Silby; Majors H. T. Hayden, M. L. Hefti, D. G. Henderson, F. X. Lawler, N. E. Pridgen, R. J. Squires, M. T. Warring, R. D. Woidyla, and R. C. Young, Jr.; Captains C. A. Braley, J. T. Conway, and J. V. DiBernardo, and Captain P. R. Hefler, USAF; Chief Warrant Officer M. H. Handelsman; Warrant Officer M. D. Labonne; Mr. H. A. Bartholomew, Mr. Peder Gustawson, and Ms. Anna C. Urband.

Indexing is by Benis Frank.

Finally, I welcome the comments, suggestions, and corrections of every reader. Is the *Guide* correct? Is it complete? Does it meet your needs? Does it answer your questions or show where to find answers? swers?

Thus, if you have corrections or suggestions, send them to the author, in care of the Naval Institute, Annapolis, Maryland, 21402. Only with your interest and assistance can *The Marine Officer's Guide* continue to meet the needs of its readers, and thereby best serve our Corps.

R. D. HEINL, Jr.

Washington, D.C.
November 1977

Contents

	vii	Foreword
	ix	Preface to Fourth Edition
	xv	Introduction to First Edition
One	3	The Marine Corps
Two	9	The Organization for National Security
Three	37	The Department of the Navy
Four	73	Missions and Status of the Marine Corps
Five	85	Organization of the Marine Corps
Six	117	The Story of the Marine Corps
Seven	153	Traditions, Flags, Decorations, and Uniforms
Eight	191	Posts and Stations
Nine	223	The Marine Corps Reserve
Ten	241	Women Marines
Eleven	251	You Become a Marine Officer
Twelve	271	Officers' Individual Administration
Thirteen	305	Pay, Allowances, and Official Travel
Fourteen	329	New Station
Fifteen	347	Managing Your Career
Sixteen	367	Leadership
Seventeen	389	On Watch
Eighteen	407	Military Courtesy, Honors, and Ceremonies
Nineteen	467	Notes on Military Justice
Twenty	497	Housekeeping
Twenty-one	509	Life in the Field
Twenty-two	525	Service Afloat
Twenty-three	591	Personal Affairs
Twenty-four	631	Marine Corps Social Life

Appendices

660	The Marines' Hymn
661	Commandants of the Marine Corps
662	The Importance of Being Inspected
667	Reading for Marines
669	Brother Marines
676	Art. 38, Marine Corps Manual, 1921
677	Birthday Ball Ceremony
681	Glossary of Marine Corps Terms
695	Index

Figures and Tables

10	Organization for National Security
23	Organization of the Department of the Army
31	Organization of the Department of the Air Force
40	Primary functions of the Navy
42	Organization of the Department of the Navy
61	Branch and corps devices of commissioned and warrant officers of the Navy
64	Specialty marks of Navy enlisted ratings
86	Organization of the Marine Corps
88	Organization of Marine Corps Headquarters
94	Organization of a Marine division
94	Typical components of Force Troops
95	Organization of a typical Marine aircraft wing
97	Typical Marine air-ground task forces
184–185	Types and combinations of uniforms for male officers
192	How a typical post or base is organized
252–254	Avenues to a career as a Marine officer
281	Distribution of commissioned officers in grades
299	Leave computation
310	Gates/flight pay entitlement
349	Typical assignment patterns for ground and aviation officers
393	Organization of a typical interior guard
409–410	Commissioned insignia of rank
411	Enlisted insignia of rank
414–415	Correct forms of address for naval and military personnel
420	Hand salute
422	Sword salutes and manual
426	Group saluting

	436	Rigging your sword
	437	Nomenclature of the sword, scabbard, and sling
	439	Correct ways in which to display the flag
	442	Folding the flag
	450–452	Table of honors for official visits
	470	Administration of military justice under UCMJ
	472	Disposition of a case under UCMJ
	478	Disposition of an Office Hours' case
	486–487	Court-martial punishments
	510	Typical nontactical camp layout
	530	Typical organization of a large ship
	551	When embarking, junior officers board first
	552	When disembarking, senior officers are first to leave the boat
	553	The coxswain and the senior officer in the boat render hand salutes

Photographs, unless otherwise indicated, are Department of Defense, U. S. Marine Corps, or U. S. Navy official photographs. Diagrams, where not original, have been reproduced from various official publications.

Introduction to the First Edition

Preparation of this book was a project which began during my tenure as Commandant of the Marine Corps, and it was one to which I lent sympathetic attention. Now that I see the results, I am well satisfied that I helped to the extent that I did.

It is high time that the Marine Corps had a work of this kind.

Within the memory not of a few senior officers on the verge of retirement, but of the bulk of our most experienced field officers, the Corps has expanded immensely and has pursued its traditional role of national force in readiness on a vastly larger stage.

Thus the Marine Corps is—or could be—in a time of transition. At such times it is all too easy to forget, depart from, or discard the well tested ways which have brought us where we are. Fortunately, those ways are still with us, and such a book as *The Marine Officer's Guide* must be of the greatest value in keeping them with us.

As I write this to readers of *The Marine Officer's Guide*, I am reminded of what must be one of the earliest surviving "fitness reports" on a young Marine officer, submitted by Captain Daniel Carmick, USMC, in April 1799: "Lt Amory is very ignorant of Military duty, as he acknowledges, but he is a smart Gentleman and far preferable to the others." For the young officer of today who (like Lt Amory) is not ashamed to admit the limits of his own experience, and is intelligent enough to profit by the experience of others, *The Marine Officer's Guide* should prove indispensable.

—L. C. Shepherd, Jr.
   General, U. S. Marine Corps (Ret.)

1

The Marine Corps

First to fight....
Retreat, hell! We just got here....
Gone to fight the Indians—will be back when the war is over....
Uncommon valor was a common virtue....
The Marines have landed, and the situation is well in hand....

Phrases like these say more about the United States Marine Corps than all the handbooks ever written. As you read this *Guide*, therefore, remember that there is far more to the Marine Corps than can ever be expressed in any manual. If you are fortunate enough to become a Marine, you will soon realize what the Corps is and what it stands for.

101 ■ What Is the Marine Corps?

Beyond the statutes and official definitions, what is the Marine Corps?

To begin with, the Marine Corps is a military anomaly—"Soldier and sailor, too."

Every world power has an army. Most powers have navies and air forces. However, only 29 countries (Argentina, Brazil, Chile, Colombia, Denmark, Great Britain, Greece, Indonesia, Iran, Republic of Korea, China, the Netherlands, Pakistan, Peru, Poland, East Germany, Spain, Taiwan, Thailand, Turkey, the USSR, Venezuela, France, Italy, Portugal, Mexico, Nigeria, the Philippines, and the United States) maintain Marine Corps.

In addition, certain other nations maintain units or forces whose functions, primarily in the field of raiding and other light amphibious operations, approximate those of Marines. Among such countries are Albania, Bulgaria, Burma, Ecuador, Israel, Yugoslavia, Paraguay, Romania, and Sweden.

Nowhere but in the United States, however, has any Marine Corps attained the status of our own. This status was not foreseen when the Continental Congress, on 10 November 1775, formed two battalions of Marines. Instead, the Corps gained its unique position through long evolution.

Much of the anomalous quality of the Marine Corps stems from the fact that the Corps possesses many individual attributes of its brother Services. As a result, you can usually discern something suggestive of the other Services in the Marine Corps, and this is only natural in a Corps which has spent most of its time spearheading, supporting, or serving beside the Army, Navy, and Air Force. But you can also see much more which belongs only to the Marine Corps.

Certainly the Marine Corps attitude is peculiar to the Corps.

Fully as important as its attitude, however, is the fundamental mission of the Corps. This primary mission—*readiness*—combined with the Marine state of mind, makes the Corps what it is today: a national force in readiness, prepared in fact, and required by law, to "perform such other duties as the President may direct"—which means "ready for anything."

Most Americans, including some who know little specific about the Corps, recognize Marines as the national force in readiness. Such tried and true phrases as "Call out the Marines!" or "The situation is well in hand," or "Tell it to the Marines," have entered American speech, and voice the country's attitude.

The existence of this nationwide feeling makes the Marine Corps a national institution.

As a Marine, you represent a national institution whose standing and reputation are in your hands.

102 ■ What the Marine Corps Stands For

The qualities that the Marine Corps stands for may seem old-fashioned. Nevertheless these attributes have shaped the Corps since 1775, from Princeton to Belleau Wood, from Trenton to Chosin Reservoir and Khe Sanh.

Here are things that the Corps stands for:

Quality and Competence. A Marine has to be good. In the Marine Corps, your best is just the acceptable minimum. It is expected, as a matter of course, that the technical performance of a single Marine or a whole Marine outfit, whether on parade or in the attack, will be outstanding.

Discipline. Of all the principles of the Marine Corps, its insistence on discipline is the most unvarying and most uncompromising.

Valor. After the seizure of Iwo Jima, Fleet Admiral Nimitz epitomized the performance of the Marines who took the island. "Uncommon valor," wrote the admiral, "was a common virtue." Three hundred Medals of Honor have been awarded to U. S. Marines. Valor and courage are the Marines' stock in trade. "Retreat, hell! We just got here," was originally uttered in 1918 by a company commander of U. S. Marines.

Esprit de corps. A Marine is intensely proud of his Corps, loyal to his comrades, and jealous for the good name of the Corps. This spirit is nowhere better expressed than in a letter, written in 1800, from William Ward Burrows, 2d Commandant of the Marine Corps, to a junior Marine officer who had been insulted by an officer in the Navy:

> Camp at Washing., Sept 22, 1800
>
> Lt. Henry Caldwell,
> Sir—
>
> When I answer'd your letter, I did not Know what Injuries you had received on board the *Trumbull*. . . . Yesterday the Secretary told me, that he understood one of the Lieutenants of the Navy had struck you. I lament that the Capt. of yr ship cannot Keep Order on board of her. . . . As to yourself I can only say, that a Blow ought never to be forgiven, and without you wipe away this Insult offer'd to the Marine Corps, you cannot expect to join our Officers.
>
> I have permitted you to leave the Ship . . . that you may be on an equal Footing with the Captain, or any one who dare insult you, or the Corps. I have wrote to Capt. Carmick, who is at Boston to call on you & be your Friend. He is a Man of Spirit, and will take care of you, but don't let me see you 'till you have wip'd away this Disgrace. It is my Duty to support my Officers and I will do it with my Life, but they must deserve it.
>
> On board the *Ganges*, about 12 mos. ago, Lt. Gale was struck by an Officer of the Navy, the Capt. took no notice of the Business, and Gale got no satisfaction on the Cruise: The moment he arrived he call'd the Lieut. out, and shot him; afterwards Politeness was restor'd. . . .
>
> Yr obdt Svt,
> W. W. BURROWS,
> LtCol Comdt, MC

Pride. Every Marine is intensely proud of Corps and Country and does his utmost to build and uphold a Corps in which a good man can take pride.

Loyalty and Faithfulness. "Semper Fidelis" ("Always Faithful") is the motto of the Corps. In addition, every honorable discharge certificate from the Marine Corps bears the phrase *"Fideli certa merces"* ("A sure reward to the faithful"). Marines know enough Latin to understand that these are not idle words. Absolute loyalty to the Corps, as well as devotion to duty, are required of every Marine. Percentages of Marines missing in action or taken prisoner by the enemy are minute. A good Marine places the interests of the Marine Corps at the top of his list.

The Individual. The Marine Corps cherishes the individuality of its members, and although sternly consecrated to discipline, has cheerfully sheltered a legion of nonconformists, flamboyant individuals, and irradiant personalities. It is a perennial prediction that colorful characters are about to vanish from the Corps. They never have, and never will. No Marine need fear that the mass will ever absorb the man.

The Volunteer. Despite occasional acceptance of drafted men in times of peak manpower demand, as in the Vietnam war, the Marine Corps is "a volunteer outfit." The Corps relies on men who *want to be Marines*. There is no substitute. In the old phrase, "One volunteer is worth ten pressed men."

The Infantry. The Corps is unique in that, no matter what military specialty, ground or aviation, he eventually pursues, every Marine is trained as a rifleman, and every officer must in addition be morally and professionally prepared to function as an infantry officer.

Relations between Officers and Men.

> The relation between officers and enlisted men should in no sense be that of superior and inferior nor that of master and servant, but rather that of teacher and scholar. In fact, it should partake of the nature of the relation between father and son, to the extent that officers, especially commanding officers, are responsible for the physical, mental, and moral welfare, as well as the discipline and military training, of the young men under their command. . . .

Thus wrote a Commandant of the Marine Corps. His words now stand as an enduring testimony to the comradeship and brotherhood among all Marines, whether officer or enlisted.

Traditions. St. Paul's injunction, "Hold the traditions which ye have been taught," could be a Marine motto. Respect for the traditions of the Corps is deeply felt. Every Marine adheres to the traditions that have shaped the Corps.

Professionalism. The U. S. Marine is a professional who stands ready to fight any enemy, any time, anywhere, whom the President or Congress may designate, and to do so coolly, capably, and in a spirit of professional detachment. He is not trained to hate, nor is he whipped up emotionally for battle or for any other duty the Corps may be called on to perform. Patriotism and professionalism are his only two "isms."

103 ■ A Commandant Writes to His Officers

Years ago, Major General John A. Lejeune, 13th Commandant of the Marine Corps, opened his heart to his officers in a collective letter.

Anyone who wants to know what the Marine Corps is and stands for need look no further than General Lejeune's "Letter No. 1":

> TO THE OFFICERS OF THE MARINE CORPS:
>
> I feel that I would like to talk to each of you personally. This, of course, it is impossible for me to do. Consequently, I am going to do the next best thing, by writing letters from time to time which will go to all the officers. In these letters, I will endeavor to embody briefly some of the thoughts which have come into my mind concerning our beloved Corps.
>
> In the first place, I want each of you to feel that the Commandant of the Corps is your friend and that he earnestly desires that you should realize this. At the same time, it is his duty to the Government and to the Marine Corps to exact a high standard of conduct, a strict performance of duty, and a rigid compliance with orders on the part of all the officers.
>
> You are the permanent part of the Marine Corps, and the efficiency, the good name, and the esprit of the Corps are in your hands. You can make or mar it.

You should never forget the power of example. The young men serving as enlisted men take their cue from you. If you conduct yourselves at all times as officers and gentlemen should conduct themselves, the moral tone of the whole Corps will be raised, its reputation, which is most precious to all of us, will be enhanced, and the esteem and affection in which the Corps is held by the American people will be increased.

Be kindly and just in your dealings with your men. Never play favorites. Make them feel that justice tempered with mercy may always be counted on. This does not mean a slackening of discipline. Obedience to orders and regulations must always be insisted on, and good conduct on the part of the men exacted. Especially should this be done with reference to the civilian inhabitants of foreign countries in which Marines are serving.

The prestige of the Marine Corps depends greatly on the appearance of its officers and men. Officers should adhere closely to the Uniform Regulations, and be exceedingly careful to be neatly and tidily dressed, and to carry themselves in a military manner. They should observe the appearance of men while on liberty, and should endeavor to instill into their minds the importance of neatness, smartness, and soldierly bearing.

A compliance with the minutiae of military courtesy is a mark of well disciplined troops. The exchange of military salutes between officers and men should not be overlooked. Its omission indicates a poor state of discipline. Similarly, officers should be equally careful to salute each other. Courtesy, too, demands more than an exchange of official salutes between officers. On all occasions when officers are gathered together, juniors should show their esteem and respect for their seniors by taking the initiative in speaking to and shaking hands with their seniors. Particularly should this be done in the case of commanding officers. The older officers appreciate greatly attention and friendliness on the part of younger officers.

We are all members of the same great family, and we should invariably show courtesy and consideration, not only to other officers, but to members of their personal families as well. Do not fail to call on your commanding officers within a week after you join a post. On social occasions the formality with which all of us conduct ourselves should be relaxed, and a spirit of friendliness and good will should prevail.

In conclusion, I wish to impress on all of you that the destiny of our Corps depends on each of you. Our forces, brigades, regiments, battalions, companies, and other detachments are what you make them. An inefficient organization is the product of inefficient officers, and all discreditable occurrences are usually due to the failure of officers to perform their duties properly. Harmonious cooperation and teamwork, together with an intelligent and energetic performance of duty, are essential to success, and these attributes can be attained only by cultivating in your character the qualities of loyalty, unselfishness, devotion to duty, and the highest sense of honor.

Let each one of us resolve to show in himself a good example of virtue, honor, patriotism, and subordination, and to do all in his power, not only to maintain, but to increase the prestige, the efficiency, and the esprit of the grand old Corps to which we belong.

With my best wishes for your success and happiness, I am, as always,

Your sincere friend,
JOHN A. LEJEUNE,
Major General Commandant.

I have just returned from visiting the Marines at the front, and there is not a finer fighting organization in the world.
—General of the Army Douglas MacArthur
21 September 1950

2

The Organization for National Security

201 ■ The Executive Office of the President

At the highest level in the Federal Government stands the President of the United States, who is, under our Constitution, Commander in Chief of the Armed Forces. The President is assisted and advised not only by the Cabinet but by several agencies which have been organized into The Executive Office of the President. Figure 2-1 shows the flow of authority and direction through the Organization for National Security from the President as Commander in Chief to the Secretary of Defense and the three military departments.

Two of these agencies are of particular importance to the Armed Services, and a third, the Central Intelligence Agency, directly under the National Security Council, must also be considered on this high level.

202 ■ The National Security Council (NSC)

Established by the National Security Act as amended, the Council has as members: the President, Vice-President, Secretary of State, and Secretary of Defense. The act provides that the Secretaries and Under Secretaries of other executive departments and of the military departments may serve as members of the Council, when appointed by the President with the advice and consent of the Senate. In addition, other Government officials attend meetings as "standing request members" (e.g., Director of the Office of Management and Budget) or on an *ad hoc* basis. The Council Staff is headed by a civilian executive secretary appointed by the President; the staff includes officers and civilian officials from the Departments of State and Defense and the four military

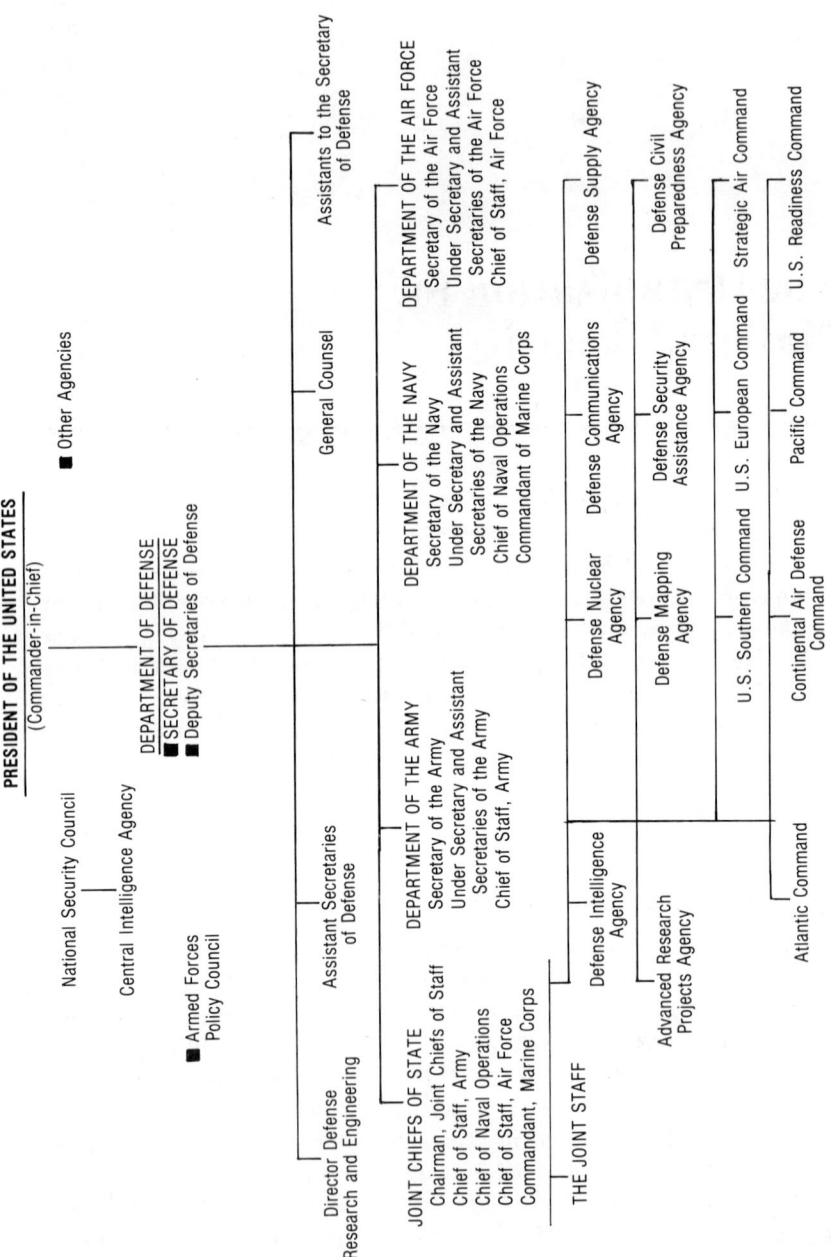

Fig. 2-1: Organization for National Security (1977).

Services. This secretariat conducts the routine business of the Council, prepares agenda for meetings, and correlates and presents for consideration a broad range of information on pertinent topics assembled by the agencies of the membership and by the Central Intelligence Agency.

The Council's function is to advise the President on domestic, foreign, and military policies and problems relating to national security, so as to enable the military Services and other departments and agencies of the Government to cooperate more effectively in matters involving national security. The duties of the Council include the following: to assess and appraise the objectives, commitments, and risks of the United States in relation to the actual and potential military power of the nation; to consider policies on matters of common interest to the departments and agencies of the Government concerned with national security; and to make recommendations to the President on subjects which may affect the national security policies of the Government.

203 ■ The Central Intelligence Agency (CIA)

This agency is administered under direction of the National Security Council by a Director appointed by the President with the advice and consent of the Senate. The Director (and his Deputy) may be either a military officer or a civilian; if an officer, while so serving he retains his Service grade and status (being carried as an extra number in grade) but is otherwise separated and exempt from normal military and administrative responsibilities.

The Agency was established to coordinate all intelligence activities of the Government. To accomplish this, CIA advises the National Security Council concerning intelligence activities of the Government which relate to national security; it makes recommendations to the National Security Council for coordination of these intelligence activities; it correlates and evaluates intelligence, and disseminates such intelligence within the government—using, where appropriate, existing agencies; it performs additional intelligence services of common concern which the National Security Council determines can be more efficiently accomplished centrally; and it performs such other functions related to intelligence as the National Security Council may direct. In carrying out the foregoing functions, one of the principal continuing tasks of the CIA is to prepare "National Intelligence Estimates," strategic intelligence on which policy decisions are based.

The CIA does not possess police, subpoena, or law-enforcement powers, or internal-security functions. The Director is responsible for protecting intelligence sources and methods from unauthorized disclosure.

204 ■ Office of Management and Budget (OMB)

Established in 1921, this office is administered by a Director. The functions of OMB are: to help the President prepare the budget and formulate the fiscal program of the Government; to supervise administration of the budget; to improve Government administrative management; to help the President bring about more efficient and economical conduct of Government service; to clear and coordinate departmental advice on proposed legislation and make recommendations as to Presidential action on legislative enactments; to assist in consideration, clearance, and preparation of Executive Orders and proclamations; to improve, develop, and coordinate Federal and other statistical services; and to inform the President of the progress of Government work proposed, initiated and completed.

From the foregoing, it will be seen that OMB is in reality far more than its title or even its mechanical functions suggest, and is in fact a civilian general staff for the President. The power of the office—the ultimate power of the purse—and the continuity of its work give OMB considerable external influence over the Defense Department and the defense policies of the Government.

205 ■ Independent Agencies

Among the numerous independent offices and establishments under direct control of the Executive are several which are of special importance in connection with national security:

Energy Research and Development Agency (ERDA) administers nuclear research and development, international cooperation, production of atomic energy and special nuclear materials. The Military Liaison Committee of ERDA, made up of representatives of Departments of the Army, Navy, and Air Force, works with the Defense Department on all military applications of atomic energy.

National Aeronautics and Space Administration (NASA) deals with problems of flight in space and in the earth's atmosphere, develops and operates space vehicles, and is charged with the exploration of space.

Selective Service System provides nationwide standby machinery for the registration and military induction—in other words, the draft—of individuals with obligated military service. As previously noted, the Marine Corps ordinarily relies on volunteers, but has been forced on certain occasions by national policy to accept draftees in times of war.

Veterans Administration (VA) administers all laws authorizing benefits for former members of the Armed Forces and their dependent beneficiaries, together with all Government insurance programs open to members of the Armed Forces. For benefits and services of the Veterans Administration, see Chapter 23.

206 ■ National Policy and National Objectives

Matters affecting national policy arise in one of three ways: within policy-making agencies of the Government; with the President or his immediate staff; or outside the Government. Such matters are customarily referred to the National Security Council, by either the President or one of the heads of departments or agencies of the Government; the Council secretariat passes the issue to members for expression of views. Among officials who have important effect upon national-policy matters are the Secretary of State, Secretary of Defense, Director of Central Intelligence, Chairman of ERDA, Joint Chiefs of Staff, and, ultimately, the National Security Council and the President. The chairmen of the Senate Committee on Foreign Affairs and the House Committee on Foreign Relations, Armed Services, Appropriations, Ways and Means, and Government Operations, as well as the senior minority members of those committees, also exercise considerable influence.

After consulting with his advisers and the National Security Council, the President makes his decision and publishes the policy to the American people and foreign nations through the press, public speeches, and White Papers. National policy as such is rarely collected and documented for public consumption; in fact, some matters of policy are considered Top Secret. Statements of policy which cannot be publicized are issued to interested Government agencies in classified form.

The national objectives of the United States materially affect the military departments. While never officially promulgated as such, the statement of national objectives that follows has been compiled from the speeches and other pronouncements of high Government officials:

> Our prime national objective is the desire of our people for peace and security, without sacrifice of either the rights of the individual or the present sovereignty we cherish.
>
> While we intend to maintain our political way of life and our form of government in our own country, our national objectives do not demand a similar political way of life or a similar form of government in other countries of the world.
>
> We intend to maintain, and to improve, if possible, our American standard of living.
>
> We desire freedom, peace, and security for the entire world, and all the benefits that these conditions can bring to all peoples everywhere.
>
> We will seek and work for an effective world organization under the United Nations, for we believe that world peace is an inseparable part of American peace.
>
> Ultimately, we hope to eliminate any sort of warfare as a means for the resolution of international disputes.
>
> Finally, if war be forced upon us, we intend to win that war, and win it in such a way that is can be followed by a stable, livable, and long-lasting peace.

THE DEPARTMENT OF DEFENSE

The Department of Defense is the largest single agency in the Government of the United States. It spends approximately 25 percent of the national budget in an ordinary fiscal year (being substantially outspent only by Health, Education, and Welfare). In the three decades since its creation, the Office of the Secretary of Defense has mushroomed from a handful of policy-makers (in 1949 the Secretary of Defense had only three special assistants) to one of the major bureaucracies of the Government, staffed by thousands of officers, enlisted men, and civilian employees.

The Defense Department includes the Office of the Secretary of Defense (OSD), the Joint Chiefs of Staff (JCS) and their supporting establishment, the Departments of the Army, Navy, and Air Force, and the four military Services (Army, Marine Corps, Navy, and Air Force) within those Departments, the unified and specified commands, and such other agencies as the Secretary of Defense establishes to meet specific requirements. The central function of the Department of Defense is to

The Pentagon, headquarters of the Department of Defense, is the world's largest office building, with three times more floor space than the Empire State Building, 32,000 employees, and 17½ miles of corridors.

provide for the military security of the United States and to support and advance the national policies and interests of the United States.

207 ■ The National Security Act

The National Security Act of 1947, as amended, is the controlling military legislation of the United States. The policy section of the Act reads: "It is the intent of Congress to provide a comprehensive program for the future security of the United States; to provide for the establishment of integrated policies and procedures for the departments, agencies, and functions of the Government relating to the national security." In so doing, the Act:

1. Provides three military departments, separately organized, for the operation and administration of the Army, the Navy (including naval aviation), the United States Marine Corps, and the Air Force, with their assigned combatant and service components;
2. Provides for coordination and direction of the three military departments and four Services under the Secretary of Defense;
3. Provides for strategic direction of the Armed Forces, for their operation under unified control, for establishment of unified and specified commands, and for the integration of the four Services into an efficient team of land, naval, and air forces, but does not establish a single Chief of Staff over the Armed Forces or an Armed Forces general staff.

Unification has been accomplished by giving the Secretary of Defense authority, direction, and virtual military control over the four military Services, although he does not administer the Departments of the Army, Navy, and Air Force. He also has authority to eliminate unnecessary duplication in procurement, supply, transportation, storage, health, and research and engineering. His greatest power of persuasion and control lies in administration of the military budgets of the Department of Defense.

The Secretaries of Army, Navy, and Air Force no longer enjoy Cabinet status, but each Secretary has the right to appeal any matter directly to the President, or make representations directly to the Office of Management and Budget or to Congress; however, he must first inform the Secretary of Defense of his intention to do so.

208 ■ Office of the Secretary

The Secretary of Defense, principal assistant to the President in all matters relating to the Department of Defense, is appointed from civil life by the President with the advice and consent of the Senate. Under the President, the Secretary exercises direction, authority, and control over the Department (see Fig. 2–1). He is a member of the National Security Council, the National Aeronautics and Space Council, and the North Atlantic Council.

The Deputy Secretary of Defense is responsible for supervision and coordination of the activities of the Department.

The Armed Forces Policy Council (AFPC) advises the Secretary of Defense on matters of broad policy relating to the Armed Forces and sometimes serves as a final clearing house or court of appeal for major administrative decisions. The members are: Secretary of Defense (Chairman); Deputy Secretary of Defense; Secretaries of the Army, Navy, and Air Force; Director of Defense Research and Engineering; Chairman of the JCS; Chief of Staff of the Army; Chief of Naval Operations; Chief of Staff of the Air Force. The Commandant of the Marine Corps, although not a Council member by statute, regularly attends and participates in meetings.

209 ■ The Office of the Secretary of Defense

Various agencies, offices, and positions created by the National Security Act, together with certain other agencies that assist the Secretary of Defense, constitute the primary staff—civil and military—of the Secretary of Defense on matters within their cognizance.

Director of Defense Research and Engineering is the principal adviser and staff assistant to the Secretary of Defense in scientific and technical matters. He supervises all research and engineering activities in the Defense Department and wields extensive coordinating and directive authority over virtually all materiel programs of the Defense Establishment.

The Assistant Secretaries of Defense, including the Comptroller and General Counsel, and certain special assistants, are responsible to the Secretary for designated functional areas of staff cognizance. In 1977, there were 12 such, embracing fields such as health, legislative affairs, program analysis, nuclear matters, installations and logistics, international security affairs, manpower and reserve, public affairs, and telecommunications (including command and control systems).

These and similar responsibilities and functions are periodically rearranged by the Secretary according to existing requirements.

210 ■ The Joint Chiefs of Staff (JCS)

To promote more personal control of the Army and Navy, to make certain of direct access to the President for his principal military advisers, and to improve coordination between the Army and Navy, President

Franklin D. Roosevelt directed organization of the Joint Chiefs of Staff in 1942, with the Chief of Staff of the Army (General Marshall) and the Chief of Naval Operations (Admiral King) as its first members. Later, Lieutenant General H. H. Arnold, Chief of the Army Air Corps, and the President's Chief of Staff, Admiral William D. Leahy, were added. In 1943, President Roosevelt proposed General Holcomb, then Commandant of the Marine Corps, as a member; this move was opposed by the Army and the Chief of Naval Operations on the ground that it would make the JCS too large and unwieldy, and there the matter rested.

The Joint Chiefs formed the American delegation in the Combined Chiefs of Staff, an Anglo-American agency consisting of the Joint Chiefs of Staff and four British flag and general officers.

During World War II, the Joint Chiefs planned and directed the broad strategy of the war. In his special wartime capacity as Chief of Staff to the Commander in Chief, Admiral Leahy kept the other Chiefs advised as to the President's wishes, and reported on their meetings to the President. While there were some disagreements, under the pressure of war unanimous decisions were nearly always reached, leaving only a few issues to be resolved by the President. With the coming of peace and passage of the National Security Act in 1947, the JCS became a permanent part of the defense organization of the United States. When the National Security Act was amended in 1949, a Chairman was authorized who would preside at JCS meetings and expedite the conduct of business; but the Chairman is not to be considered Chief of Staff to either the Secretary of Defense or the President. At the request of President Truman, before enactment of the 1949 National Security Act amendments, General Dwight D. Eisenhower served informally as first Chairman of the JCS.

Thus the Joint Chiefs of Staff now consist of a Chairman appointed by the President from one of the four Services, with the advice and consent of the Senate, the Chiefs of Staff of the Army and Air Force, and the Chief of Naval Operations. When a matter which concerns the Marine Corps is under consideration, the Commandant of the Marine Corps sits with the Joint Chiefs, and on such occasions enjoys co-equal status with the other members.

Perhaps it would be well to explain that the Joint Chiefs do not in fact operate on a voting basis. If, after discussing an issue, the Joint Chiefs agree unanimously, all Chiefs and the Chairman sign (or, as the process is called, "red-band") a paper giving to the Secretary of Defense and the Services their decision. But if, after all views are presented, disagreement remains, the Joint Chiefs come to what is known as a "split." It is the statutory duty of the Chairman to inform the Secretary of Defense and the President of such disagreement. Each Chief writes his own recommendation to the Secretary of Defense. The Chairman forwards these opinions to the Secretary of Defense and, if appropriate, the President, usually with a covering letter, whereupon the Secretary resolves the issue.

The Chairman, who takes precedence over all officers of the Armed Services, serves as presiding officer, provides agenda for meetings, and manages the Joint Staff.

As principal military advisers to the President, the National Security Council, and the Secretary of Defense, the Joint Chiefs prepare strategic plans and provide strategic direction of the military forces; they prepare joint logistic plans, and assign logistic responsibilities in accordance with such plans; they establish unified commands in strategic areas; they formulate policies for joint training of the military forces, and coordinate the education of members of the military forces; they review major material and personnel requirements of the military forces, in accordance with strategic and logistic plans; and they provide United States representation on the Military Staff Committee of the United Nations.

211 ■ Organization of the JCS

The supporting establishment of the Joint Chiefs of Staff comprises the Joint Staff and a group of other agencies outside the Joint Staff which report directly to the JCS.

The Joint Staff provides planning and staff assistance for the JCS. It is limited in size to 400 officers (having expanded four times over since it was created in 1947) who are chosen from the Army, Navy, Marine Corps, and Air Force. The Director, Joint Staff, an officer of three-star grade, attends meetings of the Joint Chiefs of Staff and serves in effect as the expediter/coordinator of the JCS organization. It is important to remember, however, that his scope of action and authority has been unequivocally limited by Congress to "authority, management, and supervision of the Joint Staff proper," and that he is not a quasi–Deputy Chairman of the JCS, as some erroneously assert. The Joint Staff is divided into directorates: Personnel (J–1), Operations (J–3), Logistics (J–4), Plans and Policy (J–5), Communications-Electronics (J–6); and various Special Assistants.

Other agencies within the JCS organization but not part of the Joint Staff include the Joint Secretariat, Special Assistant to the JCS for National Security Council Affairs, Special Assistant to the JCS for Disarmament, the Joint Meteorological Group, and the U. S. Representatives in various U.N., NATO, and inter-American military staff groups or agencies.

212 ■ Other Defense Agencies

Other than the unified and specified commands established by the JCS (see below), a number of major agencies and joint service schools come within the aegis of the Defense Department or Joint Chiefs of Staff. Certain of these agencies are of commanding size and stature and perform major functions for the Defense Department that were once considered to be within the operating and administrative purview of the military departments. Agencies included are:

Defense Nuclear Agency (DNA)
Defense Logistic Agency (DLA)
Defense Intelligence Agency (DIA)
Defense Communications Agency (DCA)
Defense Advanced Research Projects Agency
National Security Agency (NSA)
Joint Strategic Target Planning Staff

The Joint Service Schools are three in number, and all come under the Joint Chiefs of Staff, viz.: National Defense University; Industrial College of the Armed Forces; Armed Forces Staff College. In addition, the United States provides facilities and support for the Inter-American Defense College, which is located adjacent to the National Defense University. (See Section 1506.)

213 ■ Unified and Specified Commands

Coming directly under the Joint Chiefs of Staff are the unified and specified commands, predominantly located outside the United States and covering areas of greatest strategic importance. A *unified command* is a command with a broad continuing mission, under a single commander, composed of components of two or more Services. Representation on the commander's staff usually comes from all Services, and the command includes "Service component commanders" who command all units from their respective Services within the unified command. There are six unified commands, as follows:

Atlantic Command
Pacific Command
U. S. European Command
U. S. Southern Command
Continental Air Defense Command
Readiness Command

A specified command, like a unified command, has a broad continuing mission, but is normally composed of forces from but one Service. There are three specified commands: Strategic Air Command (SAC), Aerospace Defense Command (ADCOM), and Military Airlift Command (MAC).

For detailed, up-to-date information on the organization and functioning of the Joint Chiefs of Staff, its supporting organization, and the unified and specified command structure, consult *Organization and Functions of the Joint Chiefs of Staff* (JCS Publication 4) and *Unified Action Armed Forces* (UNAAF) (JCS Publication 2).

THE DEPARTMENT OF THE ARMY

214 ■ Mission of the Army

The National Security Act charges the Department of the Army with providing support for national and international policy and the security of the United States by planning, directing, and reviewing the operations of the Army Establishment to include the organization, training,

The Army, represented by this M60–A2 battle tank on the NATO front, is organized, trained, and equipped for prompt and sustained combat on land.

and equipping of land forces of the United States for prompt and sustained combat operations on land.

This mission is further interpreted and delineated in *Functions of the Department of Defense and Its Major Components*, a Defense Directive known colloquially as "The Functions Paper" or, from its origin in March 1948, as "The Key West Agreement." This document sets out functions of the Armed Forces in greater detail than, and as a supplement to, the National Security Act, which is of course controlling.

215 ■ The Structure of the Army

Command channels flow from the President, through the Secretary of Defense and the Secretary of the Army, to Army units and installations throughout the world.

A field army is composed of a headquarters, and two or more corps, each of two or more divisions. The division is the smallest unit which contains a balanced proportion of the combined arms and services and which therefore is constituted to operate independently. Below division level, units are mainly composed of the separate arms or services of the Army. The company is the smallest administrative unit in the Army.

The Army is made up of the following basic and special branches:

Basic branches	*Special branches*
Infantry	Judge Advocate General's Corps
Armor	Chaplains
Artillery	Army Medical Service:
Air Defense Artillery	Medical Corps
Corps of Engineers	Dental Corps
Signal Corps	Veterinary Corps
Adjutant General's Corps	Army Nurse Corps
Quartermaster Corps	Army Medical Specialist
Finance Corps	Corps
Ordnance Corps	
Chemical Corps	
Transportation Corps	
Military Police Corps	
Intelligence Corps	

Officers from all branches are detailed to the *General Staff Corps* and the *Inspector General's Department*.

216 ■ The Secretary of the Army and His Assistants

The Secretary of the Army heads the Department of the Army. He is responsible for all affairs of the Army Establishment. In addition, the

Secretary of the Army has certain quasi-civil functions such as: maintenance and operation of the Panama Canal and the Corps of Engineers' civil works program; supervision of U. S. battle monuments; direction of the civilian marksmanship program; and functions for the government of the District of Columbia and all aspects of the Federal Civil Defense Program assigned to the Department of Defense.

Immediately under the Secretary of the Army are the following assistants and major agencies:

Under Secretary of the Army
Assistant Secretaries of the Army (Financial Management, Research
    and Development, Installations and Logistics, Civil Works, Manpower
    and Reserve)
General Counsel
Chief of Legislative Liaison
Chief of Information
Administrative Assistant
General Staff Committees
    on Army National Guard
    and Reserve Policy

217 ■ Army Staff

The Army Staff is the staff of the Secretary of the Army and includes the Chief of Staff and his immediate assistants, the General and Special Staffs (see Fig. 2–2).

Duties of the Army Staff include: preparing plans for national security, both separately and in conjunction with the other military Services; investigating and reporting on questions affecting the efficiency of the Army and its state of preparation for military operations; execution and supervision of approved plans and instructions; and acting for the Secretary and the Chief of Staff in keeping all officers informed and coordinating the Army Establishment.

The Chief of Staff. The principal military adviser to the Secretary of the Army is the Chief of Staff. The Chief of Staff, a general, takes rank above all other officers on the active list of the Army, Marine Corps, Navy, and Air Force, except the Chairman of the Joint Chiefs of Staff, and the Chief of Naval Operations and the Chief of Staff, United States Air Force, if those latter two officers' appointments antedate his. The Chief of Staff is directly responsible to the Secretary for the efficiency of the Army, its state of preparedness for military operations, and plans therefor. He presides over the Army staff, and advises the Secretary of the Army in regard thereto.

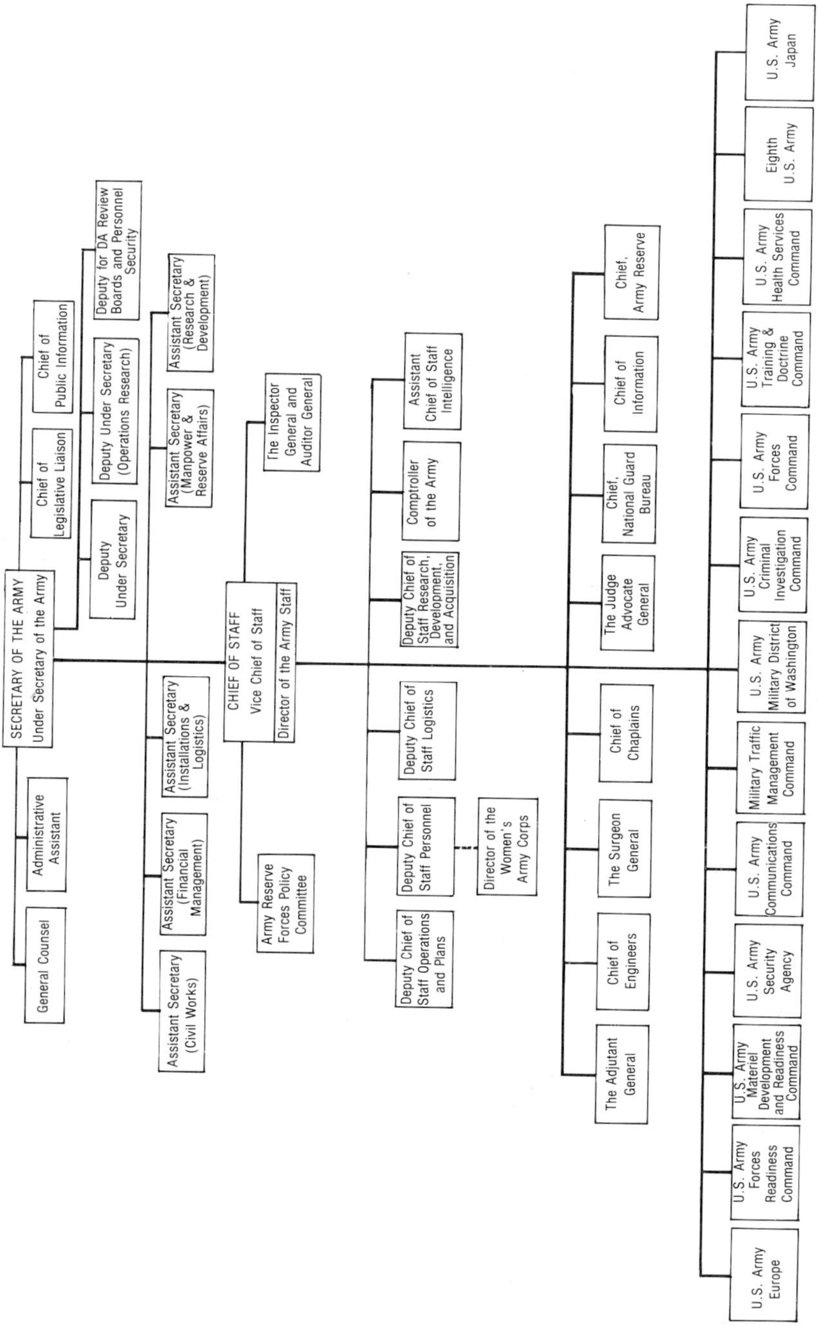

Fig. 2–2: Organization of the Department of the Army (1977).

Under the Chief of Staff, the following assistants or agencies perform staff functions for him:

Vice Chief of Staff
The Judge Advocate General
Comptroller of the Army
Chief, National Guard Bureau
Chief of Information*
The Inspector General
Deputy Chief of Staff (Operations and Plans)
Deputy Chief of Staff (Personnel)
Deputy Chief of Staff (Logistics)
Deputy Chief of Staff (Research, Development and Acquisitions)
Assistant Chief of Staff for Intelligence
Director Women's Army Corps
Secretary of the General Staff

*Also has dual direct responsibility to Secretary of the Army.

218 ■ The General Staff

Under the Chief of Staff, the General Staff renders professional advice and assistance to the Secretary of the Army; provides broad policies and plans for major Army commanders and the heads of other Department of the Army staff agencies; and prepares and issues directives in the name of the Secretary of the Army, to implement approved plans and policies. The General Staff is made up of all commissioned officers assigned to the offices of the Chief of Staff, Vice Chief of Staff, Deputy and Assistant Chiefs of Staff, Army Comptroller, and Secretary of the General Staff. The Deputy Chiefs of Staff also supervise and control the Special Staff.

219 ■ Special Staff

The Special Staff advises the Secretary of the Army and Chief of Staff, through the cognizant General Staff channel, on specialized matters within its several fields. The following officers comprise the Special Staff:

Chief of Finance
Chief of Engineers
Chief of Military History
Chief of National Guard Bureau
Chief, Army Reserve
The Adjutant General
Chief of Chaplains
Surgeon General
Judge Advocate General

220 ■ **Major Army Commands**

Materiel Development and Readiness Command performs materiel development and procurement and supply functions, as well as service test and evaluation functions, and combat development.

Training and Doctrine Command controls Army unit and individual training and provides early integration into the Army of new doctrine and organization. This command also produces the Army's comprehensive library of field and training manuals.

Forces Command (which closely corresponds functionally to our Fleet Marine Force) includes all operating forces of the Army not otherwise assigned.

Other major Army commands include Army Communications Command, Army Security Agency, Military Traffic Management Command, Army Health Services Command, Military District of Washington, Army components of unified commands, and the U. S. Military Academy (see Section 223).

221 ■ **The United States Army**

The United States Army ensures the security of the United States within its area of responsibility and particularly within the Zone of the Interior. Part of the Army is on full-time duty. Other components are ordinarily inactive in peacetime. All components may be called to active duty during an emergency declared by Congress or in the event of war.

Regular Army. The Regular Army for nearly two centuries has been the framework upon which we have built our wartime armies.

It is the duty of the Regular Army to:

1. Perform occupation duties;
2. Garrison the United States and overseas possessions;
3. Train the National Guard, Organized Reserve, and ROTC;
4. Provide an organization for the administration and supply of the peacetime military establishment;
5. Provide educated officers and men to become leaders, in event of war, of the expanded Army of the United States;
6. Expand and record the body of military knowledge so as to keep this country up to date and prepared;
7. Constitute, with the National Guard and units of the Organized Reserve, a covering force in case of a major war;
8. Cooperate with the Marine Corps, Navy, and Air Force in carrying out their missions.

National Guard. The National Guard is the militia of the United States. In time of peace, the National Guard of any State can be called to active duty by the governor of that State to perform emergency duties. Units or individual members of the National Guard can be called to active duty by the Federal Government only during war or national emergency, or with their own consent in time of peace.

In addition to augmenting the Regular Army, the Guard:

1. Trains additional volunteers and assigned selectees;
2. Supplies instructors for schools and training centers;
3. Furnishes cadres of experienced officers and men for new units;
4. Furnishes enlisted men who qualify for officer commissions.

Ready and Standby Reserve Corps. Like the National Guard, Army Reserve units train in local armories and are subject to orders to active duty under similar conditions. Individual members are assigned to Army Reserve organizations in or near their home towns. The Organized Reserve, however, is not subject to state control of any kind.

222 ■ Women in the Army

Women have served in and with the United States Army for many years and in a number of wars, at first as "contract nurses," then, as organized, in the Army Nurse Corps, the Army Medical Specialist Corps, and the Women's Army Corps.

223 ■ Education of Officer Candidates

Three school systems train candidates for commissions in the Army—the Military Academy, Officer Candidate Schools, and Reserve Officers Training Corps.

The United States Military Academy, West Point, New York, was established in 1802 for the purpose of training young gentlemen as commissioned officers. *Duty, Honor, Country*—superb motto of the Corps of Cadets—have long served to set West Point's high standard. As Colonel Archibald Henderson, 5th Commandant of the Marine Corps, remarked in 1823,

> It but rarely happens that a graduate from West Point is not a gentleman in his deportment, as well as a soldier in his education.

The Military Academy is commanded by a Superintendent, an Army general officer. Marine officers now serve on West Point's academic staff. The four-year curriculum includes cultural subjects as well as military science. The cadet graduates with a B.S. degree, and, if physically fit, is commissioned as a second lieutenant in the Army, the Marine Corps, or the Air Force.

Officer Candidate School is conducted at Fort Benning, Georgia.

"*Duty, Honor, Country*": West Point cadets carry the Colors as "the long grey line" passes in review on the Plain.

The course is usually six months. Most candidates are chosen from enlisted men and women of the Regular Army, the Reserve, and the National Guard.

The Reserve Officers' Training Corps (ROTC). Over 282 civilian universities and colleges now have the familiar Army ROTC training units, long a major source of officers for our Army (and from time to time for the Marine Corps as well); the first unit was established in 1862.

Army and Air Force ROTC are basically similar to the Naval ROTC, described in Section 340.

DEPARTMENT OF THE AIR FORCE

224 ■ Mission of the Air Force

The Department of the Air Force and the United States Air Force were established in 1947 by the National Security Act of 1947. The Department includes aviation forces, both combat and service. It organizes, trains, and equips forces of the Air Force for the conduct of prompt and sustained combat operations in the air—specifically, forces to defend the United States against air attack, to gain and maintain general air supremacy, to defeat enemy air forces, to control vital air areas, and to establish local air supremacy as required. The Air Force has primary responsibility for: developing doctrines and procedures (in coordination with the other Services) for the defense of the United States against air attack; organizing, training, and equipping Air Force forces for strategic air warfare; organizing and equipping Air Force forces for joint amphibious and airborne operations, in coordination with the other Services; furnishing close combat and logistical air support to the Army; providing air transport for the Armed Forces, except as otherwise assigned; developing, in coordination with the other Services, doctrines, procedures, and equipment for shore-based air defense, including continental United States. As in the case of the other Services, the foregoing missions derive from the National Security Act and from the Functions Paper (see Sections 214 and 301).

225 ■ Structure of the Air Force

As a result of experience during and after World War II, Korea, and Vietnam, the Air Force has evolved the following basic organizational structure:

Flight. The lowest tactical echelon recognized in the Air Force structure. Flights are not formally designated in the structure, but are subdivisions of combat squadrons. They provide the basis for combat formations, and are used for training purposes.

Squadron. The basic unit in the organizational structure of the United States Air Force. A squadron is manned and equipped to best perform a specific military function such as combat, maintenance, food service, and communications.

Group. A flexible unit composed of two or more squadrons whose functions may be either tactical or administrative in nature.

Wing. The smallest Air Force unit manned and equipped to operate independently in sustained action until replacement and resupply can take place.

Air Division. An air combat organization normally consisting of two or more wings. Divisions are operational in nature with minimum administrative or logistics responsibilities.

Numbered Air Force. The intermediate command echelon designed to control and administer a grouping of combat wings. It is flexible in organization and can vary in size. Usually, a numbered air force has one

Air Force Minuteman intercontinental ballistic missile on the way.

of three missions—strategic, tactical, or defensive. Its wings may be grouped for operational control under Air Divisions, or be directly under the numbered air force.

Major Command. A functionally titled command echelon directly below Headquarters USAF, charged with a major responsibility in fulfilling the Air Force mission.

226 ■ The Secretary of the Air Force and His Assistants

The Secretary of the Air Force heads the Department of the Air Force (see Fig. 2–3) and is responsible for all matters pertaining to its operation.

Principal assistants to the Secretary are:

The Under Secretary of the Air Force
Assistant Secretary of the Air Force (Manpower and Reserve Affairs)
Assistant Secretary of the Air Force (Financial Management)
Assistant Secretary of the Air Force (Installations and Logistics)
Assistant Secretary of the Air Force (Research and Development)
General Counsel
Assistant to the Secretary (International Affairs)
Director, Legislative Liaison
Director, Information
Administrative Assistant to the Secretary
Director, Space Systems

227 ■ The Air Staff

The Chief of Staff, U. S. Air Force, presides over the Air Staff, and supervises such personnel and organizations of the Air Force as the Secretary of the Air Force determines. Under the Chief of Staff, the Air Staff organization consists of the following:

Vice Chief of Staff
Assistant Vice Chief of Staff
Secretary of the Air Staff
Scientific Advisory Board
Chief Scientist
Chief, Operations Analysis
Surgeon General
The Inspector General
The Judge Advocate General
Chief of Air Force Chaplains
Assistant Chief of Staff, Intelligence
Assistant Chief of Staff, Reserve Forces

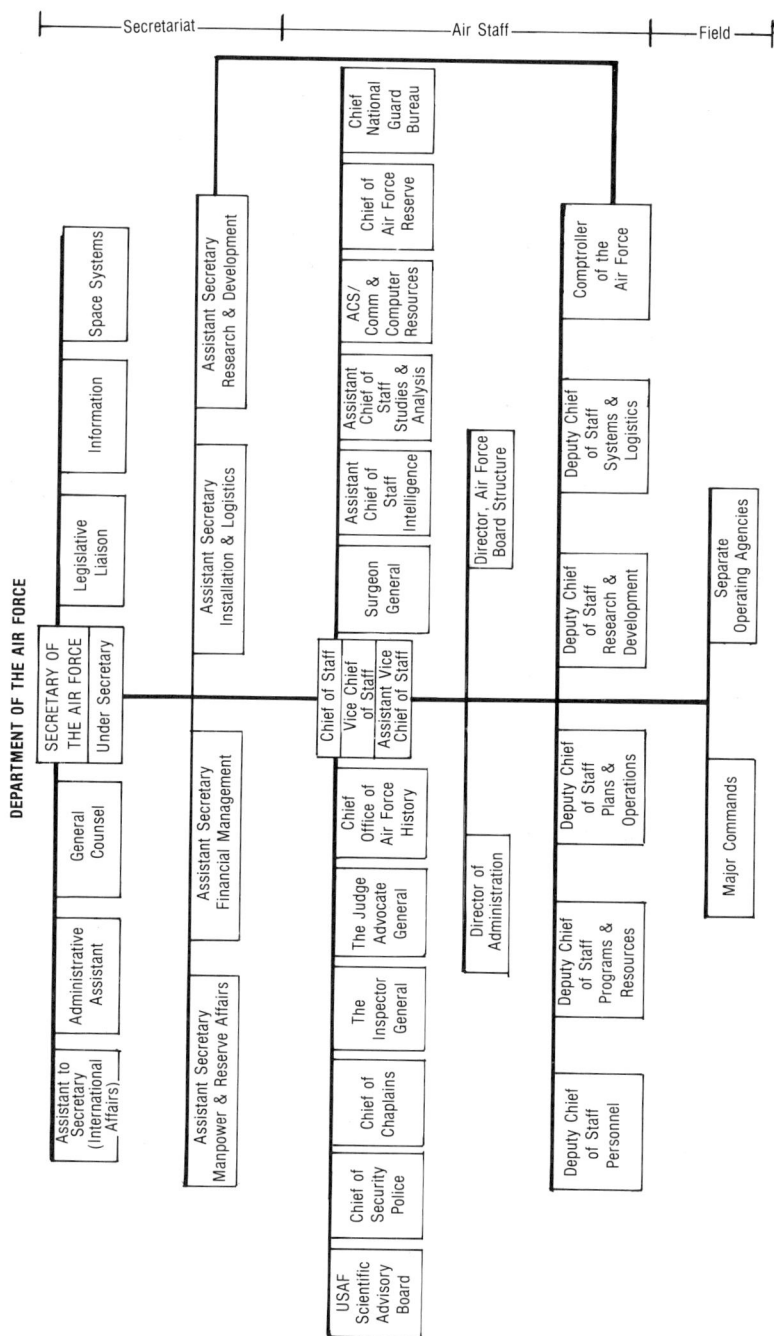

Fig. 2-3: Organization of the Department of the Air Force (1977).

Director of Administrative Services
Comptroller of the Air Force
Deputy Chiefs of Staff for Personnel, Plans and Operations, Programs and Resources, Research and Development, and Systems and Logistics

228 ■ Major Air Commands

Some 26 major air commands and separate operating agencies together represent the field organization of the Air Force. These commands are organized on a functional basis in the United States and on an area basis overseas.

32 *Major USAF Commands*
Aerospace Defense Command
Air Force Systems Command
The Air University
Headquarters Command USAF
Strategic Air Command
USAF Security Service
U. S. Air Forces Europe
Alaskan Air Command
Air Force Logistics Command
Air Training Command
Military Airlift Command
Tactical Air Command
USAF Communications Service
Pacific Air Force
USAF Southern Command

USAF Separate Operating Agencies
Air Force Accounting and Finance Center
Air Force Audit Agency
Air Force Data Automation Agency
Air Force Intelligence Service
Air Force Inspection and Safety Center
Air Force Test and Evaluation Center
Air Force Military Personnel Center
Air Force Office of Special Investigations
Air Force Reserve
Air National Guard
Air Reserve Personnel Center
Air Force Academy

Of special interest among the foregoing commands is the Military Airlift Command (MAC), which provides worldwide air transportation for the Department of Defense.

Components of the USAF. The United States Air Force is composed, in its entirety, of: the regular Air Force, Air National Guard, and Air Force Reserve.

229 ■ Women in the Air Force

Women serve in the Air Force as line officers and in enlisted ranks, and as nurses and medical specialists. Present provisions of law for Air Force women personnel are similar to those for women of the other Services.

230 ■ Education of Air Force Officer Candidates

Candidates for Air Force commissions receive their training in a number of different institutions: the Air Force Academy, Officer Training School, Air Force ROTC, and the Airman Education and Commissioning Program.

The Air Force Academy, Colorado Springs, Colorado, is one of the principal sources for regular Air Force officers. The curriculum includes both academic and military subjects, and leads to a B.S. degree and a commission as a second lieutenant in the Air Force (or the Army, Marine Corps, or Navy).

Officer Training School, located at Lackland Air Force Base, Texas, is coeducational. Male and female college graduates may apply for this three-months' course. The curriculum includes administration, organization, supply, military law, world affairs, leadership, and human rela-

High over Edwards AFB, California, the Air Force's firstline air-superiority fighter, the F–15 Eagle, shows its paces.

tions. Graduates are either assigned directly to duty or pursue additional training in an aircrew (pilot or navigator) or technical course. Graduates are given the opportunity to apply for regular commissions.

The Air Force ROTC is similar in purpose and organization to the Army and Naval ROTC.

The Airman Education and Commissioning Program provides undergraduate education, followed by officer training and a commission, for selected career-minded airmen serving on active duty.

THE DEPARTMENT OF THE NAVY

231 ■ The Department of the Navy

The Department of the Navy occupies co-equal status with the Department of the Army and the Department of the Air Force in the Organization for National Security (Fig. 2–1) and in the Department of Defense. The organization of the Department of the Navy and the place of the United States Marine Corps in the Naval Establishment are described in detail in Chapters 3, 4, and 5.

THE COAST GUARD

232 ■ The U. S. Coast Guard

The Coast Guard is a military Service within the Armed Forces at all times. Although under the Transportation Department in peacetime, Coast Guard personnel receive the same pay as Service personnel under

Coast Guard Cutter Mellon *stands ready to answer distress calls. All Coast Guard vessels are easily distinguished by their broad red and narrow blue stripes.*

the Department of Defense, and are subject to the Uniform Code of Military Justice.

Headed by an admiral as Commandant in peacetime, the Coast Guard's rank and rating structure is very similar to the Navy's. On declaration of war or when the President directs, the Coast Guard operates as a Service within the Naval Establishment.

Both in war and peace, the Coast Guard performs a wide range of functions, mainly military. In peace, most of these duties center about protecting lives and property on the seas and along the coasts, and include: manning lifeboat stations, search and rescue centers, and lighthouses, lightships, and loran stations; icebreaking, locating icebergs, and recommending safe sea lanes in northern waters; enforcing maritime pollution-control, fisheries, and safety regulations; and manning ocean "stations" to obtain weather data, assist aircraft navigation, and render on-the-spot assistance to planes and ships in trouble.

The Coast Guard is also charged with enforcing a variety of laws at sea and in Alaska for other Government agencies. In wartime it performs such duties as: escort and antisubmarine warfare, manning transports, port security, and operating landing craft.

The Service has been in continuous operation since 1790, when it was organized by Alexander Hamilton as the Revenue Marine Service. The name was later changed to Revenue Cutter Service, and, in 1915, to Coast Guard. Its motto is *Semper Paratus* ("Always Ready").

The Coast Guard in peacetime includes about 30,000 officers, men, and women. Appointments to the Coast Guard Academy at New London, Connecticut, are by competitive examination open to civilians and enlisted men and women from any Armed Service between the ages of 18 and 22. The four-year course is basically engineering; cadets receive a B.S. degree on graduation, and are then commissioned as ensigns in the U. S. Coast Guard.

The principal foundations of all states are good laws and good arms; and there cannot be good laws where there are not good arms.
—Niccolo Machiavelli, The Prince

3

The Department of the Navy

With 6,000 miles of coastline and with 71 percent of the globe covered by water, the United States has long appreciated the importance for our security of strong naval forces. Essentially, the United States is a maritime power and American strategy must always be fundamentally maritime. In two world conflicts we have successfully kept war from our shores by superior sea power. Without control of the seas, we could not have transported fighting men, equipment, and supplies to distant battle areas, nor could we ever project our fighting power from the seas onto the land.

Today, control of the seas is more important than ever. The United States depends increasingly on overseas areas for raw materials to sustain our growing industrial production. Our security is bound up with the security of friendly powers in many parts of the world. By maintaining control of the seas, we assure that our lifelines to these nations and to our far-flung advance bases will not be broken. More important, we assure control of the seas to assure our ability to use the seas for the offensive operations which victory requires. For all these reasons we must maintain the United States Fleet. In doing this, the basic policy of the Department of the Navy is as follows:

> ... to maintain the Navy and Marine Corps as an efficient, mobile, integrated force of multiple capabilities, and sufficiently strong and ready at all times to fulfill their responsibilities ... to support and defend the Constitution of the United States against all enemies, foreign and domestic; to insure, by timely and effective military action, the security of the United States, its possessions, and areas vital to its interest; to uphold and advance the national policies and interests of the United States; and to safeguard the internal security of the United States.

As you reflect on the import of this policy, remember the words of General Shepherd, 20th Commandant: "Both the functions and the future of the Marine Corps are intimately linked with those of the U. S. Navy."

301 ■ Mission of the Department of the Navy

The responsibilities of the Department of the Navy are expressed as follows in the National Security Act as amended:

Sec. 206 (a) The term "Department of the Navy" as used in this Act shall be construed to mean the Department of the Navy at the seat of government; the headquarters, United States Marine Corps; the entire operating forces of the United States Navy, including naval aviation, and of the United States Marine Corps, including the reserve components of such forces; all field activities, headquarters, forces, bases, installations, activities, and functions under the control or supervision of the Department of the Navy; and the United States Coast Guard when operating as a part of the Navy pursuant to law.

(b) The Navy, within the Department of the Navy, includes, in general, naval combat and service forces and such aviation as may be organic therein. The Navy shall be organized, trained, and equipped primarily for prompt and sustained combat incident to operations at sea. It is responsible for the preparation of naval forces necessary for the effective prosecution of war except as otherwise assigned and is generally responsible for naval reconnaissance, antisubmarine warfare, and protection of shipping.

All naval aviation shall be integrated with the naval service as part thereof within the Department of the Navy. Naval aviation consists of combat and service and training forces, and includes land-based naval aviation, air transport essential for naval operations, all air weapons and air techniques involved in the operations and activities of the Navy, and the entire remainder of the aeronautical organization of the Navy, together with the personnel necessary therefor.

The Navy shall develop aircraft, weapons, tactics, technique, organization, and equipment of naval combat and service elements. Matters of joint concern as to these functions shall be coordinated between the Army, the Air Force, and the Navy.

The Navy is responsible, in accordance with integrated joint mobilization plans, for the expansion of the peacetime components of the Navy to meet the needs of war.

(c) The Marine Corps, within the Department of the Navy, shall be so organized as to include not less than three combat divisions and three air wings, and such other land combat, aviation, and other services as may be organic therein. The Marine Corps shall be organized, trained, and equipped to provide fleet marine forces of combined arms, together with supporting air components, for service with the fleet in the seizure or defense of advanced naval bases and for the conduct of such land operations as may be essential to the prosecution of a naval campaign. In addition, the Marine Corps shall provide detachments and organizations for service on armed vessels of the Navy, shall provide security detachments for the protection of naval property at naval stations and bases, and shall perform such other duties as the President may direct. However, these additional duties may not detract from or interfere with the operations for which the Marine Corps is primarily organized.

The Marine Corps shall develop, in coordination with the Army and the Air Force, those phases of amphibious operations that pertain to the tactics, technique, and equipment used by landing forces.

The Marine Corps is responsible, in accordance with integrated joint mobilization plans, for the expansion of peacetime components of the Marine Corps to meet the needs of war.

Except in time of war or national emergency declared by Congress after June 28, 1952, the authorized strength of the Regular Marine Corps, excluding retired members, is 400,000. However, this strength may be temporarily exceeded at any time in a fiscal year if the daily average number in that year does not exceed it.

The functions of the Navy and Marine Corps were further defined in the 1948 Key West Agreement (see Fig. 3-1). The primary purpose of this paper was to amplify the functions of the Army and Air Force, both of which were treated only generally in the National Security Act. However, the Agreement did assign collateral functions to the Navy and the Marine Corps, and spelled out some basic functions in more detail. As regards Marine Corps functions, however, it should be borne in mind that their primary source is the National Security Act rather than the Functions Paper.

302 ■ The Executive Branch and the Department of the Navy

As described in Chapter 2, the Department of the Navy, as one of the military departments within the Department of Defense, comes under the President and the Secretary of Defense. Its organization and executive duties are covered later in this chapter.

303 ■ The Legislative Branch and the Department of the Navy

Under our Constitution, Congress is given the authority "to provide and maintain a navy . . . and to make rules for the government of the land and naval forces." In the words of a decision by the late Chief Justice Charles Evans Hughes, "Congress provides; the President commands." Congress therefore enacts laws governing the size, scope, functions, and authority of the Navy and the Marine Corps. Congress authorizes and provides funds for construction of ships and shore bases, and for all Navy and Marine Corps activities.

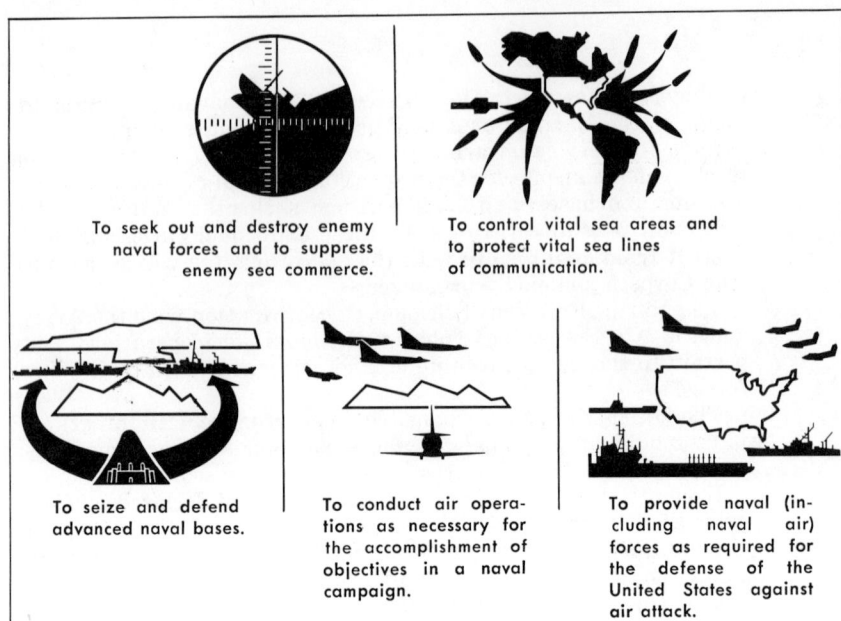

Fig. 3-1: Primary functions of the Navy, as taken from the Functions Paper.

Legislation affecting the Navy and the Marine Corps may be originated by members or committees of Congress, by the Department of the Navy, or by other offices of the Department of Defense. The Department of Defense then submits comments and recommendations on such military legislation. Officials of the Department of the Navy are frequently summoned to appear at hearings on proposed legislation, and Congressional investigations.

304 ■ The Judicial Branch and the Department of the Navy

The Department of the Navy in its official capacity may sue or be sued. It has a right to appear in its own defense or in defense of its officials, or members of the Services; it may enter briefs, and may argue and appear before the courts; and it is bound by decisions of Federal Courts.

THE DEPARTMENT OF THE NAVY

305 ■ Organization

The Department of the Navy consists of four principal parts: the Operating Forces of the Navy, the United States Marine Corps, the

Naval Material Command, and other supporting organizations (see Fig. 3-2).

The Operating Forces of the Navy include the Office of the Chief of Naval Operations, the several fleets, seagoing forces, district forces, Fleet Marine Forces and other assigned Marine Corps forces, the Military Sealift Command, and such other Navy shore (field) activities and commands as are assigned by the Secretary of the Navy.

The U.S. Marine Corps, within the Department of the Navy, includes Headquarters, U. S. Marine Corps, the Operating Forces of the Marine Corps, the Marine Corps Supporting Establishment, and the Marine Corps Reserve (see Chapters 4, 5, and 9).

The Naval Material Command includes the Office of Naval Material and subordinate functional commands in 1977 as follows: Naval Air Systems Command, Naval Electronics Systems Command, Naval Facilities Engineering Command, Naval Seas Systems Command, and Naval Supply Systems Command.

Other supporting organizations include the Bureau of Naval Personnel, the Bureau of Medicine and Surgery, Office of the Comptroller of the Navy, Office of Naval Research, Office of the Judge Advocate General, offices of the Staff Assistants to the Secretary, and shore (field) activities as assigned by the Secretary of the Navy.

306 ■ The Secretary of the Navy (SecNav)

The Secretary of the Navy is the head of the Department of the Navy. He is responsible for the policies and control of the Department of the Navy, including its organization (except as otherwise prescribed in law), administration, operation, and efficiency. As far as practicable, the Secretary discharges these responsibilities through his Civilian Executive Assistants and the other military and civilian assistants. The Secretary, however, retains personal direction over activities relating to legislation and Congress, and relationships with the Secretary of Defense and other principal Government officials and the public.

The Secretary of the Navy is the principal morale officer of the Department of the Navy, and as such directs a continuing effort to promote the welfare and morale of all hands.

The Secretary may communicate directly with any principal official of the Department of the Navy, the Shore Establishment, or the Operating Forces.

The Secretary recommends to the Secretary of Defense and the President, appointments, removals, or reassignments of the legally

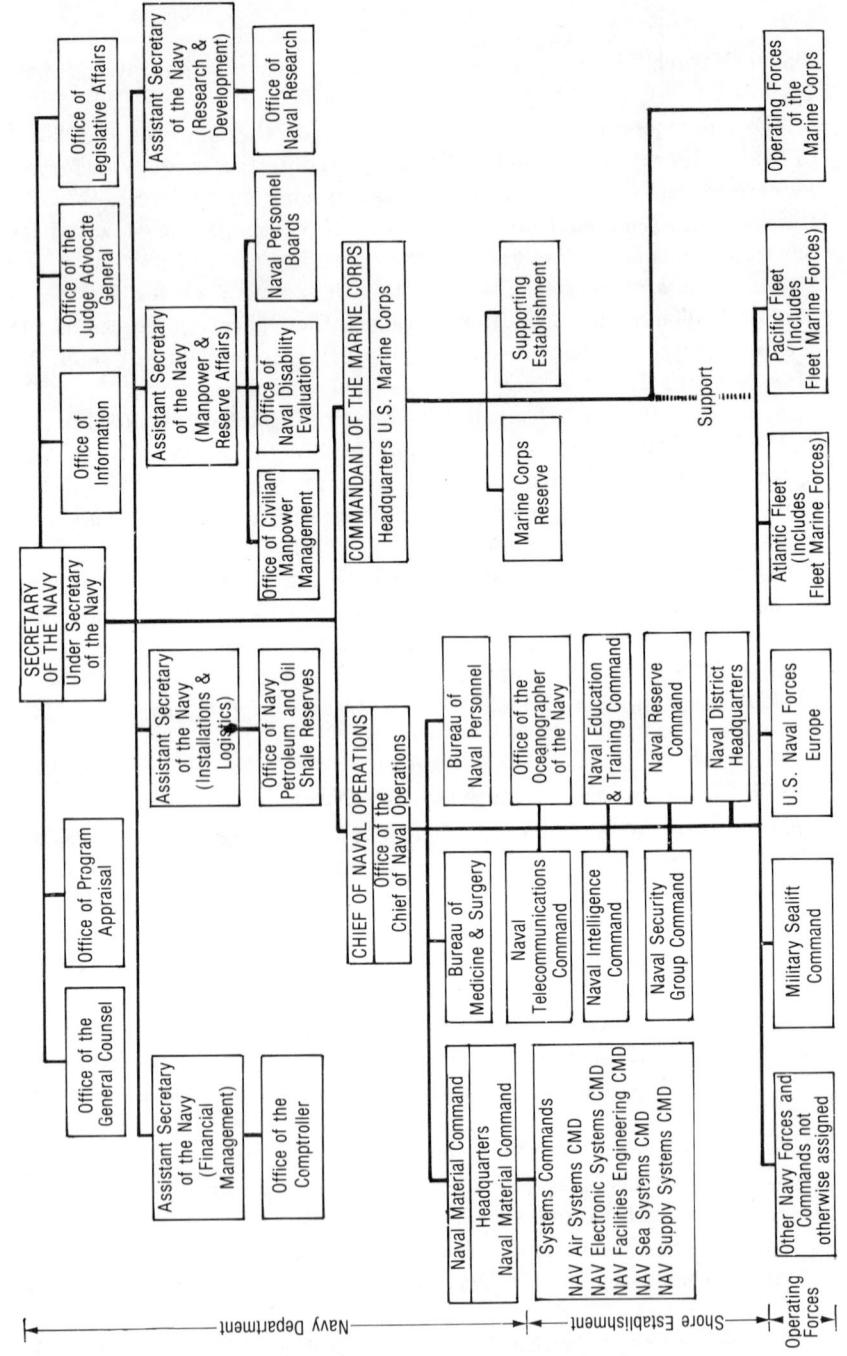

Fig. 3-2: Organization of the Department of the Navy (1977).

constituted positions of the Department of the Navy. In his own discretion he controls the selection and assignment of all other principal officials of the Department.

Policy Council. The Department of the Navy Policy Council, chaired by the Secretary and composed of the Under Secretary, Chief of Naval Operations, Commandant of the Marine Corps, and the Vice Chief of Naval Operations, is one of the Secretary's major instruments for military-civilian coordination and for timely consideration of policy.

307 ■ The Civilian Executive Assistants

The Under Secretary, Assistant Secretary (Financial Management), Assistant Secretary (Research and Development), Assistant Secretary (Installations and Logistics), and Assistant Secretary (Manpower and Reserve) are the principal Civilian Executive Assistants. They exercise top management coordination over the bureau and offices of the Navy Department. Each Civilian Executive Assistant oversees areas of responsibility assigned by the Secretary.

308 ■ The Under Secretary of the Navy

The Under Secretary of the Navy is the principal Civilian Executive Assistant to the Secretary of the Navy. In this capacity he acts for and with the authority of the Secretary of the Navy during the latter's absence or disability. In addition, the Under Secretary directly supervises:

Board of Decorations and Medals. This board sets policy regarding and passes on all awards and medals for the Naval Services.

Office of Program Appraisal. This office provides the Secretary with an independent capability to appraise progress against approved programs as well as to analyze proposed programs.

Office of Legislative Affairs. This office advises and assists the Secretary and all other principal military and civilian officials of the Department in connection with legislative affairs and Congressional relations.

Office of Information. This office performs public information and public relations functions for the Navy Department. The Chief of Information also has collateral public relations responsibility to the Chief of Naval Operations.

Office of the General Counsel. This office furnishes legal services in the field of commercial law. The General Counsel is responsible for legal aspects of procurement, contracts, property disposition, and renegotiation.

Office of the Judge Advocate General. The duties of this office are described in Section 328.

309 ■ Assistant Secretary of the Navy (R&D)

The Assistant Secretary of the Navy (Research and Development) supervises all research, development, engineering, test, and evaluation efforts within the Department of the Navy. Among other agencies under him is the Office of Naval Research (ONR). In addition, he chairs the Research and Development Committee.

310 ■ Assistant Secretary of the Navy (Installations and Logistics)

The Assistant Secretary of the Navy (Installations and Logistics) supervises all matters relating to: procurement, production, supply, distribution, alteration, maintenance, and disposal of real estate and facilities; as well as all matters relating to the acquisition, construction, utilization, improvement, alteration, maintenance, and disposal of real estate and facilities, including utilities, housing, and quarters. He supervises the Office of Naval Petroleum and Oil Shale Reserves, which is charged by law with custody of the Federal Petroleum reserves; and coordinates Department of the Navy responsibilities in connection with the Mutual Defense Assistance Program.

311 ■ Assistant Secretary of the Navy (Manpower and Reserve)

The Assistant Secretary of the Navy (Manpower and Reserve Affairs) supervises all manpower matters including personnel administration, utilization, and morale and performance. The various *Naval Personnel Boards* and *Office of Naval Disability Evaluation* function under this Assistant Secretary. These Boards, which promote and safeguard the welfare and rights of Navy and Marine Corps officers and enlisted men, include:

Board for Correction of Naval Records
Naval Clemency and Parole Board
Navy Discharge Review Board
Naval Physical Disability Review Board

312 ■ Assistant Secretary of the Navy (Financial Management)

The Assistant Secretary of the Navy (Financial Management) is responsible for supervision of all matters related to the financial management of the Department of the Navy, including budgeting, accounting, financing, progress and statistical reporting, and auditing. In addition, for the present, he is designated as Comptroller of the Navy.

In accordance with the policy of the Secretary of the Navy, this position is held by one of the Secretary's Civilian Executive Assistants, although a flag officer may be, and has in the past been, so designated.

313 ■ Order of Succession

During any temporary absence of the Secretary of the Navy, the order of succession as Acting Secretary is: Under Secretary; Assistant Secretaries in the order prescribed by the Secretary, or, if no order is prescribed, then the order in which the respective Assistant Secretaries took office; Chief of Naval Operations; and Vice Chief of Naval Operations.

314 ■ The Naval Executive Assistants to SecNav

The Chief of Naval Operations (CNO), an admiral, is the senior military officer of the Department of the Navy. He is principal naval adviser to the President and to the Secretary of the Navy on the conduct of war, and is naval executive assistant to the Secretary of the Navy on the conduct of the activities of the Department of the Navy. As Navy member of the Joint Chiefs of Staff, CNO is responsible additionally to the President and Secretary of Defense for certain duties external to the Department of the Navy.

The *Commandant of the Marine Corps*, a general, and senior officer in the Corps, commands the United States Marine Corps and is directly responsible to the Secretary of the Navy for its administration, discipline, organization, training, requirements, efficiency, and readiness, and for the total performance of the Marine Corps. The Commandant of the Marine Corps has additional responsibility to CNO for the organization, training, readiness, and performance of elements of the operating forces of the Marine Corps assigned to the Operating Forces of the Navy, and he is responsible to the Civilian Executive Assistants for matters related to the duties assigned them, as well as to the President and Secretary of Defense for certain duties external to the Department of the Navy.

In addition, the Commandant's responsibilities include:

1. Planning and determining the support needs of the Corps for equipment, weapons, materials, supplies, facilities, maintenance, and supporting services. This responsibility includes determination of Marine Corps characteristics of materiel to be procured or developed, and the training required to prepare Marines for combat.

2. Development, in coordination with the Army, Navy, and Air Force, of the tactics, techniques, doctrines, and equipment employed by landing forces in amphibious operations.

3. Plans for and determination of present and future needs, qualitative and quantitative, for regular and reserve personnel of the Marine Corps. This includes responsibility for leadership in maintaining a high degree of competence on the part of all hands in necessary fields of specialization through education, training, and equal opportunities for advancement; and for leadership in maintaining the esprit of Marines and the prestige of a Marine Corps career.

4. Planning and determining the needs for the care of the health of Marines (in coordination with the Surgeon General of the Navy).

5. Budgeting for the Marine Corps (except as otherwise directed by SecNav), and supervising the performance of its supporting establishment.

6. Formulation of Marine Corps strategic plans and policies and participation in formulation of joint and combined strategic plans and policies and related command relationships.

7. Sitting with the Joint Chiefs of Staff when matters which concern the Marine Corps are under consideration, on which occasions the Commandant enjoys co-equal status with members of the JCS. The Commandant also regularly attends meetings of the Armed Forces Policy Council.

The Commandant of the Coast Guard is a naval executive assistant to SecNav when the Coast Guard is attached to the Navy pursuant to law (see Chapter 2).

315 ■ Chief of Naval Material

The Chief of Naval Material (CNM) heads the Naval Material Command, consisting principally of the Office of Naval Material and subordinate functional commands organized from the former Bureaus of the Navy Department. The CNM, a subordinate of the Chief of Naval Operations, is responsible to both CNO and CMC for providing material support needs of the Operating Forces of the Navy and of the Marine Corps; for development and operation of the Navy Supply System; for acquisition, development, maintenance, and disposal of facilities (except maintenance or operation of Marine facilities); and for procurement, production, and contracting policies throughout the Department of the Navy.

316 ■ Naval Professional Assistants

These officials comprise the remaining Bureau Chiefs (now under CNO, however), Chief of Naval Research, and the Judge Advocate General. These functions are described later in this chapter.

OFFICE OF CHIEF OF NAVAL OPERATIONS (OPNAV)

317 ■ General

The offices, boards, and agencies reporting to and performing duties for the Chief of Naval Operations (CNO) are collectively referred to as the Office of the Chief of Naval Operations (OPNAV). The duties and composition of the principal staff units of OPNAV are discussed in the following sections.

> NOTE: Although the Chiefs of Naval Personnel, of the Bureau of Medicine and Surgery, and of Naval Material now report to and perform duties for CNO, their organizations do not comprise part of the Office of the Chief of Naval Operations.

318 ■ The Chief of Naval Operations (CNO)

The Chief of Naval Operations, as noted above in Section 314, is the senior military officer of the Department of the Navy. He is also a member of the Armed Forces Policy Council and of the Joint Chiefs of Staff. The principal responsibilities of the Chief of Naval Operations are:

1. To command the Operating Forces of the Navy.

2. To organize, train, prepare, and maintain the readiness of Navy forces for assignment to unified or specified commands. This responsibility includes determination of training required by all members of the Navy and Naval Reserve, for combat.

3. To plan and determine the material support needs of the Operating Forces of the Navy (less Fleet Marine Forces and other assigned Marine Corps forces).

4. To plan for and determine present and future needs, qualitative and quantitative, for regular and reserve personnel of the Navy. This includes responsibility for leadership in maintaining high competence on the part of all hands through education, training, and equal opportunities for promotion; and for leadership in maintaining the morale and motivation of members of the Navy, and the prestige of a naval career.

5. To plan and determine the needs for care of the health of the Navy and their dependents.

6. To budget for the operating costs of the fleets and shore activities of the Operating Forces of the Navy (except as otherwise directed by SecNav), and to supervise the performance of the shore activities assigned to the Operating Forces of the Navy.

7. To formulate Navy strategic plans and policies and participate

in formulation of joint combined strategic plans and policies and related command relationships.

8. In addition, except with respect to the Marine Corps, the CNO supervises the military administration of the Department of the Navy in such matters as security, intelligence, discipline, communications, and the customs and traditions of the Navy.

319 ■ The Vice Chief of Naval Operations (VCNO) (Op–09)

The Vice Chief of Naval Operations acts by delegated authority for CNO on all matters not specifically reserved to CNO alone, performs duties of CNO during CNO's absence, and is principal adviser to CNO. Under the Vice Chief of Naval Operations are the Plans and Programs Office, and that of the Assistant VCNO, who discharges certain CNO administrative responsibilities. In addition, the VCNO supervises the Director of Naval Intelligence; of Naval Reserve; of Naval Communications; of Research, Development, Test and Engineering; and of Naval Education and training; and the Navy Inspector General (who also has certain direct responsibilities to the Secretary of the Navy).

The *Naval Inspector General* (Op–008), under the direction of SecNav or CNO, inquires into and reports on any and all matters affecting the discipline, readiness, effectiveness, efficiency, or economy of the Department of the Navy.

As regards the Marine Corps Supporting Establishment, these inquiries and reports are limited to management efficiency and compliance with policies of the Civilian Executive Assistants and technical instructions by the Naval Professional Assistants.

Marine Corps Liaison Officer (Op–09M) provides liaison for Marine Corps matters with the Chief of Naval Operations.

The *Assistant VCNO* (Op–09B) consolidates a number of miscellaneous activities and functions, including, among others, Naval History Division, Naval Records Management Division, Organizational Management and Administrative Services Divisions.

Field activities under the VCNO include:

1. The *Oceanographic Office*, headed by the Oceanographer, which executes hydrographic surveys in foreign waters and on the high seas; collects and disseminates oceanographic and navigational information; and prepares and issues maps, charts, sailing directions, navigational publications, and radio broadcasts. The Office was originally established in 1830 as The Depot of Charts and Instruments.

2. The *Naval Observatory*, in Washington, D.C., which determines and broadcasts the correct time. These signals establish U. S. standard

The Naval Observatory in Washington determines the correct time for the United States and has made fundamental contributions to such sciences as mathematics, astronomy, and optics.

time, are needed by navigators at sea and in the air to determine their astronomical positions, and are used for many other purposes requiring exact time. The Observatory takes an active part in designing and constructing its precision optical and electrical instruments. It has made notable contributions to mathematics, celestial mechanics, optics, fundamental astronomy, and the design of astronomical instruments. The Naval Observatory's *Nautical Almanac Office* compiles publications required for fundamental positional astronomy.

3. The *Naval Weather Service,* with headquarters at Washington Navy Yard, which provides worldwide weather forecasts and information for the Fleet.

320 ■ The Deputy Chiefs of Naval Operations (DCNO)

Six Deputy Chiefs of Naval Operations perform the following functions:

DCNO (Manpower) (Op–01) is responsible for coordination of basic training, and for personnel plans and policies for the Operating Forces and the Naval Reserve. The same officer serves as Chief of Naval Personnel (see Section 326) and as DCNO (Manpower).

DCNO (Submarine Warfare) (Op–02) has cognizance of the organization, readiness, administration, and operations of the submarine forces (including strategic submarines), and of submarine manpower and training requirements.

DNCO (Surface Warfare) (Op–03) has cognizance over the organization, readiness, and administration of the surface seagoing forces (including surface warfare plans), programs, weapons, training, and ship acquisition.

DCNO (Logistics) (Op–04) plans the logistic requirements of the Operating Forces of the Navy, including ships' materiel readiness, shore facilities programming, and inspection and survey of warships.

DCNO (Air Warfare) (Op–05) is responsible for the military aspects and for logistic support of naval aviation, and for the integration of Marine Corps aviation with the naval aviation program. To accomplish this, the Deputy Chief of Staff for Aviation, Marine Corps Headquarters, also is detailed as Assistant Chief of Naval Operations (Marine Aviation) (Op–05M).

DCNO (Plans, Policy, and Operations) (Op–06) formulates naval strategic plans and policies with respect to politico-military policy, command relationships (including joint and combined matters), Pan American affairs, and foreign military assistance.

NAVAL MATERIAL COMMAND

As already stated, the Naval Material Command consists of "systems commands" which have superseded several of the former Bureaus. This command is headed by the Chief of Naval Material.

321 ▪ Air Systems Command

This command has cognizance over Navy and Marine aircraft, air-launched weapons systems (including aerial torpedoes and mines), airborne electronics (avionics), airborne underwater sound systems and minesweeping, airborne pyrotechnics, astronautics, aircraft and air weapons targets, and photographic, meteorological, and pyrotechnic gear. The command handles procurement, design, manufacture, maintenance, alteration, and material effectiveness of naval aircraft, as well as the design, construction, and maintenance of aeronautics shore establishment stations.

322 ▪ Sea Systems Command

This command coordinates all naval shipbuilding; designs, constructs, procures, and maintains all ships and craft; has cognizance over surface navigational radar and sonar and search radar; is responsible for rescue and salvage systems and for degaussing and minesweeping gear, and also for inactive-ship maintenance. Sea Systems Command also has cognizance over all shipborne weapons systems (including air-launched underwater weapons such as mines, torpedoes, and missiles); ammunition and explosives; harbor-defense gear; landing force gear; and research and development associated with all the foregoing.

323 ■ Electronics Systems Command

This command has cognizance over all ship and shore electronics, certain space programs, other radio or electronics programs and gear, and electronic standards and compatibility throughout the Navy.

324 ■ Supply Systems Command

Although this is a new title, the basic organization remains that of the former Bureau of Supplies and Accounts, and thus coordinates the Navy supply systems; procures, stores, and issues supplies, provisions, clothing, fuel, and whatever other material the technical bureaus do not procure directly; keeps the property and money accounts of the Navy; and pays invoices and Navy payrolls.

325 ■ Facilities Engineering Command

Like the Supply Systems Command, this command, too, is a virtually unchanged redesignation of a former Bureau—in this case, Yards and Docks, the organization which designs, constructs, and maintains public works and utilities at continental and outlying bases; and trains, organizes, and maintains the Construction Battalions (Seabees), which specialize in advance base development and frequently serve with the Fleet Marine Force. The Facilities Engineering Command is also responsible for maintenance of all Navy shore facilities and plant property, except insofar as Marine Corps installations are concerned.

OTHER SUPPORTING ORGANIZATIONS

326 ■ Bureau of Naval Personnel (BuPers)

BuPers procures, trains, and distributes Navy officers and enlisted personnel; supervises promotion, discipline, and welfare of Navy personnel; administers the Naval Reserve; and operates field personnel establishments. The Chief of Naval Personnel is one of the two Bureau chiefs who remain directly under the Chief of Naval Operations rather than under the Chief of Naval Material. The Chief of Chaplains, who is responsible for the spiritual care and welfare of the Navy and Marine Corps, comes under the Chief of Naval Personnel.

327 ■ Bureau of Medicine and Surgery (BuMed)
BuMed is responsible for the health, sanitation, and medical and dental care of the Navy and Marine Corps. BuMed operates the Medical Department and exercises administrative control over the Medical, Dental, Nurse, Hospital, and Medical Service Corps. The Surgeon General of the Navy heads the Bureau. Like the Chief of Naval Personnel, the Surgeon General (and BuMed) functions under direct supervision of the Chief of Naval Operations.

328 ■ Office of the Judge Advocate General (OffJAG)
This office handles all legal matters of the Department of the Navy in the field of military, administrative, legislative, and general law. This responsibility covers the entire Naval Establishment, including the Marine Corps.

329 ■ Office of Naval Research (ONR)
ONR assures the Navy a well-rounded program by coordinating research throughout the Department of the Navy. In addition, ONR is responsible for protecting the Navy's interests in patents, inventions, and related matters.

THE SHORE ESTABLISHMENT

330 ■ General
The Shore Establishment comprises all field activities of the Department of the Navy, except shore activities assigned to the Operating Forces of the Navy. The Shore Establishment includes those Operating Forces of the Marine Corps which are not assigned to the Operating Forces of the Navy or to a unified or specified command.

The activities of the Shore Establishment are generally distributed along our coasts where they can best serve the Operating Forces. However, many activities in which such close proximity is not essential (notably air, ordnance, and supply) are distributed well inside continental United States and territories.

331 ■ Status of the Marine Corps in the Shore Establishment
Within the Shore Establishment the Commandant of the Marine Corps exercises military command over the Marine Corps supporting estab-

The activities of the Shore Establishment are distributed along our coasts where they can best serve the Operating Forces of the Navy. Norfolk Naval Operating Base serves the East Coast.

lishment, over operating forces of the Marine Corps not assigned to the operating forces of the Navy or to a JCS-established command, and over special activities of the Marine Corps (Marine activities responsible for providing service or security, or those which perform duties for government activities other than those of the Department of the Navy).

332 ■ Naval Districts

The Shore Establishment includes 12 Naval Districts with headquarters as follows:

1st Naval District: (see note)
3rd Naval District: (see note)
4th Naval District: Philadelphia, Pennsylvania

The Puget Sound Naval Shipyard exemplifies the Shore Establishment on the West Coast. Above is the world's largest drydock.

5th Naval District: Norfolk, Virginia
6th Naval District: Charleston, South Carolina
8th Naval District: New Orleans, Louisiana
9th Naval District: Great Lakes, Illinois
11th Naval District: San Diego, California
12th Naval District: San Francisco, California
13th Naval District: Bremerton, Washington
14th Naval District: Pearl Harbor, Hawaii
Naval District of Washington, D.C.

NOTE: In several instances command of naval districts has been consolidated with that of co-located naval bases or other commands, or (as in the case of 1st and 3d Districts) grouped in a single headquarters, e.g., 4th Naval District.

Naval districts exercise the functions of public relations, legal, military administration, discipline, intelligence, and disaster planning.

Naval districts enable CNO to maintain coordination control over shore activities. District commandants exercise command over Shore

Establishment activities in their districts—*except* field activities under the Chief of Naval Air Training and Commander, Naval Reserve Training, and the installations of the Marine Corps Supporting Establishment. Bureaus and systems commands may delegate to a district commandant control over some of their functional responsibilities in a district. In the role of grassroots public relations, the district commandants perform important service for the Secretary of the Navy. Closely allied to this, district commandants administer the Naval Reserve program in their districts—except activities assigned to the Chief of Naval Reserve Air Training.

333 ■ Naval Operating Bases

A naval operating base centralizes in one command adjacent activities whose prime responsibility is to support the fleet. At each major naval base complex, a single officer is designated as Commander Naval Operating Base and is in command over all fleet support activities. Whenever (as is in most cases the situation) a major base complex coincides with a naval district headquarters (see Section 332), the Naval Operating Base Commander is also the Naval District Commandant. A naval operating base includes a shipyard, and may include an air station. Commanders of naval shipyards are technically qualified officers skilled in industrial management.

Commanding officers of the component activities of a naval base receive instructions on support and technical matters directly from the responsible agencies in the Department of the Navy.

334 ■ Naval Air Bases

Naval air bases commands comprise Shore Establishment activities furnishing aviation logistic support to the Operating Forces of the Navy.

The commander of naval air bases within a naval district comes under the command and coordination control of the district commandant. Commanding officers of component activities of naval air bases are subject to the command of the air bases commander. Support of such activities stems directly from the Department of the Navy.

Certain air activities in naval districts are not assigned to the Commander, Naval Air Bases. These include:

The Naval Air Training Command
The Marine Corps Air Bases Command, Cherry Point, North Carolina
The Marine Corps Air Facility, Quantico, Virginia
The Naval Air Missile Test Center
Naval Air Material Center

OPERATING FORCES OF THE NAVY

335 ■ General

The Operating Forces of the Navy comprise the four fleets, the seagoing forces, district forces, Fleet Marine Forces, the Military Sealift Command, and such shore activities of the Navy and other forces as may be assigned by the President or Secretary of the Navy. The Chief of Naval Operations commands the Operating Forces of the Navy.

336 ■ Major Components of the Operating Forces of the Navy

The composition of the Operating Forces of the Navy, as assigned by the Secretary of the Navy, is indicated in Fig. 3–2.

Fleet Organization. There are four regularly constituted fleets—the Third and Seventh Fleets in the Pacific (under Commander-in-Chief, Pacific Fleet); the Second Fleet (under Commander-in-Chief, Atlantic Fleet); and the Sixth Fleet (under Commander Naval Forces, Europe). Normally, Commander Third Fleet exercises operational control over all naval forces on the Pacific Coast and Eastern Pacific, as does Commander Second Fleet over similar forces on the Atlantic Coast. The Sixth and Seventh Fleets, in the Mediterranean and Far East respectively, represent the cutting edge of American sea power in those vital regions. Each ordinarily includes one or more embarked "floating battalions," i.e., Marine battalion landing teams (BLTs) or Marine amphibious units (MAUs). In addition, the Second Fleet now maintains floating units of the FMF in the Caribbean and West Indies.

Type Commands. The major fleets are organized into broad categories under commanders whose titles are self-explanatory and generic, such as Commander Surface Forces, Commander Submarine Forces, Commander Air Forces, and Commanding General Fleet Marine Force. The purpose of this "type organization" is to *prepare* and *provide* forces for operations—not to *conduct* operations.

Military Sealift Command (MSC) provides ocean transportation for personnel (including sick and wounded), material (including petroleum products), mail, and cargoes for agencies of the Department of Defense and, as authorized, for other agencies of the Government.

A task force directly under CNO, MSC provides ships and crews for the peacetime needs of the Services, keeps available for emergency a nucleus of auxiliary ships which are gainfully employed, and provides an operating and administrative organization capable of rapid expansion.

Nuclear-powered attack carriers, such as USS Nimitz, shown here, are the backbone of the striking power of the U.S. Fleets.

Naval Forces Europe serve as a naval component command under U. S. CinCEurope.

Naval Forces Caribbean functions as a local area subcommand under Commander-in-Chief Atlantic Fleet for Fleet units and activities in the Caribbean.

337 ■ The "Task Force Principle"

The "task force principle" is the name given to the Navy and Marine Corps system of organizing forces for given tasks while preserving a separate administrative organization for training and housekeeping. This is the fundamental organizational principle of the U. S. Fleets.

"Type Organization." All forces in the U. S. Fleets are grouped into the "type organization" of the Fleet. As its name implies, the type organization is based on types of ships or forces. Note that the Fleet Marine Force is a type command, since it comprises all Marine Corps tactical units—air and ground—assigned to the Fleet.

"Task Organization." The other facet of Fleet organization is the "task organization." The task organization conducts operations, using units prepared and provided by the type organization. Taking the Atlantic Fleet as an example, it includes several permanent *task fleets* and *task forces*. Certain of these, such as the antisubmarine warfare forces, are task-organized from aviation, surface, and submarine forces, to maintain control of the sea. The Second Fleet is a task fleet whose functions are offensive: fast carrier operations and amphibious assault.

Ballistic-missile submarines, such as USS Henry Clay, shown here, are the invulnerable maritime component of U.S. strategic nuclear capabilities.

Under the above system a flexible structure is provided, consisting of Fleets further subdivided into Forces, Groups, Units, and Elements. Each of these descending subdivisions has a numbered designation and appropriate communication call-signs. When a Task Fleet Commander receives a task from higher authority, he can then assign necessary forces under his command to do the job, creating an ad hoc organization of ships and units as needed. Such a "task organization" is adaptable to any magnitude of organization, ranging from the campaigns of entire fleets in general war to a single ship on a temporary mission. For example, an LPH, or helicopter assault ship, might be given a task designation simply to steam across Chesapeake Bay for a Navy Day visit, and, on conclusion of the job, the task designation would cease.

A typical (hypothetical) Task Fleet numbering system would be one in which the Commander Seventh Fleet would assign his major forces to numbered task forces (TF), such as Striking Force, TF 70; Amphibious Task Force, TF 71; Service Force, TF 72; and so forth.

Within each force he would then assign logical subdivisions of that force as task groups (TG), such as TG 70.1, Carrier Group; TG 70.2, Gunfire Support and Covering Group; and so forth.

Within each task group, in turn, would be found task units (TU). For example, TG 70.1, Carrier Group, noted above, might be divided into TU 70.1.1, Carrier Unit, and TU 70.1.2, Destroyer Screen Unit.

Note the fashion in which components of a task organization are designated by addition of decimal separators and successive numbers; this enables you to determine at a glance the chain of operational command in which a given unit may be functioning.

Significance. The simultaneous organization of the Fleets by types and task is obviously complex. It is, nevertheless, a system which

Destroyer USS Hull fires her lightweight rapid-fire 8-inch gun, a new-model weapon capable of hurling a 260-pound projectile 17 miles, and of being mounted on destroyers and other smaller ships for effective naval gunfire support.

adapts itself precisely to the job to be done, large or small, temporary or permanent. Moreover, it is completely flexible and innately economical of forces.

COMPONENTS OF THE UNITED STATES NAVY

338 ■ Composition
The United States Navy consists of the regular Navy and the Naval Reserve.

339 ■ The Regular Navy
Commissioned officers of the regular Navy (and Naval Reserve) are divided among the line and staff corps. (See Fig. 3–3.)

Line. Line officers exercise military command, are accountable for the exercise of their authority, and cannot divest themselves of the accompanying responsibility. Only line officers command at sea. Among line officers are several types of "line" specialists: naval constructors, naval engineers, and specialists in such fields as intelligence, oceanography, communications, and public information. Specialists cannot command at sea. In general, only line officers exercise command ashore, except that members of certain corps, such as Medical, Supply, and Civil Engineering Corps, command shore activities and units (such as Seabees) under the cognizance of their respective bureaus. Although of course not eligible to command at sea, or to command a Navy base or station, Marine officers are nevertheless line officers of the Naval Service, and have been held, legally, to be naval officers.

Medical Corps. This corps is composed exclusively of graduate doctors of medicine, who treat the "sick, lame, and lazy," and administer the hospitals, dispensaries, sickbays, and other medical units of the Naval Establishment. Medical and dental service for the Marine Corps is provided by Navy doctors, dentists, and hospital corpsmen.

Nurse Corps. Navy nurses are commissioned officers in the Nurse Corps, serving in hospitals and dispensaries at home and on foreign station and in hospital ships and transports at sea.

Dental Corps. Composed of graduate dental surgeons, this is a separate corps whose members serve at hospitals and dispensaries and

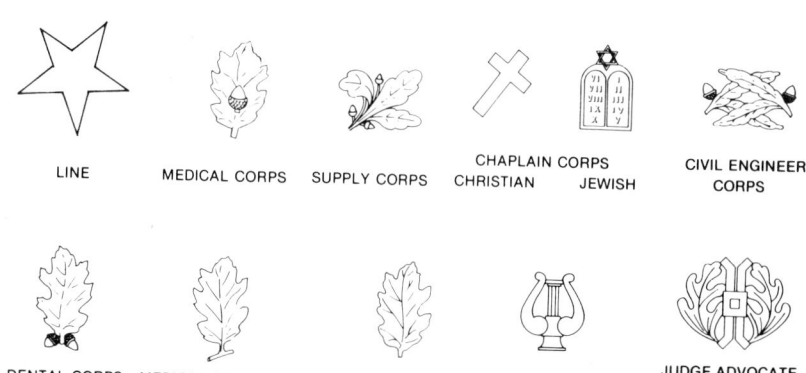

COMMISSIONED OFFICERS

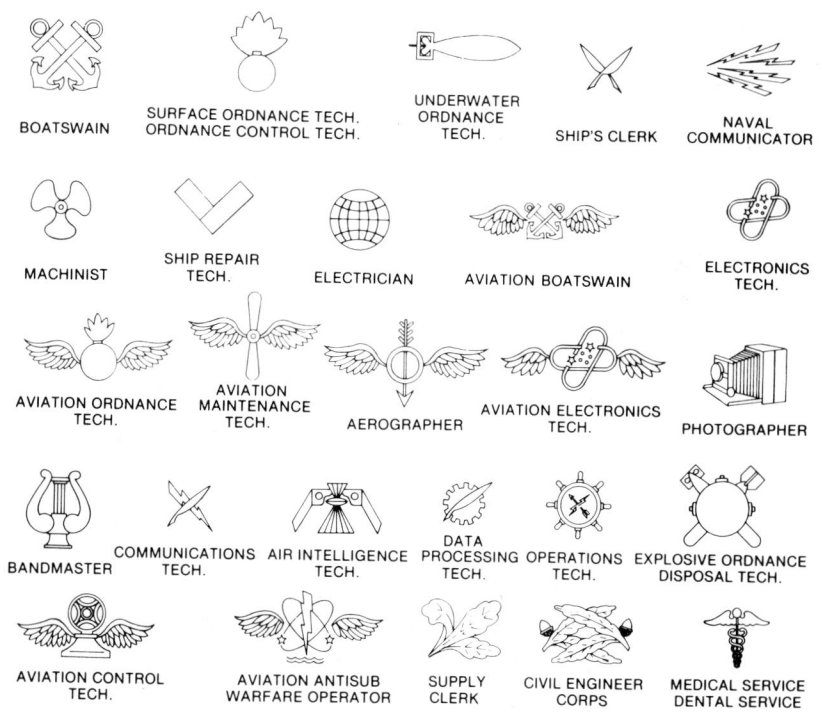

WARRANT OFFICERS

Fig. 3-3: Branch and Corps Devices of Commissioned and Warrant Officers of the Navy (1977).

on board larger ships. The Dental Corps, like the Medical, Nurse, and Medical Service Corps, comes under the Bureau of Medicine and Surgery.

Medical Service Corps. This corps is composed of specialists in the fields of optometry, pharmacy, or such allied sciences as bacteriology, biochemistry, psychology, sanitation engineering, or medical administration and statistics.

Supply Corps. This is the business branch of the Navy, which administers the Navy supply system and receives and disburses funds for supply and for pay, subsistence, and transportation.

Chaplain Corps. Ordained ministers of various denominations, officers of the Chaplain Corps conduct religious services and promote the spiritual and moral welfare of the Navy and Marine Corps. The Chief of Chaplains heads the corps.

Civil Engineer Corps. This corps is composed of graduate civil engineers, normally restricted to shore duty, who supervise buildings, grounds, and plants at shore stations, as well as construction of buildings and the layout of shore stations. This corps conceived, organized, and commanded the Navy construction battalions (Seabees), which served so illustriously beside Marines during World War II, Korea, and Vietnam.

Judge Advocate General's Corps. This corps consists of Navy lawyers who have been duly certified to practice and perform staff-legal and judicial duties under the Judge Advocate General and within the system of military jurisprudence.

Warrant Officers. Navy warrant and commissioned warrant officers possess the most detailed practical knowledge of the complicated mechanisms of our modern Navy, and thus provide an invaluable footing of technical know-how for the Fleet. In spite of recent proposals, based on administrative considerations, that the warrant grades be abolished in favor of limited-duty billets, and that no further warrants be issued, it has now been firmly decided that the Navy warrant officer is here to stay. Warrant grade titles include: Boatswain, Machinist, Electrician, Ship's Clerk, Aerographer, Photographer, Supply Clerk, Ordnance and Mine Warfare Technician, Communications and Electronics Technician, Ship Repair Technician, Equipment and Building Foreman, Bandmaster, Medical and Dental Service Warrant, Aviation Maintenance and Operations Technician, as well as others. (See Fig. 3-3.)

Enlisted Men. Basic legislation allows 500,000 enlisted men in the regular Navy. During an emergency the actual number depends on the size of the Fleet and Shore Establishment to be maintained.

Enlisted men and women of the Navy and Coast Guard are divided into rating groups as illustrated in Fig. 3-4. Just as we expect Navy officers to recognize and identify Marine noncommissioned officers, so a Marine officer should be able to identify Navy petty officers in the various ratings.

340 ■ Education of Officer Candidates

Several systems provide candidates for commissions in the Navy, as follows:

United States Naval Academy, Annapolis, Maryland. The Naval Academy was established in 1845 at old Fort Severn, at the mouth of the Severn River, to train naval officers.

Midshipmen at the U. S. Naval Academy. Frequent drills in military precision instill pride and esprit de corps in prospective officers.

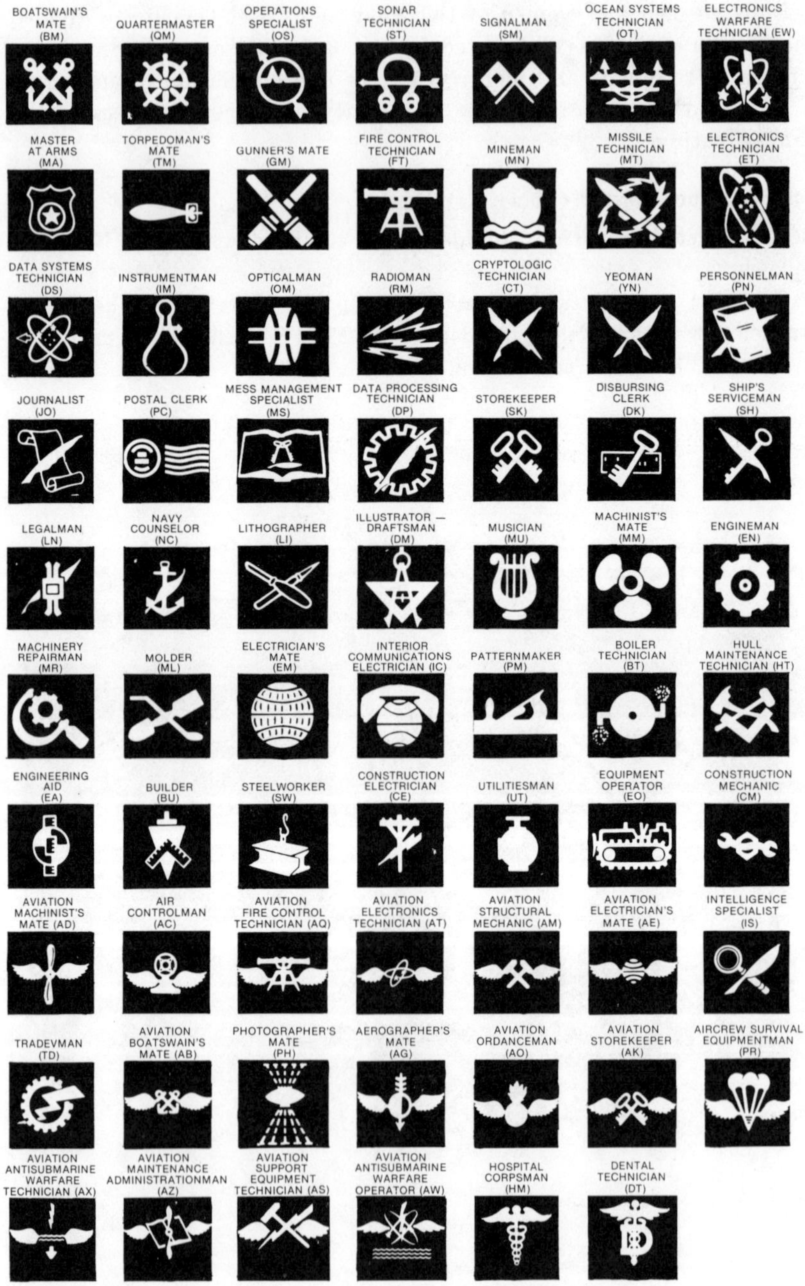

Fig. 3-4: Specialty marks of Navy enlisted ratings (1977).

Since the 1880s the Marine Corps has each year commissioned a number of Naval Academy graduates as second lieutenants, preference being given to sons of Marine officers and to former Marines. At present 16 percent of the midshipmen per graduating class are selected for Marine Corps commissions, the competition for these vacancies being high.

The strength of the Brigade of Midshipmen is maintained by appointments from Senators, Congressmen, and territorial delegates; by competitive appointments from regular and reserve enlisted men and women of the Navy and Marine Corps; and by appointments-at-large by the President and Vice-President. These so-called Presidential Appointments are made on a competitive basis from among the children of regular and reserve officers and children of any members of the Armed Forces killed or disabled in line of duty. In addition, the children of a Medal of Honor man are admitted to the Academy on passing the usual mental and physical examinations.

The Academy is commanded by a Superintendent of flag rank. The staff includes officers of all the Services, as well as civilian professors. The course is four years, and leads to a B.S. degree.

Naval Reserve Officers' Training Corps (NROTC). The NROTC offers the opportunity for young men and women to qualify for Naval and Marine Corps Reserve commissions while attending college. After World War II, NROTC graduates became eligible for regular commissions in the Navy and Marine Corps, with the end result that today the program is maintained for one purpose: to educate and train young men and women for careers as commissioned officers in the Regular Navy and Marine Corps.

Reserve midshipmen are selected competitively in each state. As NROTC students they lead the same academic life as college contemporaries, but, in addition to normal studies, they receive professional training in naval science.

Tuition and certain related expenses are wholly paid for by the Government.

Midshipmen wear uniform when engaged in naval duties, and receive the pay and allowances of a midshipman. After graduation, the NROTC graduate must serve a minimum of four years as a Marine or Navy officer.

Besides the foregoing program for regular-officer candidates, there is also a nonsubsidized NROTC program for college students who expressly wish Reserve commissions in either Marine Corps or Navy.

NROTC units are located at more than 50 institutions throughout the country. To enroll in this excellent program, you should apply

NROTC midshipmen study navigation during a training cruise. After graduation, an NROTC graduate must serve four years as a Marine or Naval officer.

either to Commandant of the Marine Corps, Washington, D.C., 20380; or Chief of Naval Personnel, Washington, D.C., 20350.

Junior NROTC units are maintained by the Navy and Marine Corps at a number of secondary schools, private and public, throughout the country. Instructors are retired officers and staff NCOs. Successful completion of three years of Junior NROTC in a Marine unit gives a year's credit on enrolling in college ROTC, or, if an enlistment is chosen, promotion to private first class.

Officer Candidate School. Located at the Naval Training Center, Newport, R.I., the Navy's Officer Candidate School trains young college graduates as Naval Reserve officers. Candidates hold the enlisted grade of officer candidate. They are obligated to serve three years on active duty as commissioned officers after passing the four-month course, and to continue a total of six years in the Reserve. Meritorious enlisted men selected for integration as Regular Navy officers also attend Officer Candidate School.

341 ■ The Naval Reserve

The purpose, classification, and organization of the Naval Reserve are generally similar to the Marine Corps Reserve, discussed in detail in Chapter 9.

The *Ready Reserve* provides trained officer and enlisted reservists, who, added to qualified personnel from other sources, complete the war organization of the Navy. Officers and men of the Ready Reserve must perform annual training and other duties to keep them ready for immediate mobilization in emergency.

The *Standby Reserve* provides a force of qualified and partially qualified officers and enlisted persons who will, except on personal application, be called to active duty only for war or a national emergency declared by Congress.

The *Retired Reserve* is liable for active duty only in time of war or emergency declared by Congress, or when otherwise authorized by law, in the event sufficient qualified personnel are not available in the Ready and Standby Reserves.

The *Naval Air Reserve* trains at naval air stations about the country, its members functioning in either of the classes of the Naval Reserve described in the preceding paragraphs.

Organization. The basic Ready Reserve unit is the division. A battalion consists of two to five surface or submarine divisions. A brigade is established in cities having more than one battalion. The squadron is the basic unit for the Naval Air Reserve. An air wing consists of a number of squadrons located in the same area. Specialist units exist in communities which have enough reserve specialists to permit such an organization.

In the Standby Reserve, volunteer training divisions and units may be organized in a wide variety of specialties, in order to keep members up to date on their specialties.

NAVAL STAFF ORGANIZATION

342 ■ Organization and Functions of an Operational Naval Staff

This discussion of the naval staff system is important to you because Marine officers serve on every naval staff of any consequence.

The operational naval staff (as in the amphibious force or group headquarters) is of primary interest to Marine officers. You will find that other naval staffs, ashore or afloat, are organized along much the same lines, with special functions added or eliminated as appropriate. For detailed, authoritative information on naval staff organization and procedure, consult *NWP-12, The Navy Staff.*

General Functions and Organization. Naval staff functions come under two headings: *administration* and *planning and operations.* To carry out these functions, the typical large naval staff is organized much like the Office of CNO (see Sections 317–320). This organization comprises:

Personal staff
Coordinating staff (equivalent to USMC executive staff)

Personal Staff. The duties of a Navy personal staff are essentially similar to those listed in Section 521 for the Marine personal staff, although titles differ somewhat.

The Chief of Staff not only carries out the functions his title implies, but is also the admiral's personal aide. He is the senior officer on the staff and coordinates all staff activities, thus serving as a member of the coordinating staff as well. In Navy commands headed by officers below flag rank, this billet bears the title of *chief staff officer* rather than "chief of staff."

The Flag Secretary serves as the admiral's administrative aide and confidential secretary. Like the chief of staff, the flag secretary also has dual status as a member of the coordinating staff, in which he heads the administrative section.

The Flag Lieutenant, in addition to personal services to the admiral, supervises salutes, honors, awards, official calls, uniform, social protocol, and transportation for the admiral and the staff (barge, staff gig, helicopter, and staff cars). The flag lieutenant keeps the staff duty officer (as well as other members of the staff and interested officers of the flagship) apprised of the movements and intentions of the admiral.

Coordinating Staff. The coordinating staff comprises four or five sections, each headed by an assistant chief of staff, as follows:

The *Administrative Section (N-1)* is headed by the flag secretary. The section combines the shoregoing functions associated with G-1, staff secretary, adjutant, and legal officer.

The *Intelligence Section (N-2)* performs naval intelligence functions analogous to those performed by the G-2 ashore. In some Navy staffs, the intelligence officer likewise has cognizance over public information. Marine officers may serve in Navy intelligence sections—primarily in connection with amphibious intelligence.

The *Operations Section (N-3)* performs plans, training, and operations duties comparable to those of the G-3 on shore. In addition, however, the naval operations section has responsibility for matters affecting readiness, and, in amphibious staffs, deals with and conducts the ship-to-shore movement. Marines frequently serve in the operations section, not only as *plans officer*, but as *military operations officer*, both of which titles are self-explanatory.

The *Logistics Section (N-4)* has all the functions of the G-4 Section in a Marine staff. In addition, this section plans the availability for overhaul of ships, screens work requests, supervises maintenance, and administers funds for repair and alterations.

The *Communications Section (N-5)* is sometimes a separate section and sometimes part of the N-1 Section. The communications officer not only supervises communications, but also controls classified publications for the staff. In amphibious staffs, a Marine communication officer usually forms part of the section.

Specialist Officers. Although the Navy organization does not include a Special Staff as such, it does, of course, have certain specialist officers who, even though integrated within and under cognizance of the respective Coordinating Staff sections, in effect comprise a special staff. While the specialties are not as numerous as on a Marine staff, duties are quite similar. The following specialist officers would be included in the typical Navy staff:

Weapons officer	Supply officer
Air officer	Chaplain
Aerological officer	Legal officer
Surgeon	

343 ■ Staff Duty Officer Afloat

Senior line officers (including Marines) take turns as staff duty officer. Their duties and status correspond somewhat to those of the field officer of the day (see Section 1703).

In port the staff duty officer takes a day's duty; he receives routine reports and acts on routine matters in the absence of officers having staff cognizance; he regulates the use of staff boats and tends the side on occasions of ceremony. In emergency, the staff duty officer must be prepared to make decisions when the admiral and chief of staff are unavailable.

Underway the staff duty officer stands watch on the flag bridge and represents the admiral in the same way that the officer of the deck (see Section 2216) represents the Captain. He makes routine reports and signals, supervises navigation and station-keeping of the force, keeps the staff log, and oversees the watch on the flag bridge. To perform efficiently, the staff duty officer must keep informed of current operations, expected hazards, conditions of readiness, launching and recovery of aircraft, joining and detaching of units, fueling and provisioning, and so on. The state of relations between the staff and flagship depends in considerable measure on the attitude and consideration of the staff duty officers.

344 ■ Marine Duties on Naval Staffs

Nowhere more than on a Navy staff is a Marine expected to be "soldier and sailor, too." Thus, when assigned to the Navy, never be surprised, regardless of what duty you find yourself performing. Your only concern should be to see that that duty is well done, so that as a Marine Corps representative you set an example to your Navy colleagues.

Subject to the foregoing, Marines are usually assigned to one or several of the following staff duties:

Staff Marine Officer. As division, squadron, force, or fleet Marine officer, you exercise staff supervision over Marine personnel and matters within your command. In practice this boils down to coordination of landing force activities involving Marines, inspection of ships' Marine detachments, advice and assistance as needed for the several Marine detachment commanders, and supervision of the flag Marine detachment (or "flag allowance"; see Chapter 22). In addition, the staff Marine officer is usually selected to maintain liaison between his staff and any Marine or Army staffs in the vicinity. The senior Marine officer on any Navy staff, regardless of other duties, performs the functions just described.

Military Operations Officer. The military operations officer has cognizance over all military operations ashore in which the naval staff may be involved. In homely language, the job of the Marine military operations officer is to keep the Navy straight in matters of land warfare and organization.

Embarkation Officer. The embarkation officer performs the duties of troop loading, billeting, and landing which are associated with his functions in the Marine staff ashore.

Other Duties. The foregoing jobs are those to which Marine representatives on Navy staffs are usually assigned. In addition, however (according to your capabilities and the needs of the organization), you may find yourself performing any of the following tasks:

Security officer
Logistics officer
Intelligence officer
Air officer
Plans officer

> NOTE: Never acquiesce in the bad old practice (still occasionally recurrent) of allowing yourself to be assigned duty in a staff section headed by a naval officer junior to you. Your lineal precedence, based on date of rank, is as binding on a Navy staff as anywhere else.

Under all circumstances, a decisive naval superiority is to be considered a fundamental principle, and the basis upon which all hope of success must ultimately depend.
—George Washington, 1780

4

Missions and Status of the Marine Corps

401 ■ Introduction

Every American respects the Marine Corps, but a surprising number of people are quite hazy on what the Corps really is and does. Thus it is up to every Marine officer to have precise knowledge of the roles, missions, and status of the Corps.

402 ■ Marine Corps Roles and Missions

The Law. The Statutes of the United States include many provisions, great and small, which affect the Corps. All these provisions have been codified under Title 10 (Armed Forces), U. S. Code.

The "charter" of the Marine Corps, however, has evolved from three laws: (1) the Act of 11 July 1798, "Establishing and Organizing a Marine Corps"; (2) the Act of 30 June 1834, "For the Better Organization of the Marine Corps"; and (3) the National Security Act of 1947 as amended.

The National Security Act, which unified the Armed Services, is the controlling military legislation of this country. It is generally discussed in Chapters 2 and 3. From our point of view as Marines, though, the Douglas-Mansfield Bill (Public Law 416, 82d Congress, 2d Session) has particular importance. This law amended the National Security Act as regards the Marine Corps. Its debates and hearings (1951–1952) contain a mine of information on the Corps.

To summarize briefly, the National Security Act as now amended (see Section 301) makes the following provisions for the Marine Corps:

1. It reaffirms the Corps's status as a Service within the Department of the Navy.

2. It provides for Fleet Marine Forces, ground and aviation.

3. It requires that the combatant forces of the Corps be organized on the basis of three Marine divisions and three air wings, and sets a 400,000-man peacetime ceiling for the regular Corps.

4. It assigns the Corps the missions of seizure and defense of advanced naval bases, as well as land operations incident to naval campaigns.

5. It gives the Marines Corps primary responsibility for development of amphibious warfare doctrines, tactics, techniques, and equipment employed by landing forces.

6. It seats the Commandant of the Marine Corps in co-equal status with members of the Joint Chiefs of Staff whenever matters of Marine Corps interest are under consideration.

7. It affords the Marine Corps appropriate representation on various joint Defense Department agencies, notably the Joint Staff.

8. It assigns the Marine Corps collateral missions of providing security forces for naval shore stations; providing ships' detachments; and performing such other duties as the President may direct.

In taking stock of Marine Corps missions found in law, it is important not to overlook the short phrase, ". . . *and shall perform such other duties as the President may direct.*" This phrase, which the Unification Act quotes directly from the 1834 Marine Corps law, stems in turn from similar language in the Act of 1798. It validates in law Marine Corps functions which transcend the Corps' purely naval missions. In mid-1951, the House of Representatives Armed Services Committee highlighted the significance of this clause, in a trenchant summary:

> It is, however, the Committee view that one of the most important statutory—and traditional—functions of the Marine Corps has been and still is to perform "such other duties as the President may direct."
>
> The campaign in Korea, in which the 1st Marine Division and the 1st Marine Air Wing are presently participating, can hardly be called a naval mission. Practically every war involving the United States has found the Marine Corps performing duties other than naval. Indeed, the first two battalions of Marines raised in this country were raised specifically for service before Boston with General Washington's army.
>
> Many Marine activities in the War of 1812 involved only land fighting; in the 1840's the Marines saw "the Halls of Montezuma" while fighting with the Army in the War with Mexico; in the early 1900's Marine activities in Central America were repeatedly entirely of a land nature; their participation in the fighting in the Boxer Uprising in China in 1900 likewise was of a land nature; certainly when in May 1917 President Wilson or-

dered the 4th Marine Brigade to serve as part of the Army's 2d Division in the Battles of Belleau Wood, Aisne-Marne, St. Mihiel, Blanc Mont, and Meuse-Argonne, and later in the occupation forces, these can hardly be described as naval missions; nor can the activities of Marine Maj. Gen. John A. Lejeune, in commanding for a time the Army's 2d Division in France, be called a naval function; nor could the service of Marine aviation in France during 1918 be accurately termed a naval activity.

It is difficult to see how the sending of Marines as the initial force to hold Iceland prior to the last war, until relieved by Army troops, could accurately be called a naval mission; how the reinforcing of Corregidor by the 4th Marine Regiment sent from China just before war broke out, could be accurately termed a naval action; and if the actions of the 1st Marine Division on Guadalcanal, commencing the first American attack of the war on August 7, 1942, can accurately be called a naval action, then in the same fashion the activities of Army divisions in this area must likewise be so termed. It further is worthy of note that on Mindanao and Luzon in the Philippines, in the last ground action against the enemy in World War II, Marine Air Groups 12, 14, 24 and 32, gave close air support to the 24th, 31st, and 41st Infantry Divisions—an activity that appears to the Committee to be only distantly related (if at all) to exclusively naval activities.

The Committee must also call attention to the fact that, after V-J Day, the V Amphibious Corps, USMC, was part of the forces sent to occupy the Japanese Home Islands; the III Marine Amphibious Corps was sent to North China to accept the surrender of Japanese troops there; that a Marine division, with other forces, was kept in China until the summer of 1947 during the attempt of the United States to settle civil war between the Chinese Government and Chinese Communists. It is a strained construction, indeed, of military activities to characterize such employment of the United States Marines as essentially naval in character.

In line with the foregoing, Marines have on several past occasions been temporarily detached by executive order of the President to service under the Secretary of War. This procedure is now less likely to occur because today's unified command arrangements enable the same operational result to be attained without the administrative complications which such a transfer entails. The last occasion on which Marines were detached to service under the Army in this way was in July 1941, when the 1st Provisional Marine Brigade in Iceland was assigned to the Army by President Roosevelt. Note the distinction between administrative transfer of Marines to Army duty (which can only be effected by order of the President) and operational attachments under unified command, which occur frequently—as was the case throughout the greater part of the Korean and Vietnamese wars.

Before we pass on, further reference to Public Law 416 is in order.

Public Law 416 set up today's procedure whereby the Commandant of the Marine Corps sits with the Joint Chiefs of Staff, and it likewise solidifies in law the modern organization of Marine combat forces into divisions and air wings. "Public Law 416," wrote General Lemuel C. Shepherd, Jr., 20th Commandant of the Marine Corps:

> ... expresses clearly the intent that the Marine Corps shall be maintained as a ready fighting force prepared to move promptly in time of peace or war to areas of trouble. It recognizes that in the future there may be a series of continuing international crises —each short of all-out war, but each requiring our nation, in response to its global responsibilities, to move shock forces into action on the shortest of notice. Finally there is evidenced in the law a determination to safeguard the amphibious force-in-readiness aspect of the Marine Corps, and action in this respect is taken ensuring the Commandant direct access to the policy-making agencies of the National Military Establishment.

So much, then, for the main provisions of law which give the Marine Corps its roles and missions. Note that, while those roles are carefully spelled out, the law nevertheless allows employment of Marines anywhere, on any service the President may desire.

Additional Missions of the Marine Corps. In addition to the missions expressly assigned by Congress, the Corps also performs several tasks either assigned by the Department of Defense or in accordance with long-standing custom.

"The Functions Paper." Originally known as "the Key West Agreement" (see Sections 214 and 301) the Defense Department directive which states the functions of that Department and of its major components is now usually spoken of as "the Functions Paper." This directive is essentially a compilation of inter-Service agreements dating from 1948 and later revised from time to time, as to how the roles-and-missions provisions of the National Security Act are to be implemented. In addition, the directive establishes a number of Service relationships and common functions within the Defense Department, which affect the Marine Corps equally with the other Services. Portions of this paper that specifically provide for Marine Corps functions are as follows:

> ... To maintain the Marine Corps, having the following specific functions:
> (1) To provide Fleet Marine Forces of combined arms, together with supporting air components, for service with the Fleet in the seizure or defense of advanced naval bases and for the conduct of such land operations as may be essential to the prosecution of a naval campaign. These functions do not contemplate the creation of a second land Army.*

(2) To provide detachments and organizations for service on armed vessels of the Navy, and security detachments for the protection of naval property at naval stations and bases.

(3) To develop, in coordination with the other Services, the doctrines, tactics, techniques, and equipment employed by landing forces in amphibious operations. The Marine Corps shall have primary interest in the development of those landing forces doctrines, tactics, techniques, and equipment which are of common interest to the Army and the Marine Corps.

(4) To train and equip, as required, Marine forces for airborne operations, in coordination with the other Services and in accordance with doctrines established by the Joint Chiefs of Staff.

(5) To develop, in coordination with the other Services, doctrines, procedures, and equipment of interest to the Marine Corps for airborne operations. . . .

Although many of the foregoing provisions stem directly from and actually include language of the National Security Act, you should never confuse the Functions Paper, only a departmental directive, with the National Security Act, which is the law and thus governs in any disagreement.

"Unified Action Armed Forces" (UNAF) is a doctrinal publication which governs the activities of two or more of the Armed Services when operating together, and prescribes joint procedures and responsibilities which apply to the Marine Corps along with sister Services. Insofar as the Corps is concerned, one of the more important portions of *UNAF* is that which delineates the amphibious and landing force developmental responsibilities of the Marine Corps.

State Department Guards. Under authority of the Foreign Service Act of 1946, the Marine Corps has a collateral mission of providing security guards for American embassies, legations, and consulates. For this duty, which demands the highest discretion and trust, the Marine Corps furnishes over a thousand officers and men who are distributed throughout more than 110 State Department overseas posts.

White House Duties. Dating from 1798, the scarlet-coated Marine Band has been styled "The President's Own" because of its privilege of providing the music for state functions at the White House. Similarly, Marines have established and guarded Presidential camps at Rapidan, Virginia, Warm Springs, Georgia, Camp David, Maryland, and elsewhere, while Marine helicopters were the first to carry a President and still routinely do so.

*The Marine Corps, though repeatedly overruled, has consistently opposed inclusion of the meaningless "second land Army" phrase at this point in the Functions Paper, since nothing could be further from the objectives or interests of the Corps.

Unwritten Missions. Nowhere do the statute books say that the Marine Corps is *the national force in readiness*, yet our history demonstrates clearly that the fundamental mission of the Corps is just that, and always has been. To quote former Assistant Secretary of the Navy John Nicholas Brown:

> Readiness, the capacity to move anywhere immediately and become effective, is always needed and at the present juncture of events is especially necessary. This is the daily bread of the Marine Corps.

In close corollary to this traditional mission is the Corps' worldwide service, in times of nominal peace, as "State Department Troops" (as during the Dominican revolt) for enforcement or protection of U.S. foreign policy, under direction of the Department of State.

403 ■ Status of the Marine Corps

"The Marine Corps is *sui generis*" ("something entirely of its own sort"), once ruled a Federal judge when construing the legal status of the Corps. This is probably the best one-sentence characterization the Marine Corps ever had.

The Marine Corps is one of the several Armed Services (Army, Marine Corps, Navy, and Air Force) which, with the Coast Guard (when attached to the Naval Establishment in time of war), comprise the Armed Forces of the United States. It is important that you be aware of this, since you may sometimes encounter the erroneous term "the three Services," which is usually a term of exclusion so far as the Marine Corps is concerned.

Side by side with the Navy, the Marine Corps is one of two military Services in the Naval Establishment, under direct control and supervision of the Secretary of the Navy. Detailed explanation of the relationship of the Marine Corps to the Navy may be found in Section 404. For the moment we can let the subject rest with the words of Representative Carl Vinson, distinguished former Chairman of the House Armed Services Committee:

> The fact is that the Marine Corps is and always has been, since its inception 175 years ago, a separate military service apart from the United States Army, the United States Navy, and the United States Air Force.

Some account of the evolution of the status of the Marine Corps is useful knowledge for you as a Marine officer.

The Act of 11 July 1798, reconstituting the Corps after its post-Revolutionary War hiatus, provided for a Corps of Marines, ". . . *in addition to* the present military establishment." In line with this thought, although the Corps' distinct status from the Navy was never questioned, it was nearly 40 years before the Corps was firmly dissociated from the Army. During this time, while on shore, Marines (like the British Marines) were promoted, paid, rationed, and disciplined under Army Regulations—practices sanctioned not only by custom but by express rulings handed down from time to time by the Attorney General of the United States.

To clarify the status of the Marine Corps, Congress in 1834 affirmed the Corps as a separate Service, but placed it unequivocally under the Secretary of the Navy and therefore under *Navy Regulations* ". . . except when detached for service with the Army, by order of the President."

For more than a century the Acts of 1798 and 1834 governed the status of the Marine Corps. In 1947 the National Security Act became law. This law as amended by Public Law 416 not only spells out the missions of our Corps today but defines the Corps in declaratory language as one of the four Services given statutory missions under the Act. In the first years of unification there was some tendency to assume that the National Security Act had intended to tri-elementalize the Armed Forces on a three-Service basis, with the Marine Corps merely a specialist branch of the Navy. This misconception was set at rest with some emphasis in the debate and hearings on Public Law 416, during which Congress avowed that the Marine Corps was not a mere appendage, but a Service in its own right.

A final and legally definitive ruling on the foregoing point is to be found in House Report 970, 84th Congress. This is the report by the House of Representatives on the codification of Title 10 (Armed Forces), U. S. Code. This report states:

> . . . the legislative history of Public Law 432, the National Security Act of 1947, and Public Law 416 of the 82nd Congress . . . clearly indicates that the Marine Corps is legally a separate and distinct military service within the Department of the Navy, with individually assigned statutory responsibilities, and that the Commandant directs and administers the Marine Corps under delegated command of the Secretary of the Navy.

The status of the Marine Corps can be summed up thus:

1. The Marine Corps is a separate military Service possessing distinct statutory roles and missions prescribed by the Unification Act;

2. The Marine Corps is a part of the Naval Establishment (or Department of the Navy) and comes directly under the Secretary of the Navy;

3. The Commandant of the Marine Corps commands the Corps as a whole, and is directly responsible to the Secretary of the Navy in a

well defined historical and legal relationship for the total performance, administration, readiness, discipline, and efficiency of the Corps.

404 ■ The Marine Corps and the Department of the Navy

The brotherhood between Marine Corps and Navy is of such long standing, so close and so smooth in operation, that the casual observer may be readily pardoned the erroneous conclusion that the Marine Corps forms part of the Navy, or vice versa.

As we have seen, this is not the case. To quote General C. B. Cates, 19th Commandant of the Marine Corps:

> The partnership between the Navy and Marine Corps had its legal birth more than 150 years ago when Congress placed both Services—which were then some 25 years of age—under a newly created Secretary of the Navy. The partnership was a close one initially, and it grew even closer with the passage of time. Today it is so close that only a handful of people—inside the Naval Services as well as outside—realize that technically the Navy and Marine Corps are separate Services under the command of the Secretary of the Navy. *Practically speaking*, the Navy and Marine Corps have lived, worked, and fought together since their inception.

To understand the place of the Marine Corps in the Naval Establishment, you must first understand exactly what constitutes the Department of the Navy, or, as it was called for many years in the past, "the Naval Establishment." As stated in *Navy Regulations*, and in Chapter 3 of this *Guide*, the Naval Establishment (i.e., the Department of the Navy) embraces all activities committed to the care of the Secretary of the Navy, and thus includes the Marine Corps. This does not make the Marine Corps a part of, but rather a partner of, the Navy proper.

The Marine Corps and Public Law 432. The law which defines the position of the Chief of Naval Operations in the Navy is Public Law 432, 80th Congress (original House of Representatives title, "H.R. 3432").

Casual reading of parts of this law by a person not conversant with the intent of Congress in framing it (or of the Navy Department in seeking it) might suggest that this act could be construed as placing the Marine Corps under command of the Chief of Naval Operations. To save confusion on this it is enough to quote from an official letter by Secretary of the Navy John L. Sullivan, on 17 December 1947, to General A. A. Vandegrift, 18th Commandant of the Marine Corps:

> The Commandant of the Marine Corps is informed that it is not the intent of the Navy Department, in seeking enactment of H.R. 3432 [Public Law 432], to alter the Commandant's direct

responsibility to the Secretary of the Navy for the administration and efficiency of the Marine Corps.

The Navy Department interprets neither Executive Order 9635 [an earlier directive defining the wartime position of the Chief of Naval Operations] nor H.R. 3432 as interposing the Chief of Naval Operations in the administrative chain of responsibility between the Secretary and the Commandant, or as otherwise modifying the historical relationship between the Secretary and the Commandant.

/s/ John L. Sullivan

The Marine Corps as a Naval Service. Attempts are sometimes made to show that the term "naval service" has a specific, technical organizational meaning which includes both the U. S. Navy and the U. S. Marine Corps, so that together they may be said to constitute one military service—the "naval service." The claim that this term has such a meaning is baseless. Historically, the phrase "naval service" originated as a matter of convenience for the Secretary of the Navy in issuing orders affecting all military personnel under his jurisdiction. It has occasionally been used for comparable specific purposes in statutes, mainly dating many years back, dealing with personnel administration or discipline, and with no general or consistent construction or definition of the term in question. That Marines, within the meaning and purposes of these statutes and regulations, are "members of the naval service" has long been accepted without dispute; but, as the codification of Title 10, U. S. Code, underscores (see Section 403), the Marine Corps is, nevertheless, a legally distinct and separate military service. Therefore, the best usage when the term "naval service" arises in connection with Marines or the Marine Corps, is to pluralize it as "the naval services," since there are always two naval services—the U. S. Navy and the U. S. Marine Corps—within the Department of the Navy, and, when the Coast Guard is assigned in time of hostilities, there are three.

Status of the Commandant of the Marine Corps. Much of the original and distinct status of the Marine Corps derives from implicit powers which vest in the office of the Commandant. As one legal authority has written:

> The word, "Commandant," was commonly used in legislation to designate an officer placed in command of a body of troops. It gave an officer so designated certain rights, duties, and power well defined by the rules and disciplines of war, customary military law, and, in the case of the Commandant of the Marine Corps, the usage of the sea. It is clear that the duties and functions were presumed by Congress to be so well understood as not to require specific mention. The principles by which one denominated Commandant was to govern himself in the execution of his office were unwritten, but long continued practice common to men

in the profession of arms had given these principles a binding force, and they were universally recognized. By the very use of the term, "Commandant," the Congress provided the Commandant of the Marine Corps a wide range of implied powers.

Navy Regulations recognize the foregoing by providing that the Commandant of the Marine Corps is a *naval executive* of the Secretary of the Navy. He is not, as is sometimes still wrongly asserted or believed, a technical or professional assistant like the Navy Department bureau chiefs.

As a naval executive, the Commandant reports directly to the Secretary of the Navy in matters pertaining to the Marine Corps, just as the other naval executive, the Chief of Naval Operations, reports to the Secretary in Navy matters. When the Secretary so directs, the Commandant may accept from the CNO orders expressly given in the name of the Secretary.

Both *Navy Regulations* and Navy Department general orders summarize the responsibilities of the Commandant as follows (see Section 314):

1. Command, administration, discipline, organization, training, efficiency, and readiness of the Marine Corps;
2. Development of tactics, techniques, doctrines, and equipment used by landing forces in amphibious operations;
3. Provision of technical advice to the Secretary of the Navy and Assistants, in the formulation of naval policies and procedures;
4. Marine Corps representation on the JCS (see Section 210).

Working Relations Between Marine Corps and Navy. Although we have emphasized the legally and intrinsically separate status of the Marine Corps in the Naval Establishment, it should not be supposed that this prevents harmonious working relations between the Marine Corps and the Navy.

Not only do individual Marines serve as part of Navy commands, and vice versa, but units are likewise freely interchanged. Every Navy staff of any consequence includes a Marine officer or officers, while all Marine units and stations have Navy doctors, dentists, chaplains, and hospital corpsmen. In addition the Fleet Marine Force includes naval gunfire liaison officers—Navy line officers who, as FMF staff officers, help to obtain gunfire support.

Each major combatant ship of the Fleet has a Marine detachment, and most Navy shore stations boast Marine barracks or detachments for security purposes. On the other hand, Navy units such as Seabees, naval beach groups, and so on, are frequently attached to the Marine Corps.

At higher levels, the Fleet Marine Force itself (see Section 505) best exemplifies the close relationship between Marine Corps and Navy. Here we have major Marine operating forces assigned by the Secretary of the Navy on a continuing basis to duty with the fleets, and, while so assigned, (operationally under the CNO) just as much part of the fleet as its ships or aircraft. Side by side with this operational relationship, however, the Commandant of the Marine Corps retains full control over the administration, readiness, and military efficiency of the units concerned. And all hands, Marine Corps and Navy, are governed alike by *Navy Regulations*.

In the words of former Secretary of the Navy Robert B. Anderson, *"They are, in every sense of the word, a team."*

405 ■ Summary

To summarize the missions and status of the Marine Corps, you may find them prescribed in or derived from the National Security Act as amended. Public Laws 432 (80th Congress) and 416 (82d Congress) supplement and affirm the National Security Act, as does the Functions Paper (even though without statutory standing). And the Defense Department manual *Unified Action Armed Forces* assigns collateral missions to the Corps and governs the mechanics of its relations with the other Services. For a detailed discussion and analysis of these and related questions by a highly qualified authority, turn to "The Legal Status of the U. S. Marine Corps," by Colonel H. M. Hoyler, in the *Marine Corps Gazette*, November 1950.

The status of the Marine Corps within the Department of the Navy can best be described in the words of Vice Admiral O. W. Colclough, while Judge Advocate General of the Navy:

> The Marine Corps has been held for years to be a separate Service, although it operates with the Navy, and under the Secretary of the Navy.

Over and above its usual status and duties within the naval framework, the Marine Corps may be, and frequently has been, assigned other duties and status elsewhere in the executive branch, under the plenary powers which the President possesses with regard to the Corps.

In the vast complex of the Department of Defense, the Marine Corps plays a lonely role.
—The Honorable John Nicholas Brown

5

Organization of the Marine Corps

501 ■ Introduction

Major General W. S. ("Bigfoot") Brown, one of the Corps' most loved old-timers, lecturing at a Service school, began with these words, "Well, gentlemen, they've given me the job of describing the organization of the Marine Corps. This surprised me somewhat, because I never knew we had any organization. . . ."

Despite that prologue, the Marine Corps does have an organization. And precise knowledge of that organization is one of the first things a Marine officer must acquire. (See Fig. 5–1.)

The Marine Corps is made up of land combat, security, and service forces; Marine Corps aviation; and the Marine Corps Reserve. In many ways the organization of the Corps resembles that of the Navy. Like the Navy, the Marine Corps is organized into three principal subdivisions:

Marine Corps Headquarters
Marine Corps Operating Forces
Marine Corps Supporting Establishment
    (including the Reserve Establishment)

Throughout these components, Marine Corps aviation is included as necessary to carry out the missions of the Corps.

MARINE CORPS HEADQUARTERS

502 ■ Headquarters, U. S. Marine Corps

Marine Corps Headquarters, in Washington, D.C., is the executive part of the Corps. It is located in Arlington Annex of the Navy De-

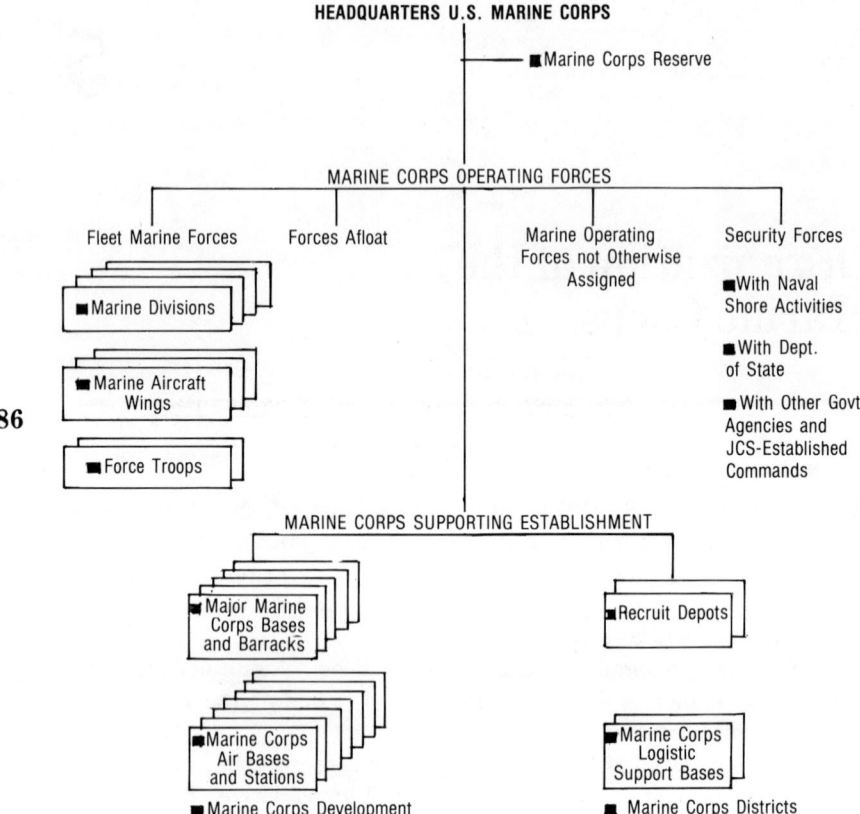

Fig. 5–1: Organization of the Marine Corps (1977).

partment. Headquarters is literally the headquarters of the Commandant, and could, in theory, take the field, as it actually did on occasion in the nineteenth and early twentieth centuries.

Although Marine Corps Headquarters exercises some technical functions like those of the Navy Department bureaus, the mistake should not be made of concluding that Headquarters is merely another bureau, or on the same level as the bureaus. An organization chart of Marine Corps Headquarters appears in Fig. 5–2.

The Commandant of the Marine Corps.

> I want each of you to feel that the Commandant of the Corps is your friend and that he earnestly desires that you should realize this. At the same time, it is his duty to the Government and to the Marine Corps to exact a high standard of conduct, a strict performance of duty, and a rigid compliance with orders. . . .

Marine Corps Headquarters occupies Arlington Annex of the Navy Department, atop a hill overlooking the Pentagon.

In those simple terms, one Commandant defined his responsibilities. Phrasing those responsibilities today, we can say that the Commandant is directly responsible to the Secretary of the Navy for the readiness, total performance, and administration of the Marine Corps as a whole, including the Reserve. He commands all Marine forces and activities except those assigned to the Naval Operating Forces or elsewhere. For the readiness and performance of those elements of the Marine Corps operating forces assigned to the Operating Forces of the Navy (i.e., the Fleet Marine Forces), the Commandant is also responsible to the Chief of Naval Operations.

The Commandant is appointed by the President, from among the active general officers of the Corps, with the advice and consent of the Senate, for a four-year term. He holds the rank of general. The Commandant may be, and frequently has been, reappointed for more than four years. Archibald Henderson, 5th Commandant, who held office more than 39 years, has the record.

The principal duties of the Commandant extend, but are not limited, to: procurement, discharge, education, training (individual and unit), and distribution of the officers and enlisted men of the Corps, and all matters of command, discipline, requirements, readiness, organization, administration, equipment, and supply of the Marine Corps. You will find a list of the Commandants in Appendix II; if you are interested in the Commandant's legal status, turn to Section 404.

The Assistant Commandant. The Assistant Commandant is a general who discharges the duties of the Commandant during the latter's absence or disability, and performs such other duties as the Commandant may direct.

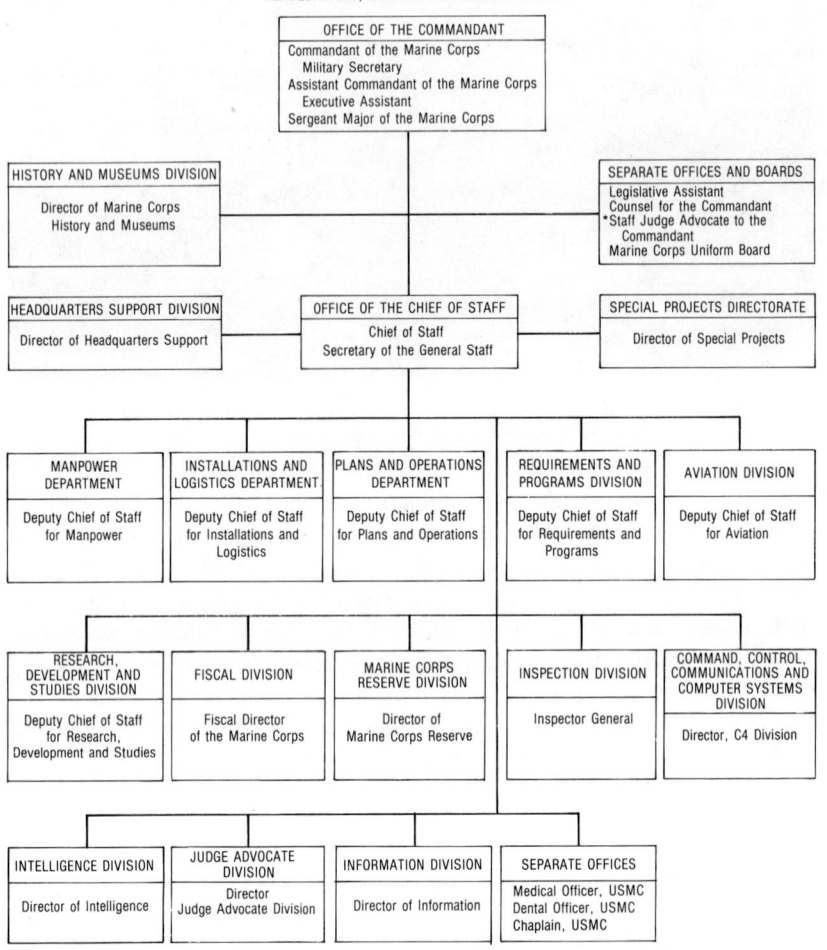

Fig. 5-2: Organization of Marine Corps Headquarters (1977).

The Chief of Staff. The Chief of Staff, a lieutenant general, is the Commandant's executive officer. He directs, coordinates, and supervises staff activities of Marine Corps Headquarters.

Deputy Chiefs of Staff. There are six Deputy Chiefs of Staff. The Deputy Chief of Staff (Plans and Operations), a lieutenant general, acts for the Chief of Staff in his absence; has cognizance over unit training and readiness, force structure, amphibious and other doctrinal matters; and represents the Marine Corps in certain joint Service matters. The Deputy Chief of Staff (Manpower), a lieutenant general, has cognizance over manpower planning, budgeting, programs, management and administration, and individual training. The Deputy Chief of Staff (Installations and Logistics), a general officer, has cognizance over matters relating to logistics policy and management, and facilities and installations. The Deputy Chief of Staff (Requirements and Programs), a general officer, has cognizance over Marine Corps programming, requirements, and systems/cost analysis. He also represents the Marine Corps in certain external functions related to his areas of cognizance. The Deputy Chief of Staff (Aviation), a lieutenant general, has cognizance over matters relating to Marine Corps aviation. In addition, as *ACNO Marine Aviation* (Op-05M), he administers, on behalf of the Deputy Chief of Naval Operations (Air), all matters pertaining to Marine Corps aviation in order to ensure maximum cooperation and integration with naval aviation as a whole. The Deputy Chief of Staff (Research, Development, and Studies), a general officer, has cognizance over all Marine Corps R&D matters.

The Military Secretary manages the immediate office of the Commandant.

The Sergeant Major of the Marine Corps is the senior noncommissioned officer of the Corps and, by virtue of his billet, is senior to all other enlisted Marines. He advises and assists the Commandant in all matters within his cognizance.

The Secretary of the General Staff assists the Chief of Staff and staff by coordinating staff action, and assures that staff matters presented to the Chief of Staff and to the Commandant are complete.

The Legislative Assistant to the Commandant is the Commandant's principal adviser in legal and legislative matters, including liaison with Congress. He prepares comments on legislative proposals referred to, or affecting, the Marine Corps (except cases in the province of the Fiscal Director).

The Director of Special Projects assists the Commandant in matters relating to briefing, symposia, foreign visits, and preparation of speeches and articles.

The Director of Headquarters Support has cognizance over administration and management services for HQMC, headquarters security, transportation, internal communications services, and military and civilian personnel for the headquarters.

The Director of Intelligence has cognizance over intelligence, counterintelligence, cryptology, and electronic warfare and (as a Service intelligence chief) maintains liaison with other Government intelligence agencies. He also disseminates intelligence information within Marine Corps Headquarters.

The Director, Judge Advocate Division serves as Staff Judge Advocate for the Commandant and has cognizance over all legal matters (except certain questions of business or budgetary law, which fall to the Counsel to the Commandant or the Fiscal Director).

The Director of Command, Control, Communications, and Computer Systems (C4) has cognizance over all Marine Corps automated data processing and programs (ADP), command and control systems, and telecommunications and communications security.

The Director of Marine Corps History and Museums. The Director of Marine Corps History and Museums has cognizance over all Marine Corps historical programs and museums, as well as maintenance of historical reference and library functions, and the Marine Corps archives.

Director of Information. The Director of Information heads the Division of Information. It is his delicate, exacting, and sometimes thankless job to represent the Marine Corps to the public. The Director of Information maintains liaison with Defense Department and Government information agencies, and with national information and news media. He also supervises Marine field activities which disseminate information.

Inspector General. The Inspector General heads the Inspection Division. It is the Inspector General's eagle-eyed responsibility to conduct inspections and investigations as directed by the Commandant; to supervise the audit of nonappropriated funds (such as recreation and exchange funds); and to maintain liaison with the inspection agencies of the other Services. It is worth comment at this point that, although an "IG Inspection" invariably begets trepidation and soul-searching, the mission of the Inspector General is to help and to improve by constructive inspection.

Fiscal Director. The Fiscal Director heads the Fiscal Division. It is his job to plan and coordinate the Marine Corps budget; to present and justify that budget to other Defense Department agencies and to Congress; to supervise the spending of appropriated funds; to disburse funds for military and civilian pay and general expenses; and to analyze, record, and report on expenditures under the Department of Defense "performance budget" procedure.

Director, Marine Corps Reserve. The Director, Marine Corps Reserve, is responsible for plans, programs, and administration of the Marine Corps Reserve (turn to Section 903).

Manpower Department. Of all agencies in Marine Corps Headquarters, the Manpower Department has more directly to do with you than any other.

"Personnel" selects and procures you (just as it recruits enlisted Marines). It gives you your commission and administers you from the moment you are sworn in until you rest beneath the trees in Arlington. The Manpower Department details you, manages your career, promotes you, and retires you. With one hand, if need be, it disciplines you, while with the other it attends to your welfare. It maintains your records at Marine Corps Headquarters, as it maintains similar records on every officer and man in the Corps. If you become a casualty, the Manpower Department notifies your next of kin, gives you your Purple Heart, and sees that you get the decorations and medals you have earned. Should you have a claim against the Government, Manpower adjudicates.

THE MARINE CORPS OPERATING FORCES

503 ■ Marine Corps Operating Forces

The Marine Corps Operating Forces fall into four categories: Marine Corps Operating Forces assigned to the Operating Forces of the Navy or to unified commands; Marine Corps Operating Forces assigned to shore activities of the Naval Establishment; Marine Corps Operating Forces assigned to the State Department; and Marine Corps Operating Forces not otherwise assigned.

504 ■ Marine Corps Operating Forces with the Operating Forces of the Navy

When Marine units are assigned to the Operating Forces of the Navy, they come under operational control of the Chief of Naval Operations, except for administration and individual and intra-unit training, which of course always remain under the Commandant. Marine units so assigned fall into two categories: *Fleet Marine Forces* and *ships' detachments*. (See Chapter 22).

505 ■ Fleet Marine Forces

The Fleet Marine Forces, or FMF, constitute the bulk of the Marines assigned to the Operating Forces of the Navy. Both the Atlantic and Pacific Fleets include Fleet Marine Forces. The FMFs are integral components in the Fleet organization and enjoy status as "type-commands" (see Section 336). This Marine expeditionary force integral

The FMF constitutes the cutting edge of America's amphibious assault deterrent.

to the Fleet was created by Major General John H. Russell, 16th Commandant.

A Fleet Marine Force is a balanced force of combined arms, including air. It consists of a headquarters, force troops (i.e., nondivisional units which reinforce and support the divisions), service and support units, one or more Marine divisions or brigades, and one or more Marine aircraft wings. Fleet Marine Forces are organized, trained, and equipped for the following jobs:

1. Service with the fleets in seizure and defense of advanced bases, and for land operations related to naval campaigns;
2. Development of amphibious tactics, technique, and equipment;
3. Training the maximum number of Marines for war or emergency expansion;
4. Immediate expeditionary service where, when, and as directed.

The FMF today includes three combat divisions and three air wings. Based on combat experience, the ratio of one air wing to support one Marine division is the fundamental proportion in the Marine air-ground team.

In addition to the divisions and air wings, Fleet Marine Force troops include extra artillery, engineers, armor, motor transport, amphibious reconnaissance, air and naval gunfire liaison company, service troops, and numerous specialized units which may be required to form balanced task forces for any kind of operations (see Fig. 5–4).

Tables of Organization ("T/O's") spell out the organization of every FMF unit, right down to the individual Marine, his duties, his rank, his specialist qualifications, and his weapons. *Tables of Equipment* ("T/E's") list the equipment required by each unit, and *Tables of Allowance* ("T/A's") give basic allowances of standard items, such as bunks, helmets, and cleaning gear—to cite examples—which vary in direct proportion to the number of men in a unit. *Know your unit's T/O inside out*, and acquire more than a nodding acquaintance with your T/E and T/A and with the organization of other FMF units with which you are in immediate contact.

The *Marine Division* (see Fig. 5-3) is the ground fighting organization of the Marine Corps. The division is a balanced force of combined arms and services, but it does not include organic aviation. The division comprises about 19,000 officers and men, more than half of whom serve in the three infantry regiments which are the division's cutting edge. To support these infantry regiments, the Marine division includes an artillery regiment, a support group, special teams to control air and naval gunfire support, engineers, shore party, motor transport, and medical, signal, and other troops normal for a force of combined arms. During World War II the Corps reached an all-time high of six divisions.

Fleet Marine Force aviation supports "grunts" of the 1st Marine Division during the assault of Peleliu, September 1944.

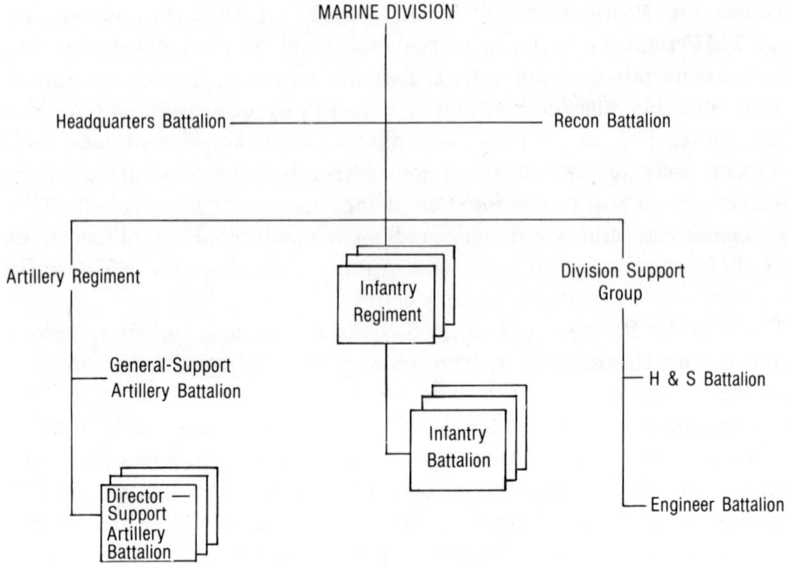

MARINE DIVISION
Headquarters Battalion
Recon Battalion
Artillery Regiment
Infantry Regiment
Division Support Group
General-Support Artillery Battalion
H & S Battalion
Infantry Battalion
Director — Support Artillery Battalion
Engineer Battalion

Fig. 5–3: Organization of a Marine Division (1977).

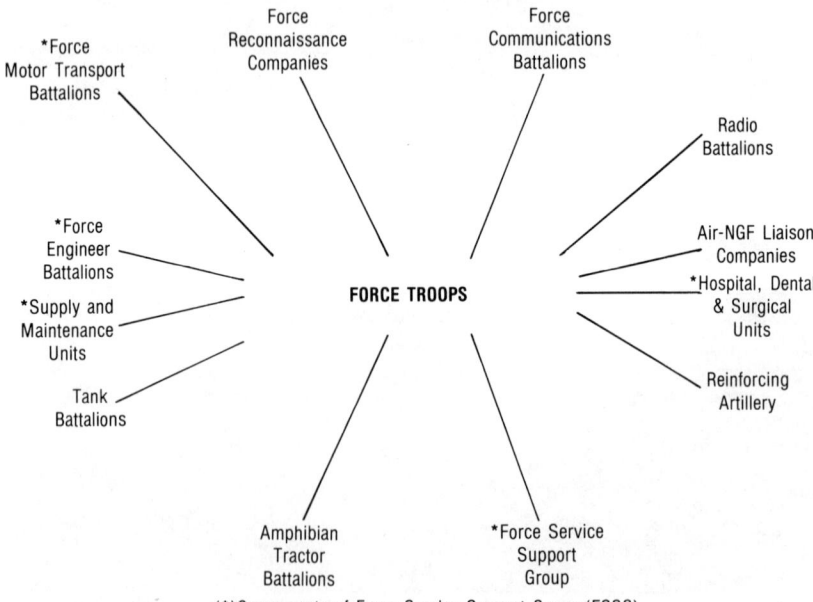

*Force Motor Transport Battalions
Force Reconnaissance Companies
Force Communications Battalions
Radio Battalions
*Force Engineer Battalions
Air-NGF Liaison Companies
FORCE TROOPS
*Supply and Maintenance Units
*Hospital, Dental & Surgical Units
Tank Battalions
Reinforcing Artillery
Amphibian Tractor Battalions
*Force Service Support Group
(*)Components of Force Service Support Group (FSSG).

Fig. 5–4: Typical Components of Force Troops (1977).

Fleet Marine Force Aviation. The basic tactical and administrative unit of FMF aviation is the squadron. Two or more tactical squadrons plus a headquarters and maintenance squadron and an air base squadron comprise the *Marine aircraft group.* Two or more groups, with appropriate supporting and service units, make up the *Marine aircraft wing* (Fig. 5–5). FMF aviation units whose aircraft permit carrier operations are so trained and serve on board aircraft carriers of the Fleet. A complete discussion of Marine aviation can be found in Sections 519–520.

506 ■ Marine Air-Ground Task Forces

For most operations, Marine ground and supporting aviation units are organized into a single integrated landing force under overall command of a single commander. A landing force built around a Marine division and a Marine aircraft wing is called a *Marine amphibious force* (MAF). A smaller landing force shaped from a regimental landing team (RLT) and an air group (MAG) is designated a *Marine amphibious brigade* (MAB), while such a force made up of an aviation squadron and a battalion landing team (BLT) is a *Marine amphibious unit* (MAU). In any case, the aviation component is organized to suit the requirements of the landing force mission and can

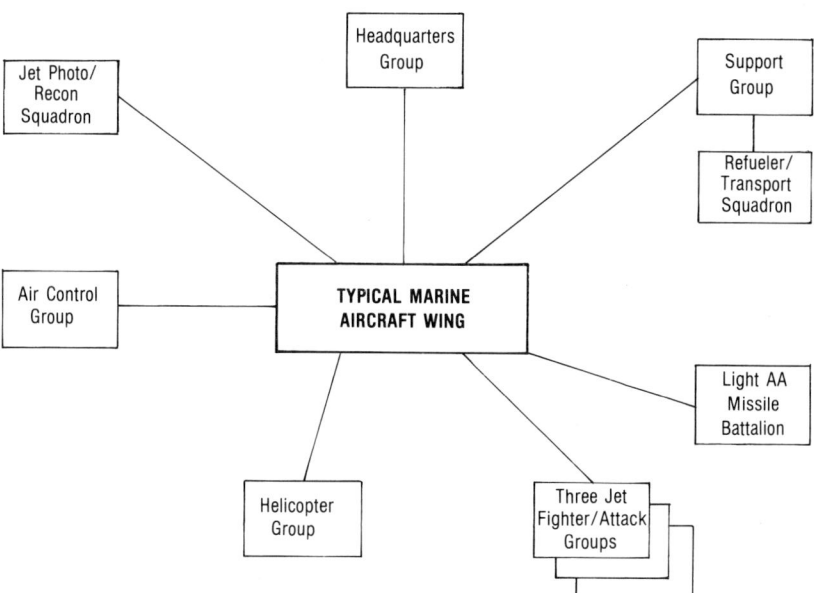

Fig. 5–5: *Organization of a typical Marine Aircraft Wing (1977).*

comprise units and aircraft of any category—helicopter, fighter/ attack, fixed-wing transport, or others. Similarly, units from Force Troops are added as may be required. For the composition of typical MAFs, MABs, and MAUs, see Fig. 5–6.

507 ■ Marine Regiments

The Corps has two kinds of regiments: infantry and artillery. They are numbered as follows:

1st–9th Marines:	Infantry
10th–15th Marines:	Artillery
16th–20th Marines:	Unassigned*
21st–29th Marines:	Infantry

*For a period during World War II, these numbers were assigned to Marine engineer regiments later disbanded.

The assignment of regiments to active Marine divisions is as follows:

1st Marine Division: 1st, 5th, 7th, 11th Marines
2nd Marine Division: 2d, 6th, 8th, 10th Marines
3d Marine Division: 3d, 4th, 9th, 12th Marines
4th Marine Division (USMCR): 23d, 24th, 25th, 14th Marines

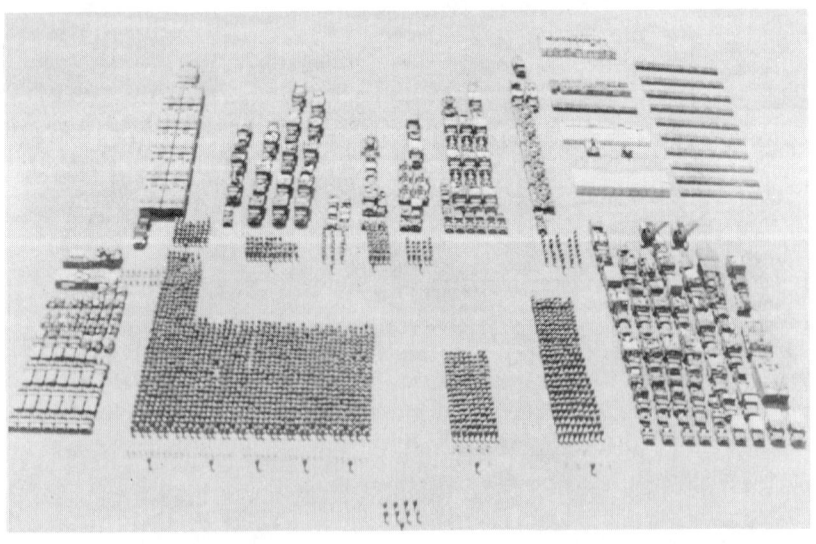

A battalion landing team (BLT) is formed by reinforcing a Marine infantry battalion with artillery, tank, engineer, amphibian tractor, shore party, and other elements.

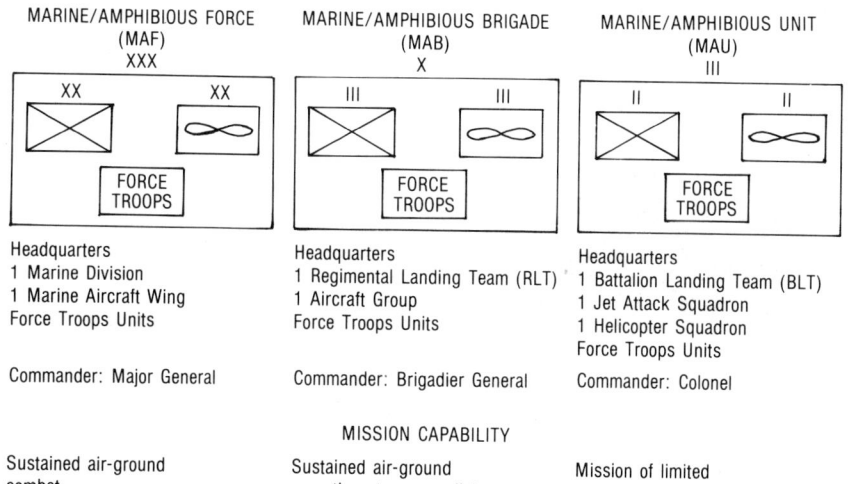

Fig. 5–6: *Typical Marine Air-Ground Task Forces (MAGTF) (1977).*

The above with one exception represents the organization for combat of those divisions in World War II. If the numbering at first sight appears illogical, you will find that there are good underlying historical reasons for it in each case.

508 ■ Seagoing Marines

Although the FMF naturally has first claim to attention among Marines assigned to the Operating Forces of the Navy, never overlook the ships' detachments, or "seagoing Marines." Every battleship (when in commission) and large aircraft carrier, and some cruisers, together with certain amphibious and depot ships, have a detachment of Marines. This is a Marine mission which dates from earliest antiquity—from the fleets of Hiram of Tyre, and of Greece and Rome, where, respectively, Marines were known as *Epibatae* and as *milites classiarii*. For more detailed information on seagoing Marines and on sea duty, turn to Chapter 22 ("Service Afloat").

509 ■ Marine Corps Operating Forces with Naval Establishment Shore Activities

Marine forces provide internal security for all major shore stations in the Naval Establishment. At such stations Marine guards perform the predominantly military activities which directly affect the internal security of the base. Many of the oldest Marine Barracks are those at naval shore stations.

Seagoing Marines, shown here aboard an aircraft carrier, form a distinct and integral part of the ship's company, and are the Corps' foremost exemplars of spit and polish.

Although Marine security forces are part of the naval stations where they serve, they also have the function of providing cadres from which the Fleet Marine Force can obtain additional trained regular Marines in a hurry. In fact, until the FMF was established, Marine Barracks were the only source of troops to form expeditionary forces. The Marine security forces therefore conduct training which is prescribed by Marine Corps Headquarters in order to keep officers and men at each Marine Barracks ready for instant field duty.

510 ▪ Marine Corps Operating Forces on Other Assignment

Since the President may assign Marines to any duty (" . . . such other duty as the President may direct"), Marine Corps Operating Forces may be, and frequently have been, detached for service outside the Naval Establishment, under unified commands, independently, or even under other executive departments (such as the Marine embassy or mail guards). Command of Marine units not otherwise assigned by the President or the Secretary of the Navy remains with the Commandant of the Marine Corps. Such forces include units in training, in a stand-by status, or any other Marine units not specifically assigned elsewhere.

THE MARINE CORPS SUPPORTING ESTABLISHMENT

511 ▪ The Supporting Establishment

The Marine Corps Supporting Establishment provides, trains, maintains, and supports the Operating Forces. The Supporting Establishment includes: Marine Corps Schools; the recruit depots; supply installations; reserve activities; certain Marine Corps bases, barracks, and air stations; Headquarters Battalion, Marine Corps Headquarters; and a number of miscellaneous small activities.

512 ■ Marine Corps Development and Education Command (MCDEC)

Stemming from World War I courses and training activities at Quantico, Virginia, and formally established by General Lejeune in 1920 as Marine Corps Schools, today's Development and Education Command is the intellectual focus of the Corps. The mission of MCDEC is twofold:

1. The formal professional education of officers and selected enlisted personnel with primary emphasis on the tactics and techniques of amphibious warfare; and,
2. The development of doctrine, tactics, techniques, and equipment for employment by landing forces in amphibious operations.

The commanding general, MCDEC, a lieutenant general, executes these missions through designated deputies and various schools and subordinate activities located at Quantico.

Schools and Educational Activities. Progressive education of each stage of an officer's career and centralization of school facilities are the concepts that have shaped the educational programs of the Marine Corps. To implement and realize these concepts, Quantico strives to:

1. Educate Marine Corps officers in command and staff functions and in the tactics and techniques of warfare with particular emphasis on amphibious operations;
2. Prepare doctrinal and related publications for employment of Marine Corps field forces with primary emphasis on doctrine, tactics, techniques, and equipment employed by landing forces in amphibious operations; and,
3. Provide nonresident correspondence instruction at the levels presented in the resident college/schools courses.

The educational activities and schools through which MCDEC accomplishes the above objectives are as follows:

Command and Staff College (C&SC) (formerly the Senior School) is a nine-months' course for majors and lieutenant colonels. The College is a high-level school. Students who attend Command and Staff College must have completed an intermediate-level school before being assigned to Command and Staff College. The curriculum is designed to prepare students for command at the regiment/group level; for staff duty at the division/wing and higher Fleet Marine Force levels; and for duties appropriate to their next higher rank with departmental, combined, joint, and high-level Service organizations.

Instruction at the College, as at other schools at Quantico, stresses planning and conduct of force-in-readiness operations by the Marine Air-Ground Team in cold, limited, or general war situations and in all phases of counterinsurgency, with due consideration to the practical demands of operating under peacetime conditions with attendant limitations on personnel, material, and money. The course covers policies, plans, and problems of Headquarters, Marine Corps, as well as the organization and functioning of the Department of Defense, the Joint Chiefs of Staff, the unified command structure, the other Services, and research and development activities. Work is also undertaken in political-military theory, foreign policy, treaty obligations, and mutual defense agreements. Emphasis is placed on executive leadership, including effective communications and conference techniques. There is a comprehensive program of required reading and book discussions, plus an opportunity to develop a conversational capability in a foreign language.

Amphibious Warfare School (AWS) (formerly the Junior School) is an intermediate-level course of instruction for captains and majors designed to prepare students for the duties of a field grade officer in the Fleet Marine Force to include command at the battalion/squadron level and staff duty at the regiment/group level.

The AWS curriculum is similar to that of Command and Staff College, but instruction is at the level of the reinforced infantry battalion. The regiment/group and other higher organizations are treated only to the extent necessary to provide understanding of command and staff functions in cold, limited, or general war, and in counterinsurgency.

Communication Officers School (COS) conducts instruction in communications for lieutenants, captains, and majors. The Basic Communication Officers Course trains selected Basic School graduates in fundamental techniques and skills and in small-unit communications. The Advanced Communication Officers Course covers all aspects of advanced communications, computer and electronic skills, and command and staff duties. COS has a Reserve Communication Course each summer.

Quantico also includes an *Instructor Training School* to ensure that all instructors become proficient in techniques of instruction. The *Extension School* is a correspondence school which prepares and administers correspondence courses based on the resident officer courses. Each course uses the same texts as the resident schools insofar as possible, supplemented with other material prepared by the staff of the Extension School.

The Amphibious Warfare Presentation Branch sponsors teams of crack instructors which travel extensively to Europe, the Far East, and to major military commands and schools in the United States, to present illustrative problems and demonstrations depicting the modern amphibious assault. The officers who are chosen for this duty have long represented Quantico's best.

Other Resident Schools. Other resident schools at Quantico include the following:

Computer Sciences School, a joint Marine Corps–Navy–Air Force school dealing with military applications of computer technology.

Marine Corps Staff Noncommissioned Officers Academy, which provides advanced NCO training for staff sergeants and for sergeants who have been selected for promotion to staff sergeant.

Officer Candidates School (Section 1107), which trains and screens male and female candidates for commissioned or warrant rank by providing basic military instruction, leadership instruction, and physical training. Successful candidates are appointed to commissioned or warrant rank in the Marine Corps or the Reserve. OCS conducts a variety of courses, including its regular course, designed primarily for college graduates, which leads to a commission and is conducted three to four times annually. Another course is a Warrant Officer Candidate Screening Course, which processes selected noncommissioned officers and graduates permanent warrant officers. During the summer OCS conducts Platoon Leaders Classes (PLC) for undergraduate students, as well as a course for NROTC midshipmen headed for the Marine Corps, who attend it instead of a summer cruise.

In many respects, however, the most important school at Quantico is *Basic School*. Basic School is the place where newly commissioned lieutenants receive their initial training and are made into Marines. It is described in detail in Section 1106.

Marine Corps Development Center. The Development Center performs research, development, test and evaluation functions, including war gaming, in order to develop tactics, techniques, and equipment for the Marine Corps and for all landing forces in amphibious operations. The traditions of the Development Center go back to Generals Lejeune and Russell, 13th and 16th Commandants, both of whom (in 1924 and 1933, respectively) suspended all instruction at Quantico in order to prosecute amphibious warfare and readiness studies.

Supporting Activities. Supporting activities at Quantico include Headquarters Battalion, Weapons Battalion, and a Support Battalion, which perform the administrative and housekeeping tasks essential to the operation of a large military base. There is also a *Marine Corps Air Facility* at Quantico, as well as a *Naval Hospital*.

513 ▪ Marine Corps Recruit Depots

The recruit depots (and Basic School) are the foundation of the Corps. "Boot Camp" transforms the average young American into a basic Marine. Picked officers and veteran enlisted drill instructors ("D.I.'s") emphasize the elements of obedience, *esprit*, and the military fundamentals every Marine must master before he takes his place in the fighting elements of the Corps.

The Marine Corps has two recruit depots—one at Parris Island, South Carolina, the other at San Diego, California. Each recruit depot trains and equips the fledgling Marine, or "boot." Recruits from the Eastern States go to Parris Island ("PI"), while those from the west

The elements of obedience, esprit, and the military fundamentals must be mastered by every Marine before he takes his place in the Corps.

go to San Diego. Often in the past, as special conditions have dictated, the Corps has trained recruits at Washington, D.C. (Marine Barracks, Eighth and Eye Streets), as well as at Quantico and Camp Lejeune. After "boot camp," the new Marine graduates to advanced individual combat training at Camps Lejeune or Pendleton.

514 ■ Marine Corps Supply

The Marine supply services provide logistic support for the Corps. Logistics comprises supply, service, transportation, and evacuation.

The supply establishment procures, stores, distributes, maintains, and repairs all materiel which passes through the Marine supply system, which coordinates and supervises procurement, stock control, and distribution of materiel.

Supply Organization. The supply organization of the Marine Corps is designed to respond to modern logistic requirements. The Marine Corps supply system heads up what is called an *inventory control point,* or "ICP," located at Marine Corps Logistic Support Base, Albany, Georgia. The main functions of the ICP are centralized procurement of virtually all materiel for the Corps, centralized processing of all requisitions, and stores accounting. In addition, however, other vital functions in support of the entire Marine Corps supply system include:

1. Cataloging. Every item entering the Marine Corps supply system is identified and cataloged and assigned a federal stock number (FSN) in accordance with the Federal Cataloging Program.

2. Provisioning. All major equipment or "end-items" entering the Marine supply system, particularly for support of the FMF, require repair parts support. This function of "provisioning" (which has nothing to do with food) is the selection and procurement of the thousands of new repair parts required each year for support of new equipment.

3. Technical Services. Users of equipment or maintenance personnel frequently encounter conditions that require engineering or technical assistance. This assistance may vary from the relatively minor determination of the exact characteristics of a repair part to development of a major modification or "engineering change."

4. Publications Support. The publications that are the bibles of the supply system are called *Marine Corps Stock Lists.* You should be as familiar with these (at least in general terms) as with the *Marine Corps Manual* and other basic administrative publications.

The principal Marine Corps logistic establishments (those that physically stock the materiel) are called *remote storage activities,* or

RSA. The two principal RSAs of the Marine Corps are the Marine Corps Logistic Support Bases at Albany, Georgia, and Barstow, California. These RSAs store and issue materiel in accordance with instructions from the ICP, conduct and supervise stores-accounting for post supply outlets, and serve as area repair centers. With one supply center at Albany and the other at Barstow, the Marine Corps system has an East Coast Complex and a West Coast Complex.

To facilitate distribution and decentralization within the limits of the system and at the same time promote responsiveness to supply needs, various smaller supply agencies support certain major posts. These smaller supply agencies, formerly known as "Stock Accounts," are now also designated as RSA. Communications and actual processing of transactions between the inventory control point and the remote storage activities are accomplished through a network of electronic transceivers and computers providing immediate response to supply needs.

Sources of Supply. Recent years have witnessed the emergence of the Defense Logistics Agency (DLA) as the overall Defense Department manager of materiel. Under DLA, Defense Logistic Centers exist for the various materiel categories, such as petroleum, subsistence, medical, industrial supplies, general supplies, communications/electronics, engineer and construction supplies, automotive and ordnance, and others. In addition, the Defense Clothing and Textile Supply Center makes or procures uniforms and clothing for all the Services. The Defense Logistics Agency and all Logistic Centers are staffed by officers from each of the Services, generally in proportion to the requirements of each. Thus, if you enter the supply field, you may reasonably expect to be detailed at some time to one of these joint agencies.

Aviation Supply. To understate, Marine aviation supply is complicated. A Marine in an aviation unit gets his clothing, individual equipment, rations, weapons, and pay from the Marine Corps. From the Naval Air Systems Command, however, he receives his vehicles, his airplane, its armament, his flight gear, and most of his training aids and manuals. His barracks, quarters, and the hangars, runways, revetments, and shops for his airplane come from the Navy's Facilities Engineering Command.

515 ■ Marine Corps Bases and Air Stations

Several Marine Corps bases, camps, barracks, and air stations exist primarily to support other Marine activities. Unlike the barracks for Marine security forces, these stations come under military command of the Commandant of the Marine Corps. These posts are:

Marine Barracks, 8th and Eye Streets, S.E., Washington, D.C.
Henderson Hall, Arlington, Virginia
Marine Corps Education and Development Command, Quantico, Virginia
Marine Corps Base, Camp Lejeune, North Carolina
Marine Corps Base, Camp Pendleton, California
Marine Corps Base, Twentynine Palms, California
Camp Elmore, Norfolk, Virginia
Camp Garcia, Vieques, Puerto Rico
Camp H. M. Smith, Oahu, Hawaii
Camp S. D. Butler, Okinawa
Marine Corps Air Station, Quantico, Virginia
Marine Corps Air Station, Beaufort, South Carolina
Marine Corps Air Station, El Toro, California
Marine Corps Air Station, Yuma, Arizona
Marine Corps Air Station, Kaneohe Bay, Hawaii
Marine Corps Air Station, Cherry Point, North Carolina
Marine Corps Air Station, Iwakuni, Japan
Marine Corps Air Station (helicopter), New River, North Carolina
Marine Corps Air Station (helicopter), Santa Ana, California
Marine Corps Air Station (helicopter), Futenma, Okinawa

In addition to the foregoing major stations, there is also the Marine Corps Mountain Warfare Training Center ("Pickel Meadows") located at Bridgeport, California.

516 ■ Marine Corps Districts

The Marine Corps divides continental United States into regional Marine Corps districts for local representation, recruiting, and officer procurement. Among a wide range of miscellaneous duties, district directors maintain liaison with corresponding agencies and headquarters of the other three Services, state adjutants-general, other Federal field agencies (particularly offices of the Veterans Administration), schools and colleges, and veterans' associations and military societies.

517 ■ The Marine Band

A unique organization in the Supporting Establishment is the U. S. Marine Band, a part of Marine Barracks, 8th and Eye Streets, Washington, D.C.

This 120-piece military band is not only the best but the oldest (1798) of the Armed Forces musical organizations. It has the privilege of providing music for all White House and official state functions in Washington, in addition to its normal duties in military parades

The U. S. Marine Band, oldest musical organization in the Armed Forces, swings up Pennsylvania Avenue, Washington, toward the Treasury.

and ceremonials. By long custom, the Director of the band is *ex officio* musical director of two of Washington's traditional dining clubs, the Gridiron Club (Washington correspondents) and the Order of the Carabao (military and naval).

518 ■ The Marine Corps Reserve

Although separate from the regular establishment, the Marine Corps Reserve (Chapter 9) forms a vital part of the Supporting Establishment.

MARINE CORPS AVIATION

519 ■ The Air-Ground Team

The role of Marine Corps aviation in the *air-ground team* is to support Fleet Marine Force operations by close and general tactical air support, and air defense. Secondarily to this main job, Marine aviation may be called on to provide replacement or augmentation squadrons for duty with the fleet air arm.

The noteworthy characteristic of Marine aviation is that it forms an inseparable part of the combined arms team with which the Corps backs up its infantry. Thus the special role of Marines in the air is to support their teammates on the ground. The kind of close air support that Marines are accustomed to demands complete integration between air and ground. Pilot and platoon leader wear the same color uniform, share the same traditions and a common fund of experience, and go to school side by side in Quantico. Battlefield and beachhead liaison between air and ground is accomplished by Marine pilots who share front-line foxholes with the riflemen while directing Marine aircraft onto targets just ahead. This makes for maximum reliance by ground on aviation, and for maximum desire by aviation to assist the infantry.

Probably the outstanding demonstration of this tradition in Marine aviation took place during the defense of Wake in 1941. Marine Fighting Squadron 211 provided a heroic air defense of Wake until there were no more airplanes left. Then the officers and men of the squadron calmly donned helmets, picked up their '03 rifles, and went down to glory as infantry. Twenty-four years later, in Vietnam, a handful of aviation mechanics made similar history when a suicide demolition-section of Vietcong sappers rushed the flight line of MAG-16 at Marble Mountain in an attempt to blow up helicopters with satchel charges. As the Vietcong charged, three mechanics downed

Marine close air support in Korea: Riflemen pause while Marine planes take out a troublesome target to the front.

Specifically designed to lift and land the Marine air-ground team, the amphibious ship (LHA), USS Tarawa, is the world's most modern and sophisticated floating base for projecting the landing force ashore.

tools, seized their rifles, and killed or wounded every attacker in one blast of well-aimed fire.

520 ■ Organization of Marine Corps Aviation

In many respects, aviation is the part of the Corps that most nearly lives up to Kipling's "Soldier and sailor, too," since Marine aviation is very closely related to naval aviation. This relationship stems not only from the long partnership between Marine Corps and Navy, but also because the preponderance of Marine squadrons are organized and equipped for carrier operations, and regularly perform tours of duty afloat. In addition, Marine pilots undergo flight training at Pensacola and win their wings as naval aviators.

The primary function of Marine Corps aviation is to participate as the supporting air component of the Fleet Marine Forces in whatever operations they may conduct. A collateral function of Marine aviation is to participate as an integral component of naval aviation in the execution of naval functions as directed by the fleet commanders.

The Commandant of the Marine Corps controls the administration, individual training, and organization of Marine aviation. The Chief of Naval Operations, however, prescribes (via the Commandant) the aeronautical training programs and standards for Marine aviation units; and the aviation materiel used by Marine squadrons comes from the same sources in the Navy as does similar materiel for Navy squadrons.

The organization of Marine Corps aviation falls into subdivisions that correspond to the organization of the Corps as a whole:

1. Office of DC/S (Aviation), Marine Corps Headquarters
2. Aircraft, Fleet Marine Forces
3. Aviation Supporting Establishment

Office of the Deputy Chief of Staff (Aviation). This office is the headquarters organization for Marine aviation. This office (which performs functions similar to those of the old Division of Aviation) plans and supervises matters relating to the organization, personnel, operational readiness, and logistics of Marine aviation. The Deputy Chief of Staff (Aviation), as stated in Section 502, is also Assistant Chief of Naval Operations for Marine aviation. This arrangement ensures that the Deputy Chief of Staff (Aviation), under one hat or another, controls all Marine aviation whether in the Marine Corps Supporting Establishment or under CNO in the Marine Corps Operating Forces (i.e., the FMF). Organized Reserve aviation training is also supervised by the Deputy Chief of Staff (Aviation).

Aircraft, Fleet Marine Forces. Aviation units in the Fleet Marine Force (FMF) comprise the combatant part of the organization. Aircraft units of both Fleet Marine Forces are under the direct command of the respective FMF Headquarters.

Operations and training of FMF air units are ultimately controlled through fleet chains of command by the Chief of Naval Operations, in whose office the ACNO, Marine aviation, actually holds the reins insofar as Marine air is concerned.

Aircraft, Fleet Marine Forces includes attack squadrons, fighter-attack squadrons (regular and all-weather), composite reconnaissance squadrons, air-control squadrons, aerial refueller squadrons, transport squadrons (helicopter and fixed-wing), and headquarters and support squadrons for the groups and wings. The wing is the basic tactical unit of Marine aviation, just as the division is the basic ground unit. You should remember, however, that the wing is a flexible, not a fixed, organization, and that different component organizations may be added or deleted.

Aircraft-carrier arresting gear installed ashore as part of Marine aviation's "SATS" system, enabled Marine squadrons to operate from short runways in Vietnam.

Aviation is represented in staff and planning billets throughout the ground organization and through the tactical air control parties (or TACP) that form part of the infantry battalion, regimental, and division headquarters of the Fleet Marine Force.

Aviation Supporting Establishment. The Marine Corps maintains several air stations and base commands in order to support aviation units operating ashore. On each coast, there is a *Marine air base command* which commands all Marine aviation shore establishments within its area that support aviation units of the FMF. The east coast command is Marine Corps Air Bases, Eastern Area, with headquarters at MCAS, Cherry Point; that on the west coast is Marine Corps Air Bases, Western Area, with headquarters at MCAS, El Toro. The air stations and facilities of the aviation supporting establishment are listed in Section 515. The Marine Air Reserve Training Command, described in Chapter 9, has its headquarters at New Orleans.

MARINE CORPS STAFF ORGANIZATION AND PROCEDURE

521 ■ Marine Corps Staff Organization

The general framework of Marine staff organization corresponds to that employed by the U. S. Army. A complete description of that organization and the staff functions it performs can be found in *Field Manual 101-5, Staff Organization and Procedure,* and the sooner you make the acquaintance of that invaluable manual, the better. Equally important is knowledge of what in effect is the Marine Corps' staff manual, *Command and Staff Action, FMFM 3-1.*

To suit differing functional needs of the Marine Corps (particularly in amphibious operations), we modify some of the staff functions described in *FM 101-5.* Moreover, the Marine Corps has evolved several special staff functions peculiar to Fleet Marine Force operations, which do not appear at all in *FM 101-5.* These latter are listed, with other staff functions, below.

Staff Organization. The Marine commander's staff consists of the following three subdivisions:

The general (or executive) staff
The special staff
The personal staff

As we discuss the staff, one principal should be kept in mind: regardless of how much help the commander receives from his staff (who serve as his eyes, ears, and agents), *the commander, and he alone, is responsible for all that his unit does or leaves undone.* This is the basic principle of command.

The General (or Executive) Staff is a coordinating staff group which plans and supervises all the basic functions of command. In units below divisional or wing level, it is known as the *executive staff;* in divisions or higher headquarters, it is the *general staff.* Except for scale, however, the functions and duties of general and executive staffs are identical.

The basic functions of command are: *personnel, intelligence, operations and training,* and *logistics.* These four functions are referred to by number, in the order just given—that is, personnel as "1," intelligence as "2," operations and training as "3," and logistics as "4." If the staff is divisional or higher (a general staff, in other words), these numbers are prefixed by the letter "G"; if the staff level is below division or wing (an executive staff), its number is prefixed by the letter "S."

The general staff, which is concerned with these command functions, is headed by a *chief of staff* (or *executive officer* in units smaller than brigade), who may be assisted by a *deputy chief of staff,* and by a *staff secretary,* who acts as office manager for the commander, the chief, and the deputy chief of staff. The remaining members of the general, or executive staff are: *personnel officer* (G-1 or S-1); *intelli-*

gence officer (G-2 or S-2); *operations and training officer* (G-3 or S-3); and *logistics officer* (G-4 or S-4). In addition to the "G's," as the four assistant chiefs of staff are referred to, the staffs of all major commands, both FMF and non-FMF, include a *comptroller*, or financial management officer, who is considered to be a member of the general staff.

You will find detailed descriptions of the duties of each of the foregoing officers both in *FM 101-5* and in corresponding Marine Corps publications.

The Special Staff. The special staff includes all the staff who are not members of either the general staff or the personal staff. The special staff is a body of specialist advisers and assistants to the commander. The job of the special staff is to provide technical advice, information, and supervision concerning all important specialized fields of military activity.

As a special staff officer, you enjoy direct access to your commander regarding any matter that relates to your particular specialty. Your routine activities are coordinated, however, by the appropriate general staff officers described above.

Special staff sections may be organized at will by the commander to fill a particular need, or existing sections may be consolidated or inactivated. Thus, the list of special staff officers which follows is typical rather than fixed (although in fact most of these appear in

Marine forward air controller, a qualified pilot serving with a ground unit, calls in a strike within yards of friendly front lines during Korean War.

tables of organization, and may therefore be considered "normal" for a major Marine headquarters):

Adjutant
Air officer (NA)
*Aircraft maintenance officer
Amphibian tractor officer
Antiaircraft artillery officer
Antimechanized officer
Artillery officer
Nuclear, biological, and chemical officer
*Aviation electronics (avionics) officer
*Aviation supply officer
*Base operations officer
Chaplain (ChC, USN))
Chief air observer
Communication-electronics officer
*Crash crew officer
Dental surgeon (DC, USN)
Disbursing officer
Embarkation officer
Engineer
*Engineering officer
Exchange officer
Fiscal officer

Explosive ordnance disposal officer
Food director
Headquarters commandant
Historian
Information services officer
Inspector
Liaison officer
Military government/civil affairs officer
Motor transport officer
Naval gunfire officer
Ordnance officer
Postal officer
Provost marshal
Shore party officer
Special services officer
Staff judge advocate
Supply officer
Surgeon (MC, USN)
Tank officer
*Utilities officer
*Weather officer
*Photographic officer

*Aviation staff function only.

Commanding officers of attached units with no special staff representation act as advisers to the commander on matters pertaining to their units.

The Personal Staff consists of the staff officers whom the commander wishes to coordinate and administer directly, rather than through his chief of staff. The personal staff thus includes such officers as aides-de-camp, and, for certain purposes, selected members of the special staff, such as the *information officer* or the *inspector*.

The relationship between the commander and his personal staff is direct, personal, and confidential. *The personal staff performs only such duties as the commander personally directs.*

522 ■ Aviation Staff Functions

Because of the inherent differences between ground and air operations, and because aviation units perform certain staff functions along naval lines (in order to keep step with naval aviation), Marine aviation has a few special staff functions not encountered in the ground organization just described. These are marked with an asterisk in Section 521. In addition, Marine Corps aviation units embody all the general staff functions covered above.

523 ■ Staff Procedure and Relationships

Although this chapter deals mainly with Marine Corps organization, it is impossible to discuss staff organization without a few words on staff procedure and relationships.

The fundamentals of staff procedure and relationships are:

Staff supervision
Completed staffwork

As a junior officer, you will probably not be assigned to staff duties for a while. Nevertheless, you ultimately face assignment on a staff, and, meanwhile, you will be on the receiving end of staff coordination and supervision. It therefore behooves you to become familiar with the following fundamentals.

Status of the Staff. No staff officer ever exercises command in his own right. The orders which a staff officer voices are those of the commander—regardless of whether or not the commander may be aware of them at the time when issued. Regardless of how much authority the commander allows his staff, *he* alone retains the responsibility. The commander holds the sack.

Staff supervision consists of advising other staff officers and subordinate commanders of the policies and desires of the commander; of interpreting those policies when necessary; and of reporting back to the commander the extent and manner in which his policies and desires are being carried out. This supervision does not extend to command.

Completed staffwork is the most important working principle of the staff. Completed staffwork has been variously defined by many commanders and by official or semiofficial publications. The definition that follows is one of the best and most generally quoted. It is well worth your time to read and ponder.

> *Completed staffwork* is the study of a problem and presentation of a solution, by a staff officer, in such form that all that remains to be done by the commander is to indicate his approval or disapproval of the completed action. The more difficult the problem is, the greater is the tendency to present the problem to the chief in piece-meal fashion. It is your duty, as a staff officer, to work

out the details. You should not burden your chief in the determination of those details, no matter how perplexing they may be. You may and should consult other staff officers. The product, whether it enunciates new policy or modifies established policy, should, when presented to the commander for approval, be worked out in finished form.

It is your job to advise the commander what he ought to do, not to ask him what you ought to do. He needs answers, not questions. Your job is to study, write, restudy, and rewrite, until you have evolved a single proposed course of action—the most advantageous course of all that you have considered. The commander then approves or disapproves.

Do not worry your commander with long explanations and memoranda. Writing a memorandum to your chief does not constitute completed staffwork, but writing a memorandum for him to send to someone else does. Your views should be placed before him in finished form, so that he can make them his own views simply by signing his name. In most cases, completed staffwork produces a single document prepared for the commander's signature, without accompanying comment. If the document stands on its own feet, it will speak for itself; if the commander wants further comment or explanation, he will ask for it.

Completed staffwork usually requires greater effort for the staff officer, but it results in greater freedom and protection for the commander. Moreover, it accomplishes two results:

(1) The commander is protected against half-baked ideas, voluminous memoranda, and immature oral presentations.

(2) The staff officer who has a valid, important proposal can more readily find receptive consideration.

The final test of completed staffwork is this:

If you yourself were the commander, would you be willing to sign the paper you have prepared? Would you stake your professional reputation on its being right?

If your answer would be "No," take the paper back and rework it, because it is not yet completed staffwork.

Fighting spirit is not primarily the result of a neat organization chart nor of a logical organizational set-up. The former should never be sacrificed to the latter.
—Ferdinand Eberstadt

6

The Story of the Marine Corps*

601 ■ Marines in the Revolution

The Marine Corps dates from 10 November 1775. On that day the Continental Congress authorized formation of two battalions of Marines.

Captain Samuel Nicholas, of Philadelphia, was commissioned captain on 28 November 1775 and charged with raising the Marines called for by Congress. Nicholas remained senior officer in the Continental Marines through the Revolution, and is properly considered our first Commandant.

The initial Marine recruiting rendezvous opened at Tun Tavern, in Philadelphia, and by early 1776, organization had progressed to the extent that the Continental Marines were ready for their first expedition. The objective was New Providence Island (Nassau) in the Bahamas, where a British fort and large supplies of munitions were known to be. With Captain Nicholas in command, 277 Marines sailed from Philadelphia in Continental warships. On 3 March 1776, Captain Nicholas landed his battalion, took the fort, and captured the powder and arms for Washington's army.

For the first time in U. S. history, the Marines had landed, and the situation was well in hand.

During the succeeding year, Nicholas, now a major, commanded a battalion of Marines which fought with distinction in the Middle Atlantic campaigns of 1776 and 1777, at Trenton, Morristown, Assanpunk, and Fort Mifflin. At sea (notably under John Paul Jones), ship-

---

*For a definitive history of the Marine Corps, see Col. R. D. Heinl, Jr., *Soldiers of the Sea* (Annapolis, Maryland, U. S. Naval Institute, 1962).

board Marines played traditional parts as prize crews, riflemen, and landing forces (such as in the Penobscot Bay expedition in 1779).

602 ■ Early Years, 1783–1811

After the end of the Revolution in 1783, both Continental Navy and Marines waned into temporary nonentity under the Confederation's unified "Department of War." Although individual Marines continued to be enlisted for and to serve in the few U. S. armed vessels of the period (such as Revenue Cutters), no Corps organization again existed until 11 July 1798, when Congress recreated the Marine Corps as a military service. Major William Ward Burrows, another Philadelphian with Revolutionary experience, was appointed Major Commandant of the Corps.

During the decade that followed, the Naval War with France (1798–1800) and the campaign against the Barbary pirates (1801–1805) provided employment for the Corps. Other noteworthy events were the movement of Marine Corps Headquarters to Washington in 1800, the physical retirement of Burrows as Commandant in 1804, and organization of the Marine Band in 1798.

The third Commandant, Lieutenant Colonel Franklin Wharton, found approximately 65 percent of his small Corps on duty in the

John Paul Jones's Marines aboard the Bonhomme Richard *during the American Revolutionary War.*

Mediterranean. Here, in 1805, the assault of the fortress of Derna, Tripoli, by a mixed force including the Marines commanded by First Lieutenant Presley Neville O'Bannon, overshadowed many less dramatic but equally important actions. O'Bannon's handful of Marines were the first U. S. forces to hoist the Stars and Stripes over territory in the Old World. The "Mameluke" sword, carried by Marine officers to this day, symbolizes O'Bannon's feat.

While events in the Mediterranean held the spotlight, Marines, together with Army and Navy forces, were active in Georgia and East Florida, and in the lower Mississippi, where, in 1804, a 106-man detachment was established at New Orleans.

603 ■ The War of 1812

During the first two years of the War of 1812 the main American achievements were at sea or on the Great Lakes. Marines distinguished themselves in the great frigate duels of the war, as well as at the Battles of Lake Erie and Ontario.

The outstanding record among seagoing Marines, however, was set by Captain John M. Gamble, captain of Marines in the USS *Essex*, the raider which virtually destroyed England's Pacific whaling trade. In April 1813, Gamble, with a crew of 14 Marines and seamen, was placed in command of a prize, the recommissioned USS *Greenwich*. After proving himself in a brilliant action against HMS *Seringapatam*, Gamble was promoted to squadron commander of a covey of prizes thinly manned by the U. S. prize crews and British prisoners. Late in 1813, Gamble and his tiny force were ordered to the Marquesas Islands to establish an advance base for further operations projected by Captain David Porter, commanding the *Essex*.

The warlike Marquesans, emboldened by Gamble's numerical weakness, began a train of depredations which culminated in a pitched battle quelled only by Gamble's naval gunfire. No sooner had this been settled, however, than mutiny reared its head among the British captives, who retook the *Seringapatam*, captured Gamble, and cast him adrift in an open boat with four loyal men. Gaining one of his ships, and despite a wound sustained in the mutiny, Gamble got her underway with no charts and a crew scarcely able to sail. He and his men made the Hawaiian Islands, only to be captured in 1814 by a British man-of-war. For all these exploits he was awarded a richly deserved brevet as lieutenant colonel.

In mid-1813, British forces under Admiral George Cockburn and Major General Robert Ross began a campaign of raids against the Middle Atlantic seaboard. A year later, in August 1814, a column of

Sharpshooting Marine riflemen dominate the action between USS Wasp *and HMS* Reindeer, *1814.*

British soldiers, sailors, and Marines advanced on Washington, D.C. On 24 August, after the Government had fled to Frederick, Maryland, an irresolute force of American soldiers and militia under Brigadier General Winder of the Army attempted to halt the much smaller British column at Bladensburg, just east of Washington. Reinforcing Winder's 6,000 soldiers were 114 Marines from Eighth and Eye Streets and a contingent of seamen gunners with five guns, the whole force being under Commodore Joshua Barney of the United States Navy.

Winder's soldiers broke and ran at the first volley from the British, who advanced unconcernedly until they hit a piece of high ground occupied by the Marines and seamen, who were standing firm. Marine volleys and Navy gunnery forced the British (seven times stronger) to halt, to deploy, and finally, three times in succession, to charge—at a cost of 249 casualties. After having suffered more than 20 percent casualties and being forced rearward by a double envelopment, the Marines and sailors withdrew in good order, with at least a moral victory to their credit.

The British, having put Washington to the torch, now determined to seize New Orleans.

Although a peace treaty was even then being signed in Europe, the British expedition forced its way up the Mississippi. On 28 December 1814, the first enemy attack spent itself against an American line generalled by Andrew Jackson, with Marines (under Major Daniel Carmick) holding the center. Less than two weeks later, on 8 January, the British tried again. This time, Jackson had distributed the Marines in small groups across the entire front. Despite a courageous assault by the redcoats, Jackson's main battle position stood unbroken. As the British commander, Pakenham, fell mortally wounded, the attack ebbed, and New Orleans was saved. Amid the general glory, Congress did the handsome thing by resolving thanks "for the valor and good conduct of Major Daniel Carmick and Marines under his command."

604 ■ Archibald Henderson Takes Over

The most important event in the history of the Marine Corps following the War of 1812 took place on 17 October 1820, when the Adjutant and Inspector, Archibald Henderson, succeeded Lieutenant Col-

Brigadier General Archibald Henderson, 5th Commandant, who held office 39 years and shaped the Marine Corps of the 19th century.

onel Anthony Gale and became fifth Commandant. Gale's term as Commandant had been cut short by a poorly timed row with the Secretary of the Navy, followed by a court-martial.

During the 39 years and 10 Presidential administrations which followed, Henderson was to dominate the Corps and impart to it the high military character which it holds to this day. Had it not been for Henderson's firmness, reinforced by a sympathetic Congress, the Corps might well have been abolished as a result of President Jackson's attempt in 1829 (with connivance from certain influential naval officers) to transfer the Marines into the Army. After the smoke of controversy had drifted clear, Congress in 1834 placed the Corps directly under the Secretary of the Navy—and increased its strength to boot. This was the first instance of Congressional redress and rescue for the Marine Corps—something which has recurred repeatedly since then.

605 ■ Actions Against the Creeks and Seminoles, 1836–1842

From 1836 through 1842, the Army, Navy, and Marine Corps bent united efforts to transfer the Creek and Seminole Indians of Georgia and Florida to new reservations. Colonel Henderson, the Commandant, spent part of this time in the field at the head of a mixed brigade of Army troops and a Marine regiment (the first organization of that size in the history of the Corps). At the Battle of Hatchee-Lustee, Florida, in 1837, Colonel Henderson won one of the few decisively successful victories of the campaign, and was thereupon brevetted brigadier general—the first general officer in the Corps. Despite this success, as well as others against the Creeks, the Seminoles continued an obstinate resistance, and, in 1842, the war ended—with most of the Seminoles still in Florida.

606 ■ To the Halls of the Montezumas

The War with Mexico included three distinct campaigns: that against Mexico City; and those against California and the west coast of Mexico. Marines took part in all three, and were, in fact, the first U. S. forces to set foot on the soil of Mexico proper (at Burrita, 18 May 1846).

A battalion of Marines formed part of General Winfield Scott's column which advanced from Veracruz toward Mexico City. The key to the capital was Chapultepec Castle, set on a crag commanding the swamp causeways into the city. In the assault on Chapultepec, the Marines were divided into storming parties to head the attack up the south approach.

"To the Halls of the Montezumas." The Marine battalion, preceded by General Quitman, enters Mexico City after having stormed Chapultepec Castle.

Under a hail of fire, the Marines moved out. Major Twiggs, the battalion commander, fell early in the attack, while Captain Terrett, a company commander, pressed home a separate assault toward the city. After a night in the outskirts of Mexico City, the Marines marched into town, in the van of their division—the first U. S. troops to enter —and occupied the palace of the Montezumas. The date was 14 September 1847, and a new phrase had been added to the annals of the Corps.

In the Pacific, Marines played important roles in the conquest of California, helping to capture Monterey, Yerba Buena (San Francisco), Los Angeles, and San Diego, while First Lieutenant Archibald Gillespie acted both as confidential agent of President Polk in the Bear State diplomatic intrigues, and subsequently as a bold combat leader.

With California uneasily at rest by 1847, Marines of the Navy's Pacific Squadron secured the Mexican west coast ports of Mazatlan, Guaymas, Muleje, and San Jose. Mazatlan was garrisoned by Marines until June 1848, when peace was concluded.

607 ■ Between the Wars

The decade that followed peace with Mexico was hardly one of peace for the Marine Corps, despite postwar reduction of the Corps to approximately 1,200.

The opening of Japan in 1853–1854 provided a historic setting for the landing of almost one-sixth of the Corps (six officers and 200 Marines, commanded by Major Jacob Zeilin, Mexican War hero and future Commandant). In the best traditions of the Corps, Major Zeilin's Marines were the first to set foot on Japanese soil.

Hardly as peaceful were the landings at Shanghai (1854) and Canton (1856). In each, the conflicts represented trials of strength between Chinese and the Americans bent on "opening" China. At Canton's "Barrier Forts," 176 Chinese cannon were taken, and 5,000 Chinese put to flight.

A hemisphere away, in Nicaragua, Panama, Paraguay, and Uruguay, Marines were scarcely less active. With discovery of gold in California, the Panamanian Isthmus assumed great importance in 1855, when a rickety U. S. railroad was finally completed across the Isthmus. Soon Panama became a hotbed of disorder, which necessitated several landings by Marines, ultimately including a brigade-sized force in 1885. In Uruguay and Paraguay, the story was the same —unsettled times, immature governments—and Marines to keep the peace.

And at home, even in Washington itself, Marines were called out in 1859 to stand off a gang of Baltimore mobsters who styled themselves the "Plug-Uglies," and who carried a loaded brass cannon for emphasis. While Marines and rioters faced each other across a downtown square, an old man, armed only with a gold-headed cane, stepped forward and placed his body across the muzzle of the mobsters' cannon. It was Brevet Brigadier General Archibald Henderson, 5th Commandant of the Marine Corps, now 74 years old. While the thugs milled about the steadfast old man, a squad of his Marines rushed the cannon, and that was that.

608 ■ The Civil War

For the Marine Corps, the opening shots of the Civil War sounded almost two years before Fort Sumter. On 17 October 1859, shortly after John Harris had succeeded Henderson as Colonel Commandant, 88 Marines were despatched by the President to Harper's Ferry, Virginia, to recapture the U. S. Arsenal, which had been seized by the insurrectionist John Brown. The Marine force reported, on his arrival, to Colonel Robert E. Lee, the senior U. S. Army officer present. When John Brown refused to surrender, the Marines, led by First Lieutenant Israel Green, smashed their way under fire into Brown's stronghold, wounded the old abolitionist, and quelled the insurrection.

After 1861, the Marine Corps—like the regular Army—was never large enough to fill the demands upon it. A Marine battalion fought in the first Battle of Manassas, and other Marine forces served ashore in the Mississippi Valley and in the defenses of Washington. All along the Confederate seaboard, from Hatteras Inlet to Hilton Head and Fort Pickens, shipboard Marines, sometimes in provisional battalions, executed successful landings, which put teeth into the Union blockade. Only at Fort Fisher did Marines share with the Navy in a bloody defeat.

By and large, the reputation of the Corps did not gain during the Civil War. Its strength was kept small—only 4,167 officers and men. And, as a Corps, Marines were not called on to perform either the readiness or amphibious tasks peculiar to the organization. As early as 1864 (and again in 1867), attempts were made to disband the Marine Corps and merge it with the Army. Both times, however, Congress stepped into the breach, and the Corps was saved.

609 ■ Post–Civil War

Although the period from 1865 to 1898 has sometimes been spoken of as one of marking time by the Marine Corps, this metaphor scarcely holds up. During that time U. S. Marines landed to protect American lives and property in Egypt, Colombia, Mexico, China, Cuba, the Arc-

Medal of Honor men, Private Purvis and Corporal Brown, pose beneath Korean battle flag captured by them during storming of the Salee River forts, 1871.

tic, Formosa, Uruguay, Argentina, Chile, Haiti, Alaska, Nicaragua, Japan, Samoa, and Panama.

In addition to these and many minor landings, a combined Marine-Navy Asiatic Fleet landing force was sent in to the west coast of Korea in 1871, where, after storming an elaborate system of Korean forts along the Salee River, it captured 481 guns and 50 Korean battle standards. In this fighting two Marines tore down the Korean flag over the enemy citadel under intense fire, and consequently were awarded Medals of Honor, the first of many to be won on the soil of the Hermit Kingdom.

Three able Commandants (Jacob Zeilin, Charles G. McCawley, and Charles Heywood) did much to sparkplug the Corps out of the Civil War doldrums, and despite its small strength (still below 3,000), the Marine Corps was in excellent shape when the United States declared war with Spain in 1898.

610 ■ War with Spain

On the night of 15 February 1898, the USS *Maine* was blown up and sunk in Havana Harbor. Twenty-eight Marines were among the 266 casualties. The gallantry of Private William Anthony, the Captain's orderly, in rescuing Captain Sigsbee despite great personal danger, made Anthony the first U. S. hero of the impending war.

The fight at Guantanamo Bay. Huntington's Marines repulse a Spanish attack, 1898.

On 1 May 1898, little more than a week after declaration of war, Commodore George Dewey destroyed the Spanish squadron in Manila Bay. Two days later Dewey landed his Marine detachments to secure Cavite Navy Yard, and settled down for the three-months' wait until the Army could get troops to the Philippines.

Just as Marines were first to land in the Philippines, so also were they the first U. S. forces to land and fight in Cuba. On 10 June 1898, an Atlantic Fleet battalion of Marines, commanded by Lieutenant Colonel R. W. Huntington, landed under cover of ships' guns at Guantanamo Bay, Cuba, and seized an advanced base for the Fleet. Four days later, at Cuzco Well, Huntington routed the remaining Spanish forces, destroyed their water supply, and completed the victory. The hero of the day was Sergeant John H. Quick, who won the Medal of Honor for semaphoring while under U. S. and Spanish shellfire, for an emergency lift of the naval bombardment.

Huntington's battalion was not the conventional ship's landing party of the nineteenth century, but rather a self-contained Marine expeditionary force, which included infantry, artillery, and a headquarters complement of specialist and service troops. The battalion formed part of the Fleet—a miniature Fleet Marine Force whose primary mission was landing on hostile shores to secure an advanced base.

611 ■ "Our Flag's Unfurl'd to Every Breeze. . . ."

Between 1899 and 1916, the Marine Corps participated in eight major expeditions or campaigns: the Philippine Insurrection; the Boxer Uprising; Panama; the Cuban Pacifications; Veracruz; Haiti; Santo Domingo; and Nicaragua.

At least as important, and possibly more so, Marine Corps developments of this period laid the foundation of American amphibious warfare techniques, and ensured the future survival and growth of the Corps. And all this was accomplished despite sustained Navy attempts to get rid of the Corps, which Congress finally squelched in 1909.

Long oppressed by Spain, the Philippines in 1899 sought to make a clean break with colonialism, and launched the Philippine Insurrection —or rather transferred their insurrection from the Spanish to the Americans. This three-year campaign included the first modern Marine brigade ever organized, as well as three exploits for which the Corps will be remembered: Major Waller's march across Samar; the storming of Sojoton Cliffs in Samar (where Captains Porter and Bearss won Medals of Honor); and the pacification of the Subic Bay area on Luzon.

As 1900 dawned, China was undergoing one of her periods of antiforeignism—the Boxer Rebellion. In Tientsin and Peking, foreign

"Our Flag's Unfurl'd to Every Breeze." A ship's guard stands by to land under cover of battleship guns. Note field hats, newly issued to Marines since 1898.

missions were soon besieged by the bloodthirsty Chinese mob. In Peking, together with other foreign garrisons, U. S. and British Marines linked arms, as so often in the past, this time to defend the beleaguered Legation Quarter throughout the summer of 1900. In the international relief column despatched to save Tientsin and Peking was a U. S. Marine force commanded by Major L. W. T. Waller and later by Major W. P. Biddle. By midsummer, Tientsin had been relieved, and on 14 August 1900, the Marines stormed the Chien Men, leading into the Tartar City, Peking, and the legations were relieved.

At almost the same time, conditions in Panama began to threaten free transit of the Isthmus. In 1903, on orders from the President, Theodore Roosevelt, a U. S. Marine brigade—led by General Elliott, the Commandant—landed at Colon to protect U. S. rights during Panama's revolt against Colombia. Marines remained in the Canal Zone until the situation became fully routine in 1911, when the Army took over. Meanwhile, in 1909, another Presidential attempt to abolish the Corps was rebuffed by Congress.

First in 1906, and again in 1912, Marine brigades were sent to Cuba to restore order under the so-called Platt Amendment, by which the United States had the right to intervene in that newly liberated country. Twenty-four towns were occupied in 1906, and 26 in 1912, and considerable fighting took place in eastern Cuba before peace was finally attained.

In 1901 the Marine Corps initiated special training and organization for the seizure and defense of advanced bases. Succeeding years saw increased use of battalions and regiments based in Navy transports as a means of projecting naval power across the shoreline. In 1910, at New London, Connecticut, Major General G. F. Elliott, 10th Commandant, established the Marine Corps Advanced Base School. For the first time in U. S. history, a school had been created to focus the thinking of an entire Service on the unsolved problems of amphibious warfare, and to perfect the already well developed expeditionary readiness of the Corps.

Hand in hand with establishment of the Advanced Base School was the organization of the Advanced Base Force, a Marine brigade containing all the necessary combined arms, maintained in readiness for immediate expeditionary service with the Fleet. The Advanced Base Force was the prototype of the Fleet Marine Force.

When President Wilson was obliged to protect American rights, property, and citizens in Mexico in 1914, Marines from the Advanced Base Force were the first to land, at Veracruz. Army forces followed in due course. Veracruz provided the first field test of the Advanced Base Force, a test that was passed with flying colors.

Marines defending the Tartar Wall, siege of the Legations, Peking, 1900.

Early in 1915, Haiti was wracked by revolution. Ships' detachments landed, but Haiti's troubles called for reinforcement by units of the Advanced Base Force. The pacification of Haiti proved to be long and arduous. Bandits were firmly established in the north of Haiti, where the rugged country gave them every advantage. Under the dynamic leadership of Colonel L. W. T. Waller and Major Smedley Butler, Marines finally brought the bandits to battle in their stronghold at Fort Rivière, in the storming of which Butler won his second Medal of Honor, having won his first at Veracruz. The resulting victory brought peace to northern Haiti. The Marines rebuilt civil government, and, for the time being, Haiti breathed easily.

In 1916, Marine forces despatched to restore order in the Dominican Republic again dramatized the ability of the Corps to take effective action on short notice. More than 2,000 Marines landed in Santo Domingo, where, until 1924, a protracted campaign was waged to suppress banditry and enable a Dominican civil government to regain control of the country.

By the end of 1916, the Corps had ended a major era of growth. It had become, in fact if not in law, a national force in readiness. In addition, the seeds of Marine amphibious development had been sown, and the leadership of the Corps had been hardened in continual combat and expeditionary experience, which was destined to pay off for years to come.

612 ■ Marines "Over There"

Although Marines served faithfully around the globe during World War I, and although Marine commitments in Haiti, Santo Domingo, Cuba, and Nicaragua remained little changed, the preeminent Marine story of World War I is that of the 4th Marine Brigade in France.

The 4th Brigade was the largest unit of Marines assembled during World War I, or ever before. It was composed of the 5th and 6th Regiments and the 6th Machine Gun Battalion, totaling some 9,444 officers and men. Of the Brigade's succession of notable actions (Belleau Wood, Soissons, Saint Mihiel, Blanc Mont Ridge, the Argonne), Belleau Wood was most significant, not only because it saved Paris from the massive German offensive in June 1918, but because it was the greatest battle up to that time in the history of the Corps. The casualties of the 4th Marine Brigade in assaulting the well-organized German center of resistance in Belleau Wood were comparable only to those later sustained in the hardest-fought beach assaults of World War II. After Belleau Wood, German intelligence evaluated the Marine Brigade as "storm troops"—the highest rating on the enemy scale of fighting men.

Fierce fighting in Belleau Wood, where Marines first defeated elite German infantry, 1918.

By 11 November 1918, the 4th Marine Brigade and Marine aviation units in France had sustained more casualties in eight months of virtually continuous combat than had the entire Corps in the preceding 143 years. The grim total was 11,366.

613 ■ Marine Aviation

Founded in 1914 as part of the Advanced Base Force, Marine Corps aviation was still in experimental stages when World War I began. During the war Marine aviation units flew in combat over France and supported the Fleet from an advanced base in the Azores. The spark plug of aviation's participation in the war was Major A. A. Cunningham, the Corps' first pilot. Starting from a strength of 7 officer pilots and 43 enlisted men in 1917, Marine aviation, at war's end, mustered 282 officers and 2,180 enlisted men. In the best traditions of the Corps, the 1st Aeronautical Company, destined for the Azores, was the first completely equipped American aviation unit to leave the U. S. for service overseas.

Marine Corps aviation was still in the experimental stage at the outbreak of World War I.

614 ■ "Beyond the Seas"

While the 4th Brigade was gaining immortal victories, the Advanced Base Force remained hard at work in Haiti, Santo Domingo, and Cuba. One regiment, the 8th Marines (later reinforced by the 9th Marines) was held in east Texas to protect the Mexican oil fields should the Germans try (as intelligence indicated they intended to do) to disrupt that source of Navy oil supply.

The 5th Marine Brigade was sent to France, but was refused permission by the Army to enter combat, and was instead parceled out in noncombatant duties—mainly provost marshal and military police in the Army communications zone.

615 ■ Expeditions Between World Wars

In addition to continuing commitments in the Caribbean, 1919 found the Corps performing occupation duty along the Rhine.

Marines were also occupying eastern Cuba, which was ultimately pacified in 1922. Two years later, in 1924, six hard-fought, campaigning years in Santo Domingo came to an end, and Marines were finally withdrawn. Haiti meanwhile was at a boil, with banditry again in full cry, this time between 1918 and 1920. Following suppression of bandit forces in 1922, Brigadier General John H. Russell, expert in Haitian affairs, was appointed U. S. High Commissioner to Haiti, to administer the American protectorate over the troubled republic. It was not until 1934 that the 1st Marine Brigade hauled down its colors in Port-au-Prince and boarded ship for home.

The year 1927 was marked by trouble in both hemispheres—in Nicaragua and China. Naturally, Marines were soon involved.

As early as 1912, Marines had landed in Nicaragua to preserve order, but were withdrawn in 1925. No sooner were they out of the country, however, than the worst civil war in the history of Nicaragua erupted, and Marines (spearheaded by the ships' detachments) were immediately despatched to Nicaragua at the mutual request of the leaders of both warring factions. Marine occupation continued after an uneasy peace, and consisted of disarming dissidents, conspicuous among whom was Augusto Sandino, a self-styled "patriot" guerrilla supported from Mexico and neighboring Honduras. A native *Guardia Nacional*, much like the *Gendarmerie d'Haiti* (also Marine-trained), was organized under the Marines to facilitate the hard job of policing a population that included thousands of demobilized revolutionary soldiers.

Marine aviation not only played a leading role in supporting ground operations in Nicaragua but also pioneered tactical and logistic air support on a scale hitherto unknown. The technique of dive-bombing (invented by Marine aviators in Haiti in 1919) was greatly refined. Virtually all the isolated patrols and outposts in the heavily jungled northern area of Nicaragua were maintained by air supply. To evacuate Marine wounded, First Lieutenant C. F. Schilt made ten landings and take-offs from a village street in Quilali in a fabric-covered scout plane under murderous fire. Lieutenant Schilt was awarded the Medal of Honor.

Fighting bandits in native village during Santo Domingo campaign.

When the Marines left Nicaragua in 1933, they turned over to the Nicaraguan government a well-organized Guardia, a military academy, a system of communications, and a first-rate public health service, plus many less obvious improvements.

As in Nicaragua, Marine embassy guards had been stationed in China for many years before 1927. But there, too, civil disturbances of increasing violence reached a peak in 1927, and additional forces—the 4th Marines to Shanghai, and a brigade to North China—were hurried in. Mainly because of these precautions, the threatening situation eased, and the brigade was withdrawn. The remaining units in Shanghai, Peking, and ultimately Tientsin as well, faced crisis after crisis with Chinese warlords and Japanese until World War II closed the ledger.

616 ■ Guarding the Mails

In 1921, after a series of violent mail robberies in the United States, the President directed the Marine Corps to guard the mails. Within a matter of hours, Marine armed guards were riding mail-cars and trucks, with orders to shoot to kill. Not a single successful mail robbery took place against a Marine guard, and in less than a year the Marines were withdrawn. Five years later, when mail robberies again broke out, Marines were called in a second time; this time as well, the robberies ended at once.

617 ■ Amphibious Pioneering

In the early 1920s it became clear to Marines that a war with Japan was inevitable, and that such a war, to be successful, would entail amphibious seizure of a chain of advanced bases across the Pacific,

The first amphibious tank being tested by the Marines at Culebra, 1924.

Prewar training for amphibious war paid off when the opening salvos of World War II rocked the world.

the world's greatest ocean. In 1921, the year after Marine Corps Schools opened at Quantico, the course of a war with Japan was forecast with prophetic accuracy by Lieutenant Colonel Earl Ellis, who subsequently "disappeared" while on an intelligence mission in the Japanese Palaus in 1923.

To Major General John A. Lejeune, 13th Commandant, the prospect of amphibious war was a bleak one. The British failure at Gallipoli had convinced orthodox military thinkers that an amphibious operation could not succeed against strong opposition. Despite the forbidding nature of the problem, General Lejeune set the Marine Corps to solving it.

Quantico was well equipped to undertake the task, and became the focal point of American fleet-centered amphibious development. Marine Corps Schools attacked the problem, and by 1934 had produced the first comprehensive U. S. manual of amphibious doctrine ever written—*Tentative Landing Operations Manual*. This historic document was adopted intact by the U. S. Navy in 1938, under the title, *FTP-167, Landing Operations Doctrine, U. S. Navy*. In 1941, when the U. S. Army issued its first amphibious publication, Quantico's book was again borrowed verbatim, even down to the illustrations, and appeared this time as *War Department Field Manual 31-5*. The tenets of *Tentative Landing Operations Manual* still constitute much of the basis for publications that are today's amphibious "bible."

To deal with materiel aspects of amphibious problems, the Marine Corps Equipment Board was brought into being in 1933. Most notable among the Board's pre–World War II achievements was the amphibian tractor, or LVT, which joined the FMF in 1940.

During these pioneering years, the Advance Base Force was redesignated in 1921 as the East Coast Expeditionary Force, and in 1933 as the Fleet Marine Force, after Major General John H. Russell, 16th Commandant, had persuaded the Secretary of the Navy that Marine expeditionary troops should form an integral part of the U. S. Fleet. The Fleet Marine Force not only constituted a force in readiness but also performed an invaluable role in testing the doctrines and materiel evolved by Marine Corps Schools and by the Equipment Board. From 1935 on, annual Fleet Landing Exercises enabled the fledgling FMF to find its footing as well as to keep the thinkers at Quantico progressing along sound lines.

Because of this pioneering in the 1920s and 1930s, the Marine Corps—ground and aviation—was ready for amphibious war, and ready also to train others to wage it, when the opening salvos of World War II rocked the world. Before World War II had run its course, seven U. S. Army divisions (including the first three Army divisions ever to receive amphibious training) were trained in landing operations by the Marine Corps, and the doctrines of Quantico girdled the globe.

618 ■ Occupation of Iceland

In 1941, with the 17th Commandant, Major General Thomas Holcomb, in office, the Fleet Marine Force was called on to demonstrate its capabilities.

Iceland, garrisoned by British forces, was critical in the Battle of the Atlantic, President Roosevelt, who well knew Iceland's strategic importance, agreed with Prime Minister Churchill that the island should be more adequately secured, and that U. S. forces should be employed. After the Army found itself unable to provide ready forces for the Iceland mission, the President turned to the Marine Corps. Less than one week later, on 22 June 1941, the 1st Provisional Marine Brigade had been organized, had embarked, and sailed. On 7 July, more than 4,000 Marines debarked at Reykjavik. Once again first on the spot, the Corps had proved itself the national force in readiness.

619 ■ "Uncommon Valor"

At the outset of World War II, in 1941, the Marine Corps totaled some 70,425, and was organized into two Marine divisions, two air wings, and seven defense battalions (advanced base artillery units). By 1944, the Corps included six divisions, four wings, and corps and force troops to support the two amphibious corps, which were the FMF's highest formations; its top strength was 471,905.

Despite differences in terrain and character of operations, both the South Pacific and Central Pacific campaigns of World War II highlighted Marine Corps attributes—the South Pacific, *readiness*; the Central Pacific, *amphibious assault virtuosity*.

The Guadalcanal campaign (1942) not only typified the South Pacific, but, more important, Guadalcanal dramatized to the American public the function of the Fleet Marine Force.

In 1942 (on the heels of valiant Marine defensive fighting at Wake, Midway, and Corregidor), it became clear that a U. S. advance base must be established in the southern Solomons. Guadalcanal, where the Japanese were already building an air strip, was the logical target. Despite high-level prophecies of disaster, and recommendations that the assault be delayed until the following year, the earliest that Army troops could be trained to participate, the 1st Marine Division was given the job of retaking Guadalcanal and adjacent Tulagi. On 7 August 1942, Marines landed.

It was the first U. S. offensive of World War II; the long road back had commenced.

Margins were never slimmer during the Pacific War than on Guadalcanal. But by late November 1942, when Army troops began to arrive in strength, battered Henderson Field was firmly secured by U. S. Marine ground and air, and, in the inner councils of Japan, it was already acknowledged that the turning point of the war had passed.

Storming Tarawa Atoll, 2d Division riflemen highlight the fighting qualities of the individual Marine.

On Iwo Jima, as on Tarawa, the fighting ability of the individual Marine came into sharp focus.

The lesson of the Guadalcanal campaign was that without a ready Marine Corps, the operation could never have taken place. Undertaken as a purely naval campaign by Fleet units and Marines, Guadalcanal demonstrated the dependence of sea power on Fleet expeditionary forces, as well as the degree to which the Marine Corps had placed itself in readiness for just such an occasion as this.

If Guadalcanal and subsequent operations in the South Pacific—New Georgia, Bougainville, Choiseul (all in 1943), and New Britain (1944)—proved that the Corps was ready for war, the campaign across the Central Pacific displayed the Marine Corps' virtuosity in amphibious assault.

The succession of Central Pacific battles—Tarawa (1943), the Marshalls, Saipan, Guam, Tinian, Peleliu (all 1944), Iwo Jima and Okinawa (both 1945)—was by hard necessity a series of frontal assaults from the sea against positions fortified with every refinement that Japanese ingenuity and pains could produce. To reduce such strongholds, wrote one historian, was "the acme of amphibious assault."

On Iwo Jima, toward the end, as on Tarawa at the beginning, the fighting ability of the individual Marine came into sharp focus. Each battle was one of frontal assault and close combat against fortified positions. Tarawa was the first combat test of the Marine Corps doctrines for amphibious assault, and Tarawa demonstrated that those doctrines would work. Two years later, at Iwo Jima—largest all-Marine battle in history—Marines reaped the benefit of Tarawa's experience (and the experience of many other hard-fought assaults) in the form of tested, combat-proven assault technique. Without Tarawa, the even greater assault on Iwo would not have been possible. Without the U. S. Marine Corps (and without the years of study, experiment, and development at Quantico), neither Tarawa nor Iwo nor the battles between could even have taken place, let alone succeeded.

But the record of the Corps in World War II was not only the great record of its seaborne assaults. Beginning on 7 December 1941, at Wake, Marine aviation was also in the war. At Midway, Guadalcanal, Bougainville, and the northern Solomons, in the Marshalls, at Peleliu, Iwo Jima, and Okinawa, Marine fliers again helped to forge the concept of the air-ground team. And just as World War II brought to fruition the long-studied amphibious assault doctrines of the Corps, so also World War II witnessed the perfection of Marine close air support. In the reconquest of the Philippines, four Marine air groups working with Marine air liaison parties on the ground reached a high point in Marine tactical air support (even though this support was for Army comrades).

By the end of the war, the Fleet Marine Force, air and ground, was poised for invasion of Japan—an invasion rendered unnecessary by U. S. sea power. The Corps which under General Holcomb, 17th Commandant, had mustered 19,354 in 1939, under Holcomb's successor, General Vandegrift of Guadalcanal, neared 500,000 in 1945. The

Marines hit the beaches in the amphibious attack on Saipan, a decisive turning point in the Pacific offensive in June 1944, but costly to the Marines.

victories in World War II had cost the Corps 86,940 casualties. In the eyes of the American public, the Marine Corps was second to none, and seemed destined for a long and useful career. Admiral Nimitz' ringing epitome of Marine fighting on Iwo Jima might very well be applied to the entire Marine Corps in World War II: *"Uncommon valor was a common virtue."*

620 ■ The Marine Corps, Postwar

After World War II, although the Corps enjoyed high public prestige and seemed indeed here to stay, the years 1946–1949 were devoted to a searching examination into the mission (and in high quarters, behind closed doors, even the need) for the Marine Corps. These doubts were inspired, as two Commandants testified before Congress, by the Army General Staff, whose long-term objective, since before World War I, has been abolition of the Marine Corps, or alternatively its reduction to a minor security and ceremonial unit. The question was firmly—and, Marines hoped, finally—resolved by the National Secur-

Marines bend the flag onto the staff, just prior to the famous flag-raising on Iwo Jima.

Marine Corsairs over Fujiyama symbolize the defeat of Japan.

ity Act of 1947, which gave the Corps firm missions and reaffirmed its status as the Service charged with primary amphibious responsibility over landing force tactics, technique, and doctrine. Subsequently the Douglas-Mansfield Bill, enacted in 1952, afforded the Commandant co-equal status with the Joint Chiefs of Staff in all matters concerning the Marines and legislated today's organization of the Corps.

While the roles, missions, and status of the Marine Corps were being debated in both executive and legislative branches of the government, the Corps maintained occupation forces in Japan and North China, and completed an orderly demobilization unmarred by indiscipline or untoward incident.

The postwar FMF comprised two major forces: Fleet Marine Force Pacific and Fleet Marine Force Atlantic—assigned respectively to the Pacific and the Atlantic Fleets. Each force embodied a Marine division and an air wing (both on peace scales), and supporting logistic units.

621 ■ Korea: "Not a Finer Fighting Organization in the World"

Like the story of the Corps in World War II, the Marines' part in the Korean War is covered by the official histories. This narrative therefore confines itself only to high spots of a gruelling three-year war, at the outset of which, to quote Hanson Baldwin, of *The New York Times*:

> The Marines were ready to fight; if they had not been, we might still be fighting in the Pusan perimeter.

On 24 June 1950, when the Russian- and Chinese-supported North Korean communists attacked South Korea, the Corps numbered ap-

Another major Marine innovation, the helicopter was first flown in battle by pilots attached to the 1st Marine Brigade, in August 1950. Here, a Marine battalion embarks for rapid movement to the front lines.

proximately 75,000. The FMF was deployed in two shrunken divisions at Pendleton and Lejeune. Aviation (which had narrowly missed transfer to the Air Force by Defense Secretary Louis Johnson) was even thinner: eleven squadrons divided into two wings, at Cherry Point and El Toro. The Chairman of the Joint Chiefs of Staff, General Omar Bradley, had predicted publicly, hardly eight months before, that the world would never see a large-scale amphibious landing.

On 2 July, faced with mounting catastrophe, General MacArthur sent his first request to the Joint Chiefs of Staff for help from the Marines: immediate assignment of a Marine regimental combat team (RCT) plus a supporting Marine air group. In the days that followed, General MacArthur sent five more pleas, culminating in one for a war-strength Marine division and a war-strength air wing.

Less than two weeks later, as soon as Navy shipping could be readied, the 1st Provisional Marine Brigade was crossing the Pacific, heading for the Pusan perimeter, into which shaken U. S. Army and Republic of Korea (ROK) units were already streaming rearward. On 3 August 1950, Marine F4Us from the USS *Sicily* scored first blood for the Corps in an air strike over Inchon; on 7 August, eight years after Guadalcanal, ground elements of the brigade were in hot action, plugging holes in the Pusan perimeter. And here for the first time helicopters were being flown in battle—by Marines.

To form this spearhead, the 1st Marine Division had been stripped, with only a cadre left behind. Now it was up to the Corps to re-form the division for the amphibious stroke that General MacArthur had visualized from the outset of the war.

For these, and other worldwide commitments, it was obvious that the 75,000-Marine peacetime Corps would not suffice. Ten days after the 1st Brigade had sailed, the President mobilized the Organized Reserve. The 2d Division, at Lejeune, was stripped of all but headquarters and a cadre; 7,000 men were transferred to Camp Pendleton. While this administrative nightmare was in progress, and the brigade was fighting desperately along the Naktong River, U. S. Marine and Navy planners were in Tokyo, translating into reality General MacArthur's plan to relieve Pusan and retake Seoul by an amphibious stroke to be delivered at the Korean west coast port of Inchon. Because of extreme tidal fluctuation, 15 September was the only suitable D-day until mid-October. Within five days after the 1st Marine Division had finally gotten together in one piece in one place, and the 1st Wing was fully in business, convoys were sailing for Inchon.

Despite unprecedented haste in preparation, and numerous calculated risks of enemy opposition, geography, and hydrography, the Inchon landing was almost anticlimactic in its success. As favorable reports poured in throughout D-day, General MacArthur signaled:

> The Navy and Marines have never shone more brightly than this morning.

On 28 September 1950, ninety days after the Communist blitz, Seoul was in the hands of the 1st Marine Division. A week later, with capture of Uijongbu, the Inchon-Seoul campaign was complete. Largely as a result of the amphibious capability and readiness of the Marine Corps, the North Korean army south of the 38th Parallel had been all but destroyed.

The strategic sequel to Inchon was another amphibious flanking maneuver, a right hook up the east coast into Wonsan, and thence

Over the seawall at Inchon, men of the 5th Marines launch still another major amphibious onslaught, 1950.

northwest—it was intended—to the Yalu. To execute this scheme, the Marines were again chosen.

Although the 1st Division's orders called for a wide-open, free-style push to the Yalu, the Marines, warned by an initial contact with the Chinese Communist Forces near Hamhung on 2 November, made haste slowly. The advance was careful, and the main supply route to the sea was painstakingly developed and secured.

On 25 November 1950, after an eerie lull, Chinese Communists hit the right wing of the Eighth U. S. Army, routed and destroyed at least one U. S. Army division, and launched an entire army group, eight divisions, against the 1st Marine Division.

The blow fell when the division's forward elements were west of Chosin Reservoir, at Yudam-Ni. It is enough to record that, in face of "General Winter" and of every weapon, artifice, and attack in overwhelmingly superior strength, the division concentrated promptly, rescued and evacuated surviving remnants of adjacent, less ready Army formations, and commenced one of the great marches of American history, from Chosin Reservoir to the sea.

On 11 December, having brought down its dead, saved its equipment, and rescued three Army battalions, the 1st Marine Division—supported by the 1st Wing—reached the sea with high morale and in fighting order. The division had shattered the Chinese Communist Forces (CCF) 9th Army Group, killed at least 25,000 Chinese, and wounded more than 12,500.

Following amphibious redeployment and a short "breather" cleaning out a North Korean guerrilla division that had infiltrated southeast Korea, the 1st Division spearheaded the IX Corps spring offensive of 1951. At the same time the 1st Wing continued to provide the

Marching down from the Reservoir, the 1st Marine Division brought out its dead, its wounded, and its weapons, and shattered an entire Chinese Communist army group in the process.

United States Marines resting in the snow in the march from Chosin Reservoir.

preponderance of all close-support sorties nominally credited to the Fifth Air Force, under whose control the wing now operated.

Although unheralded by the official publicity accorded other divisions at this time, the Marine division played a stellar role against the Chinese counteroffensive, which recoiled on the U. S. IX Corps. The division's defense of Hwachon Reservoir saved the IX Corps from disaster and largely vitiated the enemy offensive. Soon after, with initiation of armistice talks, the front stabilized, and the division found itself holding the "Punchbowl" sector on the east coast of Korea. The pattern of inconclusive, savage see-saw fighting in this sector was destined to recur here and elsewhere throughout the rest of the war.

During this time, the feats of the Marine Corps (and Royal Marines) in holding and exploiting offshore islands far up the east and west coasts of Korea began to pay off. Fifteen islands, including seven in embattled Wonsan Harbor, served as springboards for raiding, havens for aircraft in trouble, hatcheries for intelligence skulduggery, and observation posts for Marine shore fire control parties.

In spring, 1952, the 1st Division moved to the arena of its final battles in Korea: the line of the Imjin River, astride the Munsan-Ni corridor to Seoul. Here, holding a frontage greater than that of any other division in Korea, the division kept Seoul safe, anchored the United Nations' left flank, and overlooked the Panmunjom truce site. From these positions the Marines fought a series of bloody trench-warfare actions of a type and scale unheard of in the Corps since World War I: the battles for "Bunker Hill" and "The Hook," and the fighting about Outposts "Reno," "Carson," and "Vegas."

And here, still holding the Imjin, the armistice found the 1st Marine Division and its valiant supporting Marine air wing.

Korea had tested "the new Marine Corps" in readiness and fighting quality. Neither had been found wanting. Korea demonstrated to doubters, in high places and low, that amphibious operations were anything but dead. Korea proved, as had World War II before it, the high caliber and readiness of the Marine Corps Reserve. Most of all, in a time of immense upheaval in military techniques, Korea underscored the fact that the military principles, virtues, and intangibles for which the Marine Corps stood, remained as sound in 1953 as they had been in 1775.

622 ■ At Mid–Twentieth Century

As there had been no victory in Korea, so there was no demobilization. The Corps continued in the structure of three divisions and three wings that Congress had established in 1952, with one of each on the east and west coasts of the United States and in the Far East (Okinawa and Japan). From these major forces, floating battalions were maintained on station with the U. S. fleets in the Mediterranean and the Far East, and, from 1960 on, in the West Indies as well. A thorough modernization of the war-tested amphibious doctrines of the Navy and Marine Corps, originally conceived in Quantico in the late 1940s, gave these forces assault helicopters as landing craft, and specially designed helicopter carriers as transports. Blending these sophisticated and original methods with the tried and true ones of seaborne assault by landing craft and amtracs, the amphibious assault now had even greater shocking power and flexibility than before. Recognizing imitation as the sincerest flattery, the Marines who had pioneered the technique of "vertical envelopment" could watch developments with interest when, 15 years later in Vietnam, the Army gave wide publicity to its "airmobile" concept, which was simply a shore-based version of Marine-style vertical envelopment.

In the decade following Korea it became clear that, despite the ever-present threat of nuclear war, the characteristic pattern of Cold War was one of limited operations, of politico-military guerrilla warfare—in short, the type of small war in which the Marine Corps had become so thoroughly versed during the first 150 years of its existence. In the Far East, between 1955 and 1963, Marines landed in the Tachen Islands, Taiwan, Laos, Thailand, and South Vietnam, in counter-moves against communist pressure. In the Mediterranean, not only did Marines land at Alexandria to help evacuate U. S. and foreign nationals during the Suez incident of 1956, but, in July 1958, on an appeal from the Lebanese government, Marines secured Beirut against a communist coup. A Marine brigade, subsequently reinforced by Army troops, stood by for ten weeks until peaceful elections had been completed and a constitutional change of government had been duly carried out. In the Caribbean, our October 1962 confrontation with communism saw Cuba ringed with floating Marine landing forces, while other FMF units ensured that Guantanamo Bay remained safe and secure against Castro aggression. Soon afterward, in May 1963, a Marine expeditionary brigade lay off Port-au-Prince for three weeks in expectation that the regime of Haiti's dictator, Francois Duvalier, would finally topple. This was not to be, but active service loomed ahead elsewhere on Hispaniola.

623 ■ Explosion in Santo Domingo

On 24 April 1965, what had started as a military coup d'etat escalated into an attempt by Castroite communists to gain control of the Dominican Republic. In a five-day blood bath, the government of Santo Domingo ceased to exist. Foreign lives and property were in extreme danger; the American Embassy (and seven other foreign embassies) came under fire or were actually violated by leftist revolutionaries. In the circumstances, the United States government had no alternative but to step in. In the late afternoon of 28 April, President Johnson

U. S. Marine manning an improvised roadblock in Santo Domingo. During the May 1965 Dominican revolt, Marines were, as always, the first U. S. troops to land.

ordered the landing of the 3d Battalion, 6th Marines, at Ciudad Santo Domingo, from the USS *Boxer* lying offshore. It had been 39 years since U. S. Marines last landed in the Caribbean (at Bluefields, Nicaragua, in 1926).

Quickly augmented to brigade size, the landing force, despite appreciable resistance in some areas, established a demilitarized international zone protecting the American and other embassies. When, in due course, Army airborne units arrived, the Marines came under command of the Army; when pacification of the city had been effected in June, they were withdrawn. Noteworthy in this operation—aside from its dramatic demonstration that the United States was still ready to intervene when national interest so dictated—was the fact that once again ready, fully equipped seaborne Marines had reached an objective well ahead of even the fastest-moving, lightly equipped Army airborne units.

624 ■ War in Vietnam

Meanwhile, on the other side of the globe, the dragging war in Vietnam—the Nicaragua of the 1960s—heated up. Marine helicopter units (flying half the total sorties and flight hours with 20 percent of the helo-lift in the country) had been in Vietnam since 1962. So had Marine reconnaissance troops and a U. S. Marine advisory mission to train the Vietnamese Marine Corps.

In accordance with long-standing contingency plans, elements of the 3d Marine Division, supported by squadrons of the 1st Wing, deployed to a highly strategic locale in the north of the Republic of Vietnam, centered around Danang and the existing air base complex. This area

Marines slosh through flooded rice paddies during an assault on Vietcong positions, South Vietnam.

had not been chosen by chance: it had a port, independent of Saigon and others to the south; it had beaches; it commanded defiles in the coastwise road and rail net. Its main and enduring disadvantage was that it was adjacent to some of the most hard-core Vietcong regions in-country, both south and inland to the west.

On the heels of the 3d Division followed the 1st Marine Division and MAG-36, being split between Vietnam and Okinawa. By the end of the year, almost two-thirds of the combat units of the Marine Corps were thus committed to the Vietnamese war.

Adjacent Marine coastal "enclaves" were established at Danang and Chulai, with the objective not only of protecting these important air bases but of pacifying a populous and productive region, using the "oil-stain" tactics originated in this very region by France's master of colonial warfare, Marshal Lyautey. Consolidated under Headquarters, III Marine Amphibious Force, the Marines soon came to grips with the Vietcong, winning a notable action at Van Tuong near Chulai in August 1965. Subsequently, making full use of their amphibious capability, a series of surprise landing attacks, code-named *Dagger Thrust*, was executed along the Vietnamese coast, by Marines embarked with the Seventh Fleet. And late 1965 was marked by sustained patrolling, ambushes, and intermittent battles to link up the Danang and Chulai enclaves. Not only did 1966 bring more of the same but also more troops. By the end of 1966, approximately 60,000 Marines —more than one-sixth of all U. S. forces in Vietnam—were ashore and in the field. These units included troops from the newly formed 5th Marine Division, not seen on the active list since World War II, while the 1967 Marine Corps had topped Korea's peak and was surpassed in strength only by that of 1945. In hard fighting throughout I Corps area and especially along the 17th parallel demilitarized zone separating North and South Vietnam, Marines repeatedly turned back crack North Vietnamese units.

Vietnam had by now developed into two separate yet interlocking wars: the internal war of pacification, waged throughout the interior of the country against the Vietcong, at the outset authentic South Vietnamese inheritors of the old Viet Minh who had beaten the French; and the war of the DMZ, in which regular NVA (North Vietnamese Army) divisions with secure supply lines to their rear invaded the country from the north and attempted to detach the Republic's northern provinces. The first war was one of ambush, terror, and the guerrilla. The second was head-on fighting between the NVA, the Prussians of Southeast Asia, and the Marine amphibious force which stood in their path—but which, because of the skewed U. S. political strategy of the war, was prohibited from outflanking its enemies by Inchon-style attacks from the sea.

By virtue of the original deployment, it was the destiny of the 1st Division to fight the Vietcong in the "Rocket Belt" approaches to Da Nang and in such fiercely contested valleys as Que Son. The 3d, and subsequently the 5th, Divisions defended the DMZ against the North Vietnamese invaders in a war-torn battlefield defined by the DMZ on the north and Route 9, a main east-west highway, on the south. Here at Con Thien, "The Rockpile," "Leatherneck Square," and ultimately Khe Sanh (which in 1968 the NVA tried and failed to make into an American Dien Bien Phu), the Marines slugged it out with the North Vietnamese.

The Vietcong *Tet* offensive of early 1968, perceived by the U. S. public as the high-water mark of the war, actually destroyed the Vietcong. Amid hard fighting through the entire tactical zone, two Marine contributions were noteworthy. Not a single town held by Marines or by the Marine-invented "Combined Action Platoons" (Vietnamese Popular Force units stiffened by Marine squads) was lost to the communist onslaught. Hue, ancient imperial capital of Annam, lost by Vietnamese defenders, fell to the 5th Marines in one of the war's hardest fights.

Later that same year, in the most successful high-mobility regimental-scale campaign of the war, "Dewey Canyon," the 9th Marines scourged and cleaned out the major NVA Base Area 611 in the remote Da Krong Valley. Thousands of North Vietnamese were killed, large weapons caches were destroyed, and so were hundreds of tons of ammunition.

Yet even as these and other successes of combat and of pacification were being scored by all the Armed Forces, Marines included, the war was being lost at home on the pages and TV screens of the U. S. media and in the halls of Congress.

In mid-1969, as the Vietnamese Armed Forces began to take on more of their own war, American troop withdrawals commenced. By October, the first-in 3d Division and 1st Wing had been phased out. By this time the communists had reverted almost completely to the level of terror and guerrilla war. At the year's end, 90 percent of the population of the Marines' northern provinces was living in secure areas. Even so, the 1st Division still had ample work holding NVA forays at arm's length from the vital Da Nang area. More and more, however, the ARVN (Army of the Republic of Vietnam) was out front, while the Marines were in support.

One landmark Marine Corps achievement had been to forge the capable, high-spirited Vietnamese Marine Corps, whose splendid fighting during the all-out communist offensive of 1972 did so much to hold the northern provinces. Yet in the end it was all in vain: abandoned by an American Congress that cut off its weapons and supplies, Vietnam was destined to fall to communist aggression which outlasted its foes.

By that time, however (save for Marine landing forces which covered the final evacuation of American embassies in Phnom Penh and Saigon), the Marines' long war was over. In the words of the Commandant in April 1972:

> We are pulling our heads out of the jungle and redirecting our attention seaward, reemphasizing our partnership with the Navy and our shared concern in the maritime aspects of our national strategy. . . . With respect to our standards—we will maintain them: in appearance, discipline, personal proficiency, and unit performance. Without them, we would not be Marines.

625 ■ The Cutting Edge of Sea Power

An important lesson, not only in Vietnam but in Santo Domingo and the myriad lesser operations of the late 1950s and early 1960s, is that of the ability of maritime expeditionary forces, positioned in international waters, to anticipate trouble and hover indefinitely, if need be, in easy reach of the key points. While airborne forces remained tied to objectives containing major airdromes, cumbered by limits on overflight (as had happened in the Lebanon operation) and dependent even then on seaborne follow-up to give them combat staying power, the ability of the Fleet to go where it pleased, to stay there, and to project its power readily ashore made the Fleet Marine Force a cutting edge of sea power in the Cold War and in whatever history held in store to succeed it.

No-one can say that the Marines have ever failed to do their work in handsome fashion.
—Major General Johnson Hagood, USA

7

Traditions, Flags, Decorations, and Uniforms

701 ▪ "The Thin Line of Tradition"
The traditions of the Marine Corps, its history, its flags, its uniforms, its insignia—*the Marine Corps way of doing things*—make the Corps what it is and set it so distinctly apart from other military organizations and Services.

These traditions give the Marine Corps its flavor, and are the reason why the Corps cherishes its past, its ways of acting and speaking, and its uniforms. These things foster the discipline, valor, loyalty, aggressiveness, and readiness which make the term "Marine" ". . . signify all that is highest in military efficiency and soldierly virtue."

One writer on Marine traditions nailed down their importance in the following words:

> As our traditions, our institutions, and even our eccentricities—like live coral—develop and toughen, so the Corps itself develops and toughens.

And remember: whenever the Marine Corps is impoverished by the death of a tradition, *you* are generally to blame. Traditions are not preserved by books and museums, but by faithful adherence on the part of all hands—*you especially.*

MARINE CORPS TRADITIONS AND CUSTOMS

702 ▪ Globe and Anchor
When the late Major General Smedley Butler (winner of two Medals of Honor) was a lieutenant in the Philippines in 1899, he decided to get himself tattooed.

Today's emblem, the Globe and Anchor, was adopted by Brigadier General Jacob Zeilin, 7th Commandant, in 1868.

I selected an enormous Marine Corps Emblem [wrote Butler] to be tattooed across my chest. It required several sittings and hurt me like the devil, but the finished product was worth the pain. I blazed triumphantly forth, a Marine from throat to waist. The emblem is still with me. Nothing on earth but skinning will remove it.

Butler was somewhat premature in his last sentence. Within less than a year, during the storming of the Tartar Wall in Peking, a Chinese bullet struck him in the chest and gouged off part of his emblem. The rest of it accompanied him to the grave forty years later.

Whether you are a private or general is secondary compared to the privilege you share, of wearing the emblem. The Globe and Anchor is the most important insigne you have.

The Marine Emblem, as we know it today, dates from 1868. It was contributed to the Corps by Brigadier General Jacob Zeilin, 7th Commandant. Until 1840, Marines wore various devices, mainly based on the spread eagle or foul anchor. In 1840 two Marine Corps devices were accepted. Both were circled by a laurel wreath, undoubtedly borrowed from the Royal Marines' badge, but one had a foul anchor inscribed inside, while the other bore the letters "USM." In 1859 a standard center was adopted—a U. S. shield surmounted by a hunting-horn bugle, within which was the letter "M." From this time on, the bugle and letter "M," without the shield or laurel wreath, were usually worn by Marines on undress uniforms. This type of bugle was the nineteenth century symbol for light infantry or *jagers*—so called because they were recruited from the ranks of foresters, gamekeepers, and poachers, all renowned as skirmishers and riflemen.

In 1868, however, General Zeilin felt that a more distinctive emblem was needed. His choice fell on another device borrowed from the British Marines: the globe.

The Marine badge before 1868 embodied the hunting-horn symbol traditional to light infantry.

The globe had been conferred on the Royal Marines in 1827 by King George IV. Because it was impossible to recite all the achievements of Marines on the Corps Color, said the King, "the Great Globe itself" was to be their emblem, for Marines had won honor everywhere.

General Zeilin's U. S. Marine globe displayed the Western hemisphere, since the "Royals" had the Eastern hemisphere on theirs. Eagle and foul anchor were added, to leave no doubt that the Corps was both American and maritime.

703 ■ The Marine Corps Seal

The official seal of the Corps, designed by General Shepherd, 20th Commandant, consists of the Marine Corps Emblem in bronze, the eagle holding in his beak a scroll inscribed *"Semper Fidelis,"* against a scar-

The Marine Corps seal, designed by General Shepherd, 20th Commandant, was approved by the President in 1954.

let and blue background, encircled by the words, *"Department of the Navy * United States Marine Corps."*

704 ■ Marine Corps Colors

The colors of the Corps are scarlet and gold. Although associated with U. S. Marines for many years, these colors were not officially recognized until General Lejeune became 13th Commandant. Today you will see scarlet and gold throughout Marine posts—on signboards; auto tags; bandsmen's drums, pouches, and trumpet slings; MP brassards; officers' hat-cords and aiguillettes; and, it sometimes seems, everywhere in sight.

In addition to scarlet and gold, forest green enjoys at least semiofficial standing as a Marine color. During the years since 1912, when forest green was adopted for the winter service uniform, it has become standard for such equipment as vehicles, weapons, armor, and organizational chests and baggage. In addition, forest green is today virtually the distinguishing color of Marines throughout the world, being worn as a service uniform by the British, Dutch, Korean, and other corps.

Forest green comes from the same source as the light infantry bugle that was once part of the Corps badge. The costume of eighteenth century huntsmen was forest green. The riflemen recruited from that calling wore green uniforms—a green that survives not only among Marines but also in the uniforms of Britain's Rifle Brigade (the "Greenjackets"), and India's Ghurkhas.

The three colors of the Corps—scarlet, gold, and forest (or rifle) green—are the colors of the Corps necktie, designed for wear with civilian clothes.

> NOTE: Should you find it difficult to purchase the Corps necktie described above, an as yet unofficial substitute may be obtained from Brooks Brothers (New York, Washington, Boston and Atlanta). This tie consists of tiny Marine Corps emblems woven against an optional background of rifle green, maroon, or navy blue, either variation providing a distinctive, handsome, and conservative tie.

705 ■ The Marine Corps Motto

"Semper Fidelis" ("Always Faithful") is the motto of the Corps. That Marines have lived up to this motto is proved by the fact that there has never been a mutiny, or even the thought of one, among U. S. Marines.

Semper Fidelis was adopted about 1883 as the motto of the Corps. Before that, there had been three mottoes, all traditional rather than official. The first, antedating the War of 1812, was *"Fortitudine"* ("With Fortitude"). The second, "By Sea and by Land," was obviously a translation of the Royal Marines' *"Per Mare, Per Terram."* Until 1848, the third of the Marine Corps' mottoes was, "To the Shores of Tripoli," in commemoration of O'Bannon's capture of Derna in 1805. In 1848, after the return to Washington of the Marine battalion that took part in the capture of Mexico City, this motto was revised to: "From the Halls of the Montezumas to the Shores of Tripoli"—a line now familiar to all Americans. This revision of the Corps motto in Mexico has encouraged speculation that the first stanza of "The Marines' Hymn" was composed by members of the Marine battalion who stormed Chapultepec Castle.

It may be added that the Marine Corps shares its motto with England's Devonshire Regiment, the 11th Foot, one of the senior infantry regiments of the British Army, whose sobriquet is "the Bloody Eleventh" and whose motto is also *Semper Fidelis*.

706 ■ Marines' Hymn and Marine Corps March

"The Marines' Hymn" is what its name implies, the hymn of the Marine Corps. "Semper Fidelis," one of John Philip Sousa's best known works, is the Corps march.

"The Marines' Hymn," the oldest of the official songs of the Armed Services, may be found in Appendix I. Every Marine knows those words and will sing them at the drop of a field hat. The origin of the Hymn is obscure. The words date back into the nineteenth century, and the author remains unknown. The music comes from an air, "Gendarmes of the Queen," in Jacques Offenbach's opera *Geneviève de Brabant*, first performed in November 1859. Regardless of its origin, however, *all Marines get to their feet whenever "The Marines' Hymn" is played or sung.*

"Semper Fidelis" was composed by Sousa in 1888 during his tour as leader of the Marine Band. "Semper Fi," as the troops know it, is habitually rendered for parades, reviews, and march-pasts of Marines.

707 ■ Birthday of the Corps

The Marine Corps was founded by the Continental Congress on 10 November 1775. The resolution which created our Corps reads as follows:

> *Resolved.* That two Battalions of Marines be raised consisting of one Colonel, two lieutenant Colonels, two Majors, & Officers as usual in other regiments, that they consist of an equal number of privates with other battalions; that particular care be taken that no persons be appointed to office, or inlisted into said Battalions, but such as are good seamen, or so acquainted with maritime affairs as to be able to serve to advantage by sea, when required. That they be inlisted and commissioned for and during the present war with Great Britain and the colonies, unless dismissed by order of Congress. That they be distinguished by the names of the first and second battalions of American Marines, and that

they be considered as part of the number, which the continental Army before Boston is ordered to consist of.

Chapter 24 of this *Guide* tells how we celebrate the Marine Corps Birthday. Here it is enough to recount that the Birthday of the Corps has been observed for many years; and that, although the Marine Corps joins the other Services each May in observing Armed Forces' Day, November 10th remains the Marines' own day—a day of ceremony, comradeship, and celebration.

708 ■ The Mameluke Sword

The sword that Marine officers carry goes back to the *Uniform Regulations* of 1826 (with a hiatus from 1859 to 1875). Records of the day, however, indicate that swords of this pattern were worn by Marine officers before the War of 1812.

The Mameluke sword gets its name from the cross-hilt and ivory grip, both of which were used for centuries by the Moslems of North Africa and Arabia. The Marine Corps tradition of carrying this type of sword dates from Lieutenant O'Bannon's assault on Derna, Tripoli, in 1805, when he is said to have won the sword of the governor of the town.

Aside from its use on parade, many Marine Corps rituals center about your sword. You cut your wedding cake with the sword, and you wear it when you get married. At many posts, you wear it while officer of the day. Should you ever be unlucky enough to be placed under arrest, you must surrender your sword.

> P.S. Never unsheathe your sword inside a mess or wardroom. If you do, custom decrees that you must stand drinks for all present. This tradition goes back to stringent rules against dueling in the early days of the Navy and Marine Corps.

709 ■ "First on Foot, and Right of the Line"

Marines form at the place of honor—head of column, or on right of line —in any naval formation. This privilege was bestowed on the Corps by the Secretary of the Navy on 9 August, 1876.

710 ■ "First to Fight"

The slogan "First to Fight" has appeared on Marine recruiting posters ever since World War I.

Marines have been in the forefront of every American war since the founding of the Corps. Marines entered the Revolution in 1775, even before the Declaration of Independence was signed. Before declaration of the War of 1812, Marines helped to defend the USS *Chesapeake* against the British. At the outset of hostilities against Mexico, Marines

helped to raise California's Bear Flag. In the Civil War, Marines not only captured John Brown at Harper's Ferry but were among the few U. S. regulars who fought in the first Battle of Manassas in 1861. In 1898, Huntington's Fleet Marines were the first U. S. troops to occupy Cuban soil, and Admiral Dewey's Marines were the first to land in the Philippines. Marines were first to land at Veracruz (1914). In World War I, the 5th Marines formed part of the first American Expeditionary Force (AEF) contingent to sail for France. When Iceland had to be occupied in 1941, Marines, the only U.S. troops who were ready, were the first to land. In World War II, at Pearl Harbor, Ewa, Wake, Midway, Johnston Island, and Guam, Marines formed the ready forefront of our Pacific outpost line. At Guadalcanal in August 1942, Marines launched the first American offensive of the war. In the Korean War, the first reinforcements to leave continental United States were the 1st Provisional Marine Brigade. The first American troops to land in the Lebanon in 1958 were Marines. At Santo Domingo, in 1965, Marines were again the first to fight; while, in Vietnam, the first U. S. ground unit to be committed to the war was the 3d Marine Division.

On this record of readiness, "First to Fight" constitutes the Marine's pride, responsibility, and challenge.

711 ■ "Leathernecks"

The Marines' long-standing nickname, "Leathernecks," goes back to the leather stock, or neckpiece, that was part of the marine uniform from 1775 to 1875. One historian has written:

> Government contracts usually contained a specification that the stock be of such height that the "chin could turn freely over it," a rather indefinite regulation, and, as one Marine put it, one which the "taylors must have interpreted to mean with the nose pointing straight up."

Although many justifications have been adduced for the leather stock, the truth seems to be that it was intended to ensure that Marines kept their heads erect ("up and locked," the aviators would say), a laudable aim in any military organization, any time.

Descended from the stock is the standing collar, hallmark of Marine blues, whites, and evening dress. Like its leather ancestor, the standing collar regulates stance and posture and thus proclaims the wearer as a modern "Leatherneck."

712 ■ Scarlet Trouser-Stripe

Officers and noncommissioned officers have intermittently worn scarlet stripes on dress trousers ever since the early days of the Corps. It is unsubstantiated, even though oft repeated, that the right to wear scarlet stripes was conferred on the Corps as a battle honor after the Mexican war (actually the initial uniform trousers issued after reconstitution of the Corps in 1798 had scarlet piping).

713 ■ Headgear

Two Marine traditions center about headgear:
The *quatrefoil* (the cross-shaped braid atop officers' frame-type ["bar-

racks"] caps) has been worn since 1859. The design, of French origin, is a distinguishing part of the Marine officers' uniform.

The *field hat* was the rugged, picturesque expeditionary headgear of the Corps from 1898 until 1942, and became a universal favorite. As a result, although the hat became outmoded during World War II, General Cates, the 19th Commandant, authorized its use on the rifle range in 1948 and took steps to issue field hats to all medalist shooters in Marine Corps matches. Subsequently, in 1956, General Pate, the 21st Commandant, directed that field hats be worn by all recruit drill instructors, and the hat—later copied and adapted by the Army for the same purpose—has become a symbol of Marine Corps recruit training.

714 ■ Collar Emblems

Although officers have worn collar emblems since the 1870s, enlisted Marines did not rate this privilege until August 1918, when Franklin D. Roosevelt, then Assistant Secretary of the Navy, visited the 4th Marine Brigade in France, shortly after Belleau Wood. In recognition of the Brigade's victory, Mr. Roosevelt directed on the spot that enlisted Marines would henceforth wear the emblem on their collars.

715 ■ Marine Talk and Terminology

The 4th Marine Brigade's admired Army commander at Belleau Wood, Lieutenant General James G. Harbord, USA, was quick to note and record the salty Marine way of saying things:

> In the more than a month that the Marine Brigade fought in and around the Bois de Belleau, I got a good opportunity to get the Marine psychology. . . . The habitual Marine address was "Lad". . . . No Marine was ever too old to be a "lad." The Marines never start anywhere: they always "shove off." There were no kitchens: the cooking was done in "galleys." No one ever unfurled a flag—he "broke it out."

This *Guide* contains a glossary of Marine terms. Never feel selfconscious about using them. Require that subordinates use them. Accept no substitutes.

716 ■ The Canton Bell

In Quantico hangs a weathered bronze Chinese bell—the "Canton Bell" —cherished gift from the Royal Marines. This bell was taken by "the Royals" after storming the Canton Forts in South China in 1856, and for years occupied a place of honor in the Royal Marine officers' mess at Chatham. When Chatham Barracks was decommissioned after World War II, the officers of the mess voted to present their trophy to the

U. S. Marines, as a symbol of the comradeship between the two corps during this attack, and later.

717 ■ "The President's Own"
Founded in 1798 (more than a century before the bands of the other three Services), the Marine Band has performed at White House functions for every President except George Washington, and was especially sponsored by Thomas Jefferson. Because of its traditional privilege of performing at the White House, the band is spoken of as "The President's Own." President Kennedy epitomized the band's special position when he remarked in 1962, "I find that the only forces which cannot be transferred from Washington without my express permission are the members of the Marine Band, and I want it announced that we propose to hold the White House against all odds, at least for some time to come."

The Marine Band has been present at many of the most memorable and cherished moments in our nation's history, including the dedication of the National Cemetery at Gettysburg when Lincoln gave his immortal address (and his aide-de-camp was 2d Lt. H. C. Cochrane, USMC). Among the band's many traditions, including leadership for 12 years by

The U. S. Marine Band, shown here trooping the line on parade at Eighth and Eye, has performed for every President except George Washington.

Marines from Eighth and Eye form before the Marine Corps War memorial in Arlington. Besides important security duties, Marine Barracks Washington provides the top ceremonial troops for the National Capital.

John Philip Sousa, is its scarlet, full-dress blouse, the only red coat worn by American forces since the Revolutionary War. (In 1956, the Marine Corps Drum and Bugle Corps was likewise granted the privilege of wearing red coats.)

The Marine Band tours the country each fall, and has done so ever since Sousa commenced the practice in 1891, although one section of the band always remains in Washington to fulfill its traditional primary mission: "To provide music when directed by the President of the United States, the Congress of the United States, or the Commandant of the Marine Corps."

718 ■ Evening Parade

From May through October, a ceremonial Evening Parade is held each Friday evening after nightfall at the Marine Barracks, Eighth and Eye. This colorful ceremony, executed under searchlight illumination, features the Marine Band, Marine Corps Drum and Bugle Corps, a special exhibition drill platoon, and a battalion of Marines from the barracks. Evening Parades were first held in 1957 after a Marine Corps ceremonial detachment participated in the Bermuda International Searchlight Tattoo, and became a fixed Marine Corps custom following similar participation by a larger Marine detachment in the famed Edinburgh Searchlight Tattoo in Scotland in 1958. Evening Parades are open to the public, and any officer who desires to attend with a reasonable number of guests may obtain reserved seats by telephoning the Marine Barracks adjutant.

719 ■ "And St. David"

During the Boxer Uprising (1900), at Tientsin and Peking, the Marine battalion in the international relief column was brigaded with the Royal Welch Fusiliers (23d Foot), one of Britain's most renowned regiments. The resulting fellowship between the two organizations is symbolized each year on St. David's Day (1 March, the Welsh national holiday), when the Commandant of the Marine Corps and the Colonel of the Fusiliers exchange by dispatch the traditional watchword of Wales: ". . . And St. David."

720 ■ The Commandant's License Plate

If, in Washington, D.C., you ever bump a car bearing license "1775," climb out of the wreckage at attention. That license plate is set aside for the official sedan of the Commandant of the Marine Corps.

721 ■ Rum on New Year's Day

Every New Year's Day since 1804, the Marine Band serenades the Commandant at his quarters and receives a tot of hot buttered rum in return.

Serenading the Commandant on New Year's Day, Marine bandsmen get ready for their annual tot of grog.

This occasion marks the last surviving issue of "grog" in the Armed Forces.

722 ■ Marine Corps Bulldog

Ever since World War I the bulldog has been associated with the Corps. An English bulldog has been official mascot at Eighth and Eye, and therefore top dog of the Corps, since the 1920s. Prior to World War II he was always named "Jiggs." Subsequently, however, in an appropriate tribute to one of the Corps' bravest officers, the late Lieutenant General L. B. Puller, the name has been "Chesty."

723 ■ Ship's Bell

All Marine posts (and even some camps in the field) have their ship's bell, usually from a warship no longer in commission. This bell is mounted at the base of the flagpole, and the guard has the duty, between reveille and taps, of striking the bells—and also of keeping the bell in a high polish.

724 ■ Last to Leave the Ship

Marines are always or should be the last—other than the ship's captain—to leave a ship being abandoned or put out of commission. Although the tradition is an old one, it first appears in *Navy Regulations* of 1865:

> When a vessel is to be put out of commission, the Marine officer with the guard shall remain on board until all the officers and crew are detached and the ship regularly turned over to the officers of the Navy Yard or station.

725 ■ Swagger Sticks

The tradition of the swagger stick (which may be carried optionally by all officers) originated in the British Army and goes as far back as 1790. In the Marine Corps, the stick came into vogue in the latter part of the nineteenth century, and was virtually a required article of uniform until World War I. The origin of the swagger stick lay in the whips or batons carried by mounted officers of the eighteenth century. Probably the best description of the stick's function today may be quoted from the lips of a British regimental sergeant-major instructing new officers:

> Now, gentlemen, the swagger stick is not for rattling along railings, cleaning out the drains at home, or swiping the heads off poor innocent little flowers. Nor is it for poking into stomachs or for fencing duels in the Mess. No, gentlemen, *it is to make you walk like officers and above all to keep your hands out of your pockets.*

Striking (or, as the Navy phrase goes, "making") the bells is the traditional duty of the field music of the guard; so is keeping the bell highly polished.

726 ■ "Tell it to the Marines!"

In his book, *Fix Bayonets!*, Captain John W. Thomason, Jr., gives the generally accepted version of the origin of "Tell it to the Marines!":

> They relate of Charles II that at Whitehall a certain sea-captain, newly returned from the Western Ocean, told the King of flying fish, a thing never heard in old England. The King and court were vastly amused. But, the naval fellow persisting, the Merry Monarch beckoned to a lean, dry colonel of the sea regiment, with seamed mahogany face, and said, in effect: "Colonel,

this tarry-breeks here makes sport with us stay-at-homes. He tells of a miraculous fish that foresakes its element and flies like a bird over water." "Sire," said the colonel of Marines, "he tells a true thing. I myself have often seen those fish in your Majesty's seas around Barbados—" "Well," decided Charles, "such evidence cannot be disputed. And hereafter, when we hear a strange thing, we will tell it to the Marines, for the Marines go everywhere and see everything, and if they say it is so, we will believe it."

This yarn (for such it is) was for many years credited to Samuel Pepys, although scholars disclaimed it. On the other hand, the phrase, "Tell it to the Marines," is an old one and can be found in print as early as 1726.

727 ■ At Divine Service

Divine services for Marines should always include the Marine Corps Prayer (written at the suggestion of General Shepherd, 20th Commandant, by Bishop Sherrill, former Presiding Bishop of the Episcopal Church and hero in World War I):

> O Eternal Father, we commend to Thy protection and care the members of the Marine Corps. Guide and direct them in the defense of our country and in the maintenance of justice among nations. Protect them in the hour of danger. Grant that wherever they serve they may be loyal to their high traditions and that at all times they may put their trust in Thee; through Jesus Christ our Lord. *Amen.*

It is also customary for Marine Corps divine services to conclude with the traditional naval hymn, "Eternal Father, strong to save." When no chaplain is available, the commanding officer, following the traditions of the sea, may conduct divine services, hold funerals, etc. When a chaplain is present, some commanding officers may choose to read the lesson, another traditional prerogative of the CO. This is arranged beforehand with the officiating chaplain. It is a good idea to draw your chaplain's attention to the Marine Corps Prayer, above, as some chaplains are unfamiliar with it.

> NOTE: When attending divine service in uniform, or present in uniform at an occasion when prayer is offered (as at a military funeral), uncover (if not already done) and assume the *old* pre-1939 position of "Parade Rest without Arms," i.e., right foot carried six inches to the rear, left knee slightly bent, weight equally distributed on both feet; hands clasped without constraint in front of the body, left hand uppermost—and in the case of prayer, head slightly bowed. This position enables you to appear reverent and military at the same time, and was used as the traditional position for prayers at sea throughout the Old Navy.

728 ■ Conduct in Action

Over and above the competence, resolution, and courage which are expected of every Marine in battle, it is particularly expected that no wounded or dead Marine will ever be left on the field or unattended, regardless of the cost of bringing him in. As for surrender, the Marine Corps code is that expressed by Napoleon:

> There is but one honorable mode of becoming prisoner of war. That is, by being taken separately; by which is meant, by being cut off entirely, and when we can no longer make use of our arms.

729 ■ Marine Corps Museums

"The scrap-book of the Marine Corps," as it is sometimes described, is the Marine Corps Museum at the Marine Corps Historical Center, Washington Navy Yard. No summary of the traditions of the Corps would be complete without mention of this central repository of awards, battle honors, historical flags, and other objects of lasting sentimental significance to the Marine Corps. The Museum collection documents Marine Corps history from 1775 to the present day. On display is an extensive array of uniforms, weapons, artifacts, equipment, prints, and paintings giving tangible substance to the proud traditions of the Corps. Included among numerous historical flags, for example, are the famous Colors raised by the 28th Marines atop Suribachi Yama on Iwo Jima. The weapons collection is one of the finest in the world, and is backed up by over 55,000 documents covering design, patent data, and test reports on machine guns and automatic weapons alone. Every Marine officer should be thoroughly familiar with this Museum, which ranks among the best military and naval museums in the United States. In addition to the Museum at the Washington Navy Yard, there are also excellent Post Museums at Quantico and Parris Island, and the Navy and Marine Museum at Treasure Island, San Francisco; and in Philadelphia, in New Hall, a restored building from pre-Revolutionary days, may be found an outstanding collection of material dealing with the early days of the Corps and its origins in Pennsylvania.

730 ■ Marine Corps Memorial Chapel, Quantico

The Post Chapel at Quantico serves, in addition to its regular functions, at the Memorial Chapel of the Marine Corps. Here is kept a "Book of Remembrance" listing the name, rank, and date of death for Marines and members of the Navy serving with the Marine Corps who have given their lives in action. For the time being, the Book of Remembrance begins with Vietnam, but it is planned to extend it back, eventually to include by name every Marine or eligible Navy man recorded as killed in action since the Revolution.

COLORS, FLAGS, AND STANDARDS

731 ▪ Colors, Flags, and Standards

A Parris Island recruit once asked his drill instructor, "Sergeant, who carries the flag in battle?"

Came the unhesitating reply, "Son, *every* Marine carries the flag in battle!"

As the soldier's proverb says, "The flag is a jealous mistress," and any Marine will fight and die rather than permit the National Colors or a Marine Corps Color to be dishonored.

Colors or standards must never fall into enemy hands. If capture seems inevitable, they should be burned. Unserviceable colors or standards, or those from disbanded units, are turned in to the supply system. The latter in turn forwards flags of historical value to the Marine Corps Museum, which is the Corps repository for historical flags, as well as for flags and war trophies captured by Marines. Soiled, torn, or badly frayed flags, if not historical, are destroyed privately by burning.

732 ▪ Types of Flags

Marine Corps terms which deal with flags are precise and particular. As an officer, you must learn to distinguish the various kinds of flags and to speak of them in the correct terminology.

National Color or Standard. This is the American flag. When the flag is displayed over Marine or Navy posts, stations or ships, its official title is the *National Ensign.* The national flag carried by Marine organizations is made of silk or nylon and is called the *National Color (except* when borne by a mounted, mechanized, motorized, or aviation unit, when its title becomes the *National Standard).* This technical distinction between a *color* and a *standard* also applies to the battle colors and organization colors described in the following paragraphs.

The National Color is carried on all occasions of ceremony when two or more companies of a unit are present. When not in the hands of troops, the National Color is entrusted to the adjutant. With the Marine Corps Color (discussed below), the National Color is usually displayed in the office or before the tent of the commanding officer. Whenever the National Color is carried in the open, it is escorted by a *color guard* composed of selected Marines, and the Color itself is borne by an outstanding NCO, the *color sergeant.*

The National Ensign, displayed over ships and shore stations, comes in three sizes. These are:

1. *Post flag:* size 10 by 19 feet, flown in fair weather except on Sundays and national holidays;
2. *Storm flag:* size 5 by 9 feet 6 inches, flown during foul weather;

3. *Garrison flag:* size 20 by 38 feet, flown on Sundays and national holidays as provided in the *Marine Corps Flag Manual* (but never from a flagpole shorter than 65 feet).

For more information on display of the National Color or Ensign, turn to *Navy Regulations* and to *Marine Corps Flag Manual.*

Marine Corps Colors and Standards. The Commandant issues to every major Marine unit or organization a distinguishing flag, which is carried beside the National Color. These unit flags are called *Marine Corps Colors* (or Standards). A Marine Corps Color bears the emblem and motto of the Corps and the unit title, and follows the color scheme of the Corps, scarlet and gold.

The Marine Corps Color of a Fleet Marine Force unit is called the unit *Battle Color*; the Color authorized for an organization in the Supporting Establishment (such as a Marine Barracks) is called the *Organization Color.* No unit smaller than a separate battalion or regiment receives a Battle Color, nor does a temporary or provisional unit unless specially authorized by the Commandant.

Certain organized units of the Marine Corps Reserve are likewise authorized to carry Organizational Flags of the type just described, but bearing a Reserve designation.

Guidons. These are small rectangular flags, made in the Marine Corps colors, carried by companies, batteries, or detachments, or used as marker flags for ceremonies. *Organizational guidons* carry the Marine Corps emblem and the title of the unit. *Dress guidons* (used as markers) simply bear the initials "USMC."

Personal Flags. Every active general officer in command displays a *personal flag.* Marine Corps personal flags consist of a scarlet field with white stars according to the general officer's rank, arranged in the same manner as the stars on Navy personal flags. Regulations governing personal flags are in *Navy Regulations.*

Miscellaneous Flags. In addition to the ceremonial flags just described, the Corps employs several miscellaneous flags and pennants described in the *Marine Corps Flag Manual.* Examples are:

Geneva Convention flag
Church pennant
Sanitary cordon flag
Recruiting flag

733 ■ Appurtenances of Flags

The appurtenances of Marine colors, standards, flags, and guidons are:
Streamers
Bands

Cords
Tassels
Staff ornaments

Streamers denote participation in combat or award of a collective citation or decoration conferred on the unit as a whole.

A *silver band* is attached to the staff of a Marine Corps Color or Standard for each streamer awarded.

When the unit or organization does not rate streamers or bands, a *cord and tassel*, woven in the Corps colors, is substituted.

The heads of staffs bear the following *staff ornaments:*

Colors and standards: silver lance-head
Personal flag: silver halberd
Guidon: plain silver cap

734 ▪ Battle Color of the Marine Corps

The Corps as a whole has one Battle Color entitled *The Battle Color of the Marine Corps*. This Color is entrusted to the senior post of the Corps, Marine Barracks, Eighth and Eye Streets, Washington, D.C. Attached to it are all the battle honors, citations, battle streamers, and silver bands that the Corps has won since 1775. At the time of writing these honors are the following:

Presidential Unit Citation Streamer with six silver stars and two bronze stars
Navy Unit Commendation Streamer with 16 silver stars
Army Distinguished Unit Streamer with one silver oak-leaf cluster
Army Valorous Unit Award
Meritorious Unit Commendation Streamer with 24 silver and two bronze stars
Army Meritorious Unit Streamer
Revolutionary War Streamer
Naval War with France Streamer
Barbary Wars Streamer
War of 1812 Streamer
African Slave Patrol Streamer
West Indies Anti-Piracy Campaign Streamer
Indian Wars Streamer
Mexican War Streamer
Civil War Streamer
Marine Corps Expeditionary Streamer with ten silver and three bronze stars, and silver "W"
Spanish-American War Streamer

The Battle Color of the Marine Corps bears all the battle honors and citations won by the Corps as a whole since 1775.

Philippine Campaign Streamer
Cuban Pacification Streamer
Nicaraguan Campaign Streamer
Mexican Service Streamer
Haitian Campaign Streamer with bronze star
Dominican Campaign Streamer
World War I Victory Streamer with one silver and one bronze star,
    Maltese Cross, and Siberian and West Indies Clasps
Army of Occupation of Germany Streamer
Second Nicaraguan Campaign Streamer
Yangtze Service Streamer
China Service Streamer with bronze star
American Defense Service Streamer with bronze star

Asiatic-Pacific Campaign Streamer with eight silver and one bronze stars
American Campaign Streamer
European–African–Middle East Campaign Streamer with one silver and four bronze stars
World War II Victory Streamer
Navy Occupation Service Streamer with Asia and Europe Clasps
National Defense Service Streamer with bronze star
Korean Service Streamer with two silver stars
Armed Forces Expeditionary Streamer with one silver and three bronze stars
Vietnam Service Streamer with three silver and two bronze stars
French Croix de Guerre Streamer (*Fourragère*) with two palms and one gilt star
Philippine Defense Streamer with one star
Philippine Liberation Streamer with two stars
Philippine Independence Streamer
Philippine Presidential Unit Citation with two bronze stars
Republic of Korea Presidential Unit Citation
Vietnam Cross of Gallantry with palm
Vietnam Meritorious Unit Citation

DECORATIONS, MEDALS, AND UNIT CITATIONS

735 ■ Decorations and Medals

"A soldier will fight long and hard for a bit of colored ribbon," said Napoleon, who originated the awarding of personal decorations.

Napoleon's conqueror, the Duke of Wellington, in turn introduced all-hands campaign medals, the first of which went to British troops who fought at Waterloo.

Both Wellington and Napoleon realized that decorations and medals not only express national gratitude to individuals but stimulate emulation and *esprit* in battles to come.

Today, Marine Corps awards fall into three classes: personal and unit decorations; commemorative, campaign, and service medals; and marksmanship badges and trophies. The *Navy and Marine Corps Awards Manual* gives details on all these, together with guidance for anyone who wishes to originate a recommendation that an award be made.

736 ■ Personal and Unit Decorations

The United States, despite the limitations in Article I, Section 9, of the Constitution, confers numerous military decorations. These range from the Medal of Honor, at the top, to the Naval Reserve Medal in junior

RIBBONS OF DECORATIONS AND AWARDS

 Medal of Honor

 Navy Cross

 Coast Guard Distinguished Service Medal

 Silver Star Medal

 Legion of Merit

 Distinguished Flying Cross

 Coast Guard Medal

 Navy and Marine Corps Medal

 Bronze Star Medal

 Meritorious Service Medal

 Air Medal

 Joint Service Commendation Medal

 Coast Guard Commendation Medal

 Navy Commendation Medal

 Coast Guard Achievement Medal

 Navy Achievement Medal

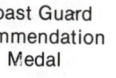 Purple Heart Medal

 Combat Action Ribbon

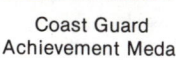 Presidential Unit Citation

 Coast Guard Unit Commendation

 Navy Unit Commendation

 Coast Guard Meritorious Unit Commendation

 Navy Meritorious Unit Commendation

 Gold Lifesaving Medal

 Silver Lifesaving Medal

 DOT Outstanding Achievement Medal

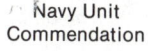 DOT Meritorious Achievement Medal

 DOT Superior Achievement Medal

 Coast Guard Good Conduct Medal

 Navy Good Conduct Medal

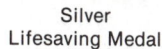 Marine Corps Good Conduct Medal

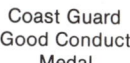 Army Good Conduct Medal

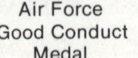  

Air Force Good Conduct Medal	Coast Guard Reserve Meritorious Service Ribbon	Naval Reserve Medal	Naval Reserve Meritorious Service Medal

Navy Expeditionary Medal	China Service Medal	American Defense Service Medal	American Campaign Medal

European-African Middle Eastern Campaign Medal	Asiatic-Pacific Campaign Medal	World War II Victory Medal	Navy Occupation Service Medal

Medal for Humane Action	National Defense Service Medal	Korean Service Medal	Antarctic Service Medal

Armed Forces Expeditionary Medal	Vietnam Service Medal	Armed Forces Reserve Medal	United Nations Service Medal

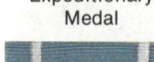   

United Nations Medal	Philippine Presidential Unit Citation	Republic of Korea Presidential Unit Citation	Republic of Vietnam Gallantry Cross Unit Citation

Republic of Vietnam Civil Actions Unit Citation	Republic of Vietnam Campaign Ribbon	Coast Guard Expert Rifleman Medal	Coast Guard Expert Pistol Shot Medal

position. In order of precedence, these personal or unit decorations are:
* ★ Medal of Honor (Navy)
* ★ Medal of Honor (Army and Air Force)
* ‡★ Marine Corps Brevet Medal
* ★ Navy Cross
* ★ Distinguished Service Cross (Army and Air Force)
* Defense Distinguished Service Medal
* Distinguished Service Medal (Navy)
* Distinguished Service Medal (Army and Air Force)
* ★ Silver Star Medal
* † Legion of Merit
* † Distinguished Flying Cross
* * Navy and Marine Corps Medal
* * Soldier's Medal
* * Airman's Medal
* * Coast Guard Medal
* † Bronze Star Medal
* Meritorious Service Medal
* † Air Medal
* Joint Services Commendation Medal
* † Navy Commendation Medal
* † Army Commendation Medal
* † Air Force Commendation Medal
* † Navy Achievement Medal
* † Coast Guard Achievement Medal
* Purple Heart
* ★ Combat Action Ribbon
* ★ Presidential Unit Citation
* ★ Distinguished Unit Emblem (Army and Air Force)
* † Navy Unit Commendation
* * Gold Life-Saving Medal (Treasury Department award)
* * Silver Life-Saving Medal (Treasury Department award)
* Marine Corps Reserve Special Commendation Ribbon
* Marine Corps Good Conduct Medal
* Navy Good Conduct Medal
* Army and Air Force Good Conduct Medal
* Coast Guard Good Conduct Medal
* Organized Marine Corps Reserve Medal
* Naval Reserve Medal

(★) Awarded for heroism only.
(†) Awarded for either heroic or meritorious acts.
(*) Awarded for heroism not in combat.
(‡) No longer awarded.

177

Among the foregoing, the *Medal of Honor* rates special mention. The Medal of Honor is the highest decoration conferred by the United States. Ordinarily, you can win it only for gallantry and intrepidity in combat, at the risk of your life, above and beyond the call of duty. Since the Civil War, when the award was created, 300 Medals of Honor have been won by U. S. Marines.

On attaining the age of 40, winners of the Medal of Honor are eligible for a special pension of $100 per month. In addition, if you hold the Medal of Honor, you are entitled to have your son(s) and daughter(s)

The Medal of Honor is the nation's highest award. Gunnery Sergeant Daniel Daly won two of them, and survived to earn the Distinguished Service Cross.

appointed to Annapolis, to West Point or to the Air Force Academy. You enjoy lifetime commissary and PX privileges; you may also travel without charge in U. S. armed forces aircraft, and, regardless of military status, you are eligible for medical and hospital care in VA hospitals. It is a tradition (though not officially recognized) that all hands salute a Medal of Honor man, regardless of rank.

737 ■ Unit Decorations

All top U. S. unit decorations, or "unit citations," as well as several foreign unit citations, have been won by Marine Corps units. If you are a member of an organization when it wins a collective citation, you are thereafter entitled to wear the citation ribbon or device as a personal decoration.

The *French Fourragère* is the senior unit award (and first collective award) won by Marines. The Fourragère dates from Napoleon's time; it was awarded to the 4th Marine Brigade in 1918 in lieu of awarding all hands the Croix de Guerre. The green and scarlet cord of the Fourragère may still be seen on the left shoulders of members of the 5th and 6th Marines, and of the few remaining Marines who were present when the 4th Brigade won the award.

The *Presidential Unit Citation* is the highest Navy and Marine Corps unit award. It was also the first American collective award, having been personally instituted by President Franklin D. Roosevelt as a citation for the defenders of Wake (1st Defense Battalion and Marine Fighting Squadron 211) in December 1941. The Presidential Unit Citation is considered to represent unit attainments that would warrant award of the Navy Cross if the recipient were an individual.

The *Distinguished Unit Emblem* is the Army and Air Force collective citation roughly equivalent to the Presidential Unit Citation. The Distinguished Unit Emblem has been awarded to several Marine ground and aviation units on detached service with the Army or Air Force.

The *Navy Unit Commendation* ranks next, in the Naval Service, after the Presidential Unit Citation. Unlike the latter, however, the Navy Unit Commendation may be won by extremely meritorious service in support of, but not participation in, combat operations. When awarded for combat performance, the NUC is comparable to the Silver Star Medal for an individual; for noncombat meritorious service, this Commendation is comparable to the Legion of Merit.

Other unit decorations awarded for lesser collective achievements of valor or merit include: Army Valorous Unit Award, Air Force Outstanding Unit Award, Meritorious Unit Commendations (Navy–Marine and Army).

738 ■ Campaign Medals

Campaign or service medals are issued to all hands who take part in particular campaigns or periods of service for which a medal is authorized. In addition to medals for specific campaigns, Marines may be awarded the *Marine Corps Expeditionary Medal* for service ashore on foreign soil, against opposition, for which no other campaign medal is authorized. Or, for similar joint operations in which the Army or Air Force may be involved, the *Armed Forces Expeditionary Medal* may be substituted. Campaign medals are often embellished by clasps or bronze stars, which denote participation in specific battles or phases of the campaign.

In addition to campaign and service medals, certain *commemorative medals* have been struck to commemorate noncombat but notable achievements, such as polar and antarctic expeditions, or pioneer flights.

739 ■ Initiating an Award

One of your responsibilities as a combat leader is to see that your men are promptly recommended for awards you believe they have earned. During active operations, it is usual for every FMF unit, from battalion up, to maintain a *board of awards*. The board evaluates and passes on recommendations for decorations that originate within the organization, but you must see that the board of awards receives recommendations promptly, and that the recommendations are accurately stated in whatever form may be required.

Few leadership derelictions are more reprehensible than failure to submit proper recommendations for awards, and then to see an award fail because you were too lazy to recommend it in the right form and with the detailed information required.

Standard Marine Corps procedure for initiating awards is described in the *Navy and Marine Corps Awards Manual.*

740 ■ Wearing Your Decorations and Medals

The Marine Corps has strict rules that govern the wearing of decorations and medals. These rules are in the *Marine Corps Uniform Regulations.* Some of them follow:

Subject to regulations, you may now accept awards from foreign nations. However, awards worth more than $50 still must be turned over to the Commandant of the Marine Corps, for retention by the Department of State pending Congressional approval, as in the past. Also, persons on duty in connection with the military assistance program may not accept awards from foreign host countries.

Most decorations, and all campaign medals, have half-size miniature reproductions known as *"miniature medals."* You wear "miniatures" with evening and mess dress, as well as with civilian full dress or dinner jacket, when appropriate. The Medal of Honor, however, is never represented in miniature, and, when miniatures are worn, the Medal of Honor is suspended in the normal fashion, about the collar.

When medals are prescribed instead of ribbons, unit citations and other ribbons for which no medal has been struck will be worn centered on the *right* breast.

Marines with eight or more ribbons of any type should wear them in rows of four rather than three, thus avoiding a top-heavy stack. Large medals may not be worn more than seven per single row; and miniatures, not more than 10 (see *Marine Corps Uniform Regulations*).

If under arrest, or suspended from rank and command, you should not wear ribbons or medals.

With every U. S. decoration (and many foreign ones, too) you receive a lapel device for wear with civilian clothes. This may be worn in the left lapel of your civilian suit when you think fit.

Decorations and medals are part of your uniform and must be worn, *except* that, when ribbons are prescribed for the shirt, you are only required to display personal decorations and unit citations; the wearing of campaign ribbons is optional.

Marksmanship badges may be prescribed for any uniform except evening and mess dress, but it is not customary to wear these badges when medals are prescribed. Nor can you wear *any* ribbon (such as Navy marksmanship ribbons) in lieu of a marksmanship or gunnery badge. Incidentally, you are limited to a ceiling of three badges of your choice, if you rate more than three.

When soiled, faded, frayed, or otherwise unserviceable, the ribbons of decorations and medals should be destroyed by burning rather than thrown away, not only to prevent reuse by unauthorized persons, but because these ribbons symbolize the bravery, devotion, and sacrifice of U. S. Marines.

Even though entitled to wear foreign decorations or medals, you must always display at least one U. S. medal or award at the same time. And remember that foreign awards take precedence *after* U. S. awards.

UNIFORMS, INSIGNIA, AND PERSONAL GROOMING

741 ■ ". . . Well Dressed Soldiers"

"It is proverbial," wrote one Commandant, "that well dressed soldiers are usually well-behaved soldiers." The Marine Corps has always set course by that axiom and has enjoyed good success and repute on both counts.

As a Marine officer, it rests squarely with you, 24 hours a day, on duty and off, to maintain the Marine Corps reputation for smart, soldierly, and correctly worn uniforms.

Marine Corps Uniform Regulations is the "bible" on uniforms, insignia, and grooming. You must:

Know those *Regulations*.
Set the example by rigid compliance.
Enforce them meticulously.

In *Uniform Regulations* you will find two essential compilations with which you should become thoroughly familiar: (1) the listing of required articles of uniform for all officers; and (2) the table (see Table 7–1) showing types and combinations of uniforms authorized for officers. This latter provides a complete checklist of articles you should (or should not) wear as part of each prescribed uniform combination. If a woman officer, you also will find instructions on uniforms and personal grooming in *Uniform Regulations*.

742 ■ Wearing the Uniform

Here are important rules that govern wearing of uniforms:

a. Uniforms designed to be buttoned *will be worn buttoned.*
b. Wear headgear whenever under arms or on watch, except when in a space where a meal is being served or divine service is being conducted; or when in quarters (if on watch) or when specifically excused from remaining covered. Remain covered at all times when out-of-doors or on topside spaces on board ship. (But see Section 2221 for shipboard ground-rules.)
c. "Mixed uniform" (components of two different uniforms, worn simultaneously—white blouse and utility trousers, for example) is strictly forbidden unless specifically authorized in *Uniform Regulations.*
d. *Full service uniform* (greens) is worn when:

   Reporting for duty ashore
   Paying official visits as defined by *Navy Regulations*
   Serving as a member of court-martial or court of inquiry
   Making boarding calls on merchantmen

e. *Undress* uniform* (blues or whites with ribbons only) is worn when:

   Reporting for sea duty
   Paying first visit to commanding officer
   Visiting foreign officers

Attending informal daytime social functions in an official capacity

f. *Dress* uniform* (blues or whites with sword, when prescribed, and medals) is worn when:

Exchanging calls or ceremonies with foreign officials, or making boarding calls on foreign men-of-war;
Exchanging official visits with U. S. civil officials, U. S. Armed Forces officers, and foreign officials;
Attending receptions tendered by or in honor of officials listed in Table of Honors, *Navy Regulations*;
Senior officer present considers the occasion appropriate.

g. *Evening dress and mess dress* are worn on formal evening occasions when civilian full dress is prescribed (such as the Marine Corps Birthday Ball) or instead of undress, on semiformal evening occasions. Evening or mess dress as appropriate will be prescribed for official or military-sponsored affairs, either formal or semiformal. Any evening function that you attend as a representative of the Corps is one for evening or mess dress. Evening or mess dress should always be prescribed for evening affairs when foreign officers, visiting officers of other Services, or distinguished civilians are to be present.

h. While the boat cloak is an optional item, you should obtain and, more important, wear this handsome, traditional garment. Specifically, the cloak should always be worn (in lieu of overcoat) with evening dress, and on any social occasion when not in line with troops, over blues. Any officer going on sea duty, other than in tropical waters, should have a boat cloak. By custom, though not prescribed in *Uniform Regulations*, you may also wear your boat cloak over civilian evening dress (dinner jacket or white tie).

i. If invited to the White House, check with the Aide-de-Camp to the Commandant, Marine Corps Headquarters, as to the correct uniform and other questions of protocol.

j. Utilities may only be worn in field work, or when it is obviously impracticable to wear service uniform. *Utilities may not be worn between home and place of duty, off Government reservations, except as specified in local regulations.*

k. You must buy and maintain in good condition all articles of uniform that the Commandant prescribes for officers, as listed in *Uniform*

---

*"Dress" (large medals in lieu of ribbons) and "undress" (ribbons and badges only) are traditional terms for these uniform categories.

Table 7-1: Types and combinations of uniforms for male officers (1977).

Uniform	Headgear	Jacket/Blouse	Shirt	Trousers
Evening Dress A (r)	Dress (n)	Evening (waistcoat)	White (c)	Evening
Evening Dress B (s)	Dress (n)	Evening (cummerbund)	White (c)	Evening
Mess Dress	Dress (n)	Mess (cummerbund)	White (e)	Bl Mess
Dress Blue A	Dress (n)	Blue	White (f)	Blue
Dress Blue B	Dress (n)	Blue	White (f)	Blue
Dress Blue C ("Seagoing")	Dress (n)	No	Khaki (h)	Blue (i)
Dress White A	Dress (n)	White	No	White
Dress White B	Dress (n)	White	No	White
Dress Blue/White A	Dress (n)	Blue	White (f)	White
Dress Blue/White B	Dress (n)	Blue	White (f)	White
Summer Service A	Svc or Garrison (j)	Green poly/wool	Khaki (k)	Green poly/wool (i)
Summer Service B	Svc or Garrison (j)	No	Khaki (k)	Green (i)
Summer Service C	Svc or Garrison (j)	No	Trop Khaki (o)	Green (i)
Winter Service A	Svc or Garrison (j)	Green	Khaki (k)	Green (i)
Winter Service B	Svc or Garrison (j)	No	Khaki (k) (o)	Green (i)

a. May be prescribed.
b. Depending on weather conditions.
c. Shirt, white, soft bosom, evening.
d. Black leather gloves shall always be worn or carried with topcoat, raincoat, or blouse in winter months.
e. Shirt, white, pleated soft bosom.
f. Shirt, white, plain soft bosom.
g. Black shoes and black socks worn when in line with troops.
h. During the summer months, commands authorized the blue uniform may, when prescribed by the Commandant, wear the shirt, khaki w/barrel cuff (with field scarf and tie clasp) or the shirt, khaki, with quarter-length sleeve with the blue trousers. The latter combination is designated Dress Blue D.
i. With web belt and brass buckle.
j. As prescribed in local regulations, but officers may substitute garrison cap when on official flights or in travel status, and garrison cap

Regulations. You must keep your full assignment of uniforms with you at all times except when in the field.

l. The law prohibits anyone not in the Armed Forces from wearing the uniform or any distinctive part thereof (Act of 3 June 1916, Sec. 125 as amended). This does not apply to honorably discharged Marines, who may continue to bear the title, and, on occasions of mili-

Footgear	Socks	Gloves	Raincoat Topcoat Boatcloak	Emblems	Medals	Badges (p)	Ribbons	Sword
Black	Black	White (d)	(a) (b) (q)	Dress	Min	No	No	No
Black	Black	White (d)	(a) (b) (q)	Dress	Min	No	No	No
Black	Black	White	(a) (b)	Dress	Min	No	No	No
Black	Black	White (d)	(a) (b) (q)	Dress	Lge	No	No	(a)
Black	Black	White (d)	(a) (b) (q)	Dress	No	(a)	Yes	(a)
Black	Black	(d)	(a) (b)	Dress	No	(a)	Yes	(a)
White	White	White	(a) (b)	Dress	Lge	No	No	(a)
White	White	White	(a) (b)	Dress	No	(a)	Yes	(a)
White (g)	White (g)	White	(a) (b)	Dress	Lge	No	No	(a)
White (g)	White (g)	White	(a) (b)	Dress	No	(a)	Yes	(a)
Black	Black	No	(a) (b)	Svc	No	(a)	Yes	(a)
Black	Black	No	(a) (b)	Svc	No	(a)	(a) (m)	(a)
Black	Black	No	(a) (b)	Svc	No	(a)	(a) (m)	(a)
Black	Black	(d)	(a) (b) (l)	Svc	No	(a)	Yes	(a)
Black	Black	(d)	(a) (b) (l)	Svc	No	(a)	(a) (m)	(a)

may be worn by FMF officers on field type duty, and by reserve officers (unless on acdu for more than 30 days).

k. With field scarf and gold clasp or gold bar.
l. Green scarf is prescribed for wear w/topcoat; optional for wear w/ raincoat.
m. As locally prescribed.
n. Company grade officers are required to possess only one frame, cap, service with black visor and it shall be authorized for wear with dress and service uniforms.
o. Shirt, tropical, khaki, w/quarter-length sleeve may be prescribed.
p. Aviation/Parachutist insignia shall be worn as prescribed.
q. Boatcloak optional for wear with Evening Dress and Blue Dress uniforms.
r. Equivalent to formal civilian full dress (white tie).
s. Equivalent to semiformal civilian dress (black tie).

tary ceremony, wear the uniform of highest rank held during war service.

m. Wearing the uniform is prohibited in connection with nonmilitary commercial or business activities, or in any circumstances that might compromise the dignity of the uniform or the Corps.

n. No Marine (including retired or reserve personnel) may wear the uniform (1) at meetings or demonstrations of organizations on the Attorney General's subversive list; or (2) while attending (unless on duty) or participating in any demonstration, assembly, or activity, the purpose of which is furtherance of personal or partisan political, social, economic, or religious issues. In other words, *demonstrations and the Marine Corps uniform or emblem don't mix.*

743 ▪ Uniform Accessories

The following rules concern the wearing of uniform accessories:

a. Belts are worn with buckle centered. Belt buckles (except on 782 equipment) must be brightly polished.
b. Leather gloves may be worn or carried with the winter service uniform, without topcoat, overcoat, or raincoat at the option of the individual. Gloves are required with winter service when wearing topcoat, overcoat, or raincoat. Local commanders designate whether gloves will be worn by troops in formation.
c. Shoes (and all leather) worn with service uniforms must be plain, without fancy stitching, and of regulation shade. Socks must match shoes in color.
d. Swords may be prescribed only with dress, undress, or service uniforms, and must be carried in line with troops in dress uniforms. Male officers must possess swords, and must have their names engraved on the blade. Sons of Marine officers, however, may carry their fathers' swords.
e. Marines may not wear jewelry, fobs, pens, or pencils exposed on the uniform, except:
Wrist watch
Regulation tie-clasp
Rings (conservative pattern)
Sunglasses (conservative design—but may not be worn in line with troops)
f. You wear the regulation black mourning band on your left arm between shoulder and elbow:
When a pallbearer or attending a military funeral in an official capacity
During prescribed official mourning
For family mourning (optional)
g. The Sam Browne belt is authorized and worn as organizational equipment in organizations specified by the Commandant. There is nothing in regulations that prohibits your buying your own Sam Browne

and wearing it as locally authorized when not in line with troops, or if on independent duty.

744 ▪ Civilian Clothes

As an officer you are expected to maintain a high standard of civilian dress. Your clothes should be conservative in cut and color, of the best quality, and well maintained. Avoid flashy shirts and neckties. Wear a hat. Never forget, incidentally, that your general neatness and grooming at all times are marked on your fitness report. This includes not only uniform but civilian clothes as well.

When off duty, wear civilian clothes. If on duty abroad, however, be sure to check local directives on wearing of plain clothes. Unless you have permission from the Commandant, you may not wear uniform when on leave outside the United States or its territories.

Here are some rules on civilian clothes:

No distinctive articles of uniform (except nonmilitary items like shoes, socks, underwear, etc.) may be worn with civilian clothing.

On board MSC transports and Government aircraft, you must wear uniform unless your orders permit plain clothes. When traveling singly by other means, you may and usually should wear mufti.

The Marine Corps mufti tie (scarlet and gold on rifle green background) may be worn with civilian clothes by all Marines, regular and reserve; by former Marines; and, honorarily, by any officer of another Service who has served as a member of a Marine Corps organization. The Corps tie marks you as a Marine. It is particularly appropriate for social functions, and you should have one in your wardrobe.

Although no official pattern is prescribed or authorized for Marine Corps blazers, such coats are frequently encountered. If you desire to have one, it should be of navy blue, single-breasted, with a Marine Corps emblem placed on the left breast pocket.

With or without the breast-pocket emblem, you can put Marine Corps dress buttons, preferably of the anodized type, on any blazer of conservative cut and color (though, as noted above, blue is preferable).

745 ▪ Grooming

a. Marines' hair will be neatly and closely trimmed; three inches is the maximum permissible length, but no officer should allow his hair to be longer than two inches. Remember Field Marshal Wolseley's dictum: "Hair is the glory of a woman, but the shame of a soldier." Get your hair cut weekly.

b. Keep clean-shaven, except for a mustache, if desired (traditionally, Marines have always rated mustaches, and naval officers, beards, but not the reverse). Eccentricities of haircut or mustache will not be tolerated.

c. *All leather must be maintained in very high polish.* The post exchange stocks regulation-color dye, and you should dye each new pair of shoes (and laces).

d. To keep shirt collars trim and scarves in place, wear a collar stay.

e. To make utility trousers look smartest with field boots, turn up the trousers so as to form interior cuffs, and place a strong rubber band, sleeve garter, or section of inner tube inside this interior cuff. This produces a neat overhang. Most post exchanges stock garters for this purpose.

f. Keep the overlap of your trousers belt within the prescribed $2\frac{2}{3}$ to $3\frac{3}{4}$ inches (and never more).

g. On promotion to first lieutenant, second lieutenants can save a tidy sum by having all their brass bars chrome- or silver-plated. This operation can be done at reasonable cost by or through any jeweler or silversmith.

h. Any officer who wishes to present a truly military appearance in blues or whites should either buy a set of anodized, high-luster buttons or have the plating stripped off a set of gold-finished buttons (and *have your blues tailored for removable buttons*). "Buffed buttons," like their anodized counterpart, present a brilliant polish far more military than gold plate, need never suffer from corrosion, and set a fine example to troops. To remove gold plate, consult a jeweler or uniform tailor (buttons with a brass base can be stripped by dipping them in nitric acid—rinse well; but be sure they are of real brass, not base metal or plastic, before any such process). You can

A Marine must maintain the Corps' reputation for smart, soldierly uniforms.

buy polishing gear for buttons at the Exchange. *If you don't have a set of buffed buttons, attend to this now, not tomorrow.* One tip: if you know any retired officer who has a spare set of buffed buttons, or even a set made of brass, see if you can talk him out of them. Aside from other advantages, the well-polished old buttons will help give you a seasoned appearance unattainable with brand-new gold plate.

i. Although your uniforms contain many pockets, the safest rule is to carry nothing in them. Specifically, you should never place anything in exterior pockets of a dress or service uniform (exceptions: pencil out of sight in a shirt pocket; notebook in hip pocket; wallet, cigarette case, and handkerchief kept flat in trousers pockets). Do not, under any circumstances, follow the sailor practice of carrying cigarettes in the cuff of a sock.
j. It is officer-like in appearance, when wearing dress shoes, to lace them so that laces are horizontal and parallel to each other, rather than criss-crossing, X-fashion.
k. At least a fortnight before the seasonal change from summer to winter uniform and vice versa, break out the forthcoming uniform, have it cleaned and pressed, and check it for completeness and repair.
l. With standing-collar uniforms (as well as evening or mess dress) it is convenient—and military—to carry your handkerchief unobtrusively tucked inside your left sleeve; this obviates convulsive dives into the interior of the uniform when a handkerchief is needed.
m. Read and follow the advice on care and marking of uniforms to be found in *Uniform Regulations*.

Old breed? New breed? There's not a damn bit of difference so long as it's the Marine breed.
—Lieutenant General Lewis B. Puller

This modern tendency to scorn and ignore tradition and to sacrifice it to administrative convenience is one that wise men will resist in all branches of life, but more especially in our military life.
—Field Marshal Wavell

8

Posts and Stations

Only the globe itself—trademark of Marines—limits the number of places where you, as a Marine, may ultimately serve.

Here, however, we are going to take a look at the permanent posts and stations of the Corps. These are the places where, between expeditions, you will spend much of your career. In addition, this chapter will describe the organization and general conditions at a typical post, as well as the facilities and services that a post or base offers to you and your dependents.

801 ■ Posts of the Corps

A number of major bases, posts, or air stations form part of the Marine Corps Supporting Establishment and are maintained exclusively for Marine Corps forces. Marines in the security forces man more than 80 Marine barracks and shore-based Marine detachments at home and abroad.

Except for posts with missions directly reflected in their titles (such as the Recruit Depots), the Corps has the following kinds of stations:

Marine Corps Bases (MCB) and *Marine Barracks (MB)* are the basic permanent posts for support of ground units of the Corps. Both are administratively autonomous and self-supporting. Marine Corps bases and camps are devoted to field training and support of major tactical units, whereas most (but not all) Marine barracks perform security missions.

Marine Corps Air Stations (MCAS) are the aviation counterparts of Marine Corps bases. Like MCB, air stations are also permanent, autonomous, and self-supporting. All MCAS have a common mission: support of Marine aviation units.

Marine Detachments (MD), as distinct from ships' detachments, are the smallest shore organizations of the Corps. An MD satellites, administratively and logistically, on some larger organization, usually Navy, and often enjoys less permanent status than other Marine activities.

A TYPICAL POST

802 ■ How a Post Is Organized

Making allowances for different missions and locations, most posts follow the same organization. Figure 8–1 shows the organization of a hypothetical post.

Command. The *commanding officer* (CO) (if a general, entitled "the commanding general") commands the post. In the fateful words of *Field Service Regulations*, he is "responsible for all that his command does or leaves undone."

The *executive officer* is the line officer next junior in rank to the CO. He is the commanding officer's *alter ego*, who relieves the commander of administrative detail and succeeds to command in the latter's absence. The extent and character of his duties vary somewhat according to the policies and peculiarities of "the Old Man." On a post commanded by a general, the executive is entitled "chief of staff," and the latter, in turn, may be assisted by a deputy.

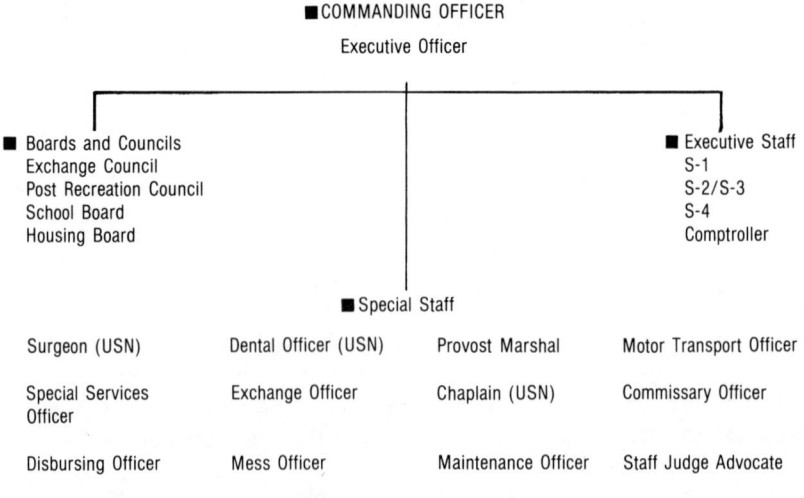

Fig. 8–1: How a typical post or base is organized.

Staff. Just as in tactical units, a post commander is assisted by an executive and special staff much like those described in Section 521. The executive staff includes a *personnel officer* (S-1); *operations and intelligence officer* (S-2/S-3); and *post supply officer* (S-4). In addition, on larger posts, the *fiscal officer* or *comptroller* occupies a status comparable to that of members of the executive staff.

The typical post special staff includes certain billets identical with those found in the FMF special staff, such as:

Communications officer
Chaplain (ChC, USN)
Adjutant (on small posts, acts as S-1)
Dental officer (DC, USN)
Disbursing officer
Exchange officer

Inspector
Mess officer
Motor transport officer
Special services officer
Staff judge advocate
Surgeon (MC, USN)

In addition to the foregoing, most posts have a few special staff functions which differ materially in scale or scope from similar FMF staff jobs, where housekeeping is not quite so important as on a post.

Provost Marshal. This is the post "chief of police" or "sheriff," responsible for public safety, traffic control, criminal investigation, internal and external security, regulation of servants and pets, and law and order in general. Frequently the provost marshal acts as *fire marshal* and thus also becomes responsible for fire protection. Law-abiding members of the post usually encounter the provost marshal in connection with licensing of vehicles or pets, and obtaining passes for guests, dependents, and servants.

Maintenance Officer. This officer is responsible for minor construction, repair, and upkeep. On small posts, the maintenance officer may also be *police officer*, and thereby responsible for the cleanliness and shipshape appearance of the post.

Public Works Officer (CEC, USN). On large stations, this is a Navy (CEC) officer who supervises new construction, improvements, and plans for post development.

Boards and Councils. To supplement the staff, most stations include one or more standing boards or councils. Some are required by regulations, while others exist to meet local needs. Typical examples are:

Exchange council
Recreation council
School board
Athletic and sports council
Housing board

803 ■ Facilities and Services

In many ways a post resembles a small community. Most if not all the facilities and services you could expect in such a town have counterparts on a Marine post. Also like small towns, however, stations of various age, locality, and mission exhibit considerable local disparities. Thus what you may find on one post may not exist, or may hardly exist, at another. Accordingly, your family fund of Service information should include "the word" on conditions at the widely separated stations where you may find yourself.

804 ■ Medical and Dental Care

Every Marine post includes medical installations for the health and sanitation of the command. These may range from a *dispensary* ("sick bay") to an *infirmary* (dispensary with limited facilities for in-patient care), or, on the largest posts, a *naval hospital* that can handle any medical or surgical emergency. Naval medical facilities in given localities are administratively grouped under what are called *Naval Regional Medical Centers*. Routine treatment and consultation are afforded daily at "sick call"—a fixed time of day when the sick bay is fully manned. Emergencies, of course, are dealt with at any time, day or night.

The Medical Department not only cares for ailments after they occur, but wages a ceaseless preventive campaign. As far as you are concerned, this manifests itself in standing requirements that all Marines undergo

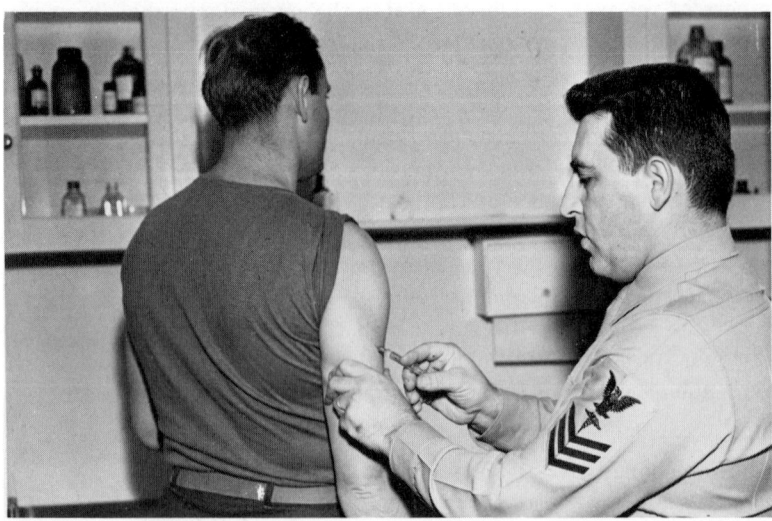

Ouch! The Medical Department wages part of its ceaseless campaign of preventive medicine.

certain immunizations, or "shots," and that every officer have a thorough physical examination once each calendar year. If for any reason you fail to take your annual physical during the year, you must explain why in writing.

805 ■ Dependents' Medical Care

Family medical care for Service dependents and for retired officers' dependents is provided by the Government, and is known as CHAMPUS (Civilian Health and Medical Program, Uniformed Services). This program permits civilian medical care and hospitalization, as well as Service medical care. A small percentage of the total annual cost of civilian medical expenses and hospitalization is borne by the individual.

Eligibility. Virtually all dependents (spouse and unmarried children under 21—subject to a few exceptions) of Marines on active duty are eligible for civilian medical care and care in Service medical facilities. Retired officers and their dependents likewise have such eligibilities, but under differing provisions (see Chapter 23). If you die, whether on active duty or after retirement, your surviving dependents remain eligible for care at Armed Forces or U. S. Public Health Service medical facilities (subject to availability of space and staff), as well as for certain civilian medical care and hospitalization.

Civilian Medical Care. Civilian semiprivate hospitalization, outpatient care by civilian facilities, routine doctor visits, prescribed medicines, laboratory and X-Ray tests, rental sick-room equipment, artificial limbs and eyes, etc., are available to active-duty dependents on a cost-sharing basis, whereby, in general, you pay only the first $50 per year per person (but not over $100 for the entire family) and 20 percent of the remaining cost; the Government pays the remaining 80 percent.

A program of financial assistance is also provided for active-duty personnel whose spouses or children are mentally retarded or physically handicapped. This program authorizes diagnostic services, treatment, and use of private nonprofit and nonmilitary institutions for such handicapped dependents with you (the sponsor) paying a varying amount according to rank and the Government paying the remaining portion of the cost up to a given maximum.

> NOTE: Two provisions will be of special interest to your wife: (1) All care received during and for a pregnancy that results in hospitalization is considered, for payment purposes, as part of that hospitalization; and (2) oral contraceptives are considered to be prescription drugs.

Medical Care at Service Facilities. When medical staff, space, and facilities are available, the Navy Medical Department will provide the following care for your dependents:

Diagnosis
Treatment of:
Acute medical conditions
Surgical conditions
Contagious diseases
Acute emergencies of any kind
Immunization
Maternity and infant care

Out-patient service, essentially of free-clinic character, is given dependents (including parents, if in fact dependent upon you, and your widow) at any naval dispensary, infirmary, or hospital that has dependent facilities. Virtually all post medical installations in the continental United States and outlying stations have dependent out-patient service. Transports that carry dependents are likewise staffed and equipped to provide full dependent medical care as required.

The extent and quality of dependent medical services vary widely with the limitations of local dispensaries, infirmaries, or hospitals, and with the medical workload as a whole. Isolated outlying stations usually have more self-sufficient medical and dental services.

On outlying stations, in addition to the Service medical care noted above, dental care can frequently be provided to dependents, subject to limitations of workload and facilities, provided adequate civilian dental services are not available. The Government will also, in appropriate cases, provide transportation for dependents from outlying stations where medical care is inadequate to centers where proper care can be provided (with round-trip expenses for attendants when they are found to be required).

A few cautionary words are in order on the subject of dependent medical care:

First and foremost, the medical needs of military personnel are the *primary* concern of the Medical Department. This means that care for dependents takes second place, and always gives way, when conflict arises, to military medicine functions.

Second, although members of the Marine Corps and Navy receive free dental care, dependents do not (except on remote, overseas stations, or as otherwise required in connection with medical or surgical treatment).

Third, to receive medical assistance from any Navy (or Armed Forces) establishment, your dependent must possess and present a Dependent Identification Card (see Section 806 below). Obtain one of these cards and have your wife carry it with her.

Fourth, if your dependents are with you, make sure, before seeking CHAMPUS treatment in a civilian hospital, that military medical treatment and hospitalization are not available (and that you have a non-availability certification, currently Form DD-1251). If in doubt, check with your medical officer or nearest sick bay.

Finally—although the professional qualifications of Navy doctors and nurses are of the first rank, care for Service dependents remains necessarily at last priority. If you wish, and can afford, the more personal attentions of a private practitioner and civilian hospital, you are free, at your own expense, or under CHAMPUS when applicable, as just noted, to obtain such services.

> NOTE: The Defense Department authorizes military medical facilities to fill prescriptions written by licensed civilian physicians and dentists for military personnel, retirees, and dependents, in reasonable quantities of non-narcotic items stocked routinely.

806 ■ Dependents' Identification and Privilege Cards

The Department of Defense issues (on application) a standard *Identification and Privilege Card* (Form DD 1173) for dependents (except children under ten) of all active-duty personnel. This card is an essential item to enable your dependents to obtain the use of medical facilities, commissary, exchange, and post theaters. It is honored not only on Marine and Navy posts and stations, but on those of the other Services as well. As soon as you acquire eligible dependent(s), you are required to apply to your commanding officer for their identification and privilege cards.

807 ■ Commissary

The military equivalent of the supermarket is the sales commissary. There, at prices equal to, or occasionally slightly lower than those charged by grocers ashore, you may buy foodstuffs from Government supplies. The privilege of making purchases is limited to regular and retired personnel; to certain reservists on active or training duty; and to certain Government civilians. Dependents of anyone entitled to commissary privileges may also use the commissary. Needless to emphasize, use of the sales commissary is a privilege, not a right, and all purchases must be for your own use and that of your household.

Everyone entitled to commissary privileges must be prepared to identify himself. Active personnel not in uniform, as well as retired personnel, are identified by the ID Card, dependents by the identification and privilege card.

Stock and services available in commissaries vary somewhat according to the size of the post and the availability of adequate civilian facilities off post. In some localities, where one Service maintains a large commissary, members of other Services stationed nearby may use this. Washington, D.C., where the Army provides commissary facilities for all four Services, is an example.

808 ■ Marine Corps Exchanges

Marine Corps Exchanges ("post exchanges" or "PX," as they are known) are maintained by all posts and by almost all FMF units of any size. Any regular Marine Corps organization may, with the approval of the Commandant, establish its own exchange.

Post exchanges go far back into U. S. military history. During the nineteenth century, when the Army pushed our frontier westward, each isolated post had its "post trader" or sutler, authorized to keep store at the post. One of the trader's perquisites was the right of trading with Indians, trappers, and hunters, and from this arose the title, "post exchange." After the frontier vanished, the name remained, carrying over from the Old Army into the Old Marine Corps. In early times, the perquisite of keeping the post trader's stores at the various Marine Barracks was awarded to the widow of some officer or senior NCO. The modern post exchange system was established by General Heywood, the 9th Commandant.

Today's exchange is the post general store. On large stations, it approximates a small department store, but the extent of what an exchange offers depends markedly on its volume of business—which in turn stems from the size of the post and the accessibility of civilian shopping centers.

The missions of Marine Corps Exchanges are: (1) to afford Service personnel (including dependents), at reasonable prices, articles necessary for health, comfort, and convenience; and (2) through reasonable profits, to afford Marines means for recreation and amusement. The latter mission is realized through donation of exchange profits to unit recreation projects and to the *Marine Corps Exchange Fund*, a nonappropriated fund maintained by Marine Corps Headquarters for the entire exchange system and for welfare or recreation.

Typical, though by no means exclusive, of goods and services obtainable in a medium-sized exchange are the following: confectionery and tobacco; toilet articles; periodicals; luggage; photographic supplies; uniforms and accessories; household supplies; soda fountain, beer hall, and lunchroom; check-cashing facilities (up to $50 in any given day for personal checks by any member of the Armed Forces); barber (and, for ladies, beauty shop); cobbler and tailor; bowling alleys and poolroom. The exchange recognizes the existence of the fair sex—both women Marines and dependents—by stocking many items of primary usefulness to women only.

Eligibility to use the exchange, like the commissary, is a privilege which extends only to active or retired Service personnel; to their dependents and widows; and to reservists on active or training duty. The *Marine Corps Exchange Manual* gives the various classes of individuals and their eligibility for exchange privileges. If in civilian clothes when making a purchase, be prepared to show either your ID Card or an Exchange Identification Card (obtainable from the exchange steward).

809 ■ Welfare Activities

In addition to welfare services provided by the chaplain, special services officer, and legal assistance officer, most large posts have representatives of the American Red Cross, Navy Relief Society, and Navy Mutual Aid Association. The assistance furnished by these groups is described in Chapter 23.

810 ■ Educational Facilities

Many posts have their own public schools for the children on the post. Because of wide variations between post school systems, however, you should investigate carefully before you assume that you will find schools at the post which meet your needs. If overseas where U. S.–owned school facilities are not available, you may claim a modest schooling allowance for each child you place in a local private school of approved standards.

Every post has a free library, open to Marines and dependents. Marine Corps Headquarters and the Bureau of Naval Personnel provide the books. A few large posts have museums.

811 ■ Recreation

Many posts include excellent on-station recreation opportunities. Some of these, such as sports facilities (golf courses, boating, athletic facilities, etc.) and hobby shops, are open to all ranks. The focus of much of your recreational and social activity, as an officer, however, is the commissioned officers' mess (see Section 2410).

At a medium-sized post, you may expect to find at least the following: commissioned officers' mess; small-bore range; swimming pool; and other athletic facilities as space and demand permit.

MAJOR POSTS AND STATIONS

812 ■ MB, Eighth and Eye Streets, S.E., Washington, D.C. 20390

"Eighth and Eye" is the senior post of the Corps, both because of its age and because it houses the Commandant. The post has been a Marine

"Eighth and Eye" is the senior post of the Corps, and includes the quarters of the Commandant (centered at upper end of the parade ground).

Barracks since 1801, and quartered Marine Corps Headquarters throughout its first century. It is the "spit-and-polish" post of the Corps, famous for its weekly Evening Parades, and constructed about a traditional barracks square in the heart of old Washington.

The Barracks provides ceremonial troops for official occasions in the nation's capital; it supports the U. S. Marine Band and the Marine Corps Drum and Bugle Corps, and the Marine Corps Institute (see Section 1509); and its officers and men are assigned to certain special security duties in and about Washington and in the Navy Yard.

Transportation. *Rail.* Amtrak via Conrail; Baltimore & Ohio; Richmond, Fredericksburg & Potomac; Chesapeake & Ohio; Southern; Seaboard Coast Line.

Bus. Greyhound; Trailways; Red Star; Virginia Stage Line.

Airports. Washington National Airport; Andrews AFB; Dulles International Airport, Chantilly, Virginia (35 miles distant).

Quarters. Center House BOQ for officers permanently attached; nearest transient officer accommodations are at Naval Facility, Andrews AFB. Family quarters are not available.

Schools. In addition to a metropolitan school system, Washington has numerous private and parochial schools at any social level, price range, and quality. Suburban public schools are adequate, with those in Montgomery County generally being considered best.

Recreation. Other than the Barracks gymnasium, the only on-post facility at Eighth and Eye is Center House, but Marine officers in the Washington area may join Messes at the Navy Yard and Naval Medical Center, Bethesda.

Commissary and Exchange. Commissary privileges are available at any of four Army commissaries in the Washington area. Excellent Marine Corps and Army exchanges are at Henderson Hall and Forts Myer and McNair, in addition to the small exchange located at Eighth and Eye.

Neighboring Marine Activities. The Washington area includes Marine Corps Headquarters in Arlington Annex, Navy Department, and Henderson Hall, Arlington, Virginia, which supports Headquarters Battalion, HQMC.

813 ■ Marine Corps Development and Education Command, Quantico, Virginia 22134

Quantico is in many ways the showplace of the Corps. It is a large (91 square miles) well developed station on the Potomac River, approximately 35 miles south of Washington. Quantico includes two major centers—the Marine Corps Educational Center, and the Marine Corps Development Center, and is the only Marine post to be the site of a U. S. National Cemetery. The base also includes a Marine Corps air facility and a naval hospital. Because of its educational and development roles, Quantico is the intellectual heart of the Corps.

Transportation. *Rail.* Amtrak via Richmond, Fredericksburg & Potomac.

Little Hall houses Quantico's post exchange, theater, and other activities.

Bus. Greyhound.

Airports. Civilian aircraft: Washington National Airport, Dulles International Airport (both more than 30 miles distant); military aircraft: MCAF, Quantico.

Highway. U.S. 1 or Interstate 95.

Quarters. Government quarters are available for all ranks, although the number is insufficient to accommodate all eligible personnel. Off-post housing is available in the surrounding area from Woodbridge south to Fredericksburg, with rapid and seasonal turnover and rentals, though still high, not as bad as in past years. Transient family accommodations are available in Harry Lee Hall, operated by the Commissioned Officers' Mess (open), and the Hostess House. Permanent and transient bachelor quarters are available in Liversedge Hall.

Schools. Post schools include all grades through high school.

Recreation. Quantico's recreational opportunities very often cause it to be called "the country club." They include: Commissioned Officers' Mess; golf (18 holes); sailing (Potomac River); equitation; Rod and Gun Club; Sky Diving Club; Skeet Club; hobby shops; swimming pools for all ranks; post gymnasium; Little Hall (theater and bowling alleys); Post Museum; and two fine libraries (James C. Breckinridge Memorial Library in Dunlap Hall and the Technical and Post Library in Little Hall).

Commissary and Exchange. Excellent.

814 ▪ Marine Activities in the Norfolk Area, 23511

Although the Norfolk-Hampton Roads area includes no major Marine Corps post, it is the location of Headquarters, FMFLant, and its supporting Camp Elmore; of Landing Force Training Command, Atlantic Fleet, at Little Creek; and of MB, Norfolk Naval Shipyard (Portsmouth)—one of the oldest barracks in the Corps—and MB, Naval Base (Hampton Roads), Norfolk. In addition, as Norfolk is the primary East Coast base of the Atlantic Fleet, many Marine officers and men serve in the area for that reason. Thus Norfolk and the surrounding area can be considered a Marine Corps station of importance.

Transportation. *Rail.* Amtrak via Chesapeake & Ohio; Norfolk & Western.

Bus. Greyhound.

Airport. Norfolk Regional Airport.

Highways. U.S. 13, 17, 58, and 60, and I-64.

Quarters. There are relatively few Government quarters available in comparison to the number of Marine officers in the area. A few quarters are to be had at NB, Hampton Roads, and a few, in excess of re-

Headquarters, Fleet Marine Forces, Atlantic, are located in the Norfolk-Hampton Roads area.

quirements, at the Marine Barracks, Navy Yard (Naval Shipyard). However, there is a wide range of excellent private housing priced reasonably and located within range of duty stations.

Schools. Norfolk has good public and parochial schools, and a few moderately good private schools.

Recreation. Naval Base, Hampton Roads, has an outstanding Commissioned Officers' Mess, and there are other excellent clubs and messes in the area, with recreational facilities of all types. Virginia Beach, adjoining Norfolk, is one of the most attractive shore resorts on the East Coast.

Commissary and Exchanges. Excellent.

815 ■ MCAS, Cherry Point, North Carolina

Commissioned in September 1942, Cherry Point is the largest Marine air station in the world and home of the 2d Marine Aircraft Wing. It is an all-weather station and site of a Naval Air Rework Facility, and of a naval hospital.

Transportation. *Rail.* Amtrak via Seaboard Coast Line (nearest main-line stop, Rocky Mount, N.C.).

Bus. Seashore Lines.

Airport. Simmons-Nott Airport, New Bern (19 miles distant).

Highway. U.S. 70.

Quarters. There are aboard the station 627 sets of quarters for married officers and five operating BOQs. A Hostess House provides

Cherry Point's headquarters, and those of a Marine aircraft wing, are housed in this administration building.

limited accommodations for military and civilian guests. Private housing is available in Havelock, Newport, New Bern, Beaufort, and Morehead City (17 miles away).

Depending on the personnel situation, waiting periods of several months for Government quarters are not uncommon. Officers reporting in should consider leaving families in comfortable and familiar surroundings elsewhere before checking into the air station or 2d Wing.

Schools. Post kindergarten and nursery. There are three public elementary schools, junior high, and high school as well as a Roman Catholic parochial grade school. The quality of the North Carolina public schools in this area is now average. Craven Community College offers off-duty courses leading to an associate degree. East Carolina University offers off-duty courses leading to a baccalaureate degree, while off-duty courses leading to a master's degree are offered by Pepperdine University and the University of Southern California. All four institutions operate a resident center at Cherry Point.

Recreation. Commissioned Officers' Mess; bowling alleys; movies; swimming pool; tennis courts; golf (18 holes); hunting and fishing; boating and sailing.

Commissary and Exchange. Excellent.

816 ■ MCB, Camp Lejeune, North Carolina 28542

Camp Lejeune is the East Coast base for the ground units of the FMF. It accommodates the 2d Marine Division plus units of Force Troops, and adjoins MCAS, New River. Its neighboring community is Jacksonville, N.C.

Transportation. *Rail.* Amtrak rail connections to Camp Lejeune via Seaboard Coast Line are Wilson, Fayetteville, and Rocky Mount, all in North Carolina.

Bus. Continental Trailways, Seashore Transportation, and Carolina Coach provide scheduled service to and from the Camp Lejeune bus terminal.

Airport. Albert Ellis Airport, Jacksonville, is 20 miles from Lejeune and like other airports at Wilmington, Kinston, and New Bern, about 45 miles distant, is served by Piedmont Airlines. Connections with major airlines can be made at Raleigh or Charlotte, both in North Carolina.

Highways. U.S. 17 and N.C. 24. U.S. 258 originates at Jacksonville and runs north into Virginia.

Quarters. A DD Form 1746 (application for quarters assignment) should be sent to Quarters and Housing Director, Camp Lejeune, prior to detachment from previous duty station; on reporting, check in with Base Housing Office. Since mandatory quarters assignment may sometimes be necessary, do not arrange for off-base housing until Base Housing gives a green light. Married company officers may (1977) anticipate up to a one-year wait for quarters. There is also an appreciable waiting period for BOQ assignments for bachelors. Limited guest or temporary accommodations may, if desired, be provided for officers with their dependents, but for these, advance reservations with the Commissioned Officers' Mess are mandatory.

Camp Lejeune is too big to fit into a single air photograph, but this one shows Division Headquarters (center foreground) and some of the many barracks on the Base. In right background lies the utility and industrial area of the Base.

Schools. Post schools from kindergarten through high school are open to children of Service families residing on Federal property. A private kindergarten is available for nominal fees. Families not on Federal property must rely on county or other local schools. East Carolina University operates a branch and an extension division at Camp Lejeune, while Pepperdine University offers degree programs at Camp Lejeune. Coastal Carolina Community College at Jacksonville offers courses leading to an associated degree.

Recreation. Commissioned Officers' Mess; 10 post theaters; hunting; fresh and salt water fishing; swimming pool; surf bathing; golf (two 18-hole courses); boating and sailing.

Commissary and Exchange. Excellent.

817 ■ Marine Corps Recruit Depot, Parris Island, South Carolina 29905

Parris Island, or "PI," is the larger and older of the two recruit depots. It trains most recruits from east of the Mississippi River, and is the only recruit depot that trains Women Marines.

Transportation. *Rail.* The nearest main-line station is Yemassee, S.C., via Amtrak on the Seaboard Coast Line, 28 miles away. Amtrak has fast sleeper service with Washington, New York, and other East Coast points.

Bus. Greyhound.

Airports. Civilian: Nearest commercial air facilities are at Travis Field, Savannah, Georgia, more than 50 miles distant. Charleston Mu-

Parris Island headquarters (right foreground) is flanked by a typical recruit barracks (left rear).

nicipal Airport is 80 miles away. Military: MCAS, Beaufort, South Carolina.

Highway. U.S. 17 or I-95, to junction with U.S. 21, which leads to Beaufort, S.C., the neighboring community. From Beaufort follow State 281 to the Main Gate at Horse Island.

Quarters. BOQ, WOQ, and married field grade quarters available on a priority basis to officers in key billets. Capehart housing is available at Laurel Bay, about 10 miles from Parris Island. Transient accommodations include a guest house for officers and families. There is civilian housing in and about Beaufort.

Schools. Grade-school children aboard the Depot attend school at Laurel Bay, which has two grade schools. The public high schools and junior high schools in Beaufort accept children from the Depot and Laurel Bay. Bus transportation is provided for all grades. On the Depot there is a Child Care Center and a Nursery School.

Recreation. Commissioned Officers' Mess; swimming pool; salt water fishing; sailing and boating; golf (18 holes); bowling; post theaters; tennis; skeet; Rod and Gun Club; hobby shop.

Commissary and Exchange. Excellent.

Neighboring Activities. Parris Island's most important Marine Corps neighbor is MCAS, Beaufort. In addition, the U. S. Naval Hospital, Beaufort, is about three miles from the Parris Island main gate.

818 ■ MCAS, Beaufort, South Carolina 29902

MCAS, Beaufort (pronounced "Bewfort") is a major jet air base capable of supporting two Marine aircraft groups and associated service units. It is close aboard Parris Island and provides military air services therefor.

Transportation. *Rail.* Amtrak via Seaboard Coast Line to Yemassee, S.C., 26 miles northwest, is the most convenient mode to or from northern or southern points.

Bus. Greyhound.

Airports. Savannah, Georgia (45 miles) and Charleston (65 miles) are the nearest commercial airports.

Highway. U.S. 17 or I-95, thence by U.S. 21 to Beaufort.

Quarters. In addition to BOQ, the station has two communities of quarters, one comprising 22 sets of on-station officers' quarters, the other, at Laurel Bay, five miles west, with over 300 sets of officers' quarters. All quarters are modern two-, three- or four-bedroom units, air-conditioned. Off-station housing conditions are the same as those described for Parris Island.

Schools. There are two Federal Government grade schools (grades K through 6) at Laurel Bay, open only to children of families occupying Government quarters. The state provides bus transportation for children attending school in Beaufort. A nursery/kindergarten is operated by Special Services for children in Government quarters.

Recreation. Commissioned Officers' Mess; tennis; swimming pool; hobby shop; fishing; hunting; boating; golf (Parris Island); outstanding ocean beaches near at hand; hobby shop.

Commissary and Exchange. The air station utilizes the Parris Island commissary, but there are two excellent seven-day stores at MCAS, one on the station, the other at Laurel Bay. The post exchange is excellent.

819 ■ Marine Corps Logistic Support Base, Atlantic (MCLSBLant) Albany, Georgia 31704

The Albany Support Base is one of the newer posts of the Corps. Its missions, all logistic, include: acquisition, storage, and disposal of materiel; technical direction of Corps stores distribution; procurement, repair/rebuild, storage, and distribution of supplies and equipment. The base has more than 3.8 million square feet of closed storage in 19 warehouses, and more than 7.7 million square feet of open storage lots.

Transportation. *Rail.* As of 1977, Amtrak has no passenger service for Albany, although the base is served by one of the largest freight rail complexes in the Southeast.

Bus. Trailways; Greyhound.

Airport. Southern Air Lines.

Highways. U.S. 82 and U.S. 19.

Quarters. Modern BOQ and married quarters are available in adequate numbers to meet the requirements; most of the MOQ are three-

Headquarters at Albany, Georgia, is one of the newest and best designed headquarters buildings in the Marine Corps. Albany serves all Marine forces and establishments on the East Coast.

MCAS Yuma, the newest Marine air station in continental United States, boasts the longest runway in the Corps and the best flying weather.

bedroom, while a few have four. There are also on-post guest accommodations.

Recreation. Commissioned Officers' Mess; golf (9 holes); swimming pools; tennis; bowling; skeet and pistol ranges; movies; fishing.

Commissary and Exchange. Excellent.

820 ■ MCAS, Yuma, Arizona 85364

One of the newest of the Marine Corps air stations, MCAS, Yuma, has a 13,300-foot main runway, an instrumented special weapons delivery range, and some of the finest flying weather to be found.

Transportation. *Rail.* Amtrak via Santa Fe at Yuma (seven miles east).

Bus. Greyhound; Trailways.

Airport. Yuma International Airport.

Highway. U.S. 80.

Quarters. Over a hundred sets of quarters and BOQ are available, together with limited transient accommodations in the Hostess House. Civilian housing is readily available in Yuma. As of 1977, however, the waiting-list for government quarters may be as long as 24 months.

Schools. The station has a child-care center, including preschool. Public and parochial elementary schools, high school, and junior college are in Yuma.

Recreation. Commissioned Officers' Mess; equitation; swimming pools; hobby shops; bowling; tennis; MARS radio station; golf (18 holes); fishing, camping, and boating at Lake Martinez station-operated recreational area.

Commissary and Exchange. Modern and adequate.

821 ■ MCB, Twentynine Palms, California 92278

With its 932 square miles of area, twice the size of Los Angeles and big enough to encompass Pendleton, Lejeune, and Quantico with room to spare, Twentynine Palms is not only the largest post in the Marine Corps but also a primary training and experimental center for Marine artillery and guided missiles. It is also the location of the Communication-Electronics School, and has facilities for the major part of Force Troops, FMF Pacific. The base also includes the Marine Corps Air/Ground Combat Training Center, which exercises the combined-arms capabilities and readiness of FMF units in a live-fire environment.

Transportation. *Rail.* Amtrak via Southern Pacific (Indio).
Bus. Local bus to Palm Springs for connections with Greyhound.
Airport. No commercial airport.
Highway. U.S. 66 to Amboy, thence by county highway to Twentynine Palms.

Quarters. Government married quarters and BOQ available.

Schools. The base has a nursery school for children from three to six, as well as a child care center. The town of Twentynine Palms has elementary, junior high, and senior high schools together with a parochial grade school (through eighth grade). Undergraduate and graduate educational opportunities are available through College of the Desert, Pepperdine, and Chapman.

Recreation. Commissioned Officers' Mess (both Open and Closed); swimming; bowling; skeet; hunting and fishing; tennis; golf (nine holes); hobby shop. The base is almost at the center of the southern

Main Gate, Twentynine Palms, stands amid typical desert cacti, palms, and boulders.

Central Repair Shop at Barstow is the largest building in the Marine Corps (10 acres).

California recreation area, and practically any kind of outdoor or indoor recreation is available.

Commissary and Exchange. Excellent.

822 ■ **Marine Corps Logistic Support Base, Pacific (MCLSBPac), Barstow, California 92311**

The Marine Corps Logistic Support base at Barstow is located to take advantage of a confluence of transportation routes and of the Mojave Desert's hot, dry climate, which inhibits deterioration of stored material. Barstow supports all Marine organizations west of the Mississippi and in the Pacific Ocean Areas, operates a central repair shop for all FMF equipment except aircraft, and manages the inventory of all supply items within the West Coast complex. The ten-acre repair shop at Yermo is the largest building in the Marine Corps, and is surrounded by a 50-acre concrete parking lot.

Transportation. *Rail.* Amtrak via Santa Fe (coastwise).

Bus. Greyhound and Trailways.

Airport. The nearest airport with scheduled service is Ontario (91 miles distant); but McCarran Field, Las Vegas, or Los Angeles International Airport is normally used.

Highway. I-15, I-40, and California 58.

Quarters. The base has over 70 married and 30 bachelor officer quarters. Officers on accompanied tours are required to occupy Government quarters, but suitable housing is available ashore for cases where off-base housing may have been approved as an exception.

Schools. Public schools are located in the Nebo Area and the towns of Barstow, Yermo, Daggett, and Hinckley, the high school as well as four parochial schools being in Barstow. In addition, an elementary school is located at Wherry Housing and provides schooling from kindergarten through sixth grade. Public school buses are provided.

Recreation. Commissioned Officers' Mess; swimming pools; tennis; golf (nine holes); equitation; bowling; skeet; hunting and fishing. Like Twentynine Palms, Barstow is in the southern California playground area, which offers extensive recreational opportunities.

Commissary and Exchange. Excellent.

823 ■ Marine Corps Recruit Depot, San Diego, California 92140

The primary mission of the recruit Depot is the training of recruit Marines, but it is also the home of Sea School, Recruiters' School, and West Coast DI School. Depot recruits come from the 8th, 9th, 12th, and portions of the 4th Marine Corps Districts. The Depot is located in Northwest San Diego on the Pacific Highway.

Transportation. *Rail.* Amtrak via Santa Fe.
Bus. Greyhound and Santa Fe Trailways.
Airport. San Diego International Airport.
Highways. U.S. 5 and 8, and I-805.

Quarters. Except for limited BOQ and six sets for key senior officers, there are no Government quarters on the Depot and no guest house. Married officers with dependents are eligible for Navy housing—mainly Capehart—operated by the 11th Naval District under agreement with the Depot. Officers with dependents must report to CO, Navy Public Works Center, San Diego, for an endorsement as to quarters availability prior to authorization of BAQ (quarters allowance).

Schools. No Depot schools, but San Diego has excellent public, private, and parochial primary and secondary schools. Also, there are three major colleges in the city, plus numerous junior colleges.

Depot Headquarters, San Diego, typifies the base's Spanish colonial architecture and atmosphere.

Recreation. Southern California supplies virtually every type of recreation from swimming to skiing. The Mess is one of the finest and oldest in the Corps.

Commissary and Exchange. There are several Navy commissaries in the San Diego area. MCRD has an outstanding Marine Exchange.

824 ■ MCB, Camp Pendleton, California 92055

"Pendleton" is the largest amphibious training base in the Corps. It serves as the major West Coast base for ground units of the Fleet Marine Force and provides facilities and support for a Marine division and some units of Force Troops. In addition, Camp Pendleton includes Edson Range, the weapons training center for the San Diego Recruit Depot. The base has a naval hospital, on the shore of Lake O'Neill, named for one of the most redoubtable and courageous Navy medical officers to serve with the Marine Corps.

At Pendleton, recruits who have successfully graduated from San Diego receive advanced individual combat training and basic specialist training. During the Pacific, Korean, and Vietnam wars, Pendleton was the training "funnel" through which the majority of Marine battle replacements passed on their way to combat.

Base headquarters at Camp Pendleton typifies the modern, functional architecture of new construction on the post.

Camp Pendleton covers 125,000 acres of terrain, which includes three mountain ranges and 17 miles of coastline.

Transportation. *Rail.* Amtrak via Santa Fe, at Oceanside, California, the adjacent civilian community.

Bus. Continental Trailways; Greyhound.

Airport. Lindbergh Field, San Diego. Camp Pendleton has an Auxiliary Landing Field, which handles military traffic.

Highways. U.S. 395, I-5.

Quarters. In addition to limited BOQ, Pendleton has approximately 475 sets of two- to four-bedroom quarters for married officers, and over 160 rooms for bachelors. There are ample transient bachelor accommodations, and transient officers with families may stay at the Hostess House, by advance reservation, up to two weeks. Private housing ashore is available in Oceanside, Carlsbad, Vista, Fallbrook, and San Clemente. Before bringing dependents to Camp Pendleton, it is nevertheless advisable, as in the case of Lejeune, to write the Base Housing Office for information. This office can post you on the quarters situation and can, if necessary, help you in finding and forecasting available civilian housing.

Schools. Educational facilities for Camp Pendleton Marines and their dependents are excellent. Public, parochial, and some private schools, together with Mira Costa and Palomar community colleges in the immediate vicinity, offer courses for qualified Marines and their families, as do other colleges in the San Diego area.

Recreation. The Commissioned Officers' Mess has three convenient locations, one with swimming pool. Surfing beaches, fresh-water and deep-sea fishing, skeet, tennis, equitation, herpetology, golf (18 holes), and several base theaters are available.

Commissary and Exchange. A modern commissary is available in the main area of the base, with an annex in the northern section. Exchanges are modern, and complete.

825 ▪ MCAS, El Toro (Santa Ana), California 92630

El Toro is the West Coast base for aviation units of the FMF. The base usually houses, supports, and trains a Marine air wing.

Transportation. *Rail.* Amtrak via Santa Fe and Southern Pacific (Los Angeles).

Bus. Greyhound and Trailways. Transit district connections to Los Angeles and San Diego.

Highway. I-5, I-405.

Airports. Orange County, Long Beach, Los Angeles International. Limousine service and transit district connections available to Santa Ana.

Battle trophies brought home by Marine aviation units adorn headquarters at El Toro.

Quarters. El Toro has a BOQ and transient officer quarters; for adequate-to-good married quarters as of 1977, there is a prolonged waiting list. Housing in Orange County is plentiful and excellent, but expensive. A family guest-house with limited facilities is available on station.

Schools. Elementary school (grades 1–6) on station. Nearest high schools are in Tustin and the community of El Toro; nearest parochial school is in Santa Ana. Several local colleges and universities conduct graduate and undergraduate programs on station.

Recreation. Commissioned Officers' Mess; swimming pool; golf course; riding; movies.

Commissary and Exchange. Excellent.

826 ■ Marine Activities in the Hawaiian Area (FPO San Francisco 96610)

The island of Oahu includes several permanent Marine Corps installations with diverse missions. The headquarters and nerve center of all Fleet Marine Force Pacific activities is at Camp H. M. Smith, overlooking Pearl Harbor from the site of the World War II Aiea Naval Hospital. In addition to Camp H. M. Smith, there is MCAS, Kaneohe, on the "windward" side of the island, home station of the 1st Marine Brigade and one of three Marine Corps air stations outside continental United States; here both ground and air units of the FMF train and operate as an integrated air-ground team. The Pearl Harbor Marine Barracks, together with detached guard companies at Barber's Point NAS, Wa-

Marine Barracks, Pearl Harbor, is a typical large Marine barracks.

hiawa communications center, and Naval Ammunition Depot, Lualualei, perform security missions. Because of its superb site and outstanding facilities, Camp H. M. Smith was chosen by Commander-in-Chief, Pacific, for his headquarters, which is a tenant activity. Camp Smith is thus the only Marine Corps station that also serves as the headquarters of a unified command. Other Marines serve at the Makalapa headquarters of Commander-in-Chief, Pacific Fleet.

Quarters. The quarters situation on Oahu has improved somewhat in recent years. Bachelor, transient, and women officers quarters are to be had. Married quarters, many of fairly new construction, are located six miles from Camp Smith and at Kaneohe. Excellent quarters are available for officers assigned to the security forces at Pearl Harbor. Up-to-date information can be obtained from individual commands concerning waiting time for quarters.

Schools. In most areas public and parochial schools of acceptable quality are located on Government property or in or near housing areas. Public high schools are also conveniently located, but may require bus transportation in some cases. Excellent private and parochial schools are to be found on Oahu. While many of these are located some distance from Marine Corps activities, they are accessible by bus.

Recreation. Oahu and its naval and military installations afford some of the best all-around recreation and liberty in the Marine Corps. In addition to several excellent clubs and messes, the Hawaiian area provides opportunity for virtually every sport or taste.

> NOTE: If you are interested in a Hawaii vacation on Waikiki Beach at affordable rates, investigate the Hotel Hale Koa (House of Warriors), a new Government-built 15-story hotel located at Fort DeRussey and open to Armed Forces personnel and their families. Reservations are basically first-come, first-served, with active-duty people having priority.

Commissary and Exchange. All Services maintain adequate commissaries and exchanges on Oahu, although prices are generally higher than on the Mainland.

827 ■ **MCB, Camp Smedley D. Butler, Okinawa (FPO Seattle 98772)**
Located at Zukeran, Headquarters, MCB, Camp Butler, is responsible for operation of all Marine facilities on the island (Camps Schwab, McTureous, Hansen, Courtney, Onna Point, and MCAS, Futenma). Headquarters, III MAF; 3d Marine Division; 1st Marine Aircraft Wing; and 3d Force Service Support Group are usually garrisoned at Camp Butler.

Transportation. Chartered commercial airlift planes use Kadena AFB, while commercial airlines use Naha International Airport. MCAS, Futenma, handles propeller-driven and helicopter military traffic for Marine units on Okinawa, while Marine jets deployed to Okinawa operate from Naval Air Facility (NAF), Kadena.

Quarters. Government quarters are available for dependents of officers assigned to the base on accompanied tours only. Waiting times may vary from 2 to 14 months, but comfortable western-style housing may be obtained in the civilian community at prices comparable to those in the United States.

Recreation. Commissioned Officers' Mess; fishing; aquatic sports; hobby shops; movies; various major sports; AFRTS radio-television; local cultural and recreational activities.

Commissary and Exchange. Excellent exchanges, operated by the Army and Air Force, and a large, well-stocked commissary are available at Zukeran.

Camp H. M. Smith, named for the Corps' greatest amphibious general, has Commander-in-Chief Pacific as a tenant, too.

828 ■ MCAS, Iwakuni, Japan (FPO Seattle 98764)
Iwakuni is the home station of all tactical jet aircraft units of the 1st Marine Aircraft Wing (whose headquarters is at Futenma, Okinawa), and of Japanese naval aviation units as well.
Transportation. *Rail.* Japan National Railways has outstanding local or high-speed service to all points in Japan except Okinawa.
Air. Semiweekly Military Aircraft Command (MAC) flights handle military traffic directly in and out of Iwakuni. Commercial flights go from Hiroshima Airport (27 miles distant), with connections for international flights at Tokyo.
Quarters. Besides BOQ, Iwakuni has Government quarters for officers on accompanied tours.
Schools. Matthew C. Perry School (kindergarten through twelfth grade) is the only western-style school available. The station provides school-busing for children whose families live off-station. Special Services operates a children's day care and preschool center (ages 3–5). Los Angeles Community College, Pepperdine University, and the University of Maryland offer college-level courses.
Recreation. Commissioned Officers' Mess; tennis; swimming; flying club; hobby shop; fishing; hunting; golf; sport parachuting; local cultural events.
Commissary. Post exchange excellent; commissary adequate.

SMALLER POSTS

829 ■ "From the Dawn to Setting Sun. . . ."
In addition to the large posts just described, the Corps maintains more than 80 Marine Barracks or Detachments at home and abroad, together with hand-picked embassy guards throughout the world. If you should be ordered to any of these posts, seek out some officer who has recently returned, or write the adjutant for local information.

830 ■ Posts in the Continental United States
Here, as of 1977, are the most important smaller Marine Corps "stateside" posts:

California:
MB, NAS, **Alameda**; MB, Fallbrook Annex, **Fallbrook**; MB, NAS, **North Island, San Diego**; MB, Naval Weapons Station, **Seal Beach**; MB, Naval Weapons Station, **Concord**; MB, NB,

Long Beach (Terminal Island); MB, NSY, Vallejo (Mare Island); MB, NAS, Moffett Field; MB, NS, San Diego; MB, NS, Treasure Island, San Francisco; Landing Force Training Command (LFTC), PhibTraPac, Coronado; Mountain Warfare Training Center, Bridgeport; MB, NAS, Lemoore.

Connecticut:
MB, Submarine Base, New London.

Florida:
MB, NAS, Jacksonville; MB, NAS, Sanford; MB, NB, Key West; MB, NAS, Cecil Field.

Maine:
MB, NAS, Brunswick.

Maryland:
MB, NS, Annapolis; MB, Fort Meade.

Missouri:
Marine Corps Finance Center, Kansas City.

Nevada:
MB, Lake Meade, Las Vegas.

New Hampshire:
MB, NB, Portsmouth; MD, Naval Disciplinary Command, Portsmouth.

New Jersey:
MB, NAS, Lakehurst; MB, Naval Weapons Center, Earle.

Pennsylvania:
MB, NB, Philadelphia.

Rhode Island:
MB, NB, Newport.

South Carolina:
MB, NB, Charleston.

Tennessee:
MB, Clarkesville.

Virginia:
MB, NSY, Norfolk (Portsmouth); MB, NB, Norfolk (Hampton Roads); MB, Naval Weapons Station, Yorktown; Landing Force Training Command (LFTC), PhibTraLant, Little Creek.

Washington:
MB, NAS, Whidbey Island; MB, Naval Ammunition Depot, Bangor; MB, Bremerton; MD, NTS, Keyport.

831 ▪ Overseas Posts

Marines serve in the overseas states and possessions of the United States, as well as in foreign countries (not including the embassy guards), as follows:

Alaska:
MB, NS, **Adak.**

Bermuda:
MB, NS.

Canal Zone:
MB, NS, **Rodman.**

Cuba:
MB, NOB, **Guantanamo Bay.**

England:
MB, U. S. Naval Activities, **London.**

Guam:
MB, NB, **Guam.**

Hawaii:
MB, **Pearl Harbor.**

Iceland:
MB, NS, **Iceland.**

The Barracks at Guantanamo Bay stands not far from ground which was first captured, then defended, by Marines during the War with Spain.

Italy:
  MB, Naval Activities, **Naples.**
Japan:
  MB, NAS, **Atsugi**; MB, Fleet Activities, **Sasebo**; MB, Fleet Activities, **Yokosuka**; MD, Camp Fuji, **Takigahara.**
Morocco:
  MB, Naval Activities, **Kenitra (Port Lyautey).**
Philippine Republic:
  MB, NB, **Subic Bay.**
Puerto Rico:
  MB, NS, **Roosevelt Roads.**
Spain:
  MB, Naval Activities, **Rota.**

832 ■ Overseas Training Missions and Embassy Guards

In addition to the foreign posts listed above, the Marine Corps maintains more than 110 State Department security detachments, commanded by NCOs, in most of the capital cities of the world. These detachments are under the immediate supervision of regional company headquarters located at key points.

To provide military advice and training for the armed forces of certain countries, the Marine Corps maintains advisory groups in Korea and Taiwan, together with individual military advisers in several other countries.

Our flag's unfurl'd to every breeze, from the dawn to setting sun...
—*"The Marines' Hymn"*

9

The Marine Corps Reserve

In many ways the Marine Corps Reserve has proved itself the backbone of today's Marine Corps. Without a high-quality, high-spirited, ready Reserve, the Marine Corps could not retain its position as the national force in readiness. The Reserve has this mission:

> To provide a trained force of qualified commissioned, warrant and enlisted personnel to meet requirements for the initial expansion of the regular Marine Corps in time of war or national emergency.

Two wars (World War II and Korea) tested the Reserve, and in each it fulfilled its mission to the hilt. In the Vietnam war, as a result of a national policy decision not to mobilize reserve forces of any Service, the Marine Corps Reserve did not play its traditional role, although many individual reservists performed outstandingly.

901 ■ History of the Reserve

The Marine Corps Reserve came into being in 1916 with an initial strength of three officers and 33 enlisted men, while Major General George Barnett was 12th Commandant. Like many other forward steps during the period, the Reserve was in fact the product of the foresight and imagination of Barnett's assistant, Colonel Lejeune, later to become 13th Commandant. Despite its eventual importance, the Reserve played no significant role in World War I. Indeed, it nearly died on its feet in the early 1920s as a result of fiscal starvation. But for the loyalty and single-mindedness of pioneer reservists of that decade, there might not be a Marine Corps Reserve.

Following enactment by Congress of the Naval Reserve Act of 1925, the Marine Corps Reserve began to come into its own. This legislation for the first time permitted individual training duty with pay, as well as the organization of drilling units in pay status. Training programs were instituted, and units sprang up in 1927. This prosperity was short-lived, however, for the depression years of 1929–1933 found the Reserve again without funds. During those lean years most units continued to drill and train without pay—even buying their own uniforms—and thus again saved the Reserve from oblivion.

It was 1935 before the Reserve was finally able to stand on its own feet. 1935, pioneering year for the Corps as a whole, was certainly a golden year for the Reserve. 1935 saw these developments:

1. Appropriations for training an organized and volunteer Marine Corps Reserve (ground and aviation) totaling almost 10,000 officers and enlisted Marines;
2. Inauguration of the now historic Platoon Leaders Classes in order to obtain a steady input of well-trained, carefully selected junior Reserve officers from colleges not participating in Army or Navy ROTC;
3. Dawn of the Reserve pilot program for Marine Corps aviation—an extra dividend of the Naval Aviation Cadet Act of 1935.

In 1938 the 1925 Naval Reserve Act was brought up to date by Congress in many aspects—perhaps the most important being the provision, for the first time, of a charter of rights and benefits for the Reserve. Among these milestones were: hospitalization; death and disability benefits; equitable promotion; retirement with pay for active service; and the right to participate in formulation of Reserve policy. Under the Naval Reserve Act of 1938, the Marine Corps Reserve has twice accomplished its job of providing a trained force in readiness.

The solid success of the peacetime Reserve was amply attested in 1939 (when individual reservists were brought to active duty after President Roosevelt's proclamation of limited national emergency in September of that year), and a year later, in 1940, when mobilization of the remainder of the Reserve brought 15,138 additional Marines to the Colors. The

Prewar Reserve Marines participate in amphibious exercises of the 1930s.

extent to which the Reserve had hewn its place in the Corps was proven, in 1945, by the fact that, of the 471,000-man Corps, largest in 170 years, approximately 70 percent were reservists.

Much of the Reserve's effectiveness throughout World War II stemmed from the philosophy behind its mobilization, a philosophy which today is stronger than ever. Although the 1940 Reserve was built around 36 hometown battalions and squadrons, each with its distinctive temper, local associations and comradeship, Major General Holcomb, then Commandant, took the position that no Marine, regular or reserve, should, while on active duty, claim any home but the Corps. Thus, as Reserve units reached mobilization points, they disbanded, and their members simply became individual Marines headed for service in the expanding regular formations of the Fleet Marine Force. To drive home the import of this decision, and to emphasize that every man privileged to wear the Globe and Anchor was a Marine, no more and no less, General Holcomb decreed that, except where required by law for administrative purposes, the word, "Reserve" and its corresponding abbreviation, "R," following the "USMC," would not be used. All hands, reserve and regular, were Marines.

That policy remains in effect and has become a tradition of the Corps.

Following World War II, the postwar buildup of its Reserves was one of the great achievements of the Corps. Through good leadership (regular and reserve), through willingness to invest capable personnel in the Reserve program, and because of the unflagging loyalty of Marine alumni—"*Who ever saw a sorehead ex-Marine?*" asked a prominent journalist—the Reserve was in unmatched readiness to back up the attenuated regular Corps when the Korean War flamed up.

In the field in Korea, as in Pacific battles before it, it was literally impossible to distinguish reservist from regular. Once again, as always, all hands were Marines. Among those who had originally started as reservists, however, it is worth noting that in World War II and Korea some 57 in all won Medals of Honor.

Soon after the end of the Korean War, Congress passed legislation (the Reserve Forces Act of 1955) that has exercised and will continue to exercise a profound effect on the reserve components of all the Armed Forces, including the Marine Corps. This law (see Section 911 for details), among other features, provided for the so-called Special Enlistment Program whereby young men, after receiving not less than 12 weeks of hard training with and by regular forces, then enter the Organized Reserve for a prescribed period of years of obligated service. At the time of writing, more than two-thirds of the Organized Reserve includes personnel with some form of legally obligated service, with a background of professional-type basic training, and therefore a degree

Off to Korea marches an Organized Reserve battalion after being mobilized by the President. Without a high-spirited combat-ready Reserve, the Inchon landings would not have been possible.

of readiness which has been greatly increased even over that traditional in earlier days. Reflecting this heightened capability, the Reserve was reorganized on 1 July 1962, to provide a distinct unit mobilization structure, embodied in the 4th Marine Division, 4th Marine Aircraft Wing, and added Force Troop units. These will be mobilized and employed as units; however, the Organized Reserve still maintains additional units whose function is to provide trained individual Marines for fleshing-out Regular and Reserve units. This extensive reorganization attests the truth of a statement by a recent Commandant:

> Never before has our dependence upon the Reserve been so great and never before has our Reserve been more worthy of that dependence.

ORGANIZATION AND COMPOSITION

902 ■ Organization of the Marine Corps Reserve

The Marine Corps Reserve today is organized and maintained under the Armed Forces Reserve Act of 1952, which superseded the 1938 Naval Reserve Act. This law incorporated the basic principles of its predecessor, but modernized the Reserve by adding features essential under mid-century world conditions and the increasingly important role that the Reserve has assumed in plans for national defense. (This and other laws bearing on the Reserve have been codified under Title 10, U. S. Code.)

Since the Reserve is a component of the Marine Corps as a whole, command and administration of the Reserve stem directly from the

Commandant. Thus the department and offices of Marine Corps Headquarters bear the same relationships and responsibilities toward the Reserve as they do toward the remainder of the Corps.

903 ■ Division of Reserve

The Division of Reserve, headed by the Director, Marine Corps Reserve, a general officer, supervises and coordinates the activities of units and members of the Reserve. The Director, Marine Corps Reserve, with headquarters now in New Orleans, advises the Commandant and staff on matters pertaining to the Reserve, and directs and exercises control over the Reserve. Under a second hat, he acts as CG, IV MAF and thus commands the 4th Marine Division and 4th Marine Air Wing, the two major Reserve formations of the FMF. This two-hatted billet is actually filled by the senior general officer commanding either wing or division, each of whom is a regular officer.

NOTE: As this edition of the *Guide* goes to press, the structure of the Marine Corps Reserve is undergoing reorganization. What is described in Sections 903–904 represents the best information available in mid-1977.

904 ■ Marine Air Reserve Training Command

Organized Reserve aviation units are supervised by the Commanding General, Marine Air Reserve Training Command, whose headquarters are at Naval Air Station, New Orleans. "ComMART," as this commander is short-titled, comes under military command of the Commandant, but, like all Marine aviation, receives certain naval logistic support from the Chief of Naval Air Reserve Training. ComMART also serves as Commanding General, 4th Marine Air Wing.

905 ■ Marine Corps Reserve Units

To a greater extent than many Marines realize, the Corps entrusts its readiness to the units of the Organized Reserve. They are the backbone of the Reserve, and constitute the mobilization backbone of the Corps.

Leaving out the volunteer training units (which will be discussed below), organized units of the Marine Corps Reserve are formed almost entirely from Class II reservists (see Section 906). Both ground and aviation units are mainly organized at or below the battalion/squadron level. Units parallel prototype units in the FMF. Thus, in the ground portion of the Organized Reserve, you will find infantry, artillery, tank, amphibian tractor, engineer, air and naval gunfire liaison (ANGLICO), and signal units—the preponderance, of course, being infantry. Organized Reserve aviation likewise follows the FMF pattern and includes, in various cities, most of the principal operating units of a Marine air-

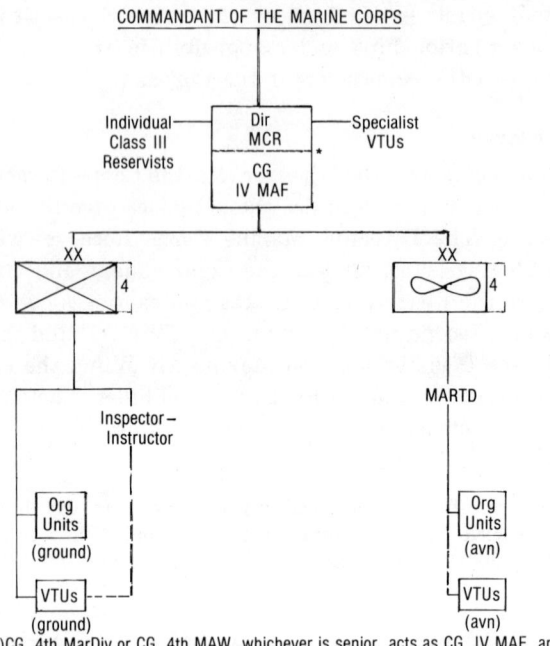

(*)CG, 4th MarDiv or CG, 4th MAW, whichever is senior, acts as CG, IV MAF, and Director, Marine Corps Reserve.

Fig. 9–1: Organization of the Marine Corps Reserve (1977).

craft wing—fighter-attack squadrons, transport squadrons, helicopter squadrons, and wing headquarters and service units as well.

Since Organized Reserve units follow the Fleet Marine Force pattern, the administration and functions of these units are carried on in the same way as in similar units in the regular establishment.

Organized ground units are commanded by Reserve officers who have been selected for their professional experience and background. Like all commanding officers, they must administer, train, and maintain the readiness of their commands. In addition, however, they must stimulate and promote whatever recruiting is needed to keep their units up to strength. Organized Reserve commanding officers usually serve for a two-year tour, which may be extended to three or more years under certain circumstances. This gives them adequate experience in command, and at the same time assures the advantages of healthy rotation.

Organized aviation units, like ground units, are commanded by Reserve officers whose responsibilities are much like those of ground unit COs in the Organized Reserve.

Because of the large amount of technical training, the paramount requirement for safe flight operations, and the quantities of expensive materiel (including airplanes) which an aviation reserve unit requires, the inspection-instruction organization for Reserve aviation outfits differs somewhat from that used with ground units.

At the home station of each Organized Reserve aviation unit, a parent Marine Air Reserve Training Detachment is located. This detachment is commanded by a regular Marine Corps aviator, usually a field officer, and includes assistant instructors and maintenance crews to support the Reserve squadron. The Reserve unit's commanding officer comes under command of the CO of the Reserve Training Detachment, or MARTD.

Staff Groups permit drill pay and Organized Reserve status for more senior Reserve officers for whom mobilization requirements exist, but who cannot train with other Organized Reserve units. Staff Groups attend paid drills and annual field training.

Volunteer Training Units (VTUs) are not part of the Organized Reserve but afford training in staff and command functions, ground and air, for Reserve officers and enlisted men ordinarily not associated with an Organized unit, who want to stay with the Corps, keep up professional training, and amass credits for Reserve retirement.

A Volunteer Training Unit may be made up of six or more members (men, women, officer or enlisted) of any military specialty or combination of specialties. Most VTUs train under a specified syllabus provided by Marine Corps Development and Education Command (MCDEC), but some specialize in given fields when all members hold the same or related military occupational specialties. Each VTU is assigned an adviser, usually the nearest I-I (see Section 908) or commanding officer, MARTD.

If you are a member of the Reserve but not affiliated with a unit, or if you live in a locality without a local unit, the district director's office can help you maintain contact with the Corps. If you wish to join the Reserve but have no hometown unit, your district director can help you in this, too.

Marine Reserve affairs outside the United States are administered by the Director, Marine Corps Reserve.

If you live outside the territorial jurisdiction of the United States and are a member of the Reserve, or desire to be, write to the Commandant of the Marine Corps, Washington, D.C. 20380.

906 ■ Composition of the Reserves

The Marine Corps Reserve includes different individual classes for mobilization planning and assignment that vary according to the preferences and background of the reservist. Thus the Reserve affords varying opportunities for activity, which can usually be adjusted to the desires of anyone qualified to be a Marine.

Fleet Marine Corps Reserve (Class I). Usually short-titled "the Fleet Reserve," Class I is composed entirely of former regular enlisted

Marines who enter this status under various laws which, in general, allow transfer to the Fleet Reserve after 20 years' service as a regular. Thus the Fleet Reserve provides a backlog of experienced enlisted Marines who may be employed without further training if the bell rings for full mobilization. Fleet reservists remain in Class I (which in peacetime amounts practically, if not legally, to semiretirement on retainer pay) until they complete 30 years' combined regular and reserve service. Then they are eligible for retirement.

Organized Marine Corps Reserve (Class II). The Organized Reserve (also known as the Selected Reserve) comprises over 30,000 officer and enlisted members of organized units that can be immediately assimilated into the FMF. For this reason Organized Reservists must measure up to physical and professional standards comparable to those required of regulars. Today, the majority of the men in the Organized Reserve are young, intelligent, highly motivated six-month trainees with a six-year military obligation; the balance are prior-service noncommissioned officers who provide the necessary hard-skill experience and seasoned leadership. Members of the Organized Reserve attend semimonthly or monthly drills, go to camp for annual training with their unit, and get paid for the drills and training they perform. If you wish to become a Class II reservist (over and above required individual standards), you must be able to attend the regular drills of your unit at its home station. This does not ordinarily present too serious a problem, however, since more than 200 U. S. cities and towns boast organized units of the Marine Corps Reserve. Today, the bulk of the members of Organized Reserve units are Marines with obligated service, and Class II of the Reserve includes only personnel in the category of the "Ready Reserve" (see Section 907).

Volunteer Marine Corps Reserve (Class III). Class III (designated as the Individual Ready Reserve) effectively speaking includes all physically qualified Marine reservists not assigned to Classes I and II. The mission of the Volunteer Reserve is to provide the Marine Corps with personnel for complete mobilization in time of war or emergency.

This means that the Individual Ready Reserve largely falls into two categories: reservists who cannot take part in an Organized unit; and people with specialist qualifications who do not need or desire the unit training and group participation that Class II entails. On the other hand, if, as a Volunteer reservist, you wish to affiliate, on a Class III basis, with your local Organized Reserve unit, you usually may do so by arrangement with the unit CO. If no Organized Reserve unit is located in your vicinity, look into the volunteer training unit situation (see Section 905).

The Individual Ready Reserve gives an opportunity for Marine Corps membership to many men and women whose personal commitments might otherwise preclude this, and the Reserve program includes voluntary training opportunities designed to suit almost any combination of individual convenience or specialization that may apply to you.

Limited Assignment (Over-age) Category. Some Reserve officers have special qualifications which are of value to the Marine Corps regardless of their age in grade. When such officers become over-age for their rank, they may be transferred to the limited assignment category and thus be retained for specific mobilization assignments.

Women Reservists. Women reservists fall into the Fleet, Organized, or Volunteer Reserve according to their individual circumstances and qualifications, and the policies prescribed by the Commandant (see Chapter 10).

907 ■ Reserve Categories for Active Duty

The reserve establishment is also broken down into three categories, in descending order of availability, of eligibility to perform active duty.

Ready Reserve. Members of the Ready Reserve may, in time of national emergency, be required to serve on active duty for a period of time not longer than 24 months. In time of war or national emergency declared by Congress, this service may be extended for the duration of the war or national emergency, plus six months thereafter. That part of the Ready Reserve which is grouped into organized units is further designated as the *Selected Reserve.*

Standby Reserve. In time of war or national emergency, Marines of the Standby Reserve may be ordered to active duty for its duration and for six months thereafter. This cannot be accomplished without their consent, however, unless the Director of Selective Service has determined that they are available for active service. Members of the so-called Inactive Status List cannot be ordered to active duty under these provisions unless the Secretary of the Navy determines that inadequate numbers of active-status personnel are available.

Retired Reserve. Those reservists who have attained retirement cannot be ordered to active duty without their own consent, even in event of war or national emergency, unless the Secretary of the Navy determines that there are insufficient reservists in an active status. The tour of duty, as above, is for the duration plus six months.

> NOTE: The President may, without declaring a national emergency, call to not more than 90 days' active duty selected organized reserve units, individuals in units who have not completed basic training, and individual reservists who have not completed two years active duty with the colors.

RESERVE TRAINING

908 ■ The Inspector-Instructor

To ensure that the Reserve has the benefit of coordinated and professional up-to-date training, advisory personnel from the regular Marine Corps are detailed to duty with the Reserve.

Each organized ground unit has a regular Marine officer (with a small staff) whose title is *"Inspector-Instructor."* This title describes the job exactly. The "I & I," as he is known, must, as instructor, provide training assistance and general guidance to the unit. As inspector, however, he must make certain that the unit is up to the standards set by Marine Corps Headquarters. Inspector-Instructors are under the direct command of the respective Commanding Generals of the 4th Division (ground I & Is) and 4th Wing (aviation). As we have seen in Section 905, functions comparable to those of the I & I are performed for aviation reserve units by the CO, air reserve training detachment.

A note for all "I & I's":
Your job (and for that matter, *any* duty in connection with the Reserve) calls for grade-A-certified leadership, imagination, and tact. Your responsibilities are heavy (much heavier than they look on paper). Your authority is slight. Nevertheless, your job presents great challenges and can give you corresponding rewards. In most cases you alone represent the active Marine Corps in your community. By your conduct, example, and loyalty to the ideals of the Corps, your fellow townsmen form their impression of the whole Marine Corps.

909 ■ Training Opportunities in the Reserve

Every Organized Reserve unit has to complete a carefully worked out annual training cycle. In addition to this unit-training program, you, as an individual reservist, may avail yourself of a wide selection of courses, volunteer periods of training duty, and gratis home study courses (both Marine Corps Schools Extension Courses and Marine Corps Institute—see Section 1509). Reserve training affords something for every individual's interests and opportunities.

Organized Reserve Training. Training of the average OMCR unit consists of 48 paid drills or flying periods, each consisting of at least four training hours, and two weeks' annual active-duty training. Certain units, such as Staff Groups and Marine Air Reserve Groups, are only authorized 24 paid drills and their two-week period of active-duty training.

Reserve training is as meaningful as human effort can make it. By holding "multiple drills" at which two or more paid drills are consolidated consecutively, Organized Reserve units now train two full days a month, frequently on a single weekend. This permits realistic field training, including overnight problems. The amount of air-ground and combined arms training is considerable, with two or more neighboring units joining in weekend exercises. Organized Reserve artillery units (generally located near Army or Marine Corps bases with range facilities) conduct live firing throughout the year.

The hometown Organized Reserve training cycle culminates annually in two weeks' training by the unit at a Marine Corps station. It follows a cycle to include desert, mountain, jungle, amphibious, air-ground, combined arms, and specialist training. When possible, the Marine Corps arranges the movement of Organized Reserve outfits to and from annual training by Marine air-lift or in amphibious shipping. This increases the adaptability, professional know-how, and experience of the unit concerned.

As a member of the Organized Reserve, you receive "drill pay" for each drill you attend, as well as for summer camp, and this pay can provide a welcome augmentation for your income.

Volunteer Reserve Training. If you are in the Volunteer Reserve, you do not have to attend drills, but may be required to perform training duty not exceeding 15 days a year. (Enlisted personnel in certain categories must perform 30 days' active duty annually.) Many opportunities are open to you to train with Organized Reserve units or Volunteer training units, on your own initiative with or without pay. If you aim to maintain your professional proficiency and to accrue credits for reserve retirement with pay (see Section 914), you can keep up-to-date by periodic spells of training duty and through the medium of MCDEC Extension School correspondence courses. A typical training cycle for a Volunteer reservist might include: completion of an Extension School course; a two-week Volunteer Reserve summer staff course at Quantico; and perhaps a few days' training duty without pay at a Fleet Marine Force post, to brush up in his specialty. If you keep professionally up to date, you may have the opportunity of performing training duty as an umpire during a large maneuver, where qualified Reserve officers often serve in this capacity.

910 ■ Categories of Reserve Training and Instruction

All training that you perform as a reservist falls into one of the following categories:

Regular drills are performed only by the Organized Reserve; they must last at least two hours, must be regularly scheduled, and must be carried out in uniform.

Equivalent instruction or *equivalent duty* is a substitute period, performed on some other day, for a regular drill that has been or will be missed. Only members of the Organized Reserve perform equivalent duty.

Appropriate duty is any type of duty, training or otherwise, authorized for you as an individual by the Commandant, as appropriate to your rank and military qualifications.

Administrative duty, over and above regular drills, is performed by Organized Reserve COs in connection with the discipline, administration, or maintenance of Government property of an Organized unit. If in the Volunteer Reserve, you may perform administrative duty without pay, but thereby accrue credit toward reserve retirement.

Annual training duty is performed by Organized Reserve units only, and lasts 15 days. It is usually conducted at a Marine Corps post, and concentrates on the mobilization mission of the unit.

Active duty, although technically not training duty, enables many reservists, especially Volunteers, to obtain much valuable training. Active duty may be either for a stated period (by contract, or "Standard Written Agreement") or of indefinite duration.

Unlimited Duty (UDR). Reserve officers who have completed their initial obligated service, plus two years' active duty, may be selected for "unlimited duty." If you come under this program, you are treated exactly as if you were a regular officer, with the same opportunity for professional schooling, duty assignments, and, eventually, retirement. You will be selected and promoted when due, as if you were a regular officer—and will not be released to inactive duty (except on your own request or refusal of further extended active duty) other than by virtue of two failures of selection, or a drastic cutback in the authorized officer strength of the Marine Corps as a whole. On the other hand, as a UDR reserve officer, you are subject to the same requirements for continuing on active duty after promotion or when in receipt of overseas orders, as apply to regular officers. This is an outstanding program which permits a suitably motivated reserve officer to have the best of both worlds.

Active duty for training (with or without pay) differs from active duty in that the avowed purpose of the duty is training. Active duty for training may be performed with or without pay, according to your request and to the state of the Marine Corps budget.

Repeated training duty without pay may be performed by any reservist, but is largely limited to aviation personnel. This category of duty confers authority to perform repeated training-duty periods (such as aerial flights), but without pay.

Extension courses are open to you from the Extension School at Quantico and from the Marine Corps Institute (MCI) in Washington. You can find out about them in Section 1509.

911 ▪ Special Enlisted Reserve Program

One provision of the 1955 Reserve Forces Act, mentioned earlier, is the special enlistment program whereby you can go on active duty for a minimum of three months and then complete your military obligation by serving six years in the Ready Reserve. In general, the program is open to young men between the ages of 17 and 26.

Once you have entered this program and have completed your 12 weeks' training with and by the regular Marine Corps (including boot camp at San Diego or Parris Island), you must remain in the Organized Reserve and complete 48 drills per year and attend up to 17 days' active duty for training, or be recalled to active duty to make up.

If you are eligible for and interested in the Marine Corps Reserve special enlistment program, you can obtain necessary detailed information from your nearest unit of the Organized Reserve or from the Director of Reserve, Marine Corps Headquarters.

MISCELLANEOUS

912 ▪ Transfer Into the Marine Corps Reserve

The Armed Forces Reserve Act permits certain members of reserve components of the Armed Forces to transfer from one Service to another. Thus, if you are a member of another reserve component and desire to complete your obligated military service in the Marine Corps Reserve, it may be possible for you to do so.

Generally speaking, if you are not on active duty, but do have a period of obligated service in your Reserve component, you may be discharged (or resign, if a Reserve officer) to accept an appointment in the Marine Corps Reserve. The Director of the Marine Corps Reserve can advise you on your eligibility for transfer, and can assist you in the administrative paperwork.

913 ▪ Privileges and Perquisites of the Marine Reservist

A Marine recruiting poster alleged to date from Revolutionary War days tells the privileges and perquisites of the Marine of 1776:

> You will receive SEVENTEEN DOLLARS BOUNTY, And on your arrival at Head Quarters, be comfortably and genteelly clothed—And spirited young boys of a promising Appearance, who are Five Feet Six Inches high, WILL RECEIVE TEN DOLLARS, and equal advantages of PROVISIONS and CLOTHING with the Men. In fact, the Advantages which the MARINE possesses, are too numerous to mention here, but among the many,

it may not be amiss to state—That if he has a WIFE or aged PARENT, he can make them an Allotment of half his PAY; which will be regularly paid without any trouble to them, or to whomsoever he may direct; that being well Clothed and Fed on Board Ship, the Remainder of his PAY and PRIZE MONEY will be clear in Reserve, for the Relief of his Family or his own private Purposes. The Single Young Man, on his Return to Port, finds himself enabled to cut a Dash on Shore with his GIRL and his GLASS, that might be envied by a Nobleman. . . .

Of course times have changed somewhat.

Today, the preeminent privilege that you gain as a member of the Marine Corps Reserve is the right to wear the Globe and Anchor and call yourself a Marine. But you do have other substantial privileges and perquisites, which are summarized below.

Pay, a short but important word, is certainly a perquisite of the reservist. For each regular drill or equivalent, you draw one day's pay, according to rank. This also applies to all active or training duty, unless you elect to perform these in nonpay status. Moreover, regardless of whether you are Volunteer, Fleet, or Organized, you accrue "fogies" (longevity credits) for *all* time spent as a member of the Reserve, active or inactive, just as if you were a regular.

Uniforms worn by reservists are, of course, regular Marine Corps uniforms. They are worn, or may be prescribed, during and when going to and from drills and instruction, and on other appropriate occasions, such as military ceremonies, military dinners or dances, and the like.

As a Reserve officer, you purchase your own uniforms, but receive an initial allowance of $200 upon performing your first 14 days' training duty (or $300 when you first report for active duty). If you have already drawn your $200 allowance, you receive $100 more when you report for active duty. For each four years' satisfactory service (*less* time spent on active duty), you receive $50 for maintenance of uniforms.

Every Reserve officer must possess a required kit, but he may in addition buy other uniforms, such as blues, whites, and evening dress, if he wants them and can find proper occasions to wear them. Marine Corps Headquarters publishes, from time to time, lists of uniforms that Reserve officers must have.

It goes without saying that the right to wear Marine Corps uniforms is a privilege which members of the Reserve have always treasured. While you wear that uniform, you are accountable to every high standard of the Corps and its discipline.

Clubs and messes (see Chapter 24) extend a hearty welcome to the Marine Corps Reserve officer, whether active or inactive. Thus you will always find a friendly greeting (and, likely as not, old comrades) in the open mess at the Marine Corps or Navy station nearest your home.

Exchange and Commissary Privileges. On drill days, reservists enjoy unlimited exchange privileges. Commissary privileges are available on a limited basis for those on duty less than 72 hours. If your period of duty exceeds 72 hours, you rate the same exchange and commissary privileges as a regular Marine.

Promotion opportunities afford reservists the chance to enhance prestige and responsibility and, of course, to be eligible for increased retirement pay (for information on Reserve retirement, see Section 914 below).

Decorations and medals are awarded to Marine reservists as to all other Marines—strictly as earned. They are worn on the uniform, or, on certain occasions, with civilian clothes, in the same way as by regulars (see *Marine Corps Uniform Regulations*).

In addition, however, the Marine Corps Reserve boasts special awards of its own:

The Organized Marine Corps Reserve Medal, with bronze stars for succeeding awards, goes to members of the Organized Reserve who maintain excellent records over a four-year period, and who, during that time, attend four annual training periods with their units, plus at least 80 percent of all scheduled drills.

The Marine Corps Reserve Ribbon is awarded to each reservist who completes a 10-year period (excluding peacetime active duty) in any class(es) of the Reserve, *provided* this time has not been counted for award of the Organized Reserve Medal.

The Marine Corps Association, which publishes the professional magazine of the Corps, the *Marine Corps Gazette,* is open to membership by officers and enlisted men of the Reserve. So also is the U. S. Naval Institute, which publishes the *United States Naval Institute Proceedings,* professional journal of the Naval Services.

Government Insurance Benefits. Since 1974, certain categories of reservists are eligible for Servicemen's Group Life Insurance (SGLI, see Section 2303), viz.:

1. Ready reservists required each year to perform at least 12 periods of inactive-duty training creditable for retirement purposes.
2. Retired reservists who have not completed 20 years' creditable service toward retirement, have not received first increment of retired pay, and are below age 61.

The new VGLI program (Section 2303) also applies to certain reservists.

Employment Protection. The Universal Military Training and Service Act (P. L. 632, 86th Congress), as amended, protects reservists against loss of seniority, status, pay, and vacation while they are away from their jobs on Reserve training duty. Also, if you should unfortunately become disabled while training and can no longer perform the duties of your civilian job, you are entitled to re-employment on other jobs whose duties you may be able to perform. If you are hospitalized incident to training duty, you may delay re-employment application for a period up to one year. On the other hand, the law requires that you request leave of absence from your employer before going on training duty, and that you must report back to work immediately on completion of training.

In addition to the foregoing, Federal employees, if in the Reserve, rate up to 15 days' extra leave with pay per year to cover periods spent on training duty, and are protected by law against "loss of time, pay or efficiency rating" while availing themselves of this additional leave for training. Government-employee reservists ordered to active duty must, by law, be restored to the job they held before being called up.

914 ■ Reserve Retirement

Retirement, both honorary (without pay, that is) and with pay, can be earned by members of the Marine Corps Reserve. In general, leaving out the Fleet Reserve, the fundamental prerequisite for Reserve retirement with pay is 20 or more years' active service (not necessarily consecutive), or 20 years' "satisfactory Federal service," not all of which need be active.

Although Reserve retirement may be effected under various provisions of law, the principal one affecting most reservists is a section of Public Law 810, 80th Congress, which makes retirement pay available to all Marine reservists who accumulate sufficient "retirement points" (credit points earned by service and training). The number of retirement points you chalk up also determines the amount of retired pay you may earn.

Here is a brief description of this retirement system.

To qualify for Reserve retirement, you must earn at least 50 points a year for a minimum of 20 years, but these years need not be consecutive.

If you served in the Reserve before 1 July 1949, you automatically accrue 50 points for each such year (other than years spent on active duty).

You get one point for each day of active duty (including training duty) served as a member of the Armed Forces before 1 July 1949.

After 1 July 1949, you must earn at least 50 points a year in order to have that year count toward Reserve retirement. The amount of retired pay you get (see below) is determined by the total number of points you accumulate. The number of points you earn depends largely on the amount of effort put into training, home study courses, and other kinds of equivalent instruction: the more you give as a reservist, the more you get.

You must serve the last 8 of your 20 years as a Reserve member of the Armed Forces.

After you have met all the requirements, you become eligible for Reserve retirement pay the first month after your sixtieth birthday.

Your retirement pay is computed as follows: Divide by 360 the sum of *all* points earned. Then multiply the result by 2½ percent. Then multiply this result by the combined annual base pay and longevity pay you would get if on active duty in the highest grade, permanent or temporary, satisfactorily held by you during your 20 years' service. The answer: the annual Reserve retired pay you will be eligible to receive on attaining the age of 60.

In addition to all the foregoing, physical disability retirement rules which govern regular Marines extend with equal force to reservists who incur Service-connected disabilities.

> NOTE: Reservists retired *with pay* are eligible to use Armed Forces medical facilities on a space-available basis and to participate in civilian out-patient care provided for retired personnel. (See Chapter 8.)

915 ■ Additional Information on Reserve Matters

The Director, Marine Corps Reserve, can supply more detailed information on the Reserve, not only to Marines, but to potential reservists and friends. The Division of Reserve is always more than glad to answer individual queries and to assist reservists in solving professional problems. However, before you write Marine Corps Headquarters, look under "United States Government" in your local telephone directory, and see if there is a local Marine Corps activity, either regular or reserve. Your local Marine officer—be he inspector-instructor, recruiter, NROTC instructor, or district director—will always be ready to help you find out what you want to know.

> *In its outstanding service to our Corps, the Marine Corps Reserve has earned the right to be called our "Secret Weapon."*
> —General L. C. Shepherd, Jr.

10

Women Marines

Although the traditional wartime mission of Women Marines was expressed in the World War II slogan, "Free a Marine to Fight!", Women Marines today perform whatever duties meet the needs of the Corps.

It is the policy of the Secretary of the Navy, as well as that expressed in Congress' charter for women in the Service (Women's Armed Services Integration Act of 1948), that Women Marines train for and perform duties in Marine Corps tables of organization; also, that their qualifications for such duties shall be in line with standards applicable to male Marines in similar jobs.

In general, Women Marines are subject to the same rules and regulations as other Marines. The military authority exercised by Women Marines on duty is the same as that authorized for male personnel of the same rank.

The professional objective of all Women Marines is thus well stated in the quotation heading this chapter: ". . . to be truly integrated in the Corps."

1001 ■ History of Women Marines

In August 1918, the first women ever to wear the Globe and Anchor enlisted in the U. S. Marine Corps. They totaled 305 in all, and were immediately nicknamed "Marinettes," an obvious derivative of the Navy's contemporary "Yoemanettes." Although the duties and scope of action of a Marinette (whose top possible rating was sergeant) were much more limited than those of today's Woman Marine, the spirit of the two was the same. The staying power of the original Marinettes is attested by the fact that seven women who wore the Marine uniform in

1918–1919 stayed on as civil servants at Marine Corps Headquarters until retirement decades later in key supervisory or executive jobs.

The "new" women's component was organized in February 1943, when Lieutenant General Thomas Holcomb, 17th Commandant, authorized creation of the Marine Corps Women's Reserve. The first officer and enlisted women were trained beside Navy WAVES in existing naval schools for women. By May 1943, seventy-five women had completed officer training, in a special course at Mount Holyoke College, and 722 had weathered recruit training. July 1943 saw establishment of the Women Reserve Schools at Camp Lejeune. Here at Lejeune were centralized recruit and officer candidate training, together with a number of specialist schools for the 18,000 enlisted women and 821 officers of the Women's Reserve, or "WR," as it was soon short-titled.

It was an emphatic tradition of the new branch of the Corps that there would be no trick nicknames for the group. As far as Women Marines were—and are—concerned, any cute, coy, or punning sobriquet,

"She is Proudest of All . . . that she is a Marine."

World War I Marinettes became, in many cases, lifetime "members" of the Corps, and several ultimately rose to key administrative positions at Marine Corps Headquarters.

official or otherwise, would merely demean the Marine Corps Emblem which they proudly wore. Today, this tradition is stronger than ever.

Much of the initial tone and standard of the Marine Corps Women's Reserve was set in ordinary course by the parent Corps, but quite as much if not more was due to the ability and effort of the first Director of the Women's Reserve, Colonel Ruth Cheney Streeter. Colonel Streeter made it her objective to integrate the women reservists into the framework of the Corps. It was her vision that lifted the World War II women from clerical specialization (the World War I role of Marinettes) into more than 200 separate occupational specialties and billets at every major Marine Corps post in continental United States, and ultimately overseas.

At the end of World War II, save for about 100 women officers and enlisted women retained on duty at Marine Corps Headquarters, under Major Julia E. Hamblet (later to become a Director of Women Marines), the Women's Reserve went home. In 1948, however, Congress passed the Women's Armed Services Integration Act, and a new chapter opened. Henceforth, each Service would have a career cadre of regulars, in addition to the reservists.

World War II: Women Marines parade at Henderson Hall.

To head the regulars ("WM," they now became), General Cates, 19th Commandant, chose Colonel Katherine A. Towle, former Assistant Dean of Women at the University of California, and wartime successor of Colonel Streeter. On 12 June 1948, the Women's Reserve went out of existence, and on the succeeding 4 November, the Women Marines came into being. Members of the former Women's Reserve were re-enlisted into the Marine Corps Reserve, and many became members of the Organized Reserve. Like the remainder of the Organized Reserve, the women found themselves mobilized only days after the onset of war in Korea, with 13 Organized Reserve women's platoons responding to the call. During the years ahead, regular Women Marines continued as an increasingly important portion of the Corps, and, during Vietnam (on a strength of some 2,000) filled many key billets which released other Marines for field service. (A symbolic watershed for the WMs came in 1965 when, during the Dominican Revolt, a woman staff sergeant on duty at the American Embassy won the first combat campaign medal ever awarded to a Woman Marine.)

Following Vietnam and the transition to all-volunteer Armed Forces, all military occupational fields save those involving direct combat or piloting aircraft were opened to women, and the billet of Director of Women Marines, with its implication of segregation of women's roles in the Corps, was abolished—the final step in a long march which commenced in 1918.

1002 ■ Organization and Composition of the Women Marines

As this is written, the strength of the WM, in round figures, approximates 3,500 officers and enlisted women. As women continue to spread into the various occupational fields, their strength is projected to exceed 5,500 by the 1980s.

In keeping with the Women Marines' desire to be fully integrated into the Corps, officers and enlisted women are assigned individually to the jobs they can best fill. While in the past, women have been grouped on each post into separate detachments commanded by the senior woman officer, this policy is on the way out (1977). Ultimately the only separate women's units will be those required for recruit training or officer-candidate indoctrination.

1003 ■ Duties and Assignments of Women Marines

All of the Marine Corps' occupational fields are open to women except those involving actual combat or piloting aircraft—in short, virtually any billet save those which might require the Marine to deploy with the assault echelon of a command in an emergency.

WMs serve in both FMF and non-FMF billets at every major Marine Corps post and station in continental United States (Washington, D.C.; Quantico and Norfolk, Virginia; Camp Lejeune and Cherry Point, North Carolina; Parris Island, South Carolina; Albany, Georgia; Yuma Arizona; San Diego, Camp Pendleton, Barstow, and El Toro, California); in Europe; in Hawaii; in the Far East (Japan and Okinawa); and in various cities throughout the United States.

1004 ■ Entrance and Initial Training

Women officer candidates are initially screened by Women Marine selection officers in a number of large cities, as well as by officer procurement offices and Marine Corps instructors at college Naval ROTC units. In addition, however, like any other Marine, the woman applicant, officer or enlisted, may begin her career at any recruiting station.

Officer candidates must have or be working toward baccalaureate degrees from accredited colleges or universities. These degrees must be in

Duties of Women Marine officers are to perform or supervise more than 100 military jobs.

fields other than medicine, dentistry, pharmacy, veterinary medicine, or theology. The minimum age for enrollment as an officer candidate is 18; the maximum, 29. Candidates must be 21 prior to being commissioned, and cannot marry during training. A limited number of college juniors are selected for the training. Enlisted women may apply for officer training if they are college graduates, or successfully complete an Officer Selection Test.

In order to become a woman officer, the candidate must successfully complete the Women Officer Candidate Course (WOCC), held annually (during the summer) by Officer Candidate School, Quantico. There are two separate phases of training for the Woman Officer candidate. The first, the Woman Officer Candidate Course, provides theoretical and practical military instruction necessary to prepare students for appointment to commissioned grade in the Marine Corps or Marine Corps Reserve. The second phase is the Officer Basic Course. This training is dependent upon successful completion of the Woman Officer Candidate Course and is conducted after the candidates are commissioned as officers. The primary purpose is to indoctrinate newly commissioned women officers in the functions and duties of company and staff officers and to develop further those qualities of leadership and standards of personal conduct and performance required of officers of the Marine Corps.

Training of women officer candidates begins at a vigorous clip and continues so, not only in classrooms, but in barracks and on the parade ground. They also undergo physical conditioning in the best Marine tradition. All instruction is backed up by a galaxy of military demonstrations such as only Quantico can present. These, plus lectures by U. S. and foreign speakers, comprise the curriculum that fits the woman candidate to don the Globe and Anchor as an officer of Marines.

> NOTE: Newly commissioned women officers may, on application, be authorized to delay active duty in order to take graduate work leading to a master's degree in an approved field.

1005 ■ Women Reservists

During World War II, all Marine women were members of the Marine Corps Women's Reserve, the famous WR. Because this term is still familiar, women Marine reservists are still occasionally spoken of collectively as "the Women's Reserve," despite the fact that women are members of the same Marine Corps Reserve classes as all other Marines, described in Chapter 9.

Women Marine reservists serve in peacetime as citizen Marines while living in their home communities. They are given the opportunity to prepare for military service, should their country need them, by their membership in an organization that does not interfere in peacetime with normal civilian life.

Women veterans of all Services are eligible for membership in the Marine Corps Reserve, provided they were not discharged for medical reasons and possess an honorable discharge. Women who have served in any of the Armed Forces other than the Marine Corps may be appointed to their former rank up through corporal, if found qualified.

There are two classes for membership for women in the Reserve—Organized and Volunteer. The woman reservist may request assignment in the Organized Reserve, in which Women Marine reservists are authorized to participate in drill pay status. Non–prior-service women are not recruited directly into Organized Reserve units. A woman reservist may request assignment to two weeks' active duty for training each year. Women will be permitted to attend annual field training with the unit, provided appropriate facilities are available at the training activity. Where the training activity does not have appropriate facilities for women, the woman so concerned may request orders for two weeks' on-the-job training at an appropriate training activity. She may further her military career through correspondence courses, and may join Volunteer Training Units located in her home community.

Women reservists, Volunteer or Organized, are subject to the same regulations as male reservists. In the event of war, Organized reservists are mobilized to meet the needs of the Marine Corps. During Korea and Vietnam, only those Volunteer women reservists were mobilized who specifically requested active duty and whose qualifications were in demand.

> NOTE: Two important billets in the Reserve program in which Women Marines may be assigned are on I & I staffs (see Section 908) and, after retirement, if they so volunteer, as instructors in the Marine Junior ROTC program (where about a fifth of the cadets are girls).

1006 ■ References

In keeping with the premise that women are an integral part of the Corps, most of the administrative instructions and information pertaining to women are embodied in corresponding or analogous material applicable to Marines of either sex. For example, much information regarding such fundamental subjects as entrance into the Corps, promotion and reduction, retirement, discharge, and civil readjustment, may be found in the *Marine Corps Manual*. For obvious reasons, however, *Marine Corps Uniform Regulations* contain much information expressly applicable to women officers.

On the social side, ground rules and fine points for Service ladies may be found in *Service Etiquette*, by Oretha D. Swartz (U. S. Naval Institute), in *Welcome Aboard*, by Jean Ebbert (U. S. Naval Institute), and *The Marine Corps Wife*, by Sally Jerome and Nancy Shea (Harpers). In reading these works, however, Women Marine officers must remember that, in most social situations, their behavior and responsibilities are first and foremost those of Marine officers, and not merely those of ladies associated with the Service.

The aim of every woman is to be truly integrated in the Corps. She is able and willing to undertake any assignment consonant with Marine Corps needs, and is proudest of all that she has no nickname. She is a "Marine."
—*Colonel K. A. Towle, USMC*

11

You Become a Marine Officer

The *Marine Corps Gazette* recounts that a young man from the hinterland, when shipping into the Corps, was asked if he intended to try for a commission.

"I don't think so," the recruit answered, "I'm not a very good shot. I'd better work on a straight salary."

Whether it is the commission that attracts you or just the salary, here are the ways in which you may become a U. S. Marine officer and what happens to you in the process.

Since the Marine Corps is not tied to any academy for its main source of officers, there are many ways in which you can obtain a commission. The variety of approaches to officer status in the Corps ensures that the base of experience, background, and education among Marine officers remains broad. It also means that, no matter what your origin, once you qualify for a Marine commission, you stand on equal footing with every other officer candidate, regardless of source or education.

1101 ■ General Requirements to Become an Officer

To be eligible for a commission in the Marine Corps, you must be a U. S. citizen, morally, mentally, and physically qualified, and your application must be approved by Marine Corps Headquarters. If you are already a veteran, you must of course have an honorable discharge, and if you are a member of the reserve component of any other Service, you must obtain a conditional release from that organization. Under certain limited conditions, regular officer transfers are authorized from the other Services into the Marine Corps; for such transactions special regulations and procedures apply which are beyond the scope of this chapter.

1102 ■ Roads to Your Commission

To obtain a commission in the Marine Corps, you may follow any one of several roads summarized in Table 11–1 and subsequently described.

Table 11-1
Avenues to a Career as a Marine Officer

Program or Source of Input	Age Limit	Education Requirements	Open To	Leads To
U. S. Naval Academy		Graduation from USNA	Midshipmen USNA[1]	2dLt USMC
U. S. Military or Air Force Academies[5]		Graduation from USMA or USAFA	Cadets USMA or USAFA[1]	2dLt USMC
Naval ROTC (Marine Option NROTC Scholarship Program)	Be 17 but not 21 years of age by 30 June of the year entering college	B.A. or B.S. degree[2]	Freshmen already in attendance, or from prospective students who have been accepted for admission as NROTC midshipmen	2dLt USMCR[3]
Naval ROTC (College Program)	Be 17 but not 21 years of age by 30 June of the year entering college	B.A. or B.S. degree[2]	Freshmen already in attendance, or from prospective students who have been accepted for admission as NROTC midshipmen	2dLt USMC
Platoon Leaders Class (Ground or Aviation)	Be at least 17 years old and less than 28 (27½ for aviation) at time of appointment to commissioned grade	B.A. or B.S. degree[2]	College freshmen, sophomores, juniors[4]	2dLt USMCR[3]
Officer Candidate Class (Grd, SNA, & SNFO)	Be at least 20 but less than 28 (27½ for aviation) at time of commissioning	B.A. or B.S. degree	Regularly enrolled senior in good standing, or a graduate, of an accredited institution	2dLt USMCR[3]

252

		granting a 4-year baccalaureate degree in a field other than medicine, dentistry, veterinary, pharmacy, chiropody, hospital administration, optometry, osteopathy, or theology[4]	
Navy Enlisted Scientific Education Program (NESEP)	Be at least 20 but less than 26 years old by 1 July of the year entering college	B.S. degree[2]	2dLt USMC
Marine Corps Enlisted Commissioning Education Program (MECEP)		USMC enlisted, male and female, with GCT 120 or higher and 6 years obligated service	
Enlisted Commissioning Program	Be at least 19½ and less than 27½ on date of application	High school graduate (or GED certificate issued by State Department of Education) and have satisfactorily completed not less than 1 year of unduplicated college work at an accredited institution. Applicants with less than a baccalaureate degree must also attain a 30th percentile standing on each of the 5 tests of the College Level Entrance Program (CLEP) General Examination	Private and above, male and female, USMC and OMCR with GCT of 120 or higher and at least 12 months remaining on current enlistment and who have completed recruit training	2dLt USMC

253

Table 11-1

Program or Source of Input	Age Limit	Education Requirements	Open To	Leads To
Limited Duty Officer	WO applicants who have a minimum of 10 and a maximum of 20 years active service and who have not reached their 46th birthday by 1 January of the fiscal year in which the appointment is to be made	GCT 110 or higher	Permanent male warrant officers in grades W-2 through W-4	1stLt USMC
Warrant Officer	Must be of an age to allow 30 years total active service by age 62	GCT 110 or higher	Sergeant or above with 5-12 years active service. COs may recommend waivers for preeminently qualified NCOs with up to 14 years of service	Warrant Officer (W-1) USMC
Woman Officer Candidate Course	Be at least 20 but less than 28 years old at time of commissioning	B.A. or B.S. degree[2]	College juniors, seniors, and graduates	2dLt USMCR[3]

NOTES:
1. Preference given to children of Marines or to former Marines.
2. Successful college graduation and completion of program is a prerequisite for commissioning.
3. Initial commission in USMCR with opportunity to augment (integrate) if qualified and selected.
4. Aviation candidates must meet flight physical standards and will be sent to flight training on completion of required ground training. Current regulations require all SNAs and SNFOs to attend Basic School prior to reporting for flight training.
5. Military and Air Force Academy graduates may by law be commissioned in the Marine Corps, but the Departments of Army and Air Force will only grant approval in exceptional cases.

(To be eligible for any program, you must be able to meet the general requirements stated in Section 1101.)

254

U. S. Naval Academy. Each graduating class from the Naval Academy at Annapolis includes a quota of midshipmen who have been selected for Marine Corps commissions. Since competition for these appointments is keen, preference is given to midshipmen who have held enlisted rank in the Marine Corps, and to children of Marines. Entrance into the Naval Academy, first step toward a commission via this route, is in turn open to civilian preparatory school and high school graduates and to qualified enlisted men from the Marine Corps and Marine Corps Reserve. Information regarding appointment to the Naval Academy may be obtained by writing to the Superintendent, U. S. Naval Academy, Annapolis, Maryland 21402. If you are already in the regular Marine Corps or Reserve, consult your commanding officer. At present, up to $16\tfrac{2}{3}$ percent of each graduating class of midshipmen may be commissioned in the Marine Corps. All USNA graduates have a five-year service obligation.

Color Parade and Color Girl highlight Annapolis graduation ("June Week"). More than 16 per cent of these midshipmen are fortunate enough to have been selected for commissions in the Marine Corps.

U. S. Military and Air Force Academies. Limited numbers of graduates of West Point and of the Air Force Academy are also eligible for regular commissions in the Marine Corps, preference going, as at Annapolis, to former Marines or children of Marines. Information regarding appointment to these academies may be obtained from the Departments of the Army and the Air Force, respectively.

From Civil Life. If you are attending an accredited college or are a college graduate, you may enter the Marine Corps as an officer, from civil life, under the following programs:

Naval ROTC. Any college man enrolled either as a scholarship midshipman in the Naval Reserve Officers' Training Corps (NROTC), or

Naval ROTC midshipmen taking training afloat. The NROTC is another route to a Marine commission.

as a "College Program Student, NROTC," can, if selected for the Marine Corps, obtain a regular or reserve commission, and NROTC is today the largest single source of regular officers in the Corps. The NROTC scholarship midshipman goes to college, when selected and approved, with his tuition, fees, and books paid for by the Navy. He also receives a retainer of $100 a month. He spends summer "cruises" afloat with the Fleet and ashore at Quantico (after selection for the Marine Corps). On graduation from college, with a B.A. or B.S. degree, he is commissioned and enters Basic School.

The NROTC "College Program Student" receives no Government financial support other than $100 a month retainer pay during his last two years in college, but takes the same Naval Science instruction in college as the scholarship midshipman. If selected for a Marine commission while in college, the contract student spends one summer on a "cruise" at Quantico. When he graduates, he is ordinarily commissioned in the Marine Corps Reserve, but may, if qualified, transfer into the regular Corps. If you are in NROTC and want to become a Marine officer, see the Marine officer attached to your Naval ROTC unit.

Platoon Leaders Class (PLC). The Platoon Leaders Class is a summer officer-candidate program designed to train college men either as ground officers or as prospective pilots or naval flight officers (NFO) in Marine aviation. PLC training is limited to two summer periods at Quantico, of six weeks each. In the case of college juniors, all training is completed in one 10-week summer period. At the completion of that training, and upon graduation, you are eligible for commission as a second lieutenant in the Marine Corps Reserve. Some graduates are appointed as regular officers. No uniforms, training, or other work is required of you during the academic year.

To enter PLC, you must be enrolled in an accredited college as a freshman, sophomore, or junior. You must be at least 17 years old on enrollment, and less than 28 on 1 July of the year in which you expect to receive your degree.

Training consists of two six-week tours (Junior and Senior Course, respectively). During each period, you receive the pay—but not the allowances—of a sergeant.

Training for both ground and aviation candidates is intensive, with initial emphasis on the basic instruction and careful screening required for all Marine officers. Aviation candidates, however, take flight examinations as part of their program, and, if found qualified on graduation, are ultimately sent to flight training on graduation from Basic School.

In addition to pay and first-class transportation to and from Quantico, you get living quarters, uniforms, and medical and dental care. And, during off-hours, you have full privileges at the library, post exchange, theater, swimming pool, and athletic field, and of course weekend liberty.

> NOTE: If you are headed for law school, you should investigate the special PLC (Law) program, under which individuals who have successfully completed PLC are allowed to remain in law school in inactive status as second lieutenants in the Reserve until they obtain their law degree and are then brought to complete their required active duty as Marine lawyers. PLCs may also be deferred from active duty to obtain a Master's degree in most recognized major fields.

Officer Candidate Course (OCC). The Officer Candidate Course is conducted for *college graduates* who are over 20 years of age and less than 28 on 1 July of the year in which commissioned.

The course provides the practical military training needed to qualify for the specialized training to be received as a second lieutenant and, in the case of aviation officer candidates, for flight training and, after Basic School, ultimate designation as a Marine pilot. It consists of ten weeks' intensive training at Quantico, Virginia. Upon successful completion of this course, you are commissioned a second lieutenant in the Marine Corps Reserve, and may be able to integrate as a regular.

In addition to pre-officer training, OCC and PLC candidates, after being commissioned, attend Basic School in Quantico before being assigned to a unit, while aviation officer candidates, if found qualified, go on from Basic School to 15 to 18 months' pre-flight and flight training in commissioned officer status.

Qualified enlisted men may also enter either the Officer Candidate or Aviation Officer Candidate courses.

From the Ranks. The Marine Corps pioneered the award of officer commissions to meritorious enlisted men long before the practice was accepted among the other three Services. In the Marine Corps the door remains open through these programs:

Navy Enlisted Scientific Program. Under this program qualified enlisted Marines with a four-year college education specializing in mathematics or science. Young enlisted men selected for this program are assigned to a special preparatory course and then to college, during which they remain on active duty. On successful completion of college, preceded by summer officer candidate training, they receive commissions as second lieutenants in the regular Marine Corps.

Meritorious Enlisted Marines. Marines meeting specified age and educational qualifications may be commissioned in the Marine Corps Reserve and thence become eligible for regular commission.

Limited Duty Officer (LDO). Warrant officers of the Marine Corps may apply for LDO commission in specialized fields such as administration, intelligence, infantry, logistics, artillery, engineers, tanks, amphibian tractors, ordnance, communications, supply, food, motor transport, and aviation. (See Section 1207.)

Warrant Officers. Long-service NCOs may obtain appointment as *warrant officers (WOs)* in specialized fields. Since qualifications for LDO and WO vary appreciably from time to time, no attempt is made to summarize the requirements for such appointments.

Temporary Officer. In addition to the established programs shown in Table 11–1 and described so far, authority exists in law to issue temporary commissions as second lieutenant and above to selected warrant officers and enlisted men in order to meet pressing or particular needs. Individuals so commissioned as *temporary officers* retain their permanent grades and status and revert thereto when the requirement for their services in the advanced rank ceases. Temporary officers have been commissioned during all twentieth century wars, including Vietnam. As in the case of LDOs and WOs, programs and requirements are variable.

Former Regular Officers. Former regular officers of the Marine Corps who have not attained their thirtieth birthday at time of appointment and who resigned from the Corps in good standing may be reappointed with the approval of the Secretary of the Navy. Former officers of the other Armed Services may, within certain limits, be appointed by transfer, in the Marine Corps Reserve, but under no circumstances in a grade higher than that held in the former Service.

Woman Officers Candidate Course (WOCC). Unmarried women college juniors, seniors, or graduates who, on 1 July of the year in which commissioned were over 18 and less than 30 years of age may attend WOCC. There they may obtain commissions in all fields except those involving actual combat. Details of this training are covered in Section 1004.

The foregoing information, so far as it goes, is current and correct (1977), although the various programs change somewhat from time to time. Should you desire a Marine Corps commission, but find nothing here which fits your situation, do not hesitate, if you are a civilian or inactive reservist, to write to the Officer Procurement Section, Marine Corps Headquarters, Washington, D.C. 20380. If you are a regular, or an organized reservist, ask your commanding officer.

1103 ■ You Become an Officer

No matter how you earn appointment as a Marine officer, the day finally arrives when you are to be sworn in as a second lieutenant.

Your commission and orders will be forwarded to your commanding officer (if you are already in the Service in some capacity), or to a Marine activity near your home, for presentation. The swearing-in ceremony and required administrative steps will ordinarily be taken care of by the presenting officer. Bear in mind, however, that pay and allowances do not begin for Regular officers until sworn in (or until commencement of active duty, in the case of reserve officers), nor can you assume title and status as an officer until you have taken your oath and formally accepted your appointment. *You should therefore be sworn in at the earliest opportunity.*

Of that oath, one of the Navy's greatest and best-loved fighting admirals, Arleigh A. Burke, once wrote:

> It is a responsibility that should not be taken easily. And its phraseology is disarmingly simple. When an officer swears "to support and defend the Constitution of the United States against all enemies, foreign and domestic"—he is assuming the most formidable obligation he will ever encounter in his life. Thousands upon thousands of men have died to preserve for him the opportunity to take such an oath. What he is actually doing is pledging his means, his talent, his very life, to his country. This is an obligation that falls to very few men.

The U. S. Supreme Court has more succinctly ruled:

> The taking of the oath of allegiance is the pivotal fact which changes the status from that of civilian to that of soldier.

As you raise your right hand and stand at attention to take your oath, you are at a turning point in your life. In a matter of seconds you will become an officer in a Corps whose valor, renown, and honor are second to none. From the moment you complete your oath *"to support and defend the Constitution of the United States of America against all enemies, foreign and domestic . . ."* you are a lieutenant of Marines responsible to the President and your superior officers, and fully amenable to military justice. It is a great moment.

1104 ■ Your Commission

After you are sworn in, you receive your commission.

A *commission* is the formal written authority, issued in the name of the President of the United States, which confers on you your rank and authority as a Marine officer. It is signed for the President, by the Secretary of the Navy, and is issued under the seal of the Department of the Navy and countersigned by an officer in the Manpower Department, Marine Corps Headquarters. Your commission states your rank and the date from which it is effective (your date of rank, so-called), and enjoins all officers, seamen, and Marines to obey any lawful order you may give. You receive a new commission for each permanent rank to which you are promoted. Since your commission is the document that attests your status and confers your authority, you should take great pains to safeguard it.

1105 ■ "Special Trust and Confidence..."

Before you file away, or even frame (as many do) your first commission, reread its opening phrase:

Know ye, that reposing special trust and confidence in....

In the foregoing words, the President of the United States certifies, via the Secretary of the Navy, that you, as a commissioned officer, have been specially set apart from your fellow citizens as one in whom "special trust and confidence" are reposed. On the basis of this special trust, you as an officer are granted special privileges; on the same basis you are subject to special responsibilities and obligations. In the words of the old French motto, *Noblesse oblige.*

In specific implementation of the foregoing, the *Marine Corps Manual* states: *The special trust and confidence which is expressly reposed in each officer by his commission is the distinguishing privilege of the officer corps.* As a commissioned officer, you should be vigilant to discharge and where necessary enforce performance of all responsibilities, and thereby guard the privileges.

1106 ■ Basic School

The orders that accompany your appointment and initial commission will direct you to proceed to Quantico, Virginia, and report as a student at Basic School. The mission of Basic School is:

> To educate newly commissioned officers in the high standards of professional knowledge, esprit de corps, and leadership traditional in the Marine Corps in order to prepare them for the duties of a company-grade officer in the Fleet Marine Force, with particular emphasis on the duties and responsibilities of a rifle platoon leader.

Ordinarily, your orders specify a date by which you must report. *It is vital that you comply carefully with your orders, and, above all, that you report on time.* There is no poorer way to start a Marine career than by being late for Basic School.

In reporting, be guided by Sections 1401–1402 of this *Guide*. These give the procedure for joining a new station. You may wear civilian clothes when first reporting to Basic School.

Hints on Reporting at Basic School. As just stated, you may report in civilian clothes, and are specifically *not* required to possess any articles of uniform upon reporting. You should, however, take care that your hair is closely trimmed and that you present a neat appearance.

You should have enough money available for living and other expenses until your first payday, which will be about three weeks after you join. Joining-instructions that accompany your orders will suggest minimum amounts of cash single or married new lieutenants should have.

Travel light and bring little baggage, as stowage space is limited. Bring two combination locks to secure lockers assigned.

Outside working hours, like any other officer, you may wear civilian clothing. This, however, must conform to accepted standards within the officer corps: clothes of eccentric fashion, design, or color are not tolerated. As a minimum you should have one suit, of conservative cut and color, as well as sports attire.

Under no circumstances should you buy uniforms or contract to buy any uniforms before you report to Basic School. Year after year, students report with uniforms and accessories bought in good faith, only to find them ill-fitted, nonregulation, and far too costly. Beware of high-pressure uniform salesmen and so-called package deals for uniforms. One of the first items of business after you join will be a uniform-orientation session conducted by the School. Only after that should you begin acquiring your uniforms.

Basic School is the oldest Marine Corps school, and is an institution whose importance to the Corps is equaled only by that of the two recruit depots. Basic School traces its history to 1 May 1891, when, as the "School of Application," it was founded at Marine Barracks, Eighth and Eye Streets, Washington, D.C., by Colonel Charles Heywood, 9th Commandant of the Corps. At various times the school has been located at Annapolis; at Port Royal, South Carolina (later to become famous as Parris Island); at Norfolk; at Philadelphia Navy Yard; and, finally, at Quantico.

Today, the Basic School is located approximately 12 miles from the main post of Quantico. The Headquarters of the Basic School is located at Camp Barrett, one of the outlying camps of the Guadalcanal Area, which comprises the greater part of the 57,000 acres of the Marine Corps Schools reservation.

No matter how you earned your commission, your first assignment as a second lieutenant will be as a student in the Officer Basic Course, a course of approximately 21 weeks' duration. In addition to the *Officer Basic Course* for newly commissioned second lieutenants, the Basic School conducts a *Marine Warrant Officer Basic Course* for newly appointed warrant officers.

Quarters for the officer students at the Basic School are found in O'Bannon Hall, which provides adjoining double rooms sharing a head and shower. This modern BOQ has space for approximately 850 resident officers and, in addition, houses the Basic School dining hall and bar, a snack bar, television lounge, library and reading room, and reception room.

Heywood Hall is the main administration building of the Basic School, wherein are located the offices of the Commanding Officer and his staff, the offices of the various instructional sections and their instructors, as well as the several student company offices. Adjoining Heywood Hall are four modern, air-conditioned classrooms with a total seating capacity of 1150 students each. These classrooms have all the necessary facilities for teaching, using various types of projectors for training films, slides, and transparencies. Conveniently located between these classrooms are two barber shops, where officer students, can, with a minimum of effort, keep their hair properly clipped at all times.

Additional facilities at Camp Barrett include a small post exchange and snack bar, post office, gymnasium, outdoor theater, chapel, armory, additional classroom facilities, gas chamber, combat conditioning facilities, "dry net" mock-up, dispensary, PX gas station, and a lighted playing-field for baseball, softball, or football.

When you report to the Basic School, you will turn in your orders to the Personnel Officer, who will then assign you to a student company. The average class is made up of two companies of about 200 students each, including young officers from various Allied countries around the world. These companies are commanded by a major with a captain as executive officer and captains and lieutenants as platoon leaders. Upon reporting to your company commander, you will receive an orientation on the course: what will be expected of you and what you can expect from the school. You will be assigned to a platoon and given a room in the BOQ. You will then be issued your field equipment, individual weapons, and textbooks for your forthcoming courses. You will do theoretical work in the classroom and then go to the field and work it out practically. About 58 percent of your training is in the field and about one-fifth of this takes place at night.

(Immediately after Basic School, every officer goes to "follow-on training" in some school or course (e.g., flight training) to qualify him in his new military specialty. Officers with an infantry (O3) MOS remain at Quantico for additional advanced infantry training. Thus, in all, your professional apprenticeship, including Basic School, lasts a minimum of six months.)

Quarters at the Basic School are not luxurious, although you will find them clean and quite comfortable. If married, you may live off the station after working hours. Basic School does not provide married quarters (nor, for that matter, does it encourage married students), although you will be given a locker in which you can secure your field equipment, rifle, and extra clothing.

The Basic School curriculum, including intense and rugged field-work, provides graduates with a foundation of leadership and professional knowledge upon which they may build their careers. All of the 24 subjects included in the Basic Course are grouped under the following major categories: Tactics, Weapons, General Military Subjects, and Physical Conditioning.

Within the foregoing academic framework, the objectives of Basic School are twofold: first, to instill the Marine Corps attitude; and, second, to teach new lieutenants the basic professional techniques which a Marine officer must know. In other words, it is up to Basic School to make a Marine officer of you. The extent to which the school succeeds, however, is largely up to you. Officer students who do not measure up to the standards set for Marine officers are dropped from the School, their commissions are revoked, and they are returned to civilian life. Remember that, at Basic School, you are under continual experienced observation. Remember, also, your first fitness report from Basic School represents a composite judgment of officers whose evaluation will receive heavy weight. Basic School is no place to "take your finger off your number."

1107 ■ Officer Candidates' School

In obtaining your commission via the Platoon Leaders Class, NROTC, NESEP, or through any of the inputs which lead you to Officer Candidate Class, you will receive pre–Basic School instruction while assigned to Officer Candidates' School. Conducted at Camp Upshur and old Brown Field, Quantico's original air station and later the immediate post–World War II site of Basic School, pre–basic training is primarily concerned with imparting to officer candidates the knowledge required of the basic enlisted Marine, while at the same time rigorously screening all candidates to be sure they are officer material. In effect, Officer Candidates' School is an officer candidates' boot camp, and a very exacting one.

HINTS FOR NEW OFFICERS

1108 ■ The Profession of Arms

On becoming a Marine officer you have entered upon one of the oldest and most honorable of the professions, the profession of arms. The profession of arms was a recognized element in ancient society, as it is in modern society.

The principles underlying all laws, rules, and conventions imposed by society to regulate the conduct of war and those who wage it are *military necessity*; *humanity*; and *chivalry*.

Once these principles are negated by a nation, that nation places in jeopardy its status in modern society and hazards any claim to the rightness of its cause.

A nation at war or planning against war rightfully expects the professional soldier (which means *you*) to protect it from its own excesses, as well as from the enemy.

Once the foregoing principles and responsibilities are forgotten by members of the Services, the profession of arms can no longer be regarded as a decent and honorable calling.

1109 ■ Do's, Don't's, and Pointers

As a Marine officer, you now represent the Corps. Conduct yourself with dignity, courtesy, and self-restraint.

Avoid any show of self-importance. Do not bluster, especially toward civilians or enlisted men.

Be wary of situations beyond your depth. A new lieutenant is not expected to be all-wise. He is expected to keep his head, and to possess enough common sense and knowledge of his own limitations to prevent him from overextending himself.

Something else will be expected of you: *not to make the same mistake twice*, particularly after having been told about it by a senior.

On joining a new organization, you will be closely looked over by all hands, officer and enlisted. The first impressions can make (or break) you. Be natural and courteous, prompt and punctilious, "squared-away" in uniform and deportment. At all costs avoid the impression of a brash young know-it-all.

From the moment you become a Marine, you should cultivate the habit of punctuality. Along with discipline, dedication, obedience, and loyalty, it should be a matter of pride never to be late. *Always be five minutes early* for any formation or official commitment.

Avoid the habit of complaining or whining, and avoid those who do. Refrain from criticizing unless you are ready and able to provide a better solution. By the same token, cultivate the habit of optimism. An optimist is like a breath of fresh air. He cheers all with whom he comes in contact. One of the great sayings of the greatest of all naval officers, Lord Nelson, was, "I am not come forth to find difficulties, but to remove them."

Be industrious and persevering, attentive to duty, and attentive to essential detail. Whether ashore or afloat, in garrison or in the field, the best officers are those who possess powers of observation, and, having those powers, know how to use them. Akin to observation is the power and habit of forethought.

Learn to control and to hide your feelings. In addition to being alert, always try to *look alert*.

Whatever you do, do thoroughly, and do it with enthusiasm and imagination. Do not confine yourself to doing only what you are told to do. Do more than you are told to do. And bear in mind that it is the smart, quick, and, if possible, *cheery* voice that gets the job done and makes the men hop.

If you are asked a question and are unfamiliar with the answer, *don't bluff*. And don't reply, "I don't know." The proper answer from a young officer in such circumstances is, *"I'll find out."*

Do not procrastinate. When you have a job to do, do it at once. If you have several items to be accomplished, *do the important thing first*. If you find yourself stymied, don't shove the matter aside or report back that you can't do it—try some other way, and keep on trying. Remember that in the Service it is *results* that count, and that if you can acquire the reputation of a "can-do," results-getting officer, you are on your way to success.

Always give thought to the Service reputation that you build and acquire day by day. An officer's reputation for character and efficiency is his invested capital. Take this away and his usefulness is gone. And remember: you cannot fool your contemporaries. Living closely together, officers soon learn the ins and outs of each other's lives and character.

Personal appearance is most important in the Service, and, although most young officers must and should economize wherever possible, purchasing inferior uniforms is a false economy of the worst kind. The only way to economize on uniforms and equipment is to get the best, and then take care of it. Economize on your bar bill rather than your tailor's. Nobody in the world looks more shabby than a shabby officer.

Keep fit. Avoid fat. The Marine Corps will help you with this by periodic physical fitness tests and by vigorous training all year, but fitness is a continuous matter which must be a continuous concern to every officer. No Marine can afford to allow himself to become fat.

Stand straight. Keep your hands out of your pockets. Never chew gum or smoke in public. Do not carry packages when in uniform. Never appear unshaven after 0800.

Be impeccable in dress and grooming. Basic School will teach you how to wear and care for your uniforms. Take similar pains with civilian clothing. Avoid cheap, flashy clothes; select conservative ties and shirts; buy and wear the Corps necktie; wear a hat.

As you acquire clothing, mark each article as laid down in *Marine Corps Uniform Regulations*. Stencil baggage and personal kit with your name, rank, and service number. *Inside* each piece of baggage, stencil or affix the same information in a permanent manner. You can buy clothing-marking sets at the post exchange.

Unscrupulous tailors sometimes impose on new officers by selling them articles that either are unnecessary or depart from approved standards. *Before you buy any uniforms, wait for advice and supervision by the staff at Basic School.* They will ensure that you spend your uniform allowance to advantage on necessary approved items.

In your relations with your fellow officers, avoid joining factions or, if there is any bad feeling between others, avoid taking sides. Don't gossip; gossip always finds its way back. Only say of a brother officer who is absent what you would say to his face.

Never, under any circumstances, speak ill of the Corps, or of your own organization, in the presence of civilians or members of the other Services. Before you voice any criticism, however merited or carefully thought-out, be sure it cannot be construed by outsiders so as to derogate the Corps. By the same token, avoid criticizing other units or Services, at least in public.

Conduct all your business through proper channels. "Channels" is a highly important word in the Service. The phrase, "Go through channels," which you will hear repeatedly, simply means, "Don't go over people's heads." In giving instructions or in doing or getting things, be careful not to go over someone's head or infringe on his areas of responsibility. This is a sure way to trouble in the Service.

Now that you are a Marine yourself, keep your eyes open for likely recruits and for potential officers among your friends. Such individual recruiting of new Marines by convinced and loyal old Marines is one of the principal ways in which the Corps maintains its quality.

Obtain suitable visiting cards, as described in Section 2409.

When the telephone rings, answer up smartly, in Marine Corps fashion, with your name and rank: "Lieutenant Burrows"—not "Hello." When you make a call, identify yourself immediately: "This is Lieutenant Wharton, Marine Corps." Be sure to add that "Marine Corps." It prevents mix-ups with the other Services.

Since you are now a member of the most professional of the Services, you should join the Marine Corps Association and subscribe to the professional journals listed in Section 1512. If you are a regular, you should by all means immediately join the Army and Navy Club in

Washington, while you can still do so (as a newly commissioned officer) without payment of initiation fees. See Section 2417 for details.

Know where to find information. Set out to go through all the basic professional publications, page by page—read *Navy Regulations, The Marine Corps Manual, Uniform Regulations, Drill and Ceremonies Manual,* and of course all the basic Field Manuals relating to Marine weapons and basic tactical principles. Basic School will direct your attention to the most important provisions in all the foregoing, but, by going through these publications on your own, you will come to know where to find information on many questions which lazy or inattentive young officers will say are not covered in the manuals.

Get into the habit of being systematic and methodical. In Lord Chesterfield's words, "Dispatch is the soul of business, and nothing contributes more to dispatch than method. Fix one certain day and hour in the week for your accounts, keep them together in their proper order, and you can never be much cheated." By so doing, you will be able to accomplish two or three times as much as an equally capable but unsystematic officer.

As an officer embarking on your new career, you should do everything possible to make your living arrangements becoming to your new station. Your pay is given you for this purpose, and you owe it to the Service to dress and live, however simply, like an officer and a gentleman.

If you don't have one already, open a checking account with a substantial bank, preferably one accustomed to handling officers' accounts on a worldwide basis. Begin systematic savings with your first pay check. Take out life insurance immediately (see Chapter 23), and begin monthly purchase of U. S. Savings Bonds by allotment.

Avoid getting into debt, and *live within your means.* If you are in debt, say, for uniforms or for a necessary vehicle, get out of it as soon as possible. In any case, see that every creditor gets at least a token remittance every month; this will go far to establish your credit and will demonstrate your sincere desire to pay your bills.

Be very slow to marry. Until World War II, officers of the Marine Corps and Navy were prohibited from marrying during their first two years' service. This wise rule protected young officers from unwise or hasty marriages and at the same time guaranteed their full attention and best efforts to their profession in the most formative period. For this very reason in particular, *do not marry, at least until you have completed Basic School.*

> NOTE: When you do marry, you are required to report this event to the Commandant of the Marine Corps. Your adjutant or first sergeant will advise you on the form of this report.

Be extremely circumspect in any kind of financial transactions with brother officers. "Neither a borrower nor a lender be" is golden advice; be not a co-signer either—many an officer has discovered that his co-signature on a "friend's" note has cost him both the friendship and the amount of the loan, too. **Other than in line of duty, you are prohibited by Navy Regulations from any pecuniary dealings with enlisted men,** and this prohibition should be strictly observed.

Do not intrude among enlisted men. They are entitled to privacy among themselves as you are. Do not enter noncommissioned officers' messes except by specific invitation of the senior NCO present. If you have been commissioned from the ranks, remember that now you are an officer. You cannot turn back the clock.

An officer is much more respected than any other man who has as little money.
—Samuel Johnson

It's not hard to be an officer, but it's damn hard to be a good officer.
—GySgt Daniel Daly

12

Officers' Individual Administration

Although a few administrators may convey the reverse impression, there is nothing inherently complicated or darkly mysterious about individual administration. This term merely comprises a number of administrative matters that concern you personally: your record, rank, promotions, retirement, official correspondence, leave, and liberty. Pay and allowances, and official travel, closely related, are covered in Chapter 13.

To give you working familiarity and ready reference in "quill driving," and to prevent its seeming a black art known only to a chosen few, read and study *Navy Regulations*, *The Marine Corps Manual*, and pertinent Marine Corps orders—and let individual administration serve and help, but never get the better of you.

OFFICERS' RECORDS

1201 ■ Your Official Record

Throughout your career, correspondence that concerns you accumulates at Marine Corps Headquarters. Adverse matter may not be placed on record without your knowledge, and must always be referred to you for statement. If you wish to give your side of the matter, you may do so; if you have nothing to say, you so state in writing. In either case, the adverse matter goes back to the Commandant of the Marine Corps, via the reporting officer and thence through normal channels. Whether favorable or unfavorable, correspondence once rightfully included in your record cannot be removed without authorization by the Secretary of the Navy.

The vital importance to you and your career of your official record cannot be overstressed. The entries in your record form the basis of your Service reputation. Without vicious intent, many a young officer, by carelessness or misdirected high spirits, has written into his record an accumulation of minor lapses that prove as detrimental to his Service reputation as a serious misstep.

Your record in Washington comprises a *correspondence jacket*, a *selection board jacket*, and certain additional files, some of which may be kept by the Judge Advocate General of the Navy.

The Correspondence Jacket has two sections:

1. *Orders*: Copies of orders issued by Marine Corps Headquarters, and all modifications thereto, regardless of origin; orders to or from active duty, retirement, etc.;

2. *Miscellaneous correspondence and forms*: Matter pertinent to your military history, such as: correspondence relative to original commissioning; medical data; pay and insurance; detailing correspondence; final security clearance; Record of Emergency Data.

The Selection Board Jacket contains only matter that bears directly on your fitness as an officer; it has four sections:

1. *Certificates and miscellaneous matter*: Diplomas and other evidence of completion of instruction while in the Marine Corps; requests for changes of orders; official photograph; court-martial order (when acquitted); and any other matter neither clearly favorable nor unfavorable but of interest to a selection board in appraising you.

2. *Favorable matter*: Anything that reflects favorably on you, such as citations and awards; recommendations for awards; letters of commendation; favorable remarks in forwarding endorsements; letters of appreciation from civilians, etc.

3. *Unfavorable matter*: Anything that reflects unfavorably on your moral, mental, or professional qualifications, such as court-martial orders (when convicted); letters of censure and related correspondence; correspondence concerning marital difficulties reflecting unfavorably on your moral attitude or conduct; indifference to indebtedness; and statements by you concerning any of the above unfavorable matters.

4. *Fitness reports*: Fitness reports; statements by you, your reporting senior, or the reviewing officer regarding a particular report; letters of commendation or censure attached to fitness reports when such letters have not previously been filed in your record.

The *Confidential File* contains any correspondence to or about an officer that must be kept in a confidential status. Most officers have no material on confidential file.

Proceedings of Courts or Boards which affect your record are filed by the Judge Advocate General (JAG) of the Navy. You may examine these records in the JAG files. If unfavorable, such matter is referred to you before being filed.

Professional examinations for promotion are also filed in the JAG Office.

Access to Record. The only persons who enjoy access to your record are you yourself (on personal application to the Officer Files Unit); your personal representative, armed with proper proof; designated Headquarters personnel; and, when authorized by the Secretary of the Navy, the courts. No person (including you or your agents) without proper authority may withdraw official records and correspondence from the files, or destroy them.

Make it a habit to review your record every time you find yourself in Washington.

1202 ∎ Your Staff Returns

Your staff returns are the administrative records that accompany you, as distinct from those filed in Washington. Staff returns comprise your *Officer's Qualification Record*, your *Personal Finance Record*, and your *Health Record*.

Your *Qualification Record* is a folder that accompanies you throughout the Corps and contains a summary of your military history; your military and civilian background, education, and skills; emergency data; and similar personal information. The qualification record enables your commanding officer to determine what you can do and what you have done in the past. The qualification record corresponds to an enlisted Marine's service record book and is used for the same purposes. The correctness of information and entries in your OQR is a responsibility which you share with your commanding officer; at least once a year you should audit the entire qualification record to be sure it is up-to-date and complete.

Your *Personal Finance Record* is a folder file of your periodic "Leave and Earnings Statements," which will show you and a paymaster how much pay you are entitled to any moment. Most Armed

Forces disbursing officers will honor a PFR, should you run short on the road between stations. It should be guarded accordingly. (See Chapter 13, "Pay, Allowances, and Official Travel.")

Your *Health Record* is opened by the Navy's Bureau of Medicine and Surgery, and very literally contains your life history, your physical description, and the medical history of every ailment that befalls you. One of the health record's most important everyday functions is to keep a record of all the "shots" and inoculations which every Marine must take.

1203 ■ Your Personal File

The day you are commissioned, start a personal file. This should contain, in one folder, all original travel orders; in another, all official correspondence from, to, and concerning you. In addition, remember that some unofficial letters you write or receive are just as important to your career as the official ones.

> NOTE: The History and Museums Division, Marine Corps Headquarters, encourages retired officers to donate personal files and papers to the Marine Corps archives.

1204 ■ Fitness Reports

Fitness reports are the periodic efficiency reports rendered on all officers in the Marine Corps and Navy. Broadly speaking, your fitness reports present a composite judgment of your military character and relative merits compared with other officers of the same rank and comparable experience. Fitness reports are the principal record of an officer's performance and conduct, and assist selection boards to determine which officers are best fitted for promotion, and provide the Commandant and staff with information as to each officer's qualifications for various types of duty.

The Marine Corps fitness report consists of: a statement of the duties you perform and your preference for future assignment; a report of your current marksmanship qualifications with required weapons; a report of results of a physical fitness test; an evaluation of your personal characteristics; an estimate of your value to the Service and your professional acceptability to your reporting senior; and a concise evaluation of your character and performance of duty.

The fitness report section of your jacket at Headquarters thus presents a running record of your performance of duty under a number of seniors in various types of service.

Types of Reports. 1. *Regular reports* are rendered semiannually; also whenever you are detached, change duty, are promoted or separated, and whenever your reporting senior changes. Your entire career must be covered by an unbroken sequence of consecutive regular reports. If any period is omitted, your record is incomplete.

2. *Concurrent reports* are in addition to regular reports, and cover additional duties performed under someone other than your regular reporting senior. Concurrent reports cover only specific periods of additional duty and need not be consecutive or submitted for less than 30 days. Completed concurrent reports are sent to your regular reporting senior to assist him in rating your regular report. He then forwards them to the Commandant.

3. *Academic reports* cover periods of duty (though not TAD) at a military or civilian school while under instruction. An academic report covers the entire period of the course (if not longer than 12 months), and permits normal semiannual regular reports to be dispensed with.

4. *Letter reports* are fitness reports in letter form rather than on the standard printed form. Letter reports are rendered only when prescribed by the Commandant, and are almost always concurrent.

5. *Special fitness reports* are required whenever an officer:

Distinguishes himself in battle;
Performs an outstanding act of valor or devotion to duty;
Displays extraordinary courage, ability, or resource in time of peril or great responsibility;
Is guilty of serious misconduct or marked inefficiency.

A Reserve officer receives a special fitness report whenever he:

Completes any active or training duty (other than periods of repeated training);
Completes his last period of repeated training duty.

In addition to the foregoing specifics, a special fitness report may be submitted at the discretion of the reporting senior whenever the individual's performance of duty—good, bad, or extraordinary—is such that Marine Corps Headquarters should be specially advised.

Reporting Seniors. Ordinarily, your reporting senior is your immediate commanding officer or the head of the staff section to which you may be attached. It is your responsibility to submit a signed fitness report, with certain entries completed, to your reporting senior, at the prescribed times. Then he fills out the remainder of the report, which, unless of an unsatisfactory character (see below), is sent to the immediate superior in command, or other designated higher authority, for review and, if the latter deems it appropriate, comment.

Although not in themselves classified, fitness reports are considered as privileged information and are handled with utmost administrative privacy. If, however, any classified papers are attached, the fitness report must be safeguarded as required for that classification.

Unsatisfactory Reports. An unsatisfactory (adverse) report is one which contains any expressly unsatisfactory rating in items 13, 14, or 15a of section B; a "Prefer Not to Have" entry in item 16; an indication in item 19 that an officer is not qualified for promotion at any time; or an unfavorable remark in Section C.

Marginal Reports. Marginal reports are those indicative of performance that may become unsatisfactory if allowed to continue uncorrected, specifically, a report which contains a "Below Average" in items 13 or 14, Section B; a mark to the immediate left of "Average" in item 15a; or an indication in item 19 that the officer is not qualified for promotion with contemporaries.

If you should be so unfortunate as to receive an unsatisfactory or marginal report, your reporting senior is obliged to refer it to you for statement (if you wish to make one) before the report goes forward. You may attach any written statement you choose, provided it is pertinent, temperate in language, and factual, and does not contain any accusations (see *Navy Regulations* in this connection). After completing and attaching your statement, sign Section D and return the report to your reporting senior. If you do not desire to make a statement (which is in effect a concession that you accept the report and that there are no mitigating or extenuating circumstances), sign the notation to this effect in Section D.

If you do submit a statement, *be extremely prudent in language* and assertion. Before you put it in, check *Navy Regulations*, and be sure that you have not forged a weapon which may turn in your hand. Many a statement dashed off in wrath or humiliation has proved more damaging to the officer concerned than the original report.

You and Your Fitness Report. Always see that your reporting senior has at least one spare report with item 22, Section D, signed by you; this covers both you and him in an emergency. But never sign item 24, Section D, in advance. When your next fitness report comes due, ask yourself, "What are my strong points? Where am I weak? What must I do to enhance my favorable characteristics and eliminate the bad ones?"

Whenever you visit Marine Corps Headquarters, take the opportunity to review the fitness reports in your Selection Board Jacket. Look for trends, and see yourself as others see you. You may be able to fool some reporting seniors some of the time, but . . . .

1205 ■ Marking Fitness Reports

Since fitness reports are decisive in the career of an officer, the preparation of a fitness report is one of the most weighty tasks you will ever perform, and an opportunity for you to contribute materially to the overall improvement of the Marine Corps. Instructions governing fitness reports are found in pertinent Marine Corps Orders, which you should review before you make out a report or a recommended report.

The suitability of an officer for future assignments, selection, or retention is based in large degree on the evaluations made by his reporting seniors. There is, in the cases of most officers in today's large Services, little else on which decisions can be based.

Officer appraisal should be a continuous process rather than an intermittent one performed only at fitness report time. Shortcomings should be pointed out as they arise—not saved up. Fitness report time can act as a convenient reminder to sit down with the officer and candidly discuss his general performance and personal qualities. This is not always an agreeable session for either party, but it is a responsibility of command.

You must therefore make out fitness reports carefully, impartially, and with a full appreciation of the task at hand and the responsibility that goes with it. A report that is unduly negative or fails to accent the positive may cost the Marine Corps a fine officer. On the other hand, your failure to point out correctable weaknesses may cause the selection and promotion of an officer unsuited for higher rank at the expense of one who is. Reporting seniors can virtually ensure selection or force the separation of a given officer. Thus, as a reporting senior, you can add luster to a career or destroy it.

Put aside prejudice or partiality as you evaluate. Compare the officer being reported on with other officers of his rank and experience. Guard particularly against the attitude of the moment; you are grading an officer's total performance during *the whole period covered*. Make initial ratings lightly in pencil, then review the report for consistency and fairness, and ink it in.

Bear in mind that any group contains a few individuals at the top and bottom, respectively, who stand out favorably or unfavorably. The majority represent a fairly level standard of performance in between. If you find that your ratings tend to put most officers at the top or at the bottom, be quite sure you can justify this departure from the normal distribution (as you are specifically required to do in the case of outstanding or unsatisfactory marks in items 16a or 19). Remember that most officers are average officers. It is the easy way out to give high ratings to all officers, rationalizing that "everyone does." But over-rating an average officer leaves no scope for the brilliant one.

Do not hesitate to use the "not observed" column for any characteristic in which your observation has been too limited to warrant sound evaluation.

Give particular thought to recommendations for the officer's next assignment. Do not feel that you must mirror his own preferences, as recorded in Part A. Make the recommendation you feel will be best for him and for the Corps. Has he had enough FMF time? Has he been to sea? Should he go to school? Where would you send him if you were the Detail Officer?

The "Remarks" section must never be left blank, and should both complement and epitomize the rest of the report. Not only should you convey what kind of officer the individual is, but you should try to do so in terms of attributes other than those covered elsewhere in the report. Whenever the report includes time in combat, the remark should so state, and should also characterize the behavior of the individual under fire, if you have observed him then. Avoid borderline remarks which may be construed as unfavorable when not so meant. If any rating in the report is unsatisfactory, state clearly, under "Remarks," the reasons why. Be quick to note and comment on improvement. And remember that realistic, conservative markings, coupled with obviously well-considered praise and insight in the remarks, are more impressive to a selection board than an overmarked report with meaningless or vapid remarks.

PROMOTION AND PRECEDENCE

1206 ■ The Marine Corps Promotion System

The Marine Corps and Navy share a system of officer promotion which has been under continuous evolution since 1915, and has in recent years served as a model for the other Services. Under this system officers who are best fitted are selected for advancement, while those least fitted are passed over and must ultimately retire. Determination of who is best fitted for promotion is accomplished by boards of senior officers, known as "selection boards."

Promotion never comes automatically. To qualify for promotion, you must not only perform effectively and loyally in your present rank, but you must develop and prove your capacity to handle the increased responsibilities of higher rank.

1207 ■ Officer Distribution

The Officer Personnel Act of 1947, with subsequent amendments, provides officer promotion machinery for the Army, Marine Corps, Navy, and Air Force. The Navy and Marine Corps provisions are generally similar. One of the most important things that this law does is to establish certain categories of officers for promotion purposes, and to regulate the number of officers who may be assigned to each grade. This is known as "officer distribution," and determines how many vacancies for promotion each rank contains.

Officer Categories. All Marine officers are line officers. Some commissioned officers, however are designated "restricted in the performance of duty," in contrast to all other Marine officers, who are described as "unrestricted officers."

Existing law provides for the following categories of *restricted officers*:

Permanent and temporary regular limited duty officers (LDO)
Permanent regular and reserve women officers
Permanent warrant officers, both male and female

All other officers in the Corps are considered for promotion and assignment purposes as not being restricted in the performance of duty.

Limited duty officers are former warrant or noncommissioned officers who have been commissioned for duty in the particular fields in which they have specialized, such as ordnance, motor transport, and so on. If he so applies, and is qualified, an LDO may be designated as an unrestricted officer. His limited-duty designation then ceases.

Like limited duty officers, warrant officers are also appointed for duty in particular fields in which they specialized as NCOs, and are thus restricted to performance of duty in the appropriate field. Warrant officers selected from certain "line duty" (i.e., Category III) MOS carry the title of "Marine Gunner" and wear the gunner's bursting bomb insignia. The MOS from which Marine gunners are appointed are: 0302 (infantry); 0802 (artillery); 1802 (tank); 1803 (amphibian tractor); 2502 (communications); and 4915 (range officer).

Reserve officers, unless on active duty with the regular establishment, are selected separately. If on active duty with the regular establishment, reserve officers are selected and promoted along with regular contemporaries. The Marine Corps has a continuing requirement for reserve officers on active duty beyond their obligated service, and now offers them a career program comparable to that of regular officers.

Authorized Commissioned Officer Strength. The authorized commissioned strength of the regular Marine Corps is 7 percent of its authorized enlisted strength. The authorized officer strength of the Organized Reserve consists of the total number of commissioned officers required to fill all billets in units of the Organized Reserve. The number of officers authorized for the Volunteer Reserve is controlled by the Commandant.

Despite the 7 percent figure provided by the 1947 Officer Personnel Act as the basis for computing the authorized regular commissioned strength of the Corps, another more recent law—the Army and Air Force Augmentation Act of 1956—enters the picture. This law gives the President authority to establish the number of permanent regular officers in each of the Services, not just the Army and Air Force. It is under this law that the number of permanent regular unrestricted commissioned officers is currently fixed.

Distribution of Officers in Grades. The number of officers who may be promoted above first lieutenant depends on the authorized numbers for each grade. These ceilings are computed by Marine Corps Headquarters and approved by the Secretary of the Navy. They are based on the *actual* number of commissioned officers on active duty at the time. The maximum percentage in each grade is:

General officer	0.75%
(not more than half may be major general or above)	
Colonel	6%
Lieutenant Colonel	12%
Major	18%
Captain	24.75%
Lieutenant (st or 2d)	38.5%

The number of *appointments to permanent rank* in any grade results from application of the foregoing percentages to the total number of permanent regular commissioned officers, and is subject to certain limitations that apply only to general officers and colonels.

The number of *appointments to temporary rank* in each grade derives from a sliding scale based on the total active-duty commissioned strength of the Corps.

If the Secretary of the Navy decides that fewer officers than those computed are required in any grade, he may establish that lesser figure as the authorized number for that grade.

Although the grade distributions shown in Fig. 12–1 represent an ideal (and also give a reasonable cross section of officer distribution for illustrative purposes), they have been superseded by more recent legislation—the Officer Grade Limitation Act of 1954—which places grade distribution on a sliding scale so that, as total authorized officer strength increases, percentages in the general and field officer grades decrease somewhat.

Limited duty officers are not "additional numbers," but the actual number of LDOs may not exceed the following percentages of the authorized number of unrestricted officers in each grade:

Lieutenant Colonel 3.64%
Major 8.62%
Captain 7.72%
Lieutenant (1st and 2d) 6.04%

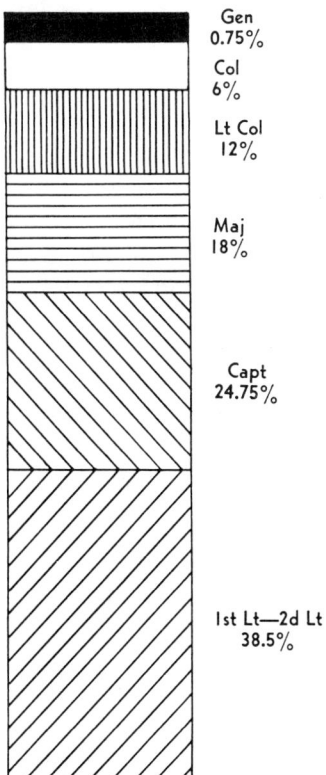

Fig. 12–1: Distribution of permanent regular commissioned officers in grades (Officer Personnel Act, 1947).

In other words, the number of limited-duty lieutenant colonels, for example, may not be more than 3.64 percent of the total number of unrestricted lieutenant colonels authorized, and so on.

General Officers. Without summarizing all provisions of the Officer Personnel Act dealing with general officers, the following are noteworthy:

The Commandant and Assistant Commandant of the Marine Corps are four-star generals. If a Marine were Chairman of the Joint Chiefs of Staff, or Chief of Staff to the President, he also would be a four-star general.

The President may appoint up to 10 percent of the authorized number of unrestricted general officers as lieutenant generals to head certain high commands or for duty of great responsibility. At present, the three-star billets of the Corps are:

Chief of Staff, HQMC
Deputy Chief of Staff (Plans and Policy), HQMC
Deputy Chief of Staff (Manpower), HQMC
Deputy Chief of Staff (Aviation), HQMC
Commanding General, MCDEC
Commanding Generals of the two Fleet Marine Forces

Brigadier generals rank with rear admirals of the lower half (that is, the junior 50 percent of the rear admirals' list) of the Navy. Major generals rank with rear admirals of the upper half, and carry the same date of rank as major general as they had in the grade of brigadier general.

For some years past, despite the various authorizations in law for flag and general officer strengths in the Armed Services, ceiling strengths in this category for each Service have been arbitrarily established each year by Congress, with the result that the number of generals in the Marine Corps (as in the other Services) has remained below that otherwise authorized.

1208 ▪ Promotion Procedure

Permanent promotion to first lieutenant is by seniority, after three years as a second lieutenant. Second lieutenants are eligible for temporary promotion on completion of 24 months' service in grade. From captain to major general, inclusive, promotion is by selection. For permanent regular officers, the standard of selection is "best fitted"; for Reserve officers, it is "qualified for continued active duty." Other than this, eligibility requirements, selection procedures, etc., are the same for all.

Eligibility for Selection. You become eligible for selection when, by the end of the fiscal year in which a selection board convenes, you will have completed service in your present grade as follows:

	Years in Grade
Brigadier General	3
Colonel	3
Lieutenant Colonel	5
Major and Captain	4
First Lieutenant	2
Chief Warrant Officer (W-3)	6
Chief Warrant Officer (W-2)	6
Warrant Officer (W-1)	3

Eligibility for selection does not necessarily mean you enter a promotion zone, as described below. Once you become eligible, though, your eligibility for promotion continues, regardless of failure of selection, as long as you remain on active duty.

The Mechanics of Selection. Each selection board receives the names and selection-board jackets (see Section 1201) of officers who are eligible for consideration. In addition, each board is informed of the names of the eligible officers who comprise the *promotion zone* for that grade. Whenever a selection board convenes, the Secretary of the Navy determines how far down the eligible list the board must go in making selections, if a satisfactory flow of promotion is to be assured. "Flow of promotion" means the progress of your career to a point at which you are considered for promotion to the next higher grade. This flow, in turn, is premised, in peacetime, on assumed *normal terms of service in grade*, as follows:

	Years in Grade	*Total Commissioned Service*
Colonel	5	30
Lieutenant Colonel	7	25
Major	6	18
Captain	6	12
First Lieutenant	3	6

Beginning with the most senior officer of the grade under consideration who has not previously failed to be selected, the promotion zone goes down to the last unrestricted officer needed to maintain the flow of promotion "up or out," as determined by the Secretary of the Navy.

A separate promotion zone is established for limited duty officers, and the selection board is allocated separate quotas of limited duty vacancies to be filled by selection. These officers are considered only upon their specialist qualifications. They do not compete for selection with unrestricted officers.

Any officer eligible for selection may forward through channels a letter inviting attention to any matter of record that he deems important in his case. This, however, is a privilege which should be exercised with the utmost prudence. Don't rock your boat.

All officers senior to or in a promotion zone fail of selection (that is, are "passed over") if not recommended for promotion. The effects of failing of selection, in terms of mandatory retirement, are covered in Section 1212.

Except for LDOs, a selection board may go below the promotion zone, among the eligibles, and select outstanding officers for accelerated promotion. But no officer below the promotion zone is considered passed over, even if an officer junior to him is selected for accelerated promotion; and not more than 5 percent of the total number of officers whom the board is authorized to select may come from below the zone, except in the case of colonels and brigadier generals being considered for the next higher grade.

Once selected, your name is submitted by the board to the Commandant, Secretaries of Navy and Defense, and finally the President (who may remove any name from a promotion list). If, as is usual, all names are approved, you are then promoted, subject to Senate confirmation, to either temporary or permanent advanced rank, according to the number of temporary and permanent vacancies in the grade ahead.

Selection Boards. Marine selection boards consist of nine active or retired officers and are convened by the Secretary of the Navy at least annually. As a matter of Marine Corps policy, the normal composition, by rank, of selection boards for the various grades is as follows:

	Number of Members by Grade:		
Promotion to:	MajGen	BrigGen	Col
Major General	9*		
Brigadier General	3*	6	
Colonel	3*	6	
Lieutenant Colonel	1	2	6
Major		1	8
Captain		1	8

*Senior selection boards virtually always include at least one LtGen.

At least three of the nine members are aviators. If reserve officers are to be considered by the board, it includes reserve membership in a proportionate number.

No officer may be a member of two successive boards for the same grade. This ensures that 18 different officers must pass on your case before you twice fail of selection.

Members of selection boards are sworn to act without prejudice or partiality, and, like members of a court-martial, may not disclose their deliberations. Specifically, their sworn duties and obligations are as follows:

To recommend the best-fitted regular officers for promotion.

To recommend for promotion reserve officers who are qualified for continued active duty in next higher grade.

To give equal weight with line duty equally well-performed, to: administrative staff duty, aviation duty, supply duty, or duty in any technical specialty.

Not to consider as prejudicial the fact that an officer under consideration may have previously been passed over.

Not to select more officers than the number set by the Secretary of the Navy (but the board need not select the full number if there are insufficient qualified names under consideration).

To recommend only officers who attain at least a two-thirds majority vote (selection boards considering active-duty officers require concurrence of at least two-thirds of the members, while boards considering inactive reserve officers select by simple majority).

To select for discharge unsatisfactory officers (if any) with less than 20 years' service (see Section 1212).

Effecting Promotions. When selected for promotion, your name goes on a promotion list in normal order of seniority. As vacancies occur, either temporary or permanent, you are then promoted. All officer promotions, regular and reserve, are, however, subject to such physical, mental, moral, and professional qualifications as the Secretary of the Navy may prescribe. On promotion, you rate the pay and allowances of the higher grade from the date of occurrence of the vacancy you are promoted to fill.

1209 ■ Women Officer Promotion

Women officers are promoted by the same basic procedures and boards as male officers. A women's quota is assigned a given selection board, which, when it considers women in the zone, contains at least one woman member. Contingent on passage of legislation under consideration in 1977, women officers may in the future be selected in direct competition with male officers under single unisex promotion zones.

1210 ■ Precedence

Precedence is your right of seniority over other officers, based on rank, and on the date of your appointment within a grade. The rank and precedence of officers on active duty is shown in numerical order in *The Combined List of Officers on Active Duty in the Marine Corps*, otherwise known as "The Blue Book." Your date of rank is also stated on your commission.

Your *Social Security Number* serves in effect as your service number instead of the "file number" every officer used to have. In addition to his Social Security Number, every Marine officer has a *number in grade*, which is his number, in order of precedence within grade, in "The Blue Book." Your number in grade gives an exact indication of your precedence.

Precedence of Officers of Different Services is in accordance with their relative grades and, within grade, in accordance with respective dates of rank, the senior in date of rank taking precedence. Among officers of different Services of the same relative grade and the same date of rank, precedence is determined according to the time each has served on active duty as a commissioned officer.

RETIREMENT AND SEPARATION

1211 ■ Introduction and Basic Definitions

Like death and taxes, it is inevitable that your Marine Corps career will eventually end with retirement or separation from the Service. The paragraphs that follow analyze retirement and separation procedures for regular officers; these procedures stem from the Officer Personnel Act and the Women's Armed Services Integration Act.

Retirement is removal from active duty after completion of certain service and longevity requirements, after which the retired officer receives retired pay. Although a retired officer is no longer on duty, he remains a member of the Marine Corps, retains his rank and status as an officer (being entitled to all military courtesies of his rank), and may under certain conditions be recalled to active duty.

Separation or discharge is an absolute termination of officer status. Depending on the character of discharge, the officer being separated may or may not receive lump-sum *severance pay. Resignation* as distinct from retirement, is total, voluntary separation from the service.

Revocation of commission may separate any officer who has been on continuing active duty for less than three years as a commissioned officer in the Marine Corps or Navy. An officer whose commission is revoked does not receive any advance pay or allowances, or severance pay.

"Total Commissioned Service." To assure that you approach both promotion and involuntary retirement in order of seniority, and to prevent amounts of actual service from conflicting with your lineal position, a standard method exists for computing what is assumed to be your "total commissioned service." This total *only* applies to your eligibility for continuation on, or separation from, the active list.

1. *If you were commissioned from the Naval Academy or pursuant to Public Law 729, 79th Congress*, your total commissioned service is computed from 30 June of the fiscal year in which you accepted your initial appointment.

2. *If commissioned from any other source*, the law assumes that you have just as much total commissioned service as the Naval Academy graduate or P.L. 729 appointee who is immediately junior to you, provided your "running mate" has not lost numbers or precedence.

This principle gives you a "running mate" who must have been appointed annually at the foot of the second lieutenants' list, and who is more likely to attain promotion after normal periods of commissioned service (see Section 1208).

1212 ▪ Involuntary Retirement and Separation

Except for physical reasons, you retire upon reaching a certain age, on completing certain periods of service, or after failure of selection for promotion.

Retirement for Age. Officers still on the active list must retire at age 62 unless the President in a particular case defers retirement. Such retirement may not be deferred beyond age 64.

Limited Duty Officers, if not otherwise subject to retirement, must retire after completing 30 years' active Marine Corps or Navy service.

Retirement Requirements

Major Generals. Major generals (as well as lieutenant generals) normally retire on 30 June of the fiscal year in which they complete 5 years' service in grade and 35 years' commissioned service. They may stay on active duty from year to year if so recommended by a board convened for the purpose and approved by the Secretary of the Navy.

Brigadier Generals. Brigadier generals retire on 30 June of the fiscal year in which they are passed over the second time.

Colonels. Colonels retire on 30 June of the fiscal year in which they complete 30 years' total commissioned service, if they have twice failed for brigadier general. After 31 years they must, unless on a promotion list, retire on the following 30 June, regardless of whether or not they have been twice passed over. Those who have lost numbers or precedence, however, may not be retired until after five years as a colonel.

A woman colonel retires at age 55, or after 30 years' active commissioned service (regular and reserve), whichever is earlier. If she reaches the age of 50 while so serving, she may retire when she ceases to serve as Director of Women Marines.

Lieutenant Colonels. Lieutenant colonels, except LDOs, retire on 30 June of the year in which they complete 26 years' total commissioned service, if they have twice been passed over for colonel. If not yet twice passed over, they do not have to retire until 30 June of the year after their second failure.

LDO lieutenant colonels must retire after 30 years' total active naval service.

A woman lieutenant colonel not on a promotion list must retire after 26 years' active commissioned service.

Majors. Majors, except LDOs, retire on 30 June of the fiscal year in which they complete 20 years' total commissioned service, if they have twice been passed over for lieutenant colonel. If they have not so failed, they do not have to retire until 30 June of the year after their second failure.

LDO majors must retire on completion of 30 years' active naval service.

Women majors not selected for lieutenant colonel retire on 30 June of the year in which they complete 20 years' active commissioned service.

Captains and Lieutenants. Male captains and first lieutenants who are twice passed over for major and captain, respectively, are honorably discharged (*not* retired) on 30 June following the second failure, with severance pay based on length of service. An LDO, however, has the option of reversion to prior enlisted status.

Women captains and first lieutenants not on promotion lists for next higher grades retire if they reach 50 while so serving; or, if below 50, they are honorably discharged with severance pay on 30 June of the year in which they complete, respectively, 13 and 7 years' active commissioned service, regular or reserve.

Warrant Officers. Warrant officers who are twice passed over for promotion to the next higher permanent warrant grade are discharged with severance pay if they have less than 18 years' active service since initial appointment as a warrant officer; if a warrant officer so passed over has 18 but less than 20 years' service since the original appointment, he will be retired (unless picked up in the interim) two months after completing 20. Any permanent regular male warrant officer who has at least 20 years' active service in the Armed Forces will be retired at age 62, or, short of that age, on completing 30 years' active service in the Armed Forces.

Officers Reported as Unsatisfactory. If at any time before completion of 20 years' service, your name comes before a selection board in normal course, and you are specifically reported by that board to be unsatisfactory in performance of your duties, you are honorably discharged with severance pay (*not* retired) on 30 June following approval of the report by the President. An LDO so reported has the normal option of reverting to his former status. Do not confuse this procedure with being passed over (which is bad enough).

Revocation of Commission. The Secretary of the Navy may revoke the commission of any regular officer who has less than three years' continuous service. Discharge of this type does not include severance pay. The most usual causes for revocation of commission are academic failure at Basic School, general low caliber or unsatisfactory performance of duty, or temperamental unsuitability.

1213 ■ Voluntary Retirement

When a regular or Reserve officer completes 20 years' active duty in the Marine Corps, Army, Navy, Air Force, or Coast Guard (including Reserve), 10 years of which must be active commissioned service, he may, in the discretion of the President, retire with the highest grade satisfactorily held, as determined by the Secretary of the Navy.

When an officer has 30 years' active service, he may, in the discretion of the President, retire with 75 percent of his active-duty pay.

When an officer has 40 years' service he shall, upon application, be retired.

Retirements take effect on the first day of the month after the Secretary of the Navy approves the request, *except* in cases of voluntary retirement in which a later date has been approved. A medical report (Form 88) must accompany all requests.

If you are considering voluntary retirement or are subject to involuntary or statutory retirement, and you have any doubt as to your physical qualification for release from active duty, obtain a preliminary physical examination *three months ahead of estimated retirement date.* If a disability is discovered at this time, you may be eligible for physical retirement. Information as to your physical condition must be received by Headquarters Marine Corps in time to modify action on your retirement papers.

After the President or Secretary of the Navy approves a request for retirement, or approves involuntary retirement proceedings, and the retirement has become effective, there is no process of law whereby the retired status may be changed except through disciplinary action, or because of physical disability incurred subsequently while serving as a retired officer on active duty.

1214 ■ Physical Retirement

Physical retirement is governed by Title 10, U. S. Code. Its important provisions are summarized below.

A *temporary-disability retired list* exists in each Service, to which are transferred individuals whose physical condition prevents proper performance of duty, but who may recover. Persons whose disability is less than 30 percent may be discharged with severance pay. Pay on the temporary-disability retired list may be either 2½ percent of active-duty pay per year for the number of years' service, *or* the percentage of disability fixed; but retired pay may not be less than 50 percent or more than 75 percent of active-duty pay. Physical-disability pay is tax-exempt.

If on the temporary-disability retired list, you must have a physical examination at least every 18 months, for not more than five years. If during that time your disability has become permanent, and is 30 percent or more, you go off the temporary disability retired list and are permanently retired for physical disability. If, upon examination, you are again found physically fit while on the temporary-disability retired list, you may, at your own consent, be reappointed to the active list in a rank not below that held when you went on the temporary-disability retired list. If you do not wish to return to active duty or are not qualified for duty but are still rated less than 30 percent disabled, you may be separated from the Service, with severance pay.

If you are hospitalized for more than three months, a clinical board (or board of medical survey) at the hospital considers your case and usually recommends that you be (1) returned to duty; (2) given further treatment; (3) ordered to limited duty and re-examination after a stated period; (4) given sick leave and re-examination thereafter; or (5) ordered before a Marine Corps physical evaluation board.

You may waive the right to appear in person, but *should be very chary of this* unless the medical evidence in your case is uncomplicated and the result certain. You should also have counsel.

Appearance before the board requires less than a day; final action is taken by the Secretary of the Navy. If you are incapacitated, and if continued hospitalization is not required, you may take leave or be assigned to temporary duty. If found qualified for continued active duty, you are discharged from the hospital and ordered back to your original organization, or to the nearest Marine Corps activity for further assignment. If retired, you must complete travel home within one year from the date of retirement, in accordance with *Joint Travel Regulations*. You may choose any residence desired, without regard to your current address of record in Headquarters.

The report of a physical evaluation board is extensively reviewed in the Navy Department. The final reviewing authority is ordinarily the Judge Advocate General of the Navy. Retirement takes effect on the first day of the month after the Secretary approves.

If you have less than eight years' active service and if disability is less than 30 percent, the physical evaluation board must determine whether your disability is the proximate result of active service. If the Board so finds, you may be discharged with severance pay. Eight years or more of active service is considered proximate result.

CORRESPONDENCE AND MESSAGES

1215 ■ Official Correspondence

Both Marine Corps and Navy employ the same forms and procedures for official correspondence. These are prescribed in *Navy Regulations*, in the *Navy Correspondence Manual*, and in the *Marine Corps Manual*. "Correspondence" embraces letters, endorsements, speedletters (shortcut, urgent communications which do not require telegraphic or radio transmission), and memoranda. Correspondence is filed in accordance with the *Navy Filing Manual*.

As an individual, you may originate official correspondence that pertains to you personally (including, however, any recommendations for improvement or innovation which may benefit the Marine Corps). Correspondence which affects a command as a whole can be originated only by, or in the name of, the commanding officer.

Except for speedletters, official correspondence must be conducted through official channels, and must be promptly forwarded. Failure to do so (if in proper form and language) is a very serious dereliction.

Correspondence with civilians or with other Government agencies should be particularly prompt, courteous, complete, and accurate.

Avoid unnecessary, verbose, imprecise correspondence. Joseph Pulitzer's rule for the staff of the old *New York World* applies with considerable force to military correspondence: *"Accuracy, brevity, accuracy!"*

During the past few years Marine Corps correspondence has grown less precise and less military. Vague expressions and superlatives have become commonplace and, in many cases, have replaced clear and understandable military terms. Complex language is used, not because it contributes to the usefulness of correspondence, but because of habit or custom. It is not consistent with the character of the Marine Corps or with the efficient conduct of its affairs to permit correspondence to become diluted by unmilitary expressions or unnecessary language.

Admirable advice regarding official correspondence, and military writing in general, was voiced some years ago by the then Chief of Staff, HQMC, one of the ablest writers of his time in the Corps:

> MEMORANDUM
> From: Chief of Staff
> To: All Officers on Duty in Headquarters Marine Corps
> Subj: Military writing in Headquarters Marine Corps
> 1. Much bad writing is done in this Headquarters. I do not expect a high literary polish on our product, but I do expect conformance to the basic rules of composition and grammar.
> 2. There is no point in trying to list the words, phrases, forms, and techniques to which I object. I ask you to consider instead a few simple ideas:
>> a. Deal with the issues. Omit irrelevant facts or opinions. It is frequently unnecessary and undesirable to tell all you know about the subject.
>> b. Develop your thoughts in logical order. Lead your reader along the short, direct route to your conclusions.
>> c. Avoid long or complicated sentences and paragraphs.
>> d. Be precise. Do not let woolly style or fancy language replace clear thought.
>> e. Exercise care in the choice and use of words. The shortest and most familiar word is usually the best. Avoid jargon.
>> f. Use qualifying words sparingly. A sound idea has no need for extravagant adjectival support.
>> g. Seek perfection in syntax and punctuation. There is no room for compromise in this regard.
>> h. Examine your writings critically. This will take a little extra time at first. We should get it back, with interest, when we all learn to write in simple, direct terms.
>
> /s/   V. E. MEGEE

1216 ■ Official Letters

According to the nature of the correspondence, official letters may follow either the *naval form* (which is used throughout the Naval Establishment) or the *business form*. Examples and detailed instructions covering both forms can be found in the *Navy Correspondence Manual*. Speedletters, however, may follow the abbreviated form and language employed in despatches and naval messages.

Hints for Official Letter-Writers. Until you are quite familiar with the prescribed forms for official correspondence, do your writing within arm's reach of the *Navy Correspondence Manual*. See that your unit clerks do likewise. Keep an up-to-date dictionary at hand (*Webster's New Collegiate Dictionary* is excellent).

Avoid pointless letters. Correspondence with higher authority should be confined to specific requests, reports, and concrete recommendations.

One letter should deal with one subject only. Cover separate subjects to the same addressee by separate letters. Answer official letters by letter, not by endorsement on the letter received, unless specifically directed to do so.

Write in terse, unadorned, direct, and clear language. Use short sentences. Psychological tests show that any sentence longer than seventeen words may stump some readers. Avoid the passive voice. Don't be afraid to use the first person. Remember Pascal's line, "I have made this letter longer than usual because I lack the time to make it shorter."

Be as temperate and courteous in correspondence as you would be in discussing the subject face to face with your correspondent.

Organize your facts and ideas before you write. The standard sequence for a staff study is a good one for almost any kind of official correspondence:

Statement of the problem
Facts bearing on the problem
Discussion
Conclusions
Recommendations

Do not send official letters to other subordinate officers in the same command. Intraunit and intraheadquarters correspondence should be by memorandum.

Block out important official letters in double-spaced rough draft. This permits you to make legible corrections and interlineations. It is good practice to compose rough drafts on a distinct color of stationery used for no other purpose.

Always capitalize the word "Marine" when used as a noun. Shun like leprosy the demeaning practice sometimes encountered in official correspondence or directives, of using the word "member" where "Marine," "officer," or "enlisted man" is meant; or "personnel" in lieu of "Marines" or "all hands." In expressing time, use the Navy 24-hour system, and never add the superfluous word "hours"—write "1230," never "1230 hours."

Here is a random list of bureaucratic gobbledygook, canned language, and trite jargon excerpted from a sampling of official correspondence. We would all be better off if most of these were never seen again:

pursuant to . . .
in conformance with . . .
the full impact . . .
pertinent facts . . .
unprecedented—as a
   substitute for "unusual"
thorough and complete
   investigation . . .
infeasible of accomplishment . . .
at the earliest practicable
   moment . . .
top management . . .
rendered mandatory . . .
salient data . . .
materially impaired
   effectiveness . . .
inter-face incompatibility
in light of the foregoing . . .
above-named personnel . . .
management—as a substitute
   for "command" or "leadership"
considered opinion . . .

VIP . . .
deobligate . . .
firm up . . .
formalize . . .
outload, offload,
   onload . . .
frame of reference . . .
effectuate . . .
definitive—where "definite"
   is meant
maximize . . .
as appropriate . . .
finalize . . .
logisticswise . . .
personnelwise . . .
appraise—where "apprise"
   is intended
forward—as a garble for
   "foreword"
marshall—as a misspelling
   of "marshal"

Finally, how would you answer the following questions about yourself as a letter writer?

1. Are most of your letters less than a page long?
2. Is your average sentence less than 17 words?
3. Do you try for short paragraphs—less than 10 lines?

4. Do you prefer active verbs (please make the following arrangements) to passive verbs (it is requested that the following arrangements be effectuated)?
5. When you have a choice, do you choose short words (pay, help, mistake) or big ones (compensation, assistance, inadvertency)?
6. Are your letters free of officialese and jargon?

1217 ■ Official Mail

Official correspondence may be mailed postage-free in what are still described as "penalty envelopes"; that is, envelopes bearing an official return address and the notation *"Official Business"* in the upper left hand corner, and, in the upper right, the phrase *"Postage and Fees Paid, Navy Department."* As an officer you are entitled to use such envelopes for correspondence that clearly involves Government business and for Government parcels within prescribed weight limits. Be scrupulous in exercising this privilege. Remember you also are a taxpayer.

1218 ■ Personal Correspondence

Because you change stations every few years, keeping correspondents advised of your correct mailing address can be quite a problem. To help out, the Navy Department has change-of-address cards, which you can get in any quantity from your mail clerk or station post office. Every time you are detached, send these to all your regular correspondents and to publications to which you are a subscriber. Should you or anyone wish to write an officer but not know his present address, simply address the letter as follows:

c/o Directory Service
Marine Corps Headquarters
Washington, D.C. 20380

> NOTE: For guidance of your family or parents at your permanent home address, Postal Laws and Regulations permit postage-free forwarding of any class of mail addressed to a member of the Armed Services, if marked, *"Change of Address Due to Official Orders, Postal Reg. 157.4."*

Fleet Post Office (FPO). Mail may be sent, at domestic postage rates, via the Fleet Post Offices, New York, San Francisco, or Seattle, as appropriate, by or to any officer or enlisted man serving beyond the continental limits of the United States on board any ship, shore station, or FMF unit. This privilege is important not only because it saves you and your correspondents appreciable money in postage (especially on publications and parcel post) but because the Fleet Postmasters keep track of the movements of mobile units and dispatch mail accordingly. A comparable system, used by the Army and Air Force for their units and stations, and for some locations overseas where there is a bulk of U. S. correspondence, is that of "APOs," via the same cities as above. Generally speaking, if a ship or unit is on the East Coast or in the Atlantic, the Caribbean, Europe, or Africa, its mail goes via FPO, New York. Mail for the West Coast, the Pacific, or Asiatic Station mainly goes via FPO, San Francisco, except for a few destinations, which are routed via Seattle.

1219 ▪ Personal Radio Traffic

If you are serving at sea or at a station outside the continental limits of the U. S. where there are no commercial cable services, you may file personal traffic for Stateside addresses via what is known as "Class E Messages." In this system your message is transmitted by naval communications to the nearest Naval Communication Station in the United States, which relays your message to its ultimate addressee via commercial telegraph. You pay for the cost of commercial relay only. If you want to send a Class E message, see your communications officer.

SECURITY OF INFORMATION

1220 ▪ Personal Security

After taking your oath and becoming a new Marine officer, in your new life and status you receive access to information not generally available to civilians. One of your most important responsibilities then becomes the safeguarding and discreet use of this information in such a way that it may never fall into the hands of enemies of the United States. Remember that, war or peace, the battle for information goes on continually. Our success in this battle will determine whether the odds of physical combat are on our side or our enemies'.

Indiscreet conversation and personal letters constitute great menaces to security. Guard against unthinking discussion of classified "shop talk," even with your family and friends. Avoid loose talk in public places. When you are on the telephone, you can never tell who may be listening. Automatic self-censorship is a responsibility of all Marines.

1221 ■ Security of Classified Matter

Classified matter is anything—either information or material—which, in the public interest, must be safeguarded against unauthorized or improper disclosure. *Navy Regulations*, as well as the *Navy Security Manual for Classified Matter* and the *Registered Publications Manual*, contain detailed instructions that must be followed to the letter when you handle classified matter. The security of classified matter is the security of the United States.

Classifications. The categories of security classification are *Top Secret, Secret,* and *Confidential*. In addition, certain information regarding nuclear weapons and related subjects is classed as *Restricted Data*; do not confuse the term with the former classification, "Restricted," which you may see on older publications but which has been replaced by "For Official Use Only." It is up to the originator of matter which should be classified—if he is so authorized—to assign it the appropriate classification, and he, as well as higher authorities, may reclassify it when appropriate. Reclassification may involve either "upgrading" or "downgrading."

Handling of Classified Matter. The precautions regarding preparation, marking, custody, handling, transmitting, stowage, disclosure, control, accounting for, and disposal of classified matter may be found in the references at the beginning of this section, and of course you must follow them to your utmost. If, however, you find yourself in a situation where you cannot physically comply with certain of these rules, you are bound simply to do your utmost, in common sense and zeal, to safeguard whatever may be entrusted to you. Should you have reason to believe classified information has been compromised, through your fault or anyone else's, you must inform your commanding officer at once.

No one, regardless of rank or position, is automatically entitled to knowledge or possession of classified matter. Such information goes only to those who *need to know*.

LEAVE AND LIBERTY

1222 ■ Leave of Absence

Subject to the needs of the Service, leave of absence provides time off for mental and physical relaxation from duty, and gives you opportunity to settle your affairs when the time comes for change of station.

Every officer on active duty accrues leave at the rate of 30 days a year (that is, 2½ days per month).

Without jeopardizing the readiness of your command, your commanding officer sometimes cannot grant every officer all the leave he rates. Whatever leave is not taken "goes on the books" until you have a maximum of 60 days' unused, or accrued, leave. Earned leave that accrues above 60 days must be automatically dropped on 1 July each year, and when you retire. As you approach retirement, it is a good idea to let leave accrue to a minimum of 30 days, since you receive a lump-sum payment for such accrued leave when you retire. Short of final years, however, take leave as you can; the days that are dropped each first of July can never come back.

Leave of absence describes authorized vacation or absence from duty, as distinguished from *liberty*, which is merely authority to be away from your place of duty for short periods, and is not charged to leave.

Accrued leave is the unused leave to your credit "on the books" each 1 July. You cannot bank up more than 60 days' accrued leave.

Annual leave is leave taken as routine vacation from duty. Annual leave is limited to your total accrued leave plus 45 days' advance leave, but may not exceed periods of 60 days.

Sick leave is given to convalescents on recommendation of the medical authorities, or to repatriated prisoners of war. Sick leave does not count against accrued leave.

Emergency leave may be granted to help alleviate some personal emergency, such as death or serious illness in the immediate family. Emergency leave is charged against accrued leave and may not exceed 105 days.

Excess leave is leave in excess of all your accrued leave plus 45 days' advance leave. Avoid taking excess leave when you can possibly do so, since your pay and allowances are checked while you are on excess leave.

Earned leave is the term used to describe the leave potential of an individual at any given date during the fiscal year. Earned leave may be calculated as follows: From the amount of accrued leave, subtract whatever leave has been taken since the outset of the fiscal year to the date in question. To that remainder add the amount of leave earned since the beginning of the fiscal year. Earned leave may exceed 60 days during the fiscal year, but will always be cut back to 60 days at the beginning of the new fiscal year.

Advance leave is an accounting term to describe leave granted in advance of accrual.

Delay in reporting is leave authorized to be taken after detachment from one permanent station and before reporting to another. Only the Commandant can grant delay to an officer.

Graduation leave is granted to officers newly commissioned from one of the Service academies (*not* to officers from any other source). It is 30 days, not chargeable to the officer's leave account, and you must take it prior to reporting to the first permanent duty station (ordinarily Basic School) or CONUS port of embarkation if ordered to permanent duty beyond the seas.

1223 ■ Computing Leave and Delay in Reporting

No small amount of low-order bookkeeping and finger-counting centers about the average officer's computations of leave and delay. Here is how to do it:

Table 12-1
Leave Computation

Amount of leave or delay	Date of departure	Must return and report not later than
10 days' annual leave	10 April (day of duty, not leave)	21 April (before forenoon quarters [0900] or beginning of working hours)
20 days' delay, with 5 days' travel time, and 4 days' "proceed" time (permanent change of station)	1 August (date of detachment, day of duty)	30 August, *before 2400* (without delay, you would be due 10 August; add 20 days' delay, making it 30 August)

Your day of departure on leave, whatever the hour, counts as a day on duty (and hence is not charged as leave).

If you return after the beginning of working hours on shore station, or after the hour of 0900 aboard ship, the calendar day of return counts as a day of leave. If, however, you return before working hours, or before 0900 aboard ship, the calendar day of return is a day of duty.

If you are reporting aboard after authorized delay, rather than from leave, you must report before midnight on the final day allowed.

Subject to the qualifications just explained about reporting before quarters or working hours, the rule to memorize is: *"Day of departure is a day of duty; day of return is a day of absence."*

One precaution: if you report back from leave or delay, outside of working hours, be sure your return is logged in by the Officer of the Day or organization duty officer.

1224 ■ Leave Requests and Records

Requesting Leave. When you want leave, it is best to give your commanding officer advance notice. Some organizations have an annual leave plan which permits all officers to book leave well in advance. After you have informal approval for your projected leave, submit a written request using the standard leave-authorization form when available (via channels, if necessary) to the officer who is authorized to grant leave—usually your battalion, squadron, or post CO. Your leave request should include: the number of days and type of leave desired; the number of days' leave you have already taken during the fiscal year; your address while on leave; whether you are a member of any court or board; and any other pertinent information or special justification for the request.

If approved, your leave request is returned by endorsement. Keep it with you throughout leave.

Address While on Leave. It is your responsibility to keep your commanding officer apprised of your address at all times while on leave. If your plans change, inform him by letter or telegram. If you are touring, set up a number of check-in points, such as hotels where you expect to stay, American Automobile Association offices, or homes of friends. You have no leg to stand on if, while on leave, your commanding officer tries to communicate with you and cannot reach you.

Your Leave Record. Every officer has a leave record (Leave and Earnings Statement, or "LES") on which all leave taken is debited and all leave earned is credited each year. This record is kept by the headquarters that administers you, but it is your responsibility to see that your leave record is correct.

1225 ■ Foreign Leave

Marine Corps Headquarters encourages officers to take foreign leave, and will assist you in passport arrangements (if required). Officers going on foreign leave may travel, on a space-available basis, in Government aircraft. Unless you are authorized by the Commandant to wear uniform while on foreign leave, you must wear civilian clothes.

It is also your responsibility to check in, preferably in person, otherwise by telegram, phone call, or letter (if you are remote from the capital city) with the U. S. Naval Attache (or, if there is no resident Naval Attache, with the resident U. S. Defense Attache). Likewise, whenever you are in a city where there is an American Consulate, you should unfailingly register your presence with the Consul. Aside from your obligations just stated, the practice of checking in gives you many potential conveniences and amenities, as well as secure forwarding addresses for mail and messages.

Permission to visit foreign countries while on leave must be granted by the Commandant, except for visits to certain foreign areas specified from time to time in current directives, for which blanket authorization is granted.

1226 ■ Liberty

Liberty in effect is local free time within limits which does not count as leave. It may be granted at any time for up to 48 hours. If the period includes a legal holiday, any commanding officer can extend a "forty-eight" to a "seventy-two." If the period includes a weekend, commanding officers so authorized by the Commandant may grant 96-hour liberty; this privilege is reserved for remote areas in which forty-eights and seventy-twos might be inadequate. But liberty may not be used as a device to extend leave.

Unless you have specific permission to the contrary, while on liberty you must remain within the general vicinity of your post. Almost all posts and units have standing orders that designate "liberty limits," beyond which ordinary liberty does not extend. The purpose of liberty limits is to prevent Marines from going so far afield that they cannot count on returning within the prescribed time.

IDENTITY DEVICES

1227 ■ "ID Card"

The "ID card," or Marine Corps identification card, is the most important identifying document you have. It identifies you as a Marine officer, and thus must be safeguarded with great care. Regular and active-duty Reserve officers have the green card, whereas inactive Reserve personnel carry the red card. Retired officers and men are issued a grey card. Loss of an ID card is a serious matter, and must be reported immediately. Carry it at all times, and never surrender it.

The ID card, however, is not a pass, but an identifying device.

> NOTE: A most objectionable practice has arisen in recent years of requiring, at such places as exchanges and entrances, and even in routine administrative transactions, that officers *in uniform* produce ID cards to prove they are officers. This in effect demeans the uniform (by subordinating it to a laminated bit of administrative paper) and puts enlisted persons (or civilians) in the position of passing on the good faith of commissioned officers. On any such occasion, therefore,

you should be aware of and enforce the *Marine Corps Manual* provision: *"Except where the security of classified material and installations impose more stringent demands, an officer's uniform will amply attest to his status as an officer."* The only exception to the provision just cited is that, because of Defense (and, ultimately, Congressional) policies, all hands, in uniform or not, are required to identify themselves as a prerequisite to use of a commissary. This demeaning requirement is said to be necessary to prevent purchases by unauthorized persons.

1228 ■ Identification Tags

Every Marine on active duty is issued two "dog tags" to identify him should he be killed or wounded in action. These tags are items of equipment. When not required to be worn, they must remain in your possession.

Note that *both* tags must be worn, when tags are required.

1229 ■ Official Photograph

As soon as you are commissioned, and prior to any consideration by a regular promotion board, you must submit an official photograph. This photo cannot be more than 12 months old when the board convenes, and must (1) be full length, one front and one side view, in service "A" uniform, uncovered; and (2) be 4"×5" prints, mounted on white bond paper with your name, Social Security number, date, height and weight stripped. This is then filed in your selection board jacket. Considering its purpose, you should take care to look your best.

1230 ■ Geneva Convention Card

This is a white card that is carried by personnel in combat or whenever capture is possible. It is designed to be turned over to a captor, should you be taken prisoner, and thus provide officially authenticated information required by the capturing power in preparation of his POW rolls. When not required to carry the Geneva Convention card, turn it in for stowage attached to your qualification record.

1231 ■ Dependents' ID Cards

Your wife and each dependent over age 10 are entitled to an Armed Forces Dependent's Card, as described in Chapter 8. Dependents under 10 are covered on your wife's card. This card, like your ID Card, is an identity device and does not in itself entitle the bearer to anything. It ordinarily serves, however, to establish identification for medical care, post exchange, and similar privileges extended to dependents. (See Section 806.)

1232 ■ Standard Vehicle Stickers

The Department of the Navy has standard bumper decals for personal vehicles of active and retired officers and enlisted people of the Navy and Marine Corps. Officer stickers are blue; enlisted, red. To obtain stickers and register your car, consult your provost marshal.

My Lord—If I attempted to answer the mass of futile correspondence that surrounds me, I should be debarred from all serious business of campaigning.

I must remind your Lordship—for the last time—that so long as I retain an independent position, I shall see that no officer under my command is debarred, by attending to the futile drivelling of mere quill driving in your Lordship's office, from attending to his first duty—which is, and always has been, so to train the private men under his command that they may, without question, beat any force opposed to them in the field.

—*Letter attributed to the Duke of Wellington*

13

Pay, Allowances, and Official Travel

It is related that, during the early days of World War II, a lofty-minded civilian visited Guadalcanal. During his tour, war aims were mentioned. Addressing Lieutenant Colonel L. B. Puller, one of the most hard-bitten professionals on the island—or for that matter, in the Marine Corps—the visitor inquired, "And what, colonel, are *you* fighting for?"

Colonel Puller paused, reflected for a moment, and answered, "$649 a month."

Whether you incline to this view, or to the sentiments of George Washington, quoted at the end of this chapter, the importance of knowing about pay and allowances need hardly be explained.

Every member of the Service is paid monthly, or twice monthly, based on his rank and length of service. Regardless of whether your rank is temporary or permanent, you are paid at the rates prescribed for that grade.

The sums you receive as "pay" in the civilian sense of the word are comprised in the term "Regular Military Compensation" (RMC), the military equivalent of a civilian salary: basic pay, quarters allowance, subsistence, and, often overlooked, the Federal income-tax advantage

NOTE: The subject matter being precise, extensive, and liable to change, this chapter is necessarily written in general terms. In case of doubt, always consult your paymaster.

deriving from tax-exempt allowances. Your pay—RMC, in other words—normally increases each October, based on comparability with pay levels in the private economy.

Besides regular pay, just described, officers whose duties or status so qualify them are entitled to incentive and special pay, including flight pay (see Section 1307).

1301 ▪ Pay System

Marines are paid through a centralized automated Joint Uniform Military Pay System (JUMPS), which constitutes part of an integrated Armed Forces pay and personnel system.

Under JUMPS a master pay account is maintained for each Marine at the Marine Corps Finance Center, Kansas City, Missouri. Monthly "leave and earnings statements" (LES) are produced from information in your master pay account; are distributed to you, your commanding officer, and your disbursing officer; and are filed at the MCFC on microfilm for permanent record. Your LES reflects what you are due, tax withholding, leave balance, and any deductions, and forecasts the amount payable for the next two paydays.

A so-called personal financial record (PFR) is also maintained for each Marine. Your disbursing officer usually keeps this, but on some occasions—as on change of station—you do.

1302 ▪ Service Creditable for Pay Purposes

In determining your length of service for pay purposes, you receive credit for all service, active or inactive, in the Marine Corps, Navy, Army, Air Force, Coast Guard, and Reserve components thereof. In addition, full time is allowed to Women Marines for service in the Army Nurse Corps, the Navy Nurse Corps, the nurse corps of the Public Health Service, and Reserve components thereof. Full time is also allowed for service as deck officer or junior engineer in the Coast and Geodetic Survey. Active service in the appointive grade as aviation cadet and service as an enlisted aviation cadet may be counted as service for pay purposes. Further, captains and lieutenants with over four years' active duty (including active duty for training) as enlisted men receive a separate, slightly higher rate of pay.

Service not creditable for longevity increases (or "fogies") is service as cadet or midshipman; service in inactive National Guard, or in State, Home, or Territorial Guard; service in ROTC; and time spent in voided fraudulent enlistment.

You count, in the computation of basic pay, the total of all periods authorized to be counted in any of the Services.

1303 ▪ Dependents

Certain of your allowances vary according to whether or not you have dependents. The law defines dependents as:

  a. Your spouse (whose dependency is presumed).
  b. Unmarried children under 21, or over 21 if handicapped and incapable of self-support.
  c. A parent (or one who has stood *in loco parentis*), if chiefly dependent on you for support.
  d. Stepchildren and adopted children, if dependent.

Except for your spouse or unmarried minor children, you must be able to prove dependency for any persons for whom you claim allowances.

1304 ▪ Subsistence

Every officer on active duty receives a basic monthly subsistence allowance, regardless of his dependency status.

1305 ▪ Quarters Allowance

With dependents you receive quarters allowance except when assigned Government quarters for yourself and dependents. If you are assigned quarters adequate for yourself but not adequate for your dependents, or if they are prevented by official orders from joining you, you continue to receive quarters allowance. While on authorized delay or in transit between permanent stations, you rate quarters allowance; also for the interval between the date when sworn in and reporting for first duty.

Without dependents you receive quarters allowance, *except*:

1. When on sea duty (unless on temporary duty not exceeding three months);
2. While in the field (unless on temporary additional duty of less than three months or required to procure quarters at own expense);
3. While occupying or assigned adequate public quarters; but officers in rank of major or above without dependents may receive allowances if they elect not to occupy available government quarters.

Officers without dependents who do not qualify for a full quarters allowance because they are on sea or field duty or living in Government quarters may be entitled to a partial quarters allowance.

In general, the status of an officer without dependents at his permanent station (as long as that duty station remains unchanged) determines his right to quarters allowance. This is true even though quarters may be occupied at some other place than his permanent duty station, including time while on temporary additional duty (TAD), leave, or hospitalization.

For quarters purposes, an officer without dependents is in general considered to be on sea duty when on temporary additional duty on board ship even though his primary duty is ashore. An officer is considered on field duty when on service with troops operating against an enemy, either actual or potential.

1306 ■ Pay on Promotion

On promotion, your increased pay takes effect as follows:

Second lieutenants permanently promoted to higher rank are entitled to the pay and allowances of the higher rank from the date stated in the letters transmitting their commissions. Second lieutenants *temporarily* promoted to first lieutenant are entitled to pay and allowances only from acceptance of appointment.

Other Marine officers are in general entitled to the pay and allowances of the rank to which promoted from the date of the occurrence of the vacancy which they are promoted to fill.

1307 ■ Incentive Pay

Incentive pay is additional pay for undertaking an aviation career, submarine duty, or performance of hazardous duty in obedience to competent orders.

The following hazardous duties currently rate extra monthly incentive pay:

a. Frequent and regular participation in aerial flights in a nonrated status or as a non–crew member
b. High or low pressure chamber inside observer
c. Parachute jumping
d. Demolition of explosives as a primary duty, including training for such duty
e. Duty involving intimate contact with lepers
f. Duty as human acceleration or deceleration experimental subject
g. Duty as a human test subject in thermal stress experiments
h. Frequent and regular participation in glider flights

The President may suspend incentive pay in wartime. Finally, you may not receive incentive pay for more than two purposes at the same time.

1308 ■ Advance Pay

Advance pay, known as a "dead horse," is designed to help pay for transportation, temporary storage of household effects, excess shipping charges and living expenses, and securing new living accommodations. Except when absolutely necessary, avoid drawing a dead horse, as it is usually difficult to repay once it is used—and liquidation is generally required within six months.

You may draw advance pay any time after receipt of orders involving detachment from permanent duty station until 30 days after reporting to a new permanent station, provided the orders are not incident to separation from the Service or trial by court-martial. Temporary duty en route is no bar to drawing a dead horse. The amount advanced normally does not exceed one month's pay, but as much as three months' basic pay (less income tax, deduction for social security, and indebtedness to the Government) may be drawn. The only possible advantage of a dead horse is the melancholy one that, should you die before it is paid back, your death gratuity cannot be checked to liquidate the overpayment, and your estate is that much ahead. This is worth considering before transfer to combat, or extensive air travel.

1309 ■ Personal Money Allowance

While occupying positions of great authority, officers above the rank of major general and rear admiral receive yearly personal money allowances as follows: lieutenant general or vice admiral, $500; general or admiral, $2,200; Chairman of the Joint Chiefs of Staff, Chief of Staff of the Army or Air Force, Chief of Naval Operations, Commandant of the Marine Corps, or Commandant of the Coast Guard, $4,000; fleet admiral or general of the army, $5,000.

1310 ■ Midshipmen and Cadets

Midshipmen and cadets from the other Academies receive base pay equal to half the base pay of a second lieutenant with less than two years' service. They receive rations in kind or commuted rations at rates periodically fixed by regulations.

1311 ■ Uniform Allowances for Officers

Regular Marine officers (except those from the ROTC, and certain temporary officers) do not receive uniform allowances. Reserve officers and ROTC graduates get: (1) a $200 initial clothing allowance payable only once; and (2) a $100 additional allowance on going on active duty for more than 90 days, and on reentry on over 90 days' active duty, following any two-year inactive period. Under some circumstances, a single entrance on active duty may qualify you for both payments.

Temporary officers get a $250 uniform allowance on initial appointment, whether regular or reserve.

Reserve officers who have not received a uniform allowance during the preceding four years, may draw $50 if during that time they have completed at least 28 days' active duty.

NOTE: The subject of uniform allowances is complex. For the final word in any individual case, consult Chapter 6, DOD *Military Pay and Allowances Entitlements Manual.*

1312 ■ Flight Pay

To qualify for flight pay (or, as it is now technically termed, "Aviation Career Incentive Pay," or ACIP) you must be designated as a student naval aviator (SNA) or student naval flight officer (SNFO), or rated as naval aviator (NA) or naval flight officer (NFO), and be assigned to an aviation unit having aircraft. During your first 12 years' service, besides having a minimum number of operational flying assignments, you must also meet annual/semiannual prescribed minimum required flight hours. After your initial 12 years in flight status, your continued qualification to fly for pay depends on satisfactory passage of "gates" set by law and listed in Table 13–1.

Table 13–1
Gates/Flight Pay Entitlement

Number of years' aviation service	*Years in operational assignment**	*Entitlement to ACIP***
12	6 or more	First 18 years of aviation service
18	More than 9 but less than 11	First 22 years of officer service
18	11 or more	First 25 years of officer service

*Operational duty assignments are listed and defined in MCO 3710.1D.
**Limited to 25 years officer service (unlimited for WOs).

1313 ■ Pay of Enlisted Personnel

The monthly basic pay of enlisted personnel of the Marine Corps, Navy, Coast Guard, Army, and Air Force may be found in current Pay Tables and is the same, grade for grade, in all of the Services.

Enlisted men may be authorized *commuted rations* in lieu of rations in kind at a current per diem rate determined by law and regulation. If rations in kind be unavailable, they are entitled to basic *allowance for subsistence* at a per diem generally larger than for *commuted rations*.

In general, enlisted Marines are entitled to a quarters allowance with or without dependents under the same conditions as officers.

Leave rations are granted to enlisted men on leave (if they are not furnished rations in kind) at a per diem rate.

1314 ▪ Clothing Allowances

An initial in-kind clothing allowance is granted to each enlisted man and enlisted woman. Six months after assignment to active duty, a monthly Basic Clothing Maintenance Allowance accrues to each enlisted person for the first three years of service; thereafter an increased Standard Clothing Maintenance Allowance continues while the individual remains on active duty.

1315 ▪ Sea or Foreign Duty (S or FD) and Family Allowance

For most types of duty in a ship or under hostile fire or beyond the continental limits of the United States (or in Alaska) as designated by SecDef, enlisted men receive S or FD pay at monthly rates depending upon the recipient's pay grade. Consult your commanding officer or paymaster.

Family separation allowance equal to "without-quarters" basic allowance for quarters is payable to a Marine *with dependents* who is on permanent duty in Alaska or outside the United States under certain conditions when dependents are not permitted to join him.

An additional allowance of $30 per month may be paid to a Marine with dependents who is away from home under certain conditions.

Both of these allowances may be paid at the same time and in addition to any other allowances or per diem to which you may be entitled.

1316 ▪ Reenlistment Bonus

The purpose of this bonus is to encourage Marines to reenlist in military skills that have high training costs and are in critical supply. Enlisted Marines with critical skills who have completed less than ten years' active service may be paid "shipping-over money" of as much as $12,000.

1317 ▪ Retired Pay

Retired pay consists of the amount of pay received by an officer on the retired list. No allowances are paid retired officers. *Severance* pay is a lump-sum payment made to an officer involuntarily discharged from the Service, based on two months' active-duty pay for each year of commissioned service, the total not to exceed two years' pay or $15,000, whichever is less. (Not paid for punitive separations.)

Computation. Generally speaking, retired pay is based on the active-duty pay of the grade in which an officer is serving at the time of retirement, plus periodic increases. Except for physical disability retirement, retired pay is computed by multiplying 2½ percent of the officer's active-duty base pay at time of retirement by the number of years' service creditable for pay purposes, the total not to exceed 75 percent of such basic pay.

Although not so counted for "fogies," six months' service counts as a year in computation of retired pay. Thus, if you retire with 29½ years' service, you receive 75 percent of your active-duty pay.

Retired Pay Accounts. Pay accounts of retired officers are maintained by the Marine Corps Finance Center, Kansas City, Missouri 64197, and retired-pay checks are mailed from that activity. Retired pay is paid monthly. All correspondence regarding retired pay should be directly addressed to the MCFC.

Income tax continues to be withheld on retired pay except for physically disabled officers wholly exempt from payment of income taxes. Unless otherwise requested, all allotments are automatically continued when you retire.

> NOTE: Both active and retired officers may have paychecks mailed directly to banks. You can have any type of account receive the money, but most officers choose a checking account. See your paymaster.

1318 ■ Settlement for Unused Leave

Each member of the Marine Corps or Marine Corps Reserve having unused leave to his credit on discharge or separation from active duty, is compensated for such unused leave on the basis of his basic pay. Payment is made for a career total up to 60 days' unused leave. Thus, as retirement approaches, it is advantageous to keep the maximum accrued leave on the books.

TRAVEL AND TRANSPORTATION OF MILITARY PERSONNEL

1319 ■ Travel Expense and Mileage

The law provides travel expenses for military personnel on a mileage or per diem basis, by rail, private conveyance, steamer, or aircraft, the allowances for travel being computed on the basis of mileage rates and/or per diem expenses allowed for persons in travel status. These rates of mileage and per diem are fixed by regulation and change frequently based on expenses, cost of living, and other such factors. Your disbursing officer will have them; basically, your options are:

a. Transportation in kind, reimbursement therefor, or a monetary allowance in lieu of cost of transportation based on distance in official mileage tables
b. The allowance in (a) plus a per diem in lieu of subsistence
c. For travel within continental United States, a mileage allowance according to current mileage tables

The actual rates of reimbursement and the regulations governing issuance of transportation are contained in the *Joint Travel Regulations* and *Navy Travel Instructions*.

1320 ■ Travel Orders

Travel Status is travel away from your duty station, under orders on official business.

When you apply for reimbursement for travel performed, or for transportation for travel to be performed, you must have travel orders.

Basic authority to issue travel orders rests with the Commandant, who delegates this authority to certain commands.

All travel orders contain the same basic information, and must:

Address the Marine by name
Reference authority to issue orders if originating from other than the Commandant
State the place or places to which the Marine is ordered to travel
State the date on which he will proceed
State the delay authorized in reporting, if any
State the modes of transportation authorized*
State the duty (official/public) to be performed
State to whom he shall report (if so required)
Give accounting data for cost of travel
State that travel is necessary in the public service*

Omission of any of these items may at least delay reimbursement, and may cause rejection of a travel claim as invalid. Be sure you understand your orders before departure, and carry them out exactly.

*Not required under certain conditions.

"Proceed" Time. The dates when you must comply with travel orders, and when you must report, depend on certain phraseology that always appears in orders. Always check to see which of the following expressions appears, and then govern yourself accordingly.

"Proceed." If your orders have no limiting date, and no haste is required in execution, you are directed simply to "proceed." You are allowed four days' "proceed time" before commencement of travel (unless you have also been granted delay in reporting, as described in Chapter 12).

"Proceed Without Delay." When haste in execution is demanded, you are directed to "proceed without delay." You are allowed only 48 hours' "proceed time" before commencement of travel.

"Proceed Immediately." When maximum haste is required, orders are worded "proceed immediately." In this case you rate only 12 hours' "proceed time" before commencement of travel.

A number of additional ground rules apply to computation and availability of "proceed time" when travel orders require temporary or temporary additional duty (TAD), or are received while on such duty. For these rules, consult Chapter 4, *Assignment, Classification, and Travel Manual.*

> NOTE: You do not rate any "proceed time" whatever for travel to first duty station after acceptance of commission; or on assignment to, or relief from, active duty.

Types of Orders. The four types of travel orders generally encountered are:

Permanent Change of Station. This includes transfer from one permanent station to another; travel to first duty station after appointment; call to active duty; change in home port or home yard of a ship (for dependents); travel home from last duty station upon retirement, separation, or relief from active duty.

Temporary Duty (TD). This is duty at a place other than permanent station, under orders which direct further assignment to a new permanent station. While on temporary duty—as distinguished from temporary *additional duty*—you have no permanent station.

Temporary Additional Duty (TAD). This includes travel away from permanent station, performance of duty elsewhere, and return to permanent station.

Blanket or Repeat Travel Orders. These are temporary additional duty orders issued to individuals for regular and frequent trips away from permanent duty stations in connection with duty.

1321 ■ What To Do About Your Orders

Here is what you do when you receive a set of orders. (This is written to apply specifically to temporary additional duty [TAD] orders as described above, as they are the most frequently encountered.)

1. On receiving orders, read them through and check the following points:

Correct rank, name, social security number, and MOS
Departure date
Place or places to be visited
Whether or not you are to report to a given headquarters or command
Mission you are to accomplish
Security clearance
Modes or options of transportation
Whether the orders are signed
First (receiving) endorsement completed

If your orders appear basically incorrect, or if it appears that you cannot carry them out as directed, return them immediately to the issuing officer with an explanation of the apparent difficulty.

2. Check out before departing and check in on return with your adjutant during working hours, or with the Officer of the Day/Staff Duty Officer at all other times.

3. If your orders so direct, you will have to report to some other headquarters or command (these are known as "reporting orders"—those not requiring you to report to anyone are known as "nonreporting orders"). When your TAD is completed at the distant place or station and before reporting, be sure your orders are endorsed and signed, stating the time and date you reported, the date your TAD was completed, and the availability or nonavailability of quarters and messing facilities.

4. If your orders *do not* direct you to report (i.e., are "nonreporting") and if you intend to claim full per diem, and if your TAD at a place is 24 hours or longer, you *must* obtain a certificate of endorsement from the CO of the post, station, base, camp, etc., at that place, to the effect that Government quarters and mess were not available for your occupancy while there on TAD. This entitles you to higher per diem.

5. If, while away from your parent command, you find you cannot carry out your orders as written without incurring additional expense, or if some unforeseen contingency arises, not provided for in the orders, request instructions by message or telephone (which can be

reimbursed at Government expense) before proceeding further. Reimbursement for unauthorized additional expenses or unauthorized travel will not be paid.

6. If orders specify travel by Government aircraft where available, you must use Government air unless you obtain a certificate or endorsement from a transportation officer to the effect that Government air transportation is not available. Under current regulations, Government air is considered "available" if there is a *scheduled* Government plane departing for your destination within 48 hours of the time you plan to leave. If no special mode of transportation—or some other option—is specified in your orders, take your choice.

7. Keep an accurate itinerary, and a record of authorized travel expenses for which you can claim reimbursement (see Section 1325).

8. Turn in your orders to your unit adjutant, complete with itinerary, on the first working day after your return to home station.

1322 ■ Travel Time

Travel time allowed in connection with permanent change of station is either the actual time required, or "constructive travel time," whichever is less.

Actual travel time is computed in whole days regardless of the length of time actually spent traveling in any given day. This requires completion of an itinerary showing all stops of one calendar day or more.

Constructive travel time for commercial transportation is one hour for each 40 miles of travel by rail or bus, with a proportionate part of one hour allowed for any fraction of 40 miles; and one hour for each 500 miles' air travel, with similar proportionate allowance for fractions of 500 miles. One day of travel time is allowed for each 18 hours of commercial constructive travel time.

Constructive travel time for travel by privately owned vehicle (POV) is based on one day for each 300 miles and for any fraction above 150 miles.

Regardless of the actual sequence of travel, constructive travel time is computed in order of POV, commercial surface, and commercial air. It is computed on the basis of the official distance between the points of duty contained in the *Official Table of Distances*.

Regardless of the mode or modes of transportation, only one day of travel time is allowed if the ordered travel is 450 miles or less.

Government and commercial vessel travel time is the actual time required to complete the trip.

Proceed, delay, and travel time are covered in detail in Chapter 4, *Assignment, Classification, and Travel Manual*. There are few parts of the *Manual* more important for a young officer to know thoroughly.

1323 ■ Reimbursement

On permanent change of station, except when performing group travel or with troops, you may choose one of the following optional allowances for travel within the United States:

1. Mileage at prescribed rate
2. Reimbursement for common-carrier travel plus per diem
3. Government transportation, if available, plus per diem
4. Common carrier transportation on Government T/R plus per diem

When you make your own travel arrangements, you may choose between mileage or reimbursement of common-carrier costs (which is limited to cost for travel over the direct route between the points of travel).

Mileage is computed via official common carrier distance or official highway distance, depending upon the mode of transportation used. These distances are given in the Official Table of Distances.

For temporary duty (TD), reimbursement for travel is the same as for permanent change of station as shown above.

On temporary additional duty orders (TAD), which permit per diem reimbursement, transportation is furnished in kind or by transportation request, and reimbursement is at specified per diem. If you choose to pay your own travel expenses rather than use a T/R, you will be reimbursed at a given prescribed rate per mile for the official distance. If travel by private automobile is authorized and used—as more advantageous to the Government—you get a different rate plus a per diem allowance. On this mode of travel, you are authorized travel time and per diem for the actual time necessary to make the trip. If the orders authorize you to travel by private conveyance you are authorized travel time at 300 miles per day, but your per diem will be based on common-carrier schedules, not to exceed time actually used. When orders direct a specific mode of transportation, but you perform travel via another mode, including privately owned conveyance, for your own convenience, you will not be entitled to reimbursement for cost of transportation, or to the monetary allowance in lieu of transportation, unless the authority responsible for furnishing the transportation requests certifies that T/Rs were not available or the mode of transportation directed was not available at the time and place required in time to comply with the orders. Travel time in excess of that authorized by the directed mode is chargeable as annual leave.

1324 ■ Per Diem Allowances

These are designed to cover hotel, meals, tips, taxi fares (other than to and from carrier terminal), laundry, and other incidental expenses. You get per diem for temporary additional duty or temporary duty, including periods of necessary delay while awaiting transportation and at ports during permanent change of station.

In the United States. Per diem rates within the United States are given in *Joint Travel Regulations.* Where Government quarters or mess are available, the allowance is reduced proportionately. If you claim maximum per diem, you must secure a certificate from the local commander that Government quarters and/or mess were not available.

Outside the United States. Per diem allowances vary widely from country to country and are subject to frequent change. They are discussed in *Joint Travel Regulations,* Chapter 4.

1325 ■ Reimbursable Expenses

Certain travel expenses (in addition to per diem) are separately reimbursable. These include:

1. Taxi fares or other local transportation between places of abode and terminals, and between terminals when free transfer is not included; also taxi fares between terminal and place of duty;
2. Tips to Pullman and baggage porters; fees for checking baggage; excess baggage, when approved;
3. Ferry fares and bridge, road, and tunnel tolls, when traveling by Government transportation;
4. *In Government aircraft*: cost of fuel, repairs, nonpersonal services, guards, and storage at other Government fields;
5. *In Government auto*: storage charges, repairs, fuel, when Government facilities are not available;
6. Telephone and telegraph charges incident to duty and arrangement of transportation, but not hotel reservations;
7. Registration fees at technical, professional, or scientific meetings, and so forth, when attendance is authorized;
8. Passport and visa fees, including cost of photographs and birth certificates required in connection therewith, and cost of traveler's checks;
9. Entry fees, port and airport taxes, and embarkation or debarkation fees upon arrival or departure from foreign countries;

10. Besides the above, any incidental expense that can be justified as necessary, and is duly approved as such.

On all the foregoing items, except tips, you may be required to produce receipts in order to support claims in excess of $15. If in doubt on any point, consult *Joint Travel Regulations* as well as your disbursing officer.

> NOTE: In case you and the disbursing officer disagree as to whether a given item is reimbursable (or if you differ on any computation of pay and allowances), you have the right to submit a claim for adjudication by the Comptroller General. The disbursing officer will explain how to go about this.

1326 ■ Travel Advance

Before departure under orders on permanent change of station, you may, if you request, draw an advance on mileage allowance, known as *Travel Advance*. An advance of per diem on temporary duty or TAD orders is also considered a Travel Advance. Do not confuse these advance payments with a "dead horse," described in Section 1308 above.

1327 ■ Travel Reimbursement

After reporting at your new station and getting your orders endorsed, present your original orders with two complete copies, to the designated administrative officer, who will help you prepare your claim for mileage and per diem on your orders. Your completed claim should then be presented to your disbursing officer within three working days. A claim must be filed even if you have drawn an advance.

1328 ■ Travel in Government Conveyance

Travel by Government aircraft or any other Government conveyance is travel in kind. You rate per diem at the prescribed rate. On extended navigational flights for proficiency purposes, if authorized at your request, no per diem is payable.

1329 ■ Travel Hints

Before departure, compute the exact time by which you must report at the next duty station (see Section 1320).

Commercial air is fast, but not always comfortable or convenient. Some financial advantage may result from traveling with your family in slack times when airlines allow reduced rates for spouses. Coach travel on *established* airlines is less expensive and no great hardship. If you are traveling with small children, the rapid completion of the trip is one of air's main appeals.

Rail travel is extremely comfortable and allows good opportunity to see the countryside. Families going by rail may consolidate individual tickets to obtain a compartment, which many regard as the best mode of travel. Amtrak has generous family-plan reduced rates, as well as a 25 percent so-called furlough fare reduction for all active-duty military personnel. (Simply produce your ID card on purchase of ticket; there is no requirement for uniform or leave papers.)

> NOTE: If you want to travel by train on official orders, you can. You may occasionally encounter old regulations or directives whose effect may seem to restrict train travel. But Congress, in P.L. 92–316, amended 22 June 1972, specifically authorizes train travel "on the same basis as travel by other authorized modes."

Travel by private automobile usually results in considerable saving, both in your transportation and that of the car; also it permits a family vacation en route, and lets you pick your own route for scenery and visits. You can take advantage also of the leisurely travel time allowed for this mode of transportation (see Section 1322).

Keep a running itinerary as you travel, including dates and local times of arrival at and departure from each stop. You will need this for your travel claim.

On temporary additional duty, jot down reimbursable expenses (see Section 1325) as you go along, even small ones; they mount up remarkably and are otherwise usually overlooked.

Last but not least—*don't mislay your original orders*. Without them not only will your reimbursement be long delayed, but you will find yourself in bad odor with your new commanding officer.

TRANSPORTATION OF DEPENDENTS AND HOUSEHOLD GOODS

1330 ■ Transportation of Dependents on Permanent Change of Station

Costs of transporting dependents on permanent change of station are paid by the Government.

If you are not traveling by private vehicle, it is simplest to obtain dependents' tickets by Government Transportation Request. You may, however, transport dependents at your own expense and claim reimbursement afterward at prescribed mileage rates within certain maximum ceilings.

In the event you plan to marry while en route to a new duty station (for example, en route to your first station after graduating from

Basic School), your proceed time, leave, and excess travel time are added to your date of detachment to determine the effective date of your orders for the purpose of entitlement to dependent's travel. *It will be to your advantage to discuss this with your disbursing officer and let him advise you.*

Claims for reimbursement must be signed by you unless you are in a casualty status. When dependents' travel is incident to your having been reported as a casualty, the claim will be signed by the senior dependent.

1331 ■ Shipment of Household Goods

Household goods include baggage, clothing, personal effects, and professional books, papers, and equipment. *Not* included, however, are vehicles and boats, wines and liquors, pets, and articles for nonmembers of your family.

Shipment can be made (including crating and drayage) at Government expense on permanent change of station, within weight allowances, under the following circumstances:

Entrance into the Service, or orders to more than 20 weeks' active duty;
Orders to sea, or duty overseas, where dependents may not follow;
Permanent change of station orders while on active duty;
Orders to duty under instruction of 20 or more weeks' duration;
Orders to or from *prolonged* hospitalization;
Honorable separation or retirement;
Death on active duty, or reported dead, missing, or interned;
Transfer between ships having different home ports;
Orders changing home port of ship to which attached;
Transfer between ship and shore station, where shore station is not ship's home port.

Shipment cannot be made:

Before receipt of orders, unless specially authorized by competent authority;
If separation is other than honorable, or if transfer is incident to trial;
For change of station by reservists on duty for less than 6 months.

> NOTE: There is also a "do-it-yourself" household-goods shipment program under which you move your own effects by commercial or rental vehicle and are paid up to 75 percent of what it would have cost the Government to ship the goods. Before doing this (which requires specific authorization), you should get the advice of your supply or transportation officer.

1332 ■ Weight Allowances

Current tables of weight allowances show the maximum weight of household goods that may be shipped by you on either permanent or temporary change of station. Remember that professional books, papers, and equipment (the adjective "professional" being construed generously) are moved without charge to your allowance.

On permanent change of station, you may ship "by expedited mode" (in most instances, simply a phrase for express shipment) up to 1,000 pounds net weight of personal property classified as *unaccompanied baggage*. This shipment should include only high-priority items necessary to permit you to carry out your duties or to prevent undue hardship to you or your dependents, and the net weight is charged against your total weight allowance. This type of shipment is invaluable for uniforms and effects required immediately after reporting.

Household goods in excess of weight allowance may be shipped, but excess costs will be charged to you. However, remember that weight allowances shown are net, i.e. they do not include packing materials.

1333 ■ Storage of Household Goods

Nontemporary Storage. There are at present some 17 situations under which an officer may be entitled to "nontemporary" storage of household effects, not exceeding prescribed weight limitations. Since the length of storage at government expense varies under different circumstances, and since you are subject to excess costs for storage beyond the authorized time-limit, you should check with the freight transportation officer holding your effects so that you will be sure where you stand. Among the most common situations under which you are entitled to nontemporary storage are the following:

Temporary duty pending detail overseas
Change of station from within U. S. to outside U. S.
Permanent change of station with temporary duty en route
Retirement, discharge with severance pay, or reversion to inactive duty
    with readjustment pay (up to one year's storage allowed)
Assignment to government quarters

Temporary Storage. You are entitled to temporary storage at Government expense for up to 90 days in connection with any authorized shipment of household goods. Under certain conditions arising from circumstances beyond your control—such as unavailability of quarters at new station, arrival of your effects before you do, early surrender of quarters, and so forth—competent authority may authorize an additional 90 days' storage. This added time in storage is not automatic; to arrange it, you should consult your traffic management officer.

Prohibited Articles. Whether in temporary or nontemporary storage, effects stored may not include automobile, inflammables, weapons, ammunition, or liquor.

1334 ■ Dislocation Allowance

When an officer with dependents has completed a permanent change of station, he gets a *dislocation allowance* to help pay the numerous extra expenses of moving. This allowance is equal to one month's quarters allowance for his rank; it is payable only once in any fiscal year, except by special authorization or when he is ordered to or from a course of instruction. It is not payable on orders to or from active duty. An officer without dependents is authorized dislocation allowance on permanent change of station if he is not assigned Government quarters at his new post.

1335 ■ Trailer Allowance

An officer on a permanent change of station is entitled to a *trailer allowance* for transportation of a "house trailer," should he own one, within the United States for use as living space. Trailer allowance therefore means the moving or transporting of a trailer at Government expense or subject to reimbursement. If you elect to claim trailer allowance, you cannot claim dislocation allowance (see Section 1334 above) or transportation of household goods.

> NOTE: Always consult your transportation or supply officer, and your disbursing officer, before taking any action in transporting your trailer. It will be to your advantage.

1336 ■ Transit Insurance

The liability that the Government will accept for a lost or damaged shipment of household goods will not exceed $15,000. In addition, the carrier's liability for loss or damage varies from 10 to 50 cents per pound on the actual weight of the shipment; but collection of claims against a carrier is a complicated, frustrating, and often fruitless process. Thus, where the value you set on your household goods exceeds the Government's liability (i.e., over $15,000), you may be wise to purchase additional protection in the form of a commercial *transit insurance policy.*

Should you take out such a policy, be careful to find out exactly what type of coverage you are getting. It is well to note, for example, that most such policies expire when your effects are delivered. Thus, when effects are delivered by van to a warehouse for temporary authorized storage (as in Section 1333 above), your policy will very likely expire as soon as the goods are accepted by the warehouse unless you have made special arrangements to extend your coverage. Further, reimbursement on such policies is computed on the ratio of the declared value of your shipment to the amount of insurance taken out. For example, if you state that your effects to be covered are worth $4,000 but only insure for $2,000, and an item worth $100 is lost or broken, the insurance company will pay only $50.

1337 ■ Check-Off List for Shipping Household Goods

Here is a summary of things you should do and think about when you ship household goods.

Have enough certified copies of your orders (usually six copies for each shipment). Then see your shipping officer at least a week before you plan to move.

Tell the shipping officer if you have professional books and papers to be shipped so that they may be weighed separately and packed without being a charge against your weight allowance.

If you plan to reach your new station before your household goods are shipped, leave or send your wife enough certified copies of orders to initiate shipment; also leave her a power of attorney or written authority to make the shipment.

If you have gold, silver, or other valuables to be shipped, inform your shipping officer in order that special arrangements may be made.

Get all possible information about your housing situation at the new station before you request shipment of your goods.

Request storage at point of origin (i.e., your old station) whenever you are in doubt as to where to ship your goods.

If goods go by van, be sure to get a copy of the inventory sheet from the driver. *Never* sign a blank "certificate of packing" which he may present you with.

If your orders are modified or canceled, or a change of destination of the shipment is desired, notify your shipping officer immediately.

Get from your shipping officer the ETA (estimated time of arrival) of your goods at destination.

Be at home on the day of the expected move. If Marines or Marine supervisory personnel are expected, have some cold beer on ice; moving is hot work.

Make arrangements for receipt of your household goods at destination. If you can't be there yourself, check with your shipping officer to find out whether storage is authorized. In cases of direct delivery by van, you or your representative must be at the new home to receive it.

> NOTE: If returning from foreign shore-duty overseas, your legitimate household effects are allowed to enter the United States duty-free.

If possible, turn over all your household goods for the same destination at the same time, except silver, gold, items of special value, or items to be shipped by express.

Let the movers know about fragile items such as chinaware and delicate glassware.

Keep groceries and food supplies together for proper packing.

Unload drawers in furniture intended for packing and crating. However, if furniture is to go by van, lightweight linens and clothing may be left in the drawers.

Arrange to have your telephone and utilities disconnected.

Label each box, showing its general contents.

DON'TS:

Don't request shipment to some place other than your new station without finding out first how much it will cost you.

Don't contract for shipment with commercial concerns unless you have been authorized in writing to do so by your shipping officer.

Don't be upset if the movers don't show up at your quarters exactly at the appointed hour. It is hard to schedule a move by the minute.

Don't try to get special services from the carrier until after you have checked with your shipping officer.

Don't pack dishes or bric-a-brac yourself. Leave this to professional packers. Usually commercial firms won't pay claims on items they didn't pack.

> NOTE: Though Service people have scant option as to when they move, the best time of year to schedule movement of household effects is from October through May (a period when only about 30 percent of all moves take place). In any given month—according to the Defense Department's Military Traffic Management and Terminal Service—the best time to move is between the 3rd and 25th. In other words, if you want better, quicker, more careful handling of household effects, don't move in the summer or at the end of a month—if you have a choice.

ALLOTMENTS AND TAXES

1338 ■ Allotments

As a matter of convenience and to facilitate regular monthly payments, you may make allotments of your pay for certain purposes.

When you make an allotment, your pay is checked that amount, and the Marine Corps mails it monthly to the designated recipient. To stop an allotment, notify your paymaster.

You may grant allotments to a bank for support of your family, for a checking account, or for savings, and you may make allotments for purchase of U. S. Saving Bonds. You cannot grant allotments to repay indebtedness, *except* for life insurance premiums on your life; for repayment of emergency loans from Navy Relief or the Red Cross; or for repayment of indebtedness to the Government.

Allotment checks are mailed on the last day of the month of checkage.

You should register an allotment for support of your family as soon as you are ordered overseas so that your dependents can rely on uninterrupted support—especially if you are headed for combat.

1339 ■ Income Tax

Your pay is taxable income and is subject to Federal and state withholding tax at its source. The following, however, are not taxable: quarters and subsistence allowances; disability retired pay; and family separation allowance. Tax exemptions for personnel serving in combat zones during hostilities are covered in Section 1341.

Marine Corps withholding tax procedure provides that the paymaster establishes your withholding rate, based on your rate of pay; this rate changes when your pay changes. Tax deductions are checked on your pay record in the same fashion as allotments. At the end of the year, the paymaster furnishes you a withholding statement to be filed with your income tax return. You in turn must inform the disbursing officer of your tax-exemption status so that he can apply the correct rate.

When hospitalized in a naval hospital for a given period, you may, if certain conditions are met, exclude from taxable income a certain portion of your pay, which is known as "sick pay." Check with your legal assistance officer (or that of the naval hospital) to determine eligibility. This can be a substantial tax benefit and should not be overlooked.

Additional tax tips regarding your personal income tax problems, responsibilities, and exemptions, will be found in Chapter 23 ("Personal Affairs").

1340 ■ Social Security Tax

Social security coverage extends to Marine officers on active duty, and requires the withholding of social security deductions from your pay.

These taxes are computed on your base pay for grade and length of service, and are deducted at rates prescribed by law.

The amount subject to withholding and the amount of tax withheld are reflected on the Internal Revenue Service Form W-2 furnished you by your paymaster at the end of each year.

1341 ■ Combat Pay and Tax Exemptions

Combat pay (technically termed "Hostile Fire Pay") is provided for all military personnel serving within geographic limits established by the Secretary of Defense during hostilities and meeting certain criteria of exposure to hostile fire or enemy action. This pay is the same for all ranks and is taxable to the same extent as other pay. Your disbursing officer can advise you as to eligibility.

Income tax exemption for officers and men serving in combat areas may be placed in effect by executive order of the President. This exemption extends to all military pay of enlisted men and warrant officers; and to the first $500 per month of taxable income received by commissioned officers. Here again, your disbursing officer can advise you as to eligibility and the precise provisions of the effective Executive Order. Note that the geographic areas of this tax exemption bear no relation to, and do not necessarily coincide with, the combat pay geographic limits mentioned above.

As to pay, I beg leave to assure the Congress that, as no pecuniary consideration could have tempted me to accept this arduous employment at the expense of my domestic ease and happiness, I do not wish to make any profit from it.
—George Washington (to Congress, on his appointment as Commander-in-Chief, 16 June 1775)

14

New Station

No statistician has ever totaled the endless adages about the importance of good beginnings, but one thing is certain: as far as your Marine Corps career is concerned, all of them are true.

The instant you show your face on a new station, you come under close observation and appraisal. And when you report for duty on your *first* station—whether in garrison, in the field, at sea, at school, or beyond the seas—you begin to lay the foundation of the service reputation which will make or break your future.

How you conduct and carry yourself, how you wear your uniforms, how you behave, how much you know, pretend to know, or don't know —by all these details you are judged.

1401 ■ Preliminaries

Toward the end of Basic School you will have received your primary MOS (see Chapter 15) and orders to your first station. In the majority of cases this will be a Fleet Marine Force unit at Lejeune, Pendleton, or Okinawa. You will have also been authorized some delay in reporting, which will give you an opportunity to catch your breath after Quantico and to square yourself away for further adventures. If you are ordered to one of the Marine divisions or air wings, your first assignment will be predetermined by your MOS, that is, by your military specialty. If, however, you are ordered to a non-FMF unit or command, it is good practice to write ahead in order to introduce yourself and to assist your new CO in deciding where and how you can best be fitted in. A good example of such a letter (which should be formal but unofficial) follows:

Chief of Staff
Marine Corps Recruit Depot
San Diego, California 92140

Dear Sir:

    I have just received orders to report to the Recruit Depot not later than 20 June 1977. My present intention is to arrive in San Diego on the 18th and report for duty at 0800 on the 19th.

    I am married but have no children, and my wife will remain on the East Coast for the time being until I have an opportunity to find suitable housing and get settled in whatever duties may be assigned me.

    Although commissioned under the NROTC program, I have had three years' enlisted service in the Marine Corps, including one year with the rifle range detachment at Camp Lejeune, and believe I could do well on the range or with any of the recruit battalions.

    Having had no previous duty at San Diego, I would appreciate it if I could be sent any orientation or general information literature as to the Base, together with any particular instructions you may have for me.

    I am looking forward to this tour of duty with pleasure and hope I may render useful service.

                              Very respectfully,
                                  Wharton Burrows,
                                  2dLt USMC

If you have a neat official photograph of yourself in uniform, it is not a bad idea to enclose it with such a letter, as this will enable your future CO to get an immediate impression and begin placing you.

> NOTE: Remember you will be reporting in to your new station or unit in uniform of the day. This requires prior planning on your part to ensure that you have a complete uniform while traveling between duty stations. You should also have enough uniforms (e.g., utilities) so you can go to work immediately if the need arises. *In no case* should you ship *all* your uniforms and assume they will be ready and waiting at your new post.

REPORTING IN GARRISON

1402 ■ Reporting on Board in Garrison

There are as many ways to report in to a new post as there are individuals. During your career you will see them all: the procrastinator, tearing out in a taxicab, five minutes before midnight on the last day his orders allow; the travel-worn parents with house-trailer, children, and wilted clothes; the careful man who arrives two days early and scouts the lay of the land before reporting.

Sentries, such as the one pictured above, stand ready to direct you to your destination on base.

Without considering the trouble in store if you miscalculate your reporting date, or if you arrive late for *any* cause, legitimate or not, you may be sure that last-minute arrival is a risky business, and one which may start you out off balance. So allow ample time, whatever else your personal logistics call for.

Let's assume, then, that you have budgeted "proceed, travel, and delay" (if any), that you've arrived at a city which adjoins your first station, and that you're there in plenty of time—at least a day to spare.

If you have a car, many problems are solved. In earlier times, for reasons of good order, public safety, and morals, junior officers were often discouraged from keeping automobiles. Those days have passed.

If you can afford a car, especially at reporting-in time, you save many steps and some Government transportation.

So here you are, at the threshold of your first station.

Put on your best civilian clothes (or, if you prefer, green service A). Get a haircut and a shoeshine. Wear headgear. Drive out to the base, show your identification card (or orders, if you have not yet been issued an ID card) to the sentry at the main gate, and ask him to direct you to the adjutant's office. The adjutant is the staff officer who traditionally receives newcomers to the command.

> NOTE: Certain large posts (such as Camp Lejeune), which include several commands, now have a "Joint Reception Center" where all company officers report initially. After leaving the JRC, you proceed as described herein, reporting to the adjutant of the command to which you have been assigned.

Before you enter the adjutant's office, remove your hat, knock (or hesitate in the doorway until invited in), and have your orders handy. Introduce yourself informally: "Sir, I'm Second Lieutenant Nicholas. I have orders to report in tomorrow, and would like to find out where and when I report, the uniform, and any information you may have on my assignment. Also, perhaps you can tell me if there's any mail for me at this station?"

The adjutant will probably have advance information of you, and will know a good deal about your immediate future. He will, in any case, look over your orders, and, unless extremely busy, chat a few minutes, if only to size you up and get you off to a proper start.

Find out the exact time and place for reporting, the name and title of the officer to whom you report, and the required uniform. This last information allows you to visit the post exchange to purchase any uniform items you may have overlooked.

After a short visit with the adjutant excuse yourself, *leaving word where you can be reached by telephone off post*, and *pick up your mail.* Depending on the organization of your new post, you may find mail in any or all of several spots: the base post office, your prospective unit (see the sergeant major, first sergeant, or mail clerk), or perhaps a unit post office. Mail is not only a pleasure to receive, but often of much importance if you have been out of touch with correspondence during travel time or leave. Getting mail and catching up with your affairs is an important part of arrival at a new station.

Usually an officer reports for duty at the commencement of office hours, or at such other time during the forenoon as the commanding officer desires. This information can be ascertained in advance of reporting. If, in order to comply with the letter of orders, you must report at night or after working hours, the officer of the day or staff duty officer will receive you, log you in, and, if necessary, provide overnight accommodations. Next day, you then report formally to your new commanding officer.

From the moment you step on board, you must be on tiptoes, fit and ready to do whatever may be required—and do it instantly.

Your greens should be freshly cleaned and pressed. Your shoes should shine, your brightwork gleam. Your field scarf should be neatly tied and securely in place. Your hair should be cut and your face newly shaven. Check to make sure that pockets are buttoned, and that all insignia are in place and correct. Look yourself over in the mirror for a last-minute check.

With you should be: your original orders, staff returns (see Section 1202), and miscellaneous papers that have to do with the day's business (baggage-checks, check-in sheet, information booklet, and so on).

If you are staying off post and have no transportation, or if you arrive by train or air, telephone the post and ask for the motor transport despatcher. Identify yourself, explain that you are reporting under orders, and ask that transportation pick you up and drive you to post headquarters. Be sure to tell where you are and who you are (so that the driver will know whom he is to look for), and emphasize that you are traveling under orders. Have the despatcher tell you when you may expect to be picked up. If, for any reason, the despatcher cannot provide for you, you are authorized to take a taxi and claim reimbursement on your travel orders.

Once on board, enter post headquarters and present yourself at the time and place previously ascertained (if you've had opportunity to conduct the preliminary reconnaissance described above); if you are in any doubt, report to the adjutant. He will take your orders and have them endorsed. He will also take up your qualification jacket. He will show you to the office of the commanding officer or executive officer (on large posts, possibly the chief of staff, the deputy chief of staff, or the G-1).

On cue from the adjutant, step smartly into the office, uncovered, halt at attention two paces before your senior's desk, and say, "Sir, Second Lieutenant Zeilin reports on board for duty." *Do not salute*; Marines do not salute uncovered or indoors (except when under arms).

The officer to whom you are reporting will then have you sit down, will put you at ease, and will chat with you a few minutes, both as a matter of courtesy and to fix you in mind. Do not get flustered; answer questions briefly and directly, and sit erect without slouching or fidgeting. Do not smoke unless invited to do so. The end of the interview will usually be indicated by instructions that you report to some lower headquarters, or commence the prescribed check-in procedure. Unless you are urgently needed, your commanding officer will almost always ask whether you have had time to "get squared away," and will allow you reasonable opportunity to attend to such personal matters as housing, administration, and the like.

After you leave "the front office," retrieve your orders with their reporting endorsement. You'll need them all day.

Your next stop probably will be the supply office, for *quarters information*. At large posts, there is usually a housing office, which not only assigns Government quarters but can give you leads on off-post housing. Because Government quarters are assigned (with few exceptions) on a first-come, first-served basis, present yourself immediately to whoever runs the quarters list, and see that your name is placed on that fateful roster, which determines when you move in. (At some posts, your control date on the quarters list is determined by the date of detachment from your previous station; it is of course in your interest to verify that this is correct.) Before making any arrangements, let alone leases, for off-post housing, *be sure to get your change-of-station orders endorsed to the effect that public quarters are not available*. If you lease ashore before getting an endorsement, you may find yourself being moved into quarters anyway and, in any case, losing your quarters allowance.

Once you've gotten yourself housed, the *disbursing office* (or *"pay office"*) should be your destination. Ask for the NCO or officer who handles officers' pay accounts. Here you'll provide certain personal statistics and here, also, you can draw pay. In addition, the disbursing office will pay you whatever *travel allowances* are due—a much-needed bonus which helps bring your pocketbook back to level after the expenses that travel entails. During their long careers, paymasters become familiar with the fact that most officers who report in are low on cash, so don't hesitate to reveal your needs and ask to be paid as soon as practicable.

With orders endorsed, quarters arranged, and pay in your pocket, you'll be ready to claim whatever *baggage and effects* you may have shipped from your home or former station. The Shipping and Receiving Section, Post Supply Office, takes care of this by holding your gear until you report in and claim it. Again, your orders are necessary. You are entitled to temporary storage of your effects as described in Section 1333. If your baggage or effects have not arrived, leave word with Shipping and Receiving where you wish to be notified when your gear gets in. Shipping and Receiving will deliver it to Government

Learn your way around as soon as you join a new post.

quarters, or to any point off post within a reasonable radius. After that, it's your problem.

You should now leave your *health record* at the sick bay (if it hasn't been mailed separately from your old station). You may find out where this is before you leave post headquarters. If your new station has a standard check-in procedure for officers newly reporting, the checksheet will tell you where to find your sick bay. With health record in hand, enter and ask for the record office. Here a Navy hospital corpsman will accept your health record, will enter you on the records, and, perhaps, will verify that you are up to date on your immunizations, or "shots." If you need dental work, or any routine medical assistance, now is the time to make your needs known. Many officers, on checking in with the sick bay, improve the occasion by picking up a small stock of medicine-chest remedies: aspirin, APC, antiseptic solution, burn ointment, and the like.

By the time you've covered the rounds just described, you will more than likely be ready for a bite to eat. This can usually be obtained at the *Commissioned Officers' Mess* (see Chapter 24), where you can meet some of your new messmates. If you're pressed for time, most post exchanges have a short-order restaurant; here you can eat on the run—less elegantly perhaps, but enough. Lunch at the Mess, however, will enable you to take care of another reporting obligation, that of *joining the Mess*. Here you receive your membership card, pay dues (if required) and ascertain the privileges and obligations of the Mess.

Last, but by no means least, if driving your own car, present yourself to the provost marshal and have your car registered. Marine Corps stations require that cars pass rigid safety examinations and that you possess liability insurance. Nothing can be more troublesome than not being able to meet post requirements for registration of a car; without post registration you park outside and walk from there. Advance precautions to have your car in good shape and fully insured may save you days of walking, plus much vexation.

> NOTE: Never park in space reserved for someone else. Such spaces are usually marked by signs or numbers painted on the curb or surface. Few events annoy a senior more than to find his space preempted by a junior.

1403 ■ Shaking Down

As you settle down in your outfit and assignment, your success will depend largely on your common sense, application, willingness to learn, and skill in human relations. Here, however, are a few tips:

Learn quickly to associate as many *names, faces, and jobs* of the officers and men about you as you can. Study your men's service records (which include photographs and background information).

Read bulletin boards—not only the current, but all the past accumulation that most bulletin boards display; it may be old hat to the plankowners, but it's background to you.

Don't forget your *official calls* (see Chapter 24).

Study the Tables of Organization and Equipment for your organization. The S-3 and S-4 can provide these.

Read Post Regulations; they can keep you out of much trouble.

Study your organization's *general orders and standing operating procedures* ("SOPs"). You can get copies from your first sergeant.

Find out the *mission of your organization*.

Learn the geography of your post by map and personal reconnaissance. "I'm a stranger here myself" is a poor reply for an officer to give.

Above all, strive to know your men. As you begin this ceaseless, vital Marine Corps task, take down your *Marine Corps Manual*, and read the article entitled "Relations Between Officers and Men."

1404 ■ Orienting Yourself

As you shake down, the sergeants in your unit can be a new lieutenant's best professional friends. Both parties observe proper military courtesy and maintain mutual respect.

An officer—especially a new one—should never be too stiff-necked or proud to learn from anybody who knows more about a particular subject.

Some commanding officers work out an informal orientation dealing primarily with internal administrative matters—mess, supply, paperwork—for new junior officers. If nothing of the kind is directed, you may find it desirable, after touching base with your captain, to do the following:

1. Ask the first sergeant to assemble copies of standing orders and manuals you ought to read—and read them carefully. Make friends with, and respect, the first sergeant.

2. Go to the supply room and find out the basics of obtaining, caring for, and accounting for supplies, equipment, and other property. There is a lot to learn about these subjects. Find out how weapons are safeguarded, and about ammunition stowage. (Incidentally, familiarize yourself with safety regulations.)

3. Visit the battalion or other mess that feeds your unit. Catch the mess sergeant at a slack time and ask him to explain how the mess and galley operate, how he draws rations, and how he handles his mess force.

Besides all the foregoing, read up on your job, using the fine professional manuals, both Marine and Army, which bear on your duty and unit. You will be surprised how soon you become recognized as professionally qualified.

REPORTING IN THE FIELD

1405 ■ Reporting in the Field

The circumstances under which you may join your first command in the field are as various as the world's geography and climate. A force on occupation duty leads a different life from a Marine division or aircraft wing at the peak of a campaign. Nevertheless, in preparing for field duty, here are some useful rules:

Get all the briefing you can, especially from those who have recently returned. Accept only guardedly advice or information from anyone who has not been on the spot recently. Find out, if you can:

The local *climate*.

The *uniforms* worn; also whether these uniforms may be procured after you arrive, or should be brought with you.

When you join an outfit in the field, like these officers in Korea, you must locate your enemy and adjacent friendly units, and learn the key terrain features.

The correct *mailing address*. This will enable correspondence to meet you rather than lag weeks behind.

Local shortages, or *hard-to-get items*. These are the things you'll want to have with you. Look through Section 2111 as a guide to what you need, and what you can afford to discard.

Don't tarry in moving forward. "The pipeline" affords many delays. Overcome them, and press forward to your destination.

Travel light enough so that you can keep all your baggage with you. For a junior officer this usually means Valpack, haversack, and little more. Keep your baggage tagged with your name and destination.

Keep your travel orders and staff returns on you. Get a notebook and pencil, and keep them handy.

Try to reach your destination with two or more hours' daylight to spare. Night is no time for strangers to be stumbling about a new unit. When you report in, follow the procedure described in Section 1402 as closely as circumstances permit. As soon as administrative formalities are complete, you should immediately:

Obtain needed uniforms and field equipment ("782 equipment") from the unit supply officer.

Find out where you wash, sleep, and eat, and when.

Ascertain the likelihood of enemy attack, and the degree of readiness being maintained. This includes blackout rules, where you wear sidearms and helmets, and when; also the password and countersign, and the presence of minefields and entanglements. Find out what you do if an enemy attack takes place immediately.

1406 ■ Taking Command in the Field

If you are taking over a command, especially one in contact with the enemy, you must:

Ascertain your mission and be sure you understand it. Know the degree of readiness required of your unit.

Meet your subordinate leaders, so that you can identify them, and they can identify you. Meet your leading NCO and keep him with you.

Walk your front lines or perimeter. Locate your unit's boundaries on the ground. Identify adjacent units. Inspect individual positions. Locate the enemy. *Show yourself to your men.*

Inspect your unit's weapons and equipment. If the situation permits, hold emergency drill.

Check your interior and exterior guard and security.

Verify your communications. Be sure you are familiar with all emergency signals.

Ascertain plans for, and amount of supporting fire available. Know how to obtain it.

Check your supply situation. This includes, at minimum, ammunition, water, and rations—"Beans and bullets," the old phrase puts it.

Make a thorough sanitary inspection, covering heads, urinals, garbage and trash disposal, and water supply. Don't overlook general policing of your area.

Know how to get medical assistance and how to evacuate casualties.

And, finally, remember that you can count on the Marines around you, just as they are depending on you.

OVERSEAS TRAVEL AND FOREIGN STATIONS

1407 ■ General

About a third of your career (other than expeditionary or war service) is spent on foreign stations. Section 213 contains a list of overseas commands where Marine officers may serve; and Chapter 8, the foreign stations where Marine Barracks or detachments are maintained. Chapter 22 contains general information on shipboard life and routine. All this will be of use as you prepare for your first trip overseas.

Probably you will be visiting a country that is new to you. You may miss some conveniences and facilities to which you are accustomed at home. Language, customs, national characteristics, and living habits may well differ markedly from your own.

Learn to view foreign usages and characteristics with understanding, and without insularity or provincialism. If only for the success of our missions overseas, Marines must earn the friendship of the people in whose countries we serve. Self-discipline, courtesy, tolerance, good humor, and generosity are the best ambassadors.

And one word more: remember Laurence Sterne's dictum on foreign travel—"An Englishman does not travel to meet English men." When abroad, meet the people and live the life of the country where you are stationed. Otherwise you might just as well never leave home.

1408 ■ Personal Effects

One thing you must bear in mind is that, even though ordered to sea or foreign service in one part of the world, you may suddenly find yourself on the way to some place quite different. Thus you must select wardrobe and personal effects which, with minimum weight and bulk, keep you prepared for duty anywhere—from the Arctic to the Caribbean, or from Vietnam to Alaska.

Your maximum baggage should comprise: regulation foot-locker, Valpack, and haversack or despatch-case. Pack by "echelon": (1) toilet articles, towel, change of clothing, flashlight, pocket flask, orders and staff returns, and a book—all in your haversack; (2) high-priority clothing and articles that will be needed immediately, in your Valpack; and (3) reserve clothing and gear for the long pull, in your foot-locker. In this way you can, if need arise, shed your luggage, allowing foot-locker (or at the bitter end, even your Valpack) to follow. To be absolutely sure none of your essential gear goes adrift, there is but one safe rule, as voiced by a well seasoned old-timer in the Corps: "Sit on your baggage and keep your orders in your pocket."

Don't overlook bedding. If possible, draw and carry a regulation sleeping bag and air mattress, unless you are sure—very sure—that you will have adequate sleeping accommodations ahead.

Common sense and the advice of officers who know your destination are the surest guides on what to take overseas. Regardless of destination, however, never be without: a complete suit of greens (with garrison cap, to save space) and accessories; at least one suit of utility clothing; field boots; field jacket; and regulation raincoat. If headed for field service or occupation duty, look through the list in Section 2111.

Do not take: valuable papers, such as insurance policies, deeds, stocks, bonds; or irreplaceable jewelry; or anything else you cannot afford to lose. Leave such items in a safe-deposit box. And be sure your wife's or next-of-kin's power of attorney contains authority to get into the safe-deposit box. Consult Section 2320 with regard to this.

1409 ■ Moving Dependents Overseas

Overseas travel of dependents on an accompanied tour via Government transportation is usually contingent on your having adequate housing (or good assurance thereof) for them at the destination.

When possible, the Marine Corps tries to arrange for you and your family to travel together to an overseas station. If there must be delay and separation before the overseas area commander allows your dependents to join you, the Corps will do its level best to get them moving quickly.

Get the latest information on living conditions in the overseas area, including climate, housing, food, educational facilities, shopping, recreation, servants, and medical care. Such data will help you decide what to take, whether children should stay home in boarding school, and similar problems. If you require particular information on these or any other matters connected with travel overseas, do not hesitate to ask the Transportation Section, Marine Corps Headquarters.

1410 ■ Preparations for Travel Overseas

Passports. You and each member of your family will require passports, unless you are ordered to one of a few areas where this rule is waived. Your change of station orders will normally include instructions that you obtain a passport, and Marine Corps Headquarters will assist in obtaining it with minimum delay. If you are anywhere near Washington, get your passport in person. If you cannot work through Washington, apply to the State Department Passport Agencies in New York, Chicago, New Orleans, Boston, or San Francisco—or, if no other source is at hand, apply to the clerk of the nearest United States Court. In any case notify Marine Corps Headquarters by telegraph when you apply. Regardless of what agency you deal with, have the following items with you when you apply.

Original orders overseas
Birth certificate and those of your dependents
Evidence of naturalization if you or your dependents are not native
   U. S. citizens
The number(s) of past U. S. passports held
Your ID card and other supporting identification
Three (3) passport-size photos, full-face and uncovered, of each person (your photo should be in uniform).

Visas. Be sure you have all visas required by countries en route, or at your destination. Again, your most reliable source of up-to-date information on this is the Manpower Department, Marine Corps Headquarters. The nearest consulates of the countries in question can of course answer your questions and issue visas when needed. Remember that you cannot get a visa until you have your passport.

Physical Examinations and Immunizations. All persons, both you and your dependents, must have physical examinations and must complete certain immunizations, or "shots," before going overseas. The requirements for both vary from time to time, according to conditions at your destination. Have your immunizations recorded and certified on your Navy Medical Department "shot card" or on an *International Certificate of Inoculation and Vaccination of the World Health Organization* (PHS Form 731); and be sure that every "shot" is recorded in your health record. Otherwise, you may find yourself getting a double dose of inoculations every time you step ashore.

You can get the necessary physical examination and shots from your medical officer. If your dependents are moving alone, from an area without a Navy surgeon nearby, have them consult the nearest Armed Forces medical activity, or, under out-patient provisions for dependents' medicare, they may receive shots and examinations from civilian sources (see Chapter 8).

Dependent medical service overseas is usually variable. So is dental service. Both you and your dependents should make every effort to be in tip-top repair before going overseas.

Baggage. Before you pack, ascertain what restrictions are in force as to weight allowances, what items may accompany you, and the precise address to which baggage must be shipped. All this information may be obtained from Marine Corps Headquarters. Mark and tag your gear clearly and indelibly. Put one copy of your basic orders *inside* each piece of baggage.

Forwarding Mail. If you can be absolutely sure, find out your new mailing address in advance, and send out official change-of-address cards to all your correspondents and to all business firms and publications with which you deal. If you are not sure, wait until you arrive, and then send the cards *at once*. Your unit mail clerk can give you as many of these cards as you need. When stationed in a foreign country, have your magazines, parcels, and any dutiable articles sent to you via the nearest Fleet Post Office (FPO) or Army Post Office (APO) (see Section 1218). This not only gets you domestic rather than appreciably higher overseas subscription rates, but enables you to receive Stateside parcels without the red tape of foreign customs. *Leave a forwarding address everywhere you stop.*

1411 ■ On Foreign Station

Arrival. Since you are traveling under orders (and with an Official or perhaps a Diplomatic Passport), you will have few if any problems with foreign customs or immigration authorities. You will probably be met by some representative of the military community that you are entering. The area or port commander will do his best to move you expeditiously to your destination; or, if delay is unavoidable, he will provide accommodations. But keep in touch with the officer you are relieving (if you know him), as he is directly interested in your safe and speedy arrival.

Language. Where your duties make it desirable, the Marine Corps makes every effort to give you language training before sending you to a foreign billet. Your wife should also enroll in some type of language instruction. Your young children will pick up the language of the country soon enough from servants and other children. To enable them to learn the country and its ways, as well as the language, give serious consideration to enrolling the children in local schools, if this is feasible. In certain overseas areas, the Navy Department will pay all or part of the cost of private schooling in eligible local schools for your children, provided requisite conditions are met. This is an important opportunity you should not overlook.

Regardless of whether you are in a U. S. Service community or on detached service alone, working with the citizens of the country where you are serving, you must perfect your command of the local language. Some commands maintain language tutors and conduct regular classes. If you cannot avail yourself of them, you can almost always hire your own tutor for a nominal fee. Have him teach the entire family.

Ability to speak the language vastly extends your opportunities, protects you against imposition, and earns the respect of all with whom you deal.

Sanitary Precautions. Sanitation and public health abroad may not attain the levels to which you have been accustomed at home. Find out the local sanitary situation as soon as you arrive. *Take nothing for granted.* Be especially careful against insect-borne and enteric diseases. Keep up your immunizations. Drink only pasteurized or boiled milk, and water that has been boiled. Avoid raw fruit and vegetables unless you are quite sure "night soil" is not used as the local fertilizer. Make your own ice at home, using pure water. Your medical officer can advise you on what you can get away with, and what you must watch.

> NOTE: As a sheet-anchor, most experienced officers traveling overseas carry a bottle of bismuth and paregoric, the tried and true Navy diarrhea mixture, which you can get from your sick bay. A word to the wise. . . .

Two other basic sanitary precautions are: know your restaurants, and watch your servants.

In China before the war, Chinese servants used to say of the "chow water" we habitually boiled and kept in the ice-box: "Melicans clazy. Cook cold water to make it hot. Put hot water in ice-box, make it cold. Take cold water from ice-box, make it hot again for coffee. Melicans clazy." But that was only one precaution. Among others, you must be sure your servants are clean, free from disease, and thoroughly checked out in whatever sanitary precautions are needed. Be sure they don't serve you overripe, spoiled food, a subject on which their views may differ from yours.

Shopping. The "bargains" you and your wife may find on foreign station will sometimes seem unbelievable. If you do not know quality, values, *and* the local market, take along some friend who does. Remember that in many foreign countries the price is "all the traffic will bear"; also that the asking price may often not be the intended selling price. Here a good knowledge of the language will help.

Look into the foreign exchange, currency, and tax situations. Find out where you can get the best *legal* rate of exchange (although your hotel will almost always change money, hotels usually charge a commission or give you a poor rate). In some countries, dollar purchasers are accorded purchase-tax exemptions; in others, some types of currency are more readily negotiable and thus get better rates of exchange. Know these fine points and take advantage of them. On the other hand, never demean your country or your Corps by black marketeering, however tempting it may appear, and regardless of what others do. This sort of thing is emphatically *not done* by Marines, and is sternly dealt with in the few cases that arise.

A good precaution is to note things you see in friends' homes and know about—*before you go overseas.* Seek the advice of longer-timers on the station whose taste you admire. One experienced officer's advice used to be—*"Don't buy any more than you can use in two lifetimes."*

But there are many bargains to be found. If you know quality, style, and value, a relatively small outlay may obtain furniture, linens, rugs, bric-a-brac, silver, chinaware, and other fine goods you might never be able to afford at home.

Take it easy at first, however. Look for a while, then buy. Don't bypass your post exchange. Purchase through an exchange gives you some assurance of quality and equity in price. Often, in fact, exchanges can do better than you as an individual, because of mass purchasing and bargaining experience.

Well begun is half done.
—*Horace*, Epistles, I

15

Managing Your Career

A second lieutenant, newly assigned to his first outfit, had the feeling that things were not quite as they should be, and sought advice of the first sergeant, a veteran of five cruises and many campaigns.

"First Sergeant," said the lieutenant, "I don't think I'm getting along with the Old Man. You always stay one jump ahead of him. How do you do it?"

"The answer is simple, Lieutenant," replied the "Top." "*I find out what the old so-and-so wants—and give it to him!*"

> NOTE: The lieutenant in question took the advice and ended up a lieutenant general.

1501 ■ A Balanced Career

If you aim for the top, you must have a balanced career. A rounded career ripens you, and guarantees decision, judgment, steadiness, and practicality at the top. Raw material for these attributes comes in average quantity among most officers. But the extent to which you develop those qualities results largely from the kind of career you pursue.

Among the ingredients of a balanced Marine career are:

Experience in command
Professional schooling
Combat experience
Staff experience
Sea and foreign shore duty

Those ingredients season a Marine. Bear them in mind. And remember, as you go from duty to duty, that while it is the Commandant who assigns you, it is still *your* career. Watch that career as anxiously as a chemist compounding a critical formula. Your career is your critical formula.

1502 ■ Assignment and Detail

The right balance in your career results largely from assignment, or "detail," as it is sometimes still known in the Marine Corps. Your assignments send you to school, to sea, and to the FMF, and determine which of the thousand-and-one Marine jobs you fill. Thus, a balanced career can be attained only through a sound pattern of assignment. It is one of the important functions of Marine Corps Headquarters to see that, during his first 20 years' service, every officer gets assignments designed to develop his potentialities, to give him equal opportunity for advancement to the limit of his capabilities and qualify him for command responsibility appropriate to his rank. (See Fig. 15–1.)

To manage your career and at the same time meet the needs of the Corps is the job of the Officer Assignment Branch, Marine Corps Headquarters. The "Detail Branch" distributes officers to all Marine commands immediately subordinate to Headquarters. These commands in turn assign you according to your experience and military specialty.

Assignments are classed as follows:

Command. Duty as commanding officer or executive officer of any Marine organization.

Staff. Duty on the general, special, or executive staff of any organization above company or squadron level.

Instructor. Duty on the staff or as instructor at any U. S. or foreign military school.

Student. Duty under instruction at any school.

Civilian Component Duty. Any duty in which an officer works directly with the civilian populace, such as inspector/instructor duty with a Reserve unit, or, if an aviator, as liaison officer with an aircraft corporation.

Special Duty. A range of varied and miscellaneous duties, such as sea duty, supply duty, duty as an aide-de-camp, and duty involving flying.

The normal tour of duty for Marine officers on duty ashore in continental United States is three years, although the demands of the Service sometimes require departure from this or any other standard duration of tour. The *geographic tour-length* in an area is five years. Geographic tours are those in which an officer is locally reassigned between major co-located commands (i.e., from 2d Marine Division to MCB, Camp Lejeune).

A tour of sea duty is two years (except for lieutenants serving as detachment executive officers, who spend only a year at sea). Foreign

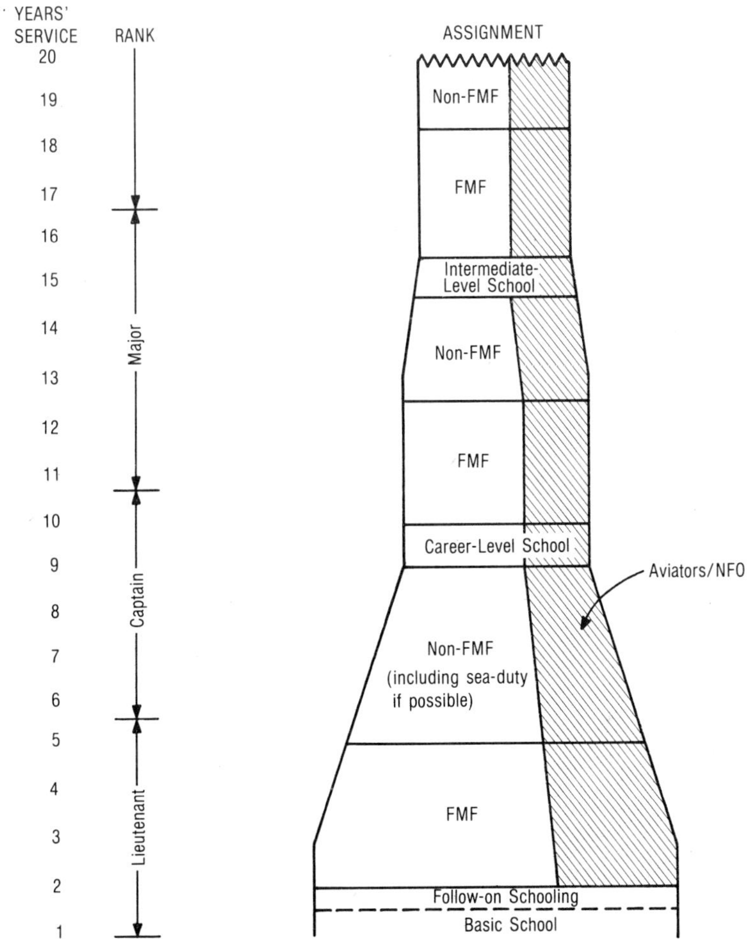

Fig. 15–1: Typical assignment patterns for ground and aviation officers.

shore duty tours run for differing periods prescribed in each locale by Marine Corps Headquarters, and officers are detailed to those duties according to their date of last return from such duty.

Overseas (and certain other) tours are basically classed as "unaccompanied" (without dependents) and "accompanied" (with dependents). You may expect one accompanied foreign tour during your career.

Although it is up to Marine Corps Headquarters, and every commanding officer, to balance your assignments, the fact remains, as has been emphasized, that it is *your* career. This is well recognized, and you have ample opportunity to put your wishes for assignment on record.

On each fitness report, you state your preference for next duty, and your reporting senior must in turn give his own recommendation as to what assignment should be made. In addition, any officer may write an official letter (or submit a "billet-preference form") to the Commandant, requesting a future assignment.

Factors involved in determining who moves where are: date of last unaccompanied overseas tour, school requirements, moves precipitated by school and/or unaccompanied tour completions, career development, time on station, and, of course, special requirements—the exceptional cases—and economy. Timing and individual availability for assignment also bear heavily on final decisions.

You can generally expect to be assigned to overseas unaccompanied tours on the basis of your date of return from previous unaccompanied duty, with respect to all other officers of your grade and military occupational specialty (see Section 1505).

The normal process of detailing officers to the various billets of the Corps revolves around annual preparation of "slates," or lists showing the assignments in each rank which are planned for the forthcoming year. Based on a normal three-year turnover, about one-third of the officer corps should be transferred each year; but, because some billets have shorter tours than three years, a larger percentage usually moves.

Slating is essentially a cross-matching of Marine Corps officer requirements (based on tables of organization, approved manning levels, school quotas, etc.) with officers available for transfer in a given calendar year. The decisions on individual assignments in each case are based on: optimum career pattern (including need for schooling); individual qualifications (such as MOS, college degree, language skill, etc.); present station (to avoid costly, time-consuming cross-country or Atlantic to Pacific moves); and of course the individual's requests, which are automatically updated by data processing on your "individual history card" (a capsulized form of your complete record used in slating and detailing).

Over and above all the foregoing, the factor of overriding importance in slating (and thus in shaping careers) is *performance*. No matter what duty you have had in the past, the singular consideration that will shape your future assignments is an established record of and reputation for consistently high performance across the range of your career.

You often hear it loosely said that "nobody pays any attention" to such requests. This is not true; contrary to all rumor, the Officer Assignment Branch does have a heart.

The literal heart of the process lies with carefully selected officers at Marine Corps Headquarters called *"monitors"* (or, in Navy parlance, "detailers"). The responsibility of a monitor is to manage the careers of officers assigned to his cognizance in the most efficient manner to meet the needs of the Marine Corps. At the same time, your monitor tries to harmonize your best interests and expressed desires with the requirements of the Corps.

To accomplish the above, the Marine Corps maintains an open door policy, in which informal, frequent contact between individual officers and their respective monitors is encouraged. You can always write, call, or, when in Washington, visit your monitor. In addition, the Manpower Department tries to have selected monitors visit major commands on the East and West coasts each year, together with a special trip to the Western Pacific (WESTPAC).

So never hesitate to make your desires known, especially when you feel your career might be broadened by some assignment which may seem professionally necessary to you, such as Fleet Marine Force or schooling. If you pass through Washington, stop by Headquarters Marine Corps and consult your monitor. Fortunately, the Corps is small enough to accommodate its own needs and the desires of its officers pretty consistently. But to accomplish this, you must do your part.

1503 ■ Initial Detail

Undoubtedly, one of your major concerns is where you will be detailed on graduating from Basic School, and here are the procedures used to determine that. These are the first steps in your career pattern.

Before you graduate, you will have a chance to submit a form showing your preference for duty. Here you indicate, in order, choices of occupational field (see Section 1505), preferences of geographic area, and any reasons to support these choices. Your company officers endorse this form with your leadership and academic grades, and an appraisal of your attitude, motivation, and general value to the Service. The form then goes to the Officer Assignment Branch, HQMC.

The ground officer assignment section, as you have seen, includes several monitors. These monitors control the detail of officers into the occupational fields under their cognizance. Just before receipt of the Basic School "Preference Statements" the monitors review their respective quotas based on future requirements. On arrival of the preference forms, the monitors consult and tentatively select candidates for their respective occupational fields. Each monitor selects more officers than he actually needs, so as to provide a margin for conflicts. If you have several talents, you may well be selected by more than one monitor.

About a month before graduation, the monitors come to Basic School and confer with your company officers. Here, differences are discussed;

on return to Headquarters, conflicts are resolved and final decisions made, and orders are then cut and disseminated.

How do the monitors make up their minds?

As to occupational field, the monitors' choice depends on the following factors:

Current requirements of the Marine Corps
Previous military education and experience
Civilian education and experience
Your desires
Basic School staff recommendations

> NOTE: In virtually all cases, lieutenants graduating from Basic School are immediately ordered to follow-on training to qualify them in their MOS. They are then ordered to their new permanent stations or organizations.

As regards geographic area, you will be ordered where the Marine Corps needs you, with your preference taken reasonably into account if possible. In any case, however, you may usually expect orders to the Fleet Marine Force.

1504 ■ Initial and Permanent Precedence

Every new second lieutenant is assigned a date of rank. Those with the same date of rank are assigned temporary initial precedence (see Section 1210), which governs their seniority until publication of the first lineal list, or "Blue Book," after they have finished Basic School. At this time their names are rearranged in permanent precedence within groups having the same date of rank.

Temporary initial precedence derives simply from alphabetic arrangement of officers with the same date of rank. To determine permanent precedence, each officer is compared with all others with the same date of rank, *appointed from the same source*. Relative standing is expressed by a percentile score derived from the individual's class standing in Service academy, NROTC, or civilian college (non–college graduates receive percentile scores based on relative standing attained in the Navy–Marine Corps standard officer selection test). Ties are resolved in favor of the officer with longest previous military training or service.

> NOTE: The foregoing rules apply only to unrestricted regular officers. Reserve, limited duty, temporary, warrant, and former officers are arranged by other procedures.

1505 ■ Your Military Occupational Specialty (MOS)

As you look down a roster of officers, you may at first be puzzled to see some such entry as this:

"1STLT JOHN HEYWOOD, 579-52-8180/0802/0840/0805"

Those mysterious numbers, you will soon realize, are individual identifying badges worn by today's Marines. Those numbers indicate the military occupational specialties, or MOS of the individual concerned.

Your *primary MOS* describes the type of unit you are considered qualified to command; your *secondary MOS* may indicate either other command qualifications or staff qualifications.

The MOS system thus provides the Marine Corps with a running inventory of talent, and indicates at a glance the professional qualifications of each officer and enlisted Marine. For example, returning to Lieutenant Heywood, just mentioned, a translation of the numbers after his name would run:

1STLT JOHN HEYWOOD, 579-52-8180: (Name and social security number)
0802: (Qualified field artillery officer)
0840: (Qualified naval gunfire spotter)
0805: (Qualified artillery air observer)

Obviously, those MOS, and the skills they represent, are extremely important to you. When you report to a new station, those numbers generally determine your assignment, because every duty, or "billet," in the Marine Corps carries the MOS appropriate to that billet.

When you report to Basic School, you are classified as a basic officer. You retain this classification only until basic training is completed and the Marine Corps has had opportunity to size you up. Before you leave Basic School, you are assigned a primary MOS by the Commandant of the Marine Corps as a basic officer in one of the following major fields of command specialization:

Infantry: Occupational field 03
Field artillery: Occupational field 08
Engineer/shore party: Occupational field 13
Armor/amphibian tractor: Occupational field 18
Communications: Occupational field 25
Supply: Occupational field 30
Disbursing: Occupational field 34
Motor Transport: Occupational field 35
Law: Occupational field 44
Military police: Occupational field 58
Aviation ground: Occupational fields 64-71
Student naval aviator: Occupational field 73

When you have enough experience and proficiency in your field of command specialization, you are reclassified as a *qualified* officer in your field. That qualification is shown by your primary MOS. If you later qualify for command in other fields, you receive secondary MOS to denote this fact. As you acquire staff specializations (for example, say, as an embarkation officer, as a naval gunfire officer, or as an operations officer), these are reflected by secondary MOS.

Every Marine officer below colonel (regular and Reserve), and every enlisted Marine, is classified under this system. Once you have your primary MOS, it can be changed only by the Commandant. Should you feel that you have not been correctly classified in accordance with your skills or experience, or wish to qualify in a new field, you may so request in an official letter to the Commandant, who then approves or disapproves the change you desire, taking into consideration the needs of the Corps, your skills, and your wishes.

One word of caution:

Although your MOS labels you for certain duties and patterns of assignment, never let that MOS be a pair of blinders. Avoid overspecialization in fact or attitude. Remember that *every Marine, regardless of MOS, must always be prepared for infantry duties* appropriate to his rank. The genius of the Corps lies in the fact that Marine officers have never let themselves be "jurisdictionalized" into competing branches or watertight professional cliques.

You are, first and foremost, a line officer of Marines; secondary to that, you pursue a major professional specialty. Command and leadership are the only universal military occupational specialties of *all* Marine officers and NCOs.

PROFESSIONAL SCHOOLING

> *Knowledge is of two kinds. We know a subject ourselves, or we know where we can find information upon it.*
> —*Samuel Johnson*

1506 ■ Professional Schooling

You cannot overemphasize the importance of professional schooling.

In school you learn from the hard-won experience of others, and you develop your own professional bent. The schooling you pursue, and your application to professional studies, probably exercise more immediate leverage on your career than any other factor.

Types and Levels of Schools. Professional schools are classified by instructional level as being resident or correspondence; as being of general curriculum, or specialized; as being Marine schools, or schools of other Services.

The levels of Marine officer schooling are: Basic (followed immediately by follow-on training in MOS); Career-Level; Intermediate; and Top. You will find examples in Section 1507.

So far as numbers permit, it is Marine Corps policy that every permanent unrestricted officer goes to school at each level through Intermediate. You may not attend two schools on the same level. For the Top level schools (such as the National Defense University), only specially qualified officers are selected. If you aspire to attend such a school, you must buck stiff competition from your equally ambitious brother officers.

In addition to resident schooling, the Marine Corps encourages all officers to enroll in correspondence or extension courses. You can pursue these on your own time, thus adding to your professional knowledge and better preparing yourself for resident instruction when the time comes. It is the advice of many successful officers that, during your first ten years' service, you should *always* be enrolled in some correspondence course (see Section 1509).

Schooling Outside the Corps. Many Marine officers attend schools conducted by other U. S. Services and foreign nations. This ensures a wide base of professional thinking throughout the Corps, and fosters insight by Marines into every aspect of the profession of arms—land, naval, and air.

It is a Marine tradition, which you should never forget, that when you attend the school of another Service or country, you should return at or near the head of the class, or—as is said half-jokingly—not return at all.

> NOTE: In addition to other overseas training opportunities, Marine officers with the necessary age and educational qualifications are eligible to apply for selection as Rhodes Scholars at Oxford University, England. If chosen, you may have two years at Oxford while remaining on the active list. Apply for information to the Chief of Naval Personnel, via the Commandant of the Marine Corps.

Marine Corps Schools. Although many Marines attend school outside the Corps, Marine Corps Development and Education Command, at Quantico, provides the bulk of the professional education for Marines. At Quantico are schools from basic through intermediate level, together with several specialist schools. Quantico's students come from all officer ranks of the Corps, from other U. S. Services, and from numerous foreign countries. Quantico is the goal of every Marine officer who wants to make the most of his career. For more information about "the Schools," read Sections 512 and 1106.

1507 ■ Resident Schools

A "resident school" is one that you attend in person, as distinguished from a "nonresident school," whose instruction is conducted by correspondence. Marine officers attend resident schools at Quantico, the majority of the schools conducted by the other Services, a number of civilian schools, and a few schools conducted by foreign countries. The list varies from time to time, and Marine Corps Headquarters periodically lists each course or school open to officer or enlisted Marines.

Here are certain courses and schools that you might attend. You may find names of still more from the current directive on this subject.

Basic Level:
The Basic School, Quantico, Va.

Career-Level:
Amphibious Warfare School, Quantico, Va.
Communication Officer School, Quantico, Va.
The Infantry School (USA), Fort Benning, Ga.
The Field Artillery School (USA), Fort Sill, Okla.
The Engineer School (USA), Fort Belvoir, Va.
The Armored School (USA), Fort Knox, Ky.
The Signal School (USA), Fort Monmouth, N.J.

Intermediate-Level:
Marine Corps Command and Staff College, Quantico, Va.
Command and General Staff College (USA), Fort Leavenworth, Kan.
The Naval War College, Newport, R.I.
Air Command and Staff College, Maxwell AFB, Ala.
Armed Forces Staff College, Norfolk, Va.
British Joint Services Staff College, Latimer, England
NATO Defense College, Rome, Italy

Top Level:
Naval War College, Newport, R.I.
National Defense University, Washington, D.C.
Army War College, Carlisle Barracks, Pa.
Air War College, Maxwell AFB, Ala.
Armed Forces Industrial College, Washington, D.C.
British Imperial Defense College, London, England

Generally speaking, the duration of resident courses increases with the level of the school. In quiet times, instruction is more leisurely, whereas during war or emergency, courses are compressed to the maximum extent. The *average* length of a peacetime resident course is from seven to nine months.

The resident schools of the Marine Corps, at Quantico, are open without distinction to officers from the line and from aviation.

1508 ■ Flight Training

If you have a yen to fly, if you can pass the searching battery of physical and psychological tests, if you are a lieutenant with less than three years' service and under 27 when you apply, and a Basic School graduate or student, you may apply to Marine Corps Headquarters and be duly assigned as a student naval aviator or naval flight officer. Flight training is conducted at the Naval Air Training Center, Pensacola, Florida. It does not count as schooling on any of the levels described in Section 1507, and, whether you succeed or not, does not debar you from normal schooling to which your rank and length of service otherwise entitle you. Once you complete flight training and are given your wings, you will be assigned to duty in an aviation unit, probably in the Fleet Marine Force.

1509 ■ Correspondence Courses

Two Marine correspondence schools are open to you: the Extension School, Quantico; and the Marine Corps Institute (MCI) in Washington. All Marine Corps correspondence schooling is free of charge to Marines and Navy personnel serving with the Corps. In addition, Marine officers are eligible to take correspondence courses conducted by the U. S. Naval War College, Newport, R.I., and, if no equivalent Marine correspondence course exists, correspondence instruction offered by the Army and Air Force.

Extension School. The Extension School parallels by correspondence the tactical instruction in the resident schools of the Marine Corps Educational Center. The Extension School also serves as a clearinghouse for applications by Marines to enroll in correspondence courses conducted by the other Services.

The courses of the Extension School are open to all Marines, regular, retired, or reserve. Successful completion of extension work accrues credit toward Marine Corps Reserve retirement, and should be carefully considered by every reserve officer.

All texts and written assignments are furnished by the Extension School, whose staff grades and returns your papers as you submit them. In addition, if you have professional questions relating to the study material issued, the School staff will be glad to help you find your answer.

You may enroll in the Extension School by written request, via your commanding officer, to the Director, Extension School, Marine Corps Educational Center, MCDEC, Quantico, Va. Before enrolling, you should discuss your plans with your commanding officer, and, if your unit is so provided, with your educational officer. Such discussion will enable you to select the course best-suited to your needs and capabilities, and the time available.

The Marine Corps Institute (MCI), located at Marine Barracks, 8th and Eye Streets, Washington, D.C., is the oldest correspondence school in the Armed Forces, having been founded in 1920 to permit World War I Marine veterans to complete interrupted education. Over the years, the MCI has changed from a general, semiacademic correspondence school to one that focuses on the professional development of enlisted Marines.

The following personnel are eligible for enrollment in the Marine Corps Institute:

Marines of any rank on active duty;
Marines of any rank in the Organized and Volunteer Reserve (provided that the courses requested are commensurate with the rank of, and are appropriate for, the individual reservist) ;
Retired Marines, members of the Fleet Marine Corps Reserve, and disabled former Marines;
Eligible members of other Armed Services (as determined by the Service concerned) ;
Civilian employees of the Marine Corps who, in the opinion of their military supervisors, would improve their efficiency and service to the Marine Corps by completing a Marine Corps Institute course related to their specific duties.

You may apply for an MCI course through your commanding officer, or if retired, directly to: Director, Marine Corps Institute, Marine Barracks, Box 1775, Washington, D.C. 20013.

In all, the Marine Corps Institute has some 150 courses, in which there are approximately a hundred thousand enrollments yearly.

Naval War College. The Naval War College extends its correspondence courses to Marine officers. Typical but not inclusive are courses in Intelligence, International Law, and Strategy and Tactics. In addition, the College Library makes loans of certain professional books, and promulgates annual lists of recommended professional reading for Navy and Marine officers. If you wish to enroll, or if you seek further information, address the President, Naval War College, Newport, R.I., via your commanding officer *and* the Director, Extension School, MCDEC, Quantico, Va.

Other Service Courses. If you want to take a correspondence course offered by the Army, Navy, or Air Force, for which the Marine Corps does not have an equivalent, apply through official channels to the Director, Extension School, MCDEC, Quantico. The Extension School will then process and forward your application.

> NOTE: For younger officers of marked ability, the White House Fellows Program is worth looking into. Officer college graduates (age 23–36), as well as civil servants, educators, and journalists, are given in this program a year of first-hand, high-level experience in the workings of the U. S. Government at White House and Cabinet level.

1510 ■ College Degree Program

One of the basic educational goals of the Corps is that any officer who does not have a baccalaureate degree will be offered the opportunity to win one.

Officers who have completed two or more years' undergraduate work are eligible for the *College Degree Program* (sometimes referred to as "Operation Bootstrap") for not more than 21 months' residential instruction at an approved institution. In this program you are ordered in a duty status to the institution in question, receiving normal pay and allowances but in turn meeting the various academic fees and expenses out of pocket. (Most tuition and other academic expenses can normally be funded through the individual officer's GI Bill of Rights entitlements.) To be selected, you may be in any grade, as a permanent regular officer, from warrant rank to that of lieutenant colonel. Applications must reach Marine Corps Headquarters at least six months before the start of the college term specified by you. Special consideration is given officers who have demonstrated interest and application in off-duty educational programs.

1511 ■ Special Education and Advanced-Degree Programs

The Corps has three programs in which officers can be encouraged and assisted to obtain advanced degrees above baccalaureate level.

The *Off-Duty Education Program* is a strictly do-it-yourself arrangement on your own time, but in which the Marine Corps pays 75 percent of tuition costs.

Special Education Program (SEP) officers study in a variety of disciplines for periods up to 24 months. Chosen curricula are tailored to billet-needs of the Corps and may or may not lead to an advanced degree. (Officers, however, are at liberty to take on extra course-loads which, taken with the prescribed curriculum, would earn a degree.) All tuition and related expenses are paid by the Marine Corps, while you continue in pay status.

Advanced-Degree Program (ADP) is essentially an advanced-level version of "Bootstrap" (see Section 1510), in which an officer, paying his own way, gets 18 months in which to gain his degree. Again, in most cases GI Bill entitlements cover academic expenses.

Typical, though by no means inclusive, of fields open to officers in the SEP and ADP are: applied mathematics (statistics); aeronautical engineering; communications engineering; computer engineering; computer science (technical); defense systems analysis; education, curriculum, and instruction; electronics engineering; financial management; management; operations analysis; and public relations/journalism.

Selection for SEP and ADP is made on application to the Commandant, by a Headquarters Marine Corps selection-board.

> NOTE: In addition to all the formal, organized schooling just described, there still remains a place for the traditional battalion or squadron officers' school, conducted by the CO and best-qualified officers of the unit. A well-tested arrangement is to hold it each Friday afternoon, followed immediately by a Happy Hour.

PROFESSIONAL READING AND WRITING

> *Read and re-read the campaigns of Alexander, Hannibal, Caesar, Gustavus Adolphus, Turenne, Eugene, and Frederick. Make them your models. This is the only way to become a great general and to master the secrets of the art of war.*
> —*Napoleon Bonaparte*

1512 ■ Professional Reading

A few officers attain high rank without having mastered the history of war, but they are few indeed. The habit of systematic, planned reading of history, biography, and literature enables you to live up to your profession and to apply the lessons of the past to the future. Remember Metternich's remark: "The past is chiefly useful to me as the eve of tomorrow—my soul wrestles with the future."

Professional reading means more than studious application to field manuals and the regulations by which we steer our course. That type of reading should be taken for granted. So should the reading of service journals. Professional reading even transcends military matters (remember Clemenceau's barbed dictum: "War is too important a matter to be left to the generals"). Your professional reading ought to embrace: military and naval history and biography; U. S. history; maritime history; international affairs; economics; and psychology.

That sounds like a large order. It will not seem so large once you begin. The most important part of a professional reading program can be summed up in one verb: *Read!*

Right now, subscribe to your professional journals: The *Marine Corps Gazette*; the *Naval Institute Proceedings*; and—for Service news—the authoritative, widely read, and well-informed *Navy Times*. If you have a bent for military history, take *Military Affairs* (the only military historical journal in the country).

In Appendix IV, look over the list of Marine books that every officer should know, and start reading them. Better still, start acquiring them. Both the Naval Institute and the *Marine Corps Gazette*, incidentally, sell current books to subscribers at discounts of 10 percent or more. If you want a more extensive compilation of outstanding professional books, the Naval War College, Newport, R.I., will be glad to send you their recommended reading list for Marine and Navy officers.

A final *must* are the high-quality Marine Corps operational histories prepared by the History and Museums Division, Marine Corps Headquarters. Every officer should know them and own them. You can buy them through the *Gazette* Bookshop.

So now it's time to build your professional library and get the habit of reading professional journals.

> P.S.—The Marine Corps packs and ships your professional library from station to station at no charge against your weight allowance and at no cost to you.

1513 ■ The Service Author

Every officer with professional ideas worth expressing should support his Service journals by contributing. There are four good reasons why you should do this:

First, you support the publications that spread military knowledge and raise professional standards.

Second, you give readers the benefit of your ideas and experience.

Third, you acquire a Service reputation for professional keenness.

Fourth—and many officers still don't realize this—you can earn much-needed extra money. Both the *Gazette* and the *Naval Institute Proceedings* pay generously, at rates that compare favorably with civilian magazines.

It is widely believed that elaborate and drastic regulations hamper an officer who chooses to write for publication. This is by no means the case *if* you confine yourself to semiofficial Service journals, such as the *Gazette* or *Proceedings*. Material prepared by active officers for outside publication, however, does require clearance by the Secretary of the Navy, and usually by the Department of Defense as well. And you may not commit yourself to furnish a manuscript for outside publication without prior clearance.

In any case, you may not reveal classified information in an article for publication any more than you can disclose the same information in a letter or in careless conversation. Nor can you represent that you are an official spokesman of the Marine Corps or the Department of the Navy, or give such an impression. Finally, whatever you write for publication should constitute a constructive contribution to the primary missions of the Department of Defense.

Short of the foregoing rules, seasoned by the canons of ordinary good taste and propriety, an active military writer is on an exact parity with outside professional writers.

If, however, you are in any doubt as to the classification or general propriety of an article or manuscript, you may always—in fast, you should—submit it to the Director of Information, Marine Corps Headquarters, who will advise you as to its suitability for publication.

And if you confine yourself to the established professional magazines such as the *Gazette* and *Proceedings*, their editorial staffs will spot any risky passages, and, for your own protection, will either edit them or secure whatever official clearance may be in order.

PUBLIC SPEAKING

1514 ■ Public Speaking

The Importance of Speaking Ability. Distinct, forceful speech is an essential quality for a successful officer. Public speaking ability is a primary tool of leadership. Since a large part of your career will be devoted to explaining, announcing, and teaching, you should learn at least the fundamentals of speaking technique—and the sooner, the better.

Some believe that public speaking ability is a magic gift, bestowed on some but denied to others. This is far from true. Public speaking, like the technique of shooting a rifle, can readily be, and has to be, *learned*. True, aptitude varies; some officers will never be brilliant speakers, others will be. But everyone—you included—can improve his speaking technique with application and practice. *Good speakers are made, not born.*

A highly recommended text (*Basic Principles of Speech*, by Sarett and Foster) has this to say:

> The most effective speaking has all the intimacy of good conversation, but has in addition the power, dignity, range, and force which are required for a formal occasion and a large audience. Any effective speaker—one who gets results—is first a fine human being, speaking honestly; and second—but always unobtrusively—a craftsman. . . .

The Qualities of an Effective Speaker. What are the qualities of effective military speakers?

Be purposeful. You must have a clear view of your objective—of what you are trying to put across. You must be able to balance the time spent against achievement of your objective. Avoid digression into irrelevancy.

Know your stuff. You must have a thorough grasp of your subject. This grasp should be backed up, if possible, by practical experience. Conversely, avoid, if you can, having to talk about matters in which you lack experience.

Take pains. Even if you have all the knowledge, skill, and experience needed to put your subject across, there is no short-cut in preparation. Choose the right approach and method of presentation; arrange your material in logical phases, each one followed by a summary; if you are instructing, use visual aids to help your audience *see* your points. Psychological studies show that 75 percent of all we learn is taken in through the eye, whereas only 13 percent comes through hearing.

Be enthusiastic. Enthusiasm is as catching as boredom. It is the driving force of a good speech or lecture. Your enthusiasm must be *balanced* and *seasoned.* If you make your hearers feel that you are a fanatic with a wild gleam in his eye, they will discount what you say, and soon become bored. And enthusiasm unseasoned by intelligence and humor soon exhausts the hearer.

Cultivate dramatic sense. Don't be content with dull, stodgy presentations; cultivate what show-business people call "sense of staging." Get in touch with the mood of your hearers. Use variety of pace, surprise, emotional and dramatic appeal to drive home your points. If you can tell a funny story well, don't be afraid to use it. But don't indulge in a gag just for the gag's sake. Remember also that there are other (and usually more effective) ways to introduce a subject than by telling an irrelevant "funny" story.

Have a confident, easy manner. Give your hearers confidence in what you say by having that confidence yourself. Speak clearly and distinctly. Remember that distinctness of speech stems from distinctness of ideas. Don't be flowery.

Have the right approach. Your particular approach should be determined by the nature of the audience. Regardless of what kind of audience you face, however, *avoid:*

Flippancy. A flippant speaker displays disrespect for his subject, and usually for his audience also.

Cheap Humor and Vulgarity. Don't play the clown to get a cheap laugh. Vulgarity offends most hearers and, if repeated, bores all. By cheapening your remarks you cheapen yourself and the Marine Corps.

Slangy Diction. Judiciously used, slang can be quite effective. But if you use slang indiscriminately, you lose your effect and grate on the audience's nerves.

"Big Words" and Pedantry. Overuse of technical terms, or of involved, long-winded constructions, wraps your subject in a fog. Five-dollar words used merely for effect do not impress your listeners with anything but your stuffiness.

Make the most of your voice and body. Voice is your basic weapon. To exploit your voice, develop *power, distinctness,* and *variety of delivery.* Your body supports your voice through erect, confident *posture,* natural *movement,* meaningful *gesture,* and *eye-contact* with every listener. Make it a rule to look every hearer in the eye while you speak to a group.

References for Military Speakers. You can improve yourself as a speaker, instructor, and officer by study of a few basic works:

How to Prepare and Conduct Military Training, Army Field Manual 21-6
Army Instruction, Army Technical Manual 21-250
Basic Principles of Speech, Sarett and Foster, 1936
Principles of Speech, Monroe, 1943.

PUBLIC RELATIONS

> *The future success of the Marine Corps depends on two factors: first, an efficient performance of all the duties to which its officers and men may be assigned; second, promptly bringing this efficiency to the attention of the proper officials of the Government, and the American people.*
> —*John A. Lejeune*

1515 ■ Public Relations

Intelligent and candid relations with the public form an important part of every Marine's career, from private to general.

Do not confuse "public relations" with public information. Public information is the technical term for that segment of public relations which has to do with the press, radio, and other information media. Public relations transcends public information.

The keystone of good relations with the public is to deserve the public's esteem. As long as the Marine Corps, seen through the citizen's unsparing eye, measures up to high standards of discipline, devotion to duty, individual smartness, and valor, which are traditional with the Corps, our relationship with the public will be what it should be.

The Marine Corps has never depended on any publicity apparatus except the heartfelt pride and loyalty of every Marine toward his Corps. The best advertisement the Marine Corps displays is the individual Marine.

In season and out, the Corps has drawn its strength and its support from the American public. Deal truthfully, pleasantly, and respectfully with every member of that public, and Marine Corps public relations will continue to take care of themselves.

NOTE: A highly interesting and penetrating short book, *This High Name* (Robert Lindsay, University of Wisconsin Press, 1956), deals with the subject of Marine Corps public relations. It is well worth your time to read.

Where duty calls, or danger, be never wanting there.
—*Hymn, "Stand Up, Stand Up for Jesus"*

16

Leadership

Leadership is a heritage which has passed from Marine to Marine since the founding of the Corps. It has been defined as the art of influencing and directing men so as to obtain their obedience, respect, confidence, and loyal cooperation.

Leadership is mainly acquired by observation, experience, and emulation. Working with other Marines is the Marine leader's school. Although some individuals possess greater instinctive gifts of leadership than others, anyone can sharpen his leadership faculties if he tries.

This chapter discusses attributes common to most successful Marine Corps leaders. It also describes techniques, procedures, and situations which contribute to or demand effective leadership.

As you peruse these paragraphs (or any other so-called text on leadership), remember, however, that the royal road to leadership is not merely to read, but rather to *lead*.

THE MARINE LEADER

1601 ■ Principles of Leadership

The late Douglas Southall Freeman, biographer of Lee and Washington, and master of war history, used to say that leadership boils down to three fundamentals:

Know your stuff.
Be a man.
Look after your men.

A World War I officer of the regular Army ably defined leadership in terms not very different:

Know your men.
Know your business.
Know yourself.

Using language only slightly changed since the original *Articles for the Government of the Navy*, today's *Navy Regulations* cover much the same ground in words which apply to every leader and commander:

> All commanding officers and others in authority in the naval service are required to show in themselves a good example of virtue, honor, patriotism, and subordination; to be vigilant in inspecting the conduct of all persons who are placed under their command; to guard against and suppress all dissolute and immoral practices, and to correct according to the laws and regulations of the Navy, all persons who are guilty of them; and to take all necessary and proper measures . . . to promote and safeguard the morale, physical well-being, and the general welfare of the officers and enlisted persons under their command or charge.

1602 ■ Attributes of a Marine Leader

> The young American responds quickly and readily to the exhibition of qualities of leadership on the part of his officers. Some of these qualities are industry, energy, initiative, determination, enthusiasm, firmness, kindness, justness, self-control, unselfishness, honor, and courage.

So Major General John A. Lejeune summarized the attributes of a Marine leader. Although Lejeune's list can scarcely be improved, it can be enlarged on.

The contagion of example is the central thought in General Lejeune's passage. It is not enough that you merely know a leader's qualities, and not enough that you proclaim them. You must *exhibit* them. To exact discipline, you must first possess self-discipline. As you demand unsparing attention to duty, you must not spare yourself. Example is better than precept.

Much of the power of example, in turn, stems from "command presence," or the kind of military appearance you make.

Command presence is the product of dignity; military carriage; neat, well-fitting uniforms; firm and unhurried speech; and self-confidence. Command presence is one useful adjunct of leadership which can be systematically cultivated. "Spit and polish" is a handmaiden of command presence.

The young American responds quickly to the qualities of leadership displayed by his officers. Leadership is a heritage passed from Marine to Marine. This picture, on Guadalcanal in 1942, catches one Commandant (Holcomb, left) with his two immediate successors (Vandegrift, right, and Cates, third from left), the Corps's collective leadership from 1936 to 1952.

Resolution and tenacity—"an unfaltering determination to achieve the mission assigned to you"—is the fuel of leadership.

Ability to teach and speak usually denotes an effective leader, and enhances whatever latent leadership talents you possess. Cultivate this gift at every opportunity. It is a lever that can decisively influence your career. See Chapter 15 for hints on public speaking and military instruction.

Loyalty downward distinguishes Marine Corps leaders. Loyalty downward means loyalty to protect and foster your subordinates, to assume responsibility for their actions (their mistakes, too), and to see that they receive all credit due them. Above all, *loyalty downward means looking out for your people.*

Encouragement of subordinates is a tradition of Marine leadership. Give subordinates all the initiative and latitude they can handle. Encourage them in professional studies and reading. Make them seek professional schooling.

Professional competence may not make the men like you but will surely elicit their respect. "You can't snow the troops" is an old Marine saying. If you are professionally able, your enlisted Marines will be the first to get the word. Conversely, they will be mercilessly quick to spot a fraud. Demonstrate competence and keenness as an officer, and your men will be content to be led by you. Never be ashamed to be known as a "hard charger."

Here are some classic remarks by one of the Corp's hardest-charging generals, the late Graves B. Erskine:

> The first thing, a man should know his business. He should know his weapons, he should know the tactics for those weapons, and he should not only be qualified for the grade he is assigned to, but at least for the next higher grade.

Education contributes to professional competence. Education and study give you technical proficiency, help you think clearly, enable you to express yourself, and command respect from all with less education.

Physical readiness, though not an end in itself, is essential for every Marine, and thus doubly so for every leader. Unless you can confidently face your physical fitness test, you are not fit for active command.

The spirit of "can-do" and "make-do" is as old as the Corps itself. To do the best you can with what you have, to do it promptly, cheerfully, and confidently, marks you as a leader in the best traditions of the Marine Corps. The world is divided into "can-do" and "can't-do" types. Be sure you fall into the former class.

Adaptability marks a seasoned Marine. As a leader, keep loose; roll with the punches. Cultivate that most admirable trait, "grace under pressure."

Devotion to the Marine Corps and its traditions begets equal earnestness and devotion from subordinates. Take the Marine Corps and its time-honored ways with full seriousness, and so will your command. That is the Marine Corps attitude.

As both summary and comment on the foregoing, here is a thought-provoking list of attributes, chosen by a panel of psychologists as desirable in a unit-commander. How do you measure up?

Serious	Stubborn
Disciplined	Proud
Loyal	Resolute
Authoritative	Hard
Courageous	Inventive
Tough	Austere
Competent	Purposeful
Aggressive	Compassionate

YOU AND YOUR SUBORDINATES

1603 ■ Dealing with Subordinates

Whether your subordinates are officers or enlisted Marines, support and back them to the hilt. They will turn to you for encouragement, guidance, and material support. Never let them down. Nothing should ever be "too much trouble" if it is needed for your men or your outfit. Protect, shelter, and feed them before you think of your own needs.

Demand the highest standards and never let those standards be compromised. Field Marshal Rommel stated this in slightly different words:

> A commander must accustom his staff to a high tempo from the outset, and continually keep them up to it. If he once allows himself to be satisfied with norms, or anything less than an all-out effort, he gives up the race from the starting post, and will sooner or later be taught a bitter lesson. . . .

Live, lead, and exercise command "by the book." Let this be understood by your men.

Keep *responsibility* centralized—in *you*. Decentralize *authority*. Give subordinates wide authority and discretion. Tell them what results you want, and leave the "how" to them. Never oversupervise.

Avoid overfamiliarity of manner or address. If you have feet of clay—and most humans do—overfamiliarity with subordinates is the surest way to advertise it. Field Marshal Rommel remarked on this also:

> The commander must try, above all, to establish personal and comradely contact with his men, *but without giving away an inch of his authority.*

Develop genuine human interest in your men as individuals. Study each man's personality. Seek out his background from his service record. Learn his name, and address him by it. Never let any Marine picture himself as "a mere cog" in the machine. No Marine is a cog.

In your daily exercise of command, avoid the "hurry-up-and-wait" tendency that characterizes ill-run commands. That is to say, think twice before you apply pressure to speed up something if the net result is simply that your people will have to stand around waiting at some further stage. Don't get them out unduly ahead of time for formations and parades, especially if every other echelon has added its few minutes of anticipation, too. And always be on time and on schedule as far as you yourself are concerned. One of the most basic rules of military courtesy is, *never keep the troops waiting.*

Respect the skill and experience of your NCOs. Remember Kipling's line, "The backbone o' the Army is the Noncommissioned man!" Learn from the wisdom of NCOs, but never let them snow you. Do everything

"*A commander must accustom his staff to a high tempo from the outset, and continually keep them up to it.*" Here on Iwo Jima in 1945, a Marine division commander (Major General G. B. Erskine) demonstrates in combat the truth of Field Marshal Rommel's precept.

in your power to enhance the prestige and authority of NCOs, except at the expense of your own prestige and authority. In public, address NCOs by name and rank. In private, you may call them by their last names only. *Never address an enlisted man by his first name or nickname.*

Be accessible to any subordinate who wishes to see you. It is a tradition of the Corps that any enlisted man who desires an interview with the commanding officer must obtain the first sergeant's permission. It is equally a tradition of the Corps that permission is unhesitatingly given unless the man is drunk or flagrantly out of uniform. In connection with such requests, you should give your first sergeant direct and positive instructions that he must report to you, the commanding officer, every complaint he receives from an enlisted man. Most of these need never come to your attention otherwise or in any official form, but this rule helps to avert trouble before it becomes serious.

1604 ■ Issuing and Enforcing Orders

> Promulgation of an order represents not over 10 per cent of your responsibility. The remaining 90 per cent consists in assuring through personal supervision on the ground, by yourself and your staff, proper and vigorous execution.

So wrote General George S. Patton on the subject of orders. Issuing and enforcing orders comprises one of the main functions of an officer.

Before you issue an order, ask yourself if it can be reasonably carried out. If, in the circumstances, an order cannot be executed as given, it should not be given.

Never give an unlawful order; that is, an order which contravenes law or regulations, or demands that your subordinates break the rules. A good test of a lawful order is, "Could a subordinate be court-martialed for failing to comply?"

Issue as few orders as necessary. Keep them concise, clear, and unmistakable in purpose. Anything that can be misunderstood, will be.

Never contravene the orders of another officer or NCO without clear and pressing reason. If possible, make this reason evident when you countermand the order in question. If orders to you conflict, obey the last one.

When you have once given an order, be sure it is executed as you gave it. Your responsibility doesn't end until you have assured yourself that the order has been carried out. Never shrug off half-hearted, perfunctory compliance. "If anyone in a key position appears to be expending less than the energy that could properly be demanded of him," wrote Rommel, "that man must be ruthlessly removed."

An order received from above should be passed on as *your* order, and should be enforced as such. Never evade the onus of an unpopular directive by throwing the blame on the next higher echelon.

It cannot be too often repeated that when you issue an order, make it clear what you want done, and who is to do it—but *avoid telling subordinates how it is to be done.* Remember the old promotion-examination question for lieutenants, in which the student is told that he has a ten-man working party, headed by a sergeant, and must erect a 75-foot flagpole on the post parade-ground. Problem—How to do it?

Every student who works out the precise calculations of stresses, tackle, and gear, no matter how accurately, is graded wrong. The desired answer is simple: The lieutenant turns to the sergeant, and says, *"Sergeant, put up that flagpole."*

When you have given an order, you must be sure it is executed as you gave it. As you can see from this picture of Colonel L. B. Puller in the field on New Britain, issuing and enforcing orders is one of the main functions of an officer.

1605 ■ "R.H.I.P."

As an officer you are entitled to take precedence ahead of your juniors and all enlisted men. This privilege is admitted in the Service proverb "R.H.I.P."—"Rank has its privileges." Just when and where you "pull rank," though, is a matter of some delicacy.

Generally speaking, you should assert your privilege when your time is circumscribed by duty, or when failure to do so would demean your status as a commissioned officer. For example, an officer should not waste his own time and the Government's by falling in line behind privates in a clothing storeroom, or hesitate to claim the attention of an administrative functionary hemmed in by enlisted men. Conversely, in situations where all men are equal, take your place with the others regardless of rank. In the mess, at the barber shop (unless there is an officer's chair), at the post exchange, or at games, avoid taking advantage of rank.

Finally, every Marine officer pulls rank in reverse when it comes to looking out for his men. In the field, before you yourself eat, every enlisted man must have had a full ration. Before you take shelter, your Marines must have shelter. "There is no fatigue the soldiers go through," said Baron von Stueben in 1779, "that the officers should not share."

MILITARY DISCIPLINE

1606 ■ The Object and Nature of Discipline

Effective performance by men in combat is the direct result and primary object of military discipline. Discipline may be defined as prompt and willing responsiveness to orders and unhesitating compliance with regulations. Since the ultimate objective of discipline is efficient performance in battle, discipline may in a very real sense spell the difference between life and death (or, more important to the Marine, between victory and defeat). It is that standard of deportment, attention to duty, example, and decent behavior which, once indoctrinated, enables men, alone or in groups, to press home the accomplishment of their missions.

To many persons, discipline simply means punishment, that is, something to be afraid of. In fact, discipline is a matter of people working well together and getting along well together—and, even if there be a lack of harmony among them, discipline is a means of cementing them as a fighting organization. In the Marine Corps, as in any military organization, it is necessary for people to do certain things in prescribed ways and at given times. If they do so, we say they are well disciplined; if not, we say they are badly disciplined.

Discipline in everyday life is what causes people to obey traffic lights, pay greens fees, go in through entrances and out through exits, return on the expiration of shore leave, and knock before entering. Nevertheless, military discipline differs fundamentally from the disciplines of civilian life. No one is less deluded by the claim that civilian personnel techniques ("management" is the word in vogue) can be used as a basis for military discipline than the man who has had professional military experience in command. Basically, the difference between the discipline of a Marine and that of a worker in industry is that the former, having accepted certain duties and responsibilities on oath, cannot quit tomorrow.

1607 ■ The Basis of Discipline

The best discipline is self-discipline. Self-discipline amounts to the Marine having control of himself and doing what is right because he wants to. To be really well-disciplined, a unit must be made up of men who are self-disciplined. In the ultimate test of combat the leader must be able to depend on his men to do their duty correctly and voluntarily whether anyone is checking on them or not. If time and the situation permit, you should make known to your men the reasons for a given order, since this knowledge will increase the desire of your people to do the job, and will enable them to do it intelligently. You must know what you want of your men, let them know, and then demand it of them. Their discipline in response to your leadership must be based on knowledge, reason, sense of duty, and loyalty.

1608 ■ Characteristics of Effective Discipline

Until men are severely tried there is no conclusive test of their discipline. Troops remain relatively undisciplined until physically and mentally conditioned to unusual exertion (a factor that shapes much of the program to be found in all varieties of recruit and officer candidate training). No body of men could possibly enjoy the dust, the heat, the blistered foot and aching back of a road march. Nevertheless, hard road marching is a necessary and sound foundation for the discipline of fighting troops. The rise in spirit within any unit, which is always marked when the men rebound from a hard march or after record day, does not come from a feeling of physical relief but from the sense of accomplishment of a goal.

Another key factor in sound discipline is consistency and firmness. You cannot wink at an infraction one day and put a man on the report for the same offense tomorrow. You must establish and make known your standards of good discipline, and be consistent, firmly consistent, every day.

Discipline imposed by compulsion and fear of punishment will inevitably break down in combat or any other severe test. If you threaten your men, discipline will also break. Discipline will not break, however, under the stress of testing demands, for troops will endure hard going when it serves an understandable end.

Military discipline is tangible in terms of its results. It allows no room for familiarity, which truly breeds contempt. On the other hand, honest comradeship is a by-product of discipline and a pillar of unit *esprit*.

PRAISE AND REPRIMAND

1609 ■ Occasions for Praise

A basic rule is to *praise in public and reprimand in private.*

Never let a praiseworthy occasion pass unmentioned. This means more than occasional back-pats. Here are ways in which you can make the most of opportunities to praise subordinates:

Promotion. When an officer is promoted (although regulations no longer so require), he should be sworn in at Office Hours by the senior Marine officer present. Administration of the officer's oath adds greatly to the solemnity of the occasion and enables him to reaffirm the original oath, which he took on receiving his first commission. If practicable, his wife and children should be invited. All brother officers who can be spared should attend. The officer administering the oath should always give a set of insignia to the individual being promoted—if possible, a set of his own insignia from an earlier rank, a gift that is always appreciated.

Enlisted promotions are effected by presentation of the man's warrant for the next higher rank. This should be accomplished at a formation. If a parade or other formation cannot be arranged, the man should receive his warrant from the commanding officer at Office Hours, in the presence of his immediate commanding officer and first sergeant. If enlisted offenders are to appear at the same Office Hours, parade them in the rear, in order to give them occasion to reflect on "the other side of the coin."

Under no circumstances should a Marine be called into the company office and receive his warrant from the first sergeant or clerk. This is the wrong way, and reflects directly on you if you permit such procedures.

Presentation of Decorations. The *Drill and Ceremonies Manual* describes the ceremony for presenting decorations. Even at some inconvenience to the unit, decorations—particularly those earned in combat or awarded for heroic action—should be presented with utmost formality at a parade or review, as laid down in the book. Avoid the easy solution of calling in the officer or man to Office Hours, and giving him his medal with a handshake.

The fundamental purpose of awards is to *inspire emulation.* To do this, you must present medals or commendations where the maximum number of other Marines know about it.

In combat, when an award can be made immediately, it is sometimes effective for a senior commander to visit the recipient at his unit, call together his comrades, and give the medal on the spot. With decorations, even more than other rewards, "he gives thrice who gives quickly." As a combat leader, be alert for every deserving act, especially by an enlisted Marine. Know the criteria and Marine Corps standards for every award, and how to initiate proper recommendations. (See Section 739.)

> NOTE: A modified version of the awards ceremony can serve equally well for such occasions as presentation of Good Conduct Medals, civilian commendatians, commissioning of meritorious NCOs as warrant officers or second lieutenants, and so on.

Retirement. The honorable retirement or transfer to the Fleet Reserve of any officer or enlisted Marine should be habitually effected at a parade or review. In the case of an officer, it is also appropriate for the officers of his unit to "dine him out" at a mess night, as described in Section 2403.

Completion of Correspondence Course. Any Marine who completes an MCI or Marine Corps Development and Education Command correspondence course should receive his diploma from the commanding officer at Office Hours.

Reenlistment. When a number of men ship over on the same day, arrange a formation in their honor. Otherwise, individuals should be shipped over at Office Hours. If practicable, make this the occasion for a furlough—and, if warranted and possible, there is no better moment to effect a promotion. Nothing starts a new cruise so handsomely as another chevron.

1610 ■ Reprimand

One basic rule of reprimand has already been stated—*do it in private.*

A second rule is found in the Marine proverb "Never give a man a dollar's worth of blame without a dime's worth of praise."

And avoid collective reprimands, let alone collective punishments. Nothing so rightly infuriates an innocent man as to be unfairly included in an all-hands blast or all-hands punishment.

Before you issue reprimand or censure, be sure that an offense or dereliction of some kind has been committed. This is basic. You cannot call down a Marine just because you don't like the color of his eyes. Before telling off any individual, ask yourself if what he has done, pushed to the limit, would sustain charges under any article in the *Uniform Code of Military Justice.* This can save you much embarrassment and injured innocence at the hands of sea-lawyers, while it sometimes cuts the other way to protect a subordinate against hasty rebuke when not warranted.

Know what you intend to say before you launch into reprimand. A sputtering, inconclusive rebuke only makes an officer look silly.

Avoid uncontrolled anger, profanity, or abuse. Many experienced Marines, officer and NCO, know how to valve off anger into indignation. Make this your object, but at all costs avoid "acting tough."

Never make a promise or threat which you are not capable of fulfilling, or which you do not intend to fulfill. Never bluff, or you will be called in short order.

Like reward, the effectiveness of reproof is in direct proportion to its immediacy. When you spot something amiss, take corrective action at once. Never let a wrongdoing Marine slide by with the thought, "Well, he's not one of *my* men. Let his own outfit handle it." *Every* U. S. Marine is one of *your* men.

If you have occasion to call down a Marine not under your command, find out who he is, and see that his commanding officer knows about it. This will be appreciated by his unit, who are just as anxious as you are to have their Marines up to snuff. Moreover, the derelictions of an individual are the responsibility of his immediate senior. A Marine with a dirty rifle is a black eye for his squad and fire-team leader; a man in your platoon who fails to salute is a discredit to your leadership. Napoleon's dictum "There are no bad regiments—only bad colonels," applies with equal force to fire teams, squads, platoons, companies, and battalions as well.

1611 ■ Office Hours

Office Hours, the Marine Corps equivalent of Mast, is the occasion when the commanding officer awards formal praise or blame, hears special requests, and awards nonjudicial punishment. Detailed treatment of Office-Hours procedure and nonjudicial punishment is contained in Chapter 19.

Remember that Office Hours is a ceremony, and that much of the desired effect depends upon the manner in which it is conducted. And when you hold Office Hours, do so with the greatest respect for each man's individuality. Not only must the punishment fit the *crime*, it must fit the *man*. Never let a man leave Office Hours with a sense of injustice or frustrated misunderstanding.

A special and important variation of Office Hours is Request Mast, an occasion set aside for individuals who may have special requests or grievances that they wish to present to the commanding officer. It is one of the responsibilities of command to keep this opportunity open to any Marine who, *in good faith*, wishes to avail himself of it. In holding Request Mast, one important point to remember is that the individual is entitled to complete privacy. Unless he requests otherwise, you should see him alone, and should take all necessary steps to avoid any prejudice to his interests which might arise out of a bona fide complaint or special request.

INSPECTIONS

1612 ■ Inspections

Inspection is one of the most important tools of command. Throughout your Marine Corps career you will continually be inspected or inspecting. Inspections serve two purposes: first, to enable commanding or superior officers to find out conditions within an organization; and second, to impart to an organization the standards required of it.

There are several types of inspection, varying from inspection of personnel in ranks to inspections of materiel, supplies, equipment, records, and buildings. Each inspection has a particular purpose, which the inspecting officer will keep foremost in mind. Thus it is up to you to ascertain or forecast the object of the inspection, and to prepare yourself and your command accordingly. For example, if the inspection is to deal with the crew-served weapons and transportation in your unit, it does no great good to emphasize clean uniforms and haircuts at the expense of materiel upkeep. On the other hand, good-looking vehicles do not excuse greasy, worn clothing at a personnel inspection.

1613 ■ Prepare for Inspection

Once you know the purpose of an inspection, you must prepare your outfit. The best way to do this is by putting yourself in the inspector's shoes. Be sure your leading NCOs also understand the "why" of the inspection so that they can cooperate intelligently in getting tuned to concert pitch. Many an inspection crisis has been averted by a quick-witted, loyal NCO with a ready answer.

Inspection is an important tool of command. Each inspection serves a particular purpose, and every man has a personal stake in the success of each inspection.

While your unit prepares for inspection, move about with a leading NCO, usually your police sergeant, and/or first sergeant. This enables you to see that preparations are what you want, and it reminds your men that you have direct interest in the hard work they are engaged in. It also lets you discover weak spots in good time.

Time preparation for inspection so that everything is ready about 30 minutes before the appointed hour. This gives your men a final opportunity to get themselves ready. It also gives you a margin to handle last-minute emergencies.

Ten minutes beforehand, have your responsible subordinates standing by their respective posts, or, if the inspection is to be in formation, have your troops paraded, steady and correct. You yourself should be either at the head of your men or at the entrance to your area, poised to meet the inspecting party. As a platoon leader, you should have your platoon sergeant and guide assist you in the inspection. If you are the company commander, you should have your first sergeant, gunnery sergeant, and police sergeant in your inspection party. The "top" should have a notebook and pencil ready to take notes. The police sergeant should have a flashlight. All rooms, compartments, sheds, etc., should be unlocked and open. Tents should be rolled, unless the weather is foul.

When the inspecting party arrives, salute and report your unit ready for inspection. Post yourself at the left rear of the inspecting officer, and follow along with him. Answer questions calmly and with good humor. Avoid alibis. Do not reprimand your men during inspection for shortcomings the inspection brings out. It is *your* outfit; the shortcomings are *yours*. Be alert for the inspecting officer's comments. These fore-arm you for the next inspection.

Afterward, if results have been notably good or notably poor, assemble your people and tell them about it. Give every man a personal stake in the success of each inspection.

NOTE: For a useful checklist in preparing a command for inspection, turn to Appendix III.

1614 ■ Inspecting

Nothing else can raise the standards of a command like an intelligent program of inspection, carefully followed up. Some officers unwisely discount the value of formal inspections, saying that these result in unbalanced, artificial impressions, and that COs ought to observe informally in order to find out "real" conditions. While it is certainly true that every CO must keep on the move and keep his eyes open, the periodic formal inspection is vital because it requires all hands to overhaul their areas of responsibility. Moreover, formal inspection is the only way to determine accurately the degree of progress being made by a unit.

Before you inspect, you, like the unit being inspected, must also make careful preparations. As an inspector, you should:

1. Know what you intend to concentrate on—in other words, the purpose of the inspection.

2. Have a planned route and sequence of inspection designed to cover the entire unit and area.

3. Organize your inspecting party. This should include one man to take notes, one man with flashlight, plus the requisite specialist talent (such as hospital corpsman, technicians, and so forth) needed to advise and assist.

4. See that you and your party are perfectly turned out and neatly uniformed. Inspections also operate in reverse.

5. Be up-to-date on details of maintenance and function of any materiel you are to inspect. If ordnance is on the program, leaf through the appropriate Technical Manual, which will contain a checklist for inspection. When you inspect, do so impartially and pleasantly. Avoid a fault-

finding spirit; the object of inspections is to help, not to antagonize. Praise individuals when you properly can. As you uncover defects, be sure that the responsible individuals understand what you have discovered and why it constitutes a defect. Avoid a dead level of criticism or complaint.

Inspect in cadence and at attention. Have a leading NCO—sergeant major or first sergeant—precede you.

Finally, regardless of the purpose of the inspection, never overlook the individual Marine. See that he is smart and military. Look him in the eye. Make him feel that he is the ultimate object, and that you are deeply interested in him as a man and a Marine.

1615 ■ Inspection Follow-Up

An inspection loses value if you fail to follow it up. This is the main reason for keeping careful notes on the comments of the inspecting officer.

Inspection notes should be disseminated to everyone concerned, broken down into items for corrective action, so that they can serve as a checklist. When you re-inspect, review previous inspection notes as a guide for follow-up. On the receiving end, you can use past notes to prepare for future occasions. *It is a grave reflection on you as a leader if the same defects continue to show up on consecutive inspections.*

Maintain high proficiency with your weapons. Enlisted Marines respect a good shot and an officer handy with small arms.

1616 ■ IG Inspections

Via his longstanding previous title, "The Adjutant and Inspector," the Inspector General of the Corps ("the IG") can trace his roots back to 1798. Today, the IG's job is to assist and examine by periodic inspections the effectiveness of Marine Corps commands in terms of ability to carry out their missions; unit leadership, economy, policies, and doctrine; work and health conditions; and discipline.

After a visit to a command by an IG team, one of four grades is awarded: noteworthy, satisfactory, satisfactory with discrepancies; and unsatisfactory. The first two are made intentionally hard to attain. Although it may seem difficult for a unit under such searching inspection to believe, the IG is basically there to help: his inspections are always a search for causes, not an inventory of symptoms.

There is no single test short of combat that can do you more good as a unit commander than a superior showing for the IG.

OTHER ASPECTS OF LEADERSHIP

1617 ■ Weapons Proficiency

A Marine leader has few better ways of setting the right example to his men than by maintaining high proficiency with infantry weapons—notably, the rifle and pistol. Marines respect a good shot and an officer who is handy with small arms. Cultivate your weapons, and do your level best each year when you go to the range. Every enlisted Marine will be watching to see how you do. Make yourself a model of marksmanship technique. Demand no special favors: behind the business end of a rifle, on the firing line, all Marines are equals. Clean and maintain your own weapon, pick up your own brass, keep your own scorebook, and keep your mouth shut.

Never violate a safety precaution. Remember the shooter's proverb: "There is no such thing as an accidental discharge."

> NOTE: When you go to the rifle range, don't forget that your qualifications and score will be entered on your next fitness report.

1618 ■ Looking Out for Your Men

In the final analysis, the essence of Marine leadership is looking out for your people.

For the sake of your men, you must be tireless, you must be imaginative, you must be willing to shoulder responsibility. Their good must be your first preoccupation. Their interest and advancement must be always on your mind.

Are they comfortably clothed, housed, and sheltered?
Are they well fed?
Are they getting their mail?
If sick and wounded, can they rely on help?
Are they justly treated?
Are you available to every man who needs counsel?
Are you alert to help each one better himself in his career?

As an officer, you demand a great deal of your men. But they in fact demand much more of you. If you let down one of your Marines, you are letting down the entire Corps. *Noblesse oblige* is the private motto of every officer of Marines.

MAXIMS OF LEADERSHIP AND COMMAND

Learn to obey before you command.—Solon
Respect yourself and others will respect you.—Confucius
The superior man is firm in the right way, and not merely firm.—Confucius.
An army of deer led by a lion is more to be feared than an army of lions led by a deer.—Philip of Macedon
He that ruleth over men must be just.—II Sam. 23
Everyone is bound to bear patiently the results of his own example.—Phaedrus
Be swift to hear, slow to speak, slow to wrath.—James 1:19
The wise man, before he speaks, will consider well what he speaks, to whom he speaks, and where and when.—St. Ambrose
You may pardon much to others, nothing to yourself.—Ausonius
Hereafter, if you should observe an occasion to give your officers and friends a little more praise than is their due, and confess more fault than you can justly be charged with, you will only become the sooner for it, a great captain.—Benjamin Franklin (to John Paul Jones)
I shall desire all and every officer to endeavor by love and affable carriage to command his soldiers, since what is done for fear is done unwillingly, and what is unwillingly attempted can never prosper.—The Earl of Essex
There is great force hidden in a sweet command.—George Herbert

The qualities of leadership are expressed in the faces of the officers who led the 1st Marine Division on Guadalcanal. Three future Commandants are in the front row: Vandegrift, Cates, and Pate (fourth, sixth, and seventh from left). General Thomas, original co-author of the Guide, sits fifth from left.

Self confidence is the first requisite to great undertakings.—Samuel Johnson

Discipline is the soul of an army. It makes small numbers formidable; procures success to the weak, and esteem to all.—George Washington

There are no bad regiments—only bad colonels.—Napoleon Bonaparte

Correction does much but encouragement does more. Encouragement after censure is as the sun after a shower.—Goethe

He never errs who sacrifices self.—Bulwer-Lytton

Impossible is a word that I never utter.—Colin d'Harleville

The man who trusts men will make fewer mistakes than he who distrusts them.—Conde di Cavour

The eye and head of the master will do more work than both of his hands.—Benjamin Franklin

The reward of the general is not a bigger tent, but command.—Justice O. W. Holmes

Historically, good men with poor ships are better than poor men with good ships.—Commander J. K. Taussig

Superiority of material strength is given to a commander gratis. Superior knowledge and superior tactical skill he must himself acquire. Superior morale, superior cooperation, he must himself create.—Admiral J. M. Reeves

I had learnt before I left school that of the many attributes necessary for success two are vital—hard work and absolute integrity. To these two I would now add a third—courage. I mean moral courage—not afraid to say or do what you believe to be right.—Montgomery of Alamein

Men are neither lions nor sheep. It is the man who leads them who turns them into lions or sheep.—Jean Dutourd

An army cannot be administered. It must be led.—Franz-Joseph Strauss

Death is as light as a feather; duty heavy as a mountain.—Emperor Meiji

What the Marine Corps teaches at Quantico is valuable in all of life. Responsibility, discipline, self-control, self-confidence, co-operation, self-reliance, accuracy, modesty, inventiveness, enthusiasm, decisiveness, devotion to duty, are the firm pillars upon which the Marine Corps has been built.—Samuel W. Meek

---

The general must know how to get his men their rations and every other kind of stores needed in war. He must have imagination to originate plans, practical sense and energy to carry them through. He must be observant, untiring, shrewd, kindly and cruel, simple and crafty, a watchman and a robber, lavish and miserly, generous and stingy, rash and conservative. All these and many other qualities, natural and acquired, must he have. He should also, as a matter of course, know his tactics; for a disorderly mob is no more an army than a heap of building materials is a house.
—Socrates

If the trumpet give an uncertain sound, who shall prepare himself to the battle?
—I Corinthians 14:8

17

On Watch

One of the most stirring guard orders ever received by U. S. Marines was issued on 11 November 1921, by Navy Secretary Edwin Denby, himself a former Marine. The nation was in the grip of a crime wave, which had been highlighted by armed robberies of the U. S. Mails. Four days before Secretary Denby penned his letter of instruction, the President had directed that the Marine Corps take over the job of safeguarding the mails, and 53 officers and 2,200 enlisted Marines were already on watch in post offices, railway mail cars, and postal trucks throughout the country. *"To the Men of the Mail Guard,"* wrote Edwin Denby:

> I am proud that my old Corps has been chosen for a duty so honorable as that of protecting the United States mail. I am very anxious that you shall successfully accomplish your mission. It is not going to be easy work. It will always be dangerous and generally tiresome. You know how to do it. Be sure you do it well. I know you will neither fear nor shirk any duty, however hazardous or exacting.
> This particular work will lack the excitment and glamor of war duty, but it will be no less important. It has the same element of service to the country.
> I look with proud confidence to you to show now the qualities that have made the Corps so well-beloved by our fellow citizens.
> You must be brave, as you always are. You must be constantly alert. You must, when on guard duty, keep your weapons in hand

Marine mail guards protected the U. S. Mails from banditry in 1921 and 1926. Not a piece of mail was lost while Marines stood watch.

and, if attacked, shoot and shoot to kill. There is no compromise in this battle with the bandits.

If two Marines, guarding a mail car, are suddenly covered by a robber, neither must hold up his hands, but both must begin shooting at once. One may be killed, but the other will get the robber and save the mail. When our men go in as guards over mail, that mail must be delivered or there must be a Marine dead at the post of duty.

To be sure of success, every Marine on this duty must be watchful as a cat, hour after hour, night after night, week after week. No Marine must drink a drop of intoxicating liquor. Every Marine must be most careful with whom he associates and what his occupations are off duty. There may be many tricks tried to get you, and you must not be tricked. Look out for women. Never discuss the details of your duty with outsiders. Never give up to another the trust you are charged with.

Never forget that the honor of the Corps is in your keeping. You have been given a great trust. I am confident you will prove that it has not been misplaced.

I am proud of you and believe in you with all my heart.

/s/   Edwin Denby

Mail robberies declined and ceased within a matter of days after Secretary Denby penned his order, and not a single piece of mail was lost to a robber while Marines stood watch.

1701 ▪ Watchstanding

The Importance of Guard Duty.   In the Marine Corps and Navy, the safety and good order of the entire command depend on those who stand guard. Thus, watchstanding is your strictest routine duty.

The importance of guard duty is underscored by the fact that sleeping on watch can be punished by death in time of war, and in peace by heavy penalties.

In addition to combat missions, the Marine Corps is the security force for the Naval Establishment, and is thus not only responsible for the good order and protection of its own posts, but also of all stations and ships where Marines are assigned. As a Marine officer, you are therefore expected to be an authority on watchstanding and guard duty, as well as a model watch officer, ashore or afloat.

Semper Fidelis never demands more than when you are on guard.

Types of Guards.   Marines maintain four kinds of guard:

Exterior guard ashore
Interior guard ashore
Ship's guard afloat
Special guards

In addition, Marines frequently perform military police and shore patrol duties for the regulation and assistance of Marines and seamen on liberty.

An *exterior guard* is maintained only in combat or when danger of attack exists. An exterior guard protects the command against outside attack. An exterior guard is organized and armed according to the tactical situation, in line with the principles laid down in *Field Service Regulations (FM 100–5)* covering security in combat.

Interior guards have the threefold mission of protecting life, preserving order and enforcing regulations, and safeguarding public property. In addition, the interior guard sometimes operates the brig.

Ships' guards carry out the same general missions afloat as interior guards do ashore, but differ in details of organization and duty because shipboard conditions and routine differ from those ashore. (See Chapter 22.)

Special guards include all guards organized for special purposes (e.g., train or boat guards, brig guard, and so on). In addition, most posts having custody of special weapons have a separate main guard for that purpose alone, leaving other normal security functions at the post to the station main guard.

Status of Marines on Watch. **Any Marine on guard, whether officer or enlisted, represents the commanding officer.** In the execution of orders or the enforcement of regulations, his authority is complete. When you receive a lawful order from a member of the guard, comply without hesitation and ask your questions afterward. Remember that an armed sentry has full authority to *enforce* his instructions.

THE INTERIOR GUARD

1702 ■ The Interior Guard

You will find detailed instructions covering interior guard duty in the *Interior Guard Manual*. Remember that these instructions, although similar in many respects to those prescribed for the Army in various manuals which you may encounter, nonetheless differ in a number of essentials. The *Interior Guard Manual* is "the bible" on Marine Corps guard duty.

The interior guard—established to preserve order, protect property, and enforce regulations—derives its authority directly from the commanding officer. Figure 17–1 shows the organization of a typical interior guard. The guard is composed of a *main guard*, and, when needed,

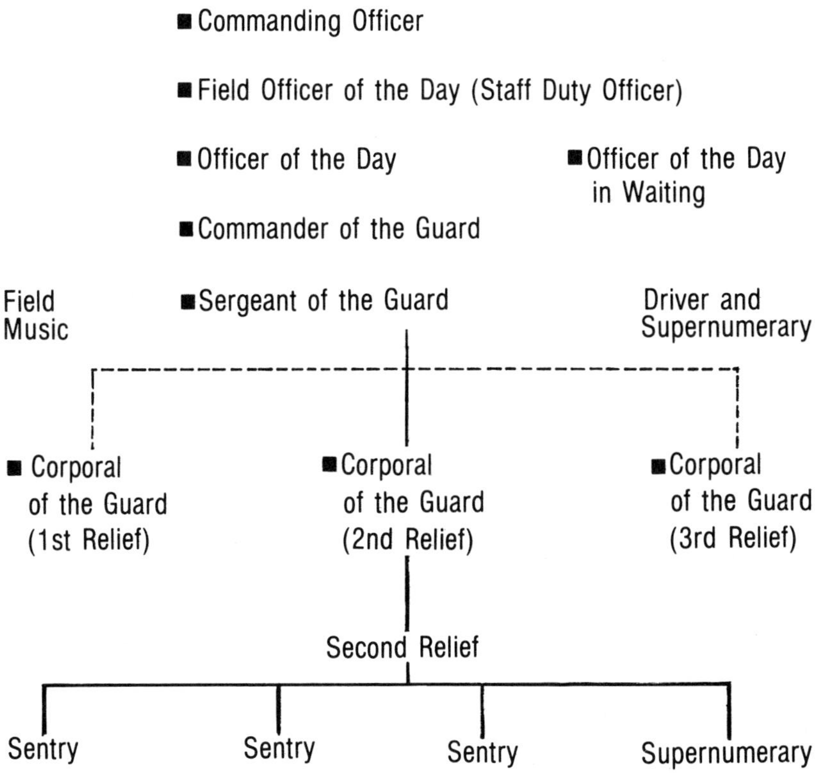

```
■ Commanding Officer

■ Field Officer of the Day (Staff Duty Officer)

■ Officer of the Day              ■ Officer of the Day
                                      in Waiting
■ Commander of the Guard
```

| Field Music | ■Sergeant of the Guard | Driver and Supernumerary |

```
■ Corporal              ■ Corporal              ■ Corporal
  of the Guard            of the Guard            of the Guard
  (1st Relief)            (2nd Relief)            (3rd Relief)
                              |
                        Second Relief

Sentry      Sentry      Sentry      Supernumerary
```

Fig. 17–1: Organization of a typical interior guard. There are as many sentries in each relief as there are posts. On a small station, the field officer of the day may be omitted.

special guards and *brig guards*—or "chasers," as the last are known in the Corps.

1703 ■ Duties of the Guard

The duties of the guard (and the CO's responsibilities in connection with the guard) are as follows:

The Commanding Officer establishes the guard and sees that it functions properly. Either the CO or his representative (usually the executive officer or adjutant) receives the daily reports from, and relieves, the officers of the day, examines the guard book, and issues whatever special instructions may be needed.

The *Field Officer of the Day* may be required on a large post where subordinate commands maintain separate guards. The field officer of the day coordinates subordinate guards, and acts for the post commander in emergency. On some stations, the field officer of the day is entitled "staff duty officer."

The *Officer of the Day (OD)* supervises the main guard (and, when directed, the brig guard), executes all orders that pertain to the guard, and is responsible that his guard performs effectively. While officer of the day, you are the direct representative of the commanding officer. Your duties are discussed in Sections 1707–1711.

The *Officer of the Day in Waiting* is the officer next on the roster for duty as officer of the day. He must be prepared to relieve the OD should the latter become ill or for any reason be unable to continue his watch.

The *Commander of the Guard*, an officer junior to the officer of the day, is responsible for the proper instruction, discipline, and performance of the guard. A commander of the guard is usually required only for a large guard, or to afford new officers practical experience before detail as officers of the day. Your first tour on watch ashore will probably be as commander of the guard, under an experienced officer of the day. Your duties as commander of the guard are covered in Section 1710.

The *Sergeant of the Guard*, whatever his actual rank, is the senior NCO of the guard. The sergeant of the guard assists the commander of the guard, or, if the guard does not include one, performs the latter's duties. The sergeant of the guard supervises the enlisted members of the guard and is responsible for Government property charged to the guard.

Nonrated members of the guard are organized into three *reliefs*, each of which includes a sentinel for each post and one supernumerary, and is commanded by a *Corporal of the Guard*. The corporal of the guard instructs and supervises his relief, which takes its successive turn on guard throughout the tour of duty.

1704 ■ Duties and General Orders for Sentinels

The sentry is the workhorse of the guard. The universal respect accorded a U. S. Marine sentinel is based on his high military efficiency and the fact that he is habitually armed and prepared to defend his post and person in the execution of orders. A sentinel's duties are to carry out the *general orders for a sentinel on post*, as well as special orders applicable to his particular post. Every Marine, officer or enlisted, must know the general orders by heart. They are:

GENERAL ORDERS FOR A SENTINEL ON POST
1. To take charge of this post and all Government property in view.

2. To walk my post in a military manner, keeping always on the alert, and observing everything that takes place within sight or hearing.
3. To report all violations of orders I am instructed to enforce.
4. To repeat all calls from posts more distant from the guardhouse than my own.
5. To quit my post only when properly relieved.
6. To receive, obey, and pass on to the sentinel who relieves me, all orders from the commanding officer (field officer of the day), officer of the day, and officers and noncommissioned officers of the guard only.
7. To talk to no one except in line of duty.
8. To give the alarm in case of fire or disorder.
9. To call the corporal of the guard in any case not covered by instructions.
10. To salute all officers, and all colors and standards not cased.
11. To be especially watchful at night, and, during the time for challenging, to challenge all persons on or near my post, and to allow no one to pass without proper authority.

In addition to routine sentry duties, nonrated members of the guard are assigned to certain special duties, such as the following:

Guardhouse Sentinel (Post No. 1): If one is assigned, he assists the corporal of the guard in carrying on guardhouse routine, and takes charge of prisoners in the brig (if administered by the guard). Your guardhouse sentinel should be picked for his intelligence, reliability, and smartness.

Main Gate Sentinel: This sentinel ensures that only authorized persons enter or leave the post through the main gate; he also directs traffic and assists visitors. Your main gate sentry stands his watch in the show-window of the station; therefore select him for soldierly appearance, judgment, and thorough knowledge of the post. *The main gate is a spot for outstanding Marines.*

Field Music: The field music (when one is available) stays at the guardhouse. He sounds the prescribed bugle calls, strikes the bells, and acts as "time-keeper" and officer of the day's messenger. The music also polices the quarters of the officer of the day and officers of the guard. In absence of a music, the latter's nonmusical duties fall to the supernumerary of the relief on duty.

Supernumerary: One additional man stands by as a supernumerary to replace any man who must be relieved. Like the music, the supernumerary can be kept busy as a messenger and general factotum in the guardhouse.

Orderlies: When orderlies are needed, they are selected from outstanding privates of the guard on the basis of soldierly appearance and correct performance of duty. When selected as an orderly, a Marine is relieved of routine guard duty and reports to the officer to whom assigned.

Driver: A motor transport man is assigned to the guard to drive the guard truck. Always keep the driver up to standard in uniform and appearance, since it is a notorious failing among guard drivers to lag behind the rest of the guard in this respect.

Color Sentinel: Whenever the unit Colors are broken out for airing —usually each Sunday and national holiday—a color sentinel is posted over them. If the Colors can be seen directly and close at hand from the guardhouse, a color sentinel is not mandatory, as the guardhouse sentinel can keep them under observation. The color sentinel guards the Colors, cases and protects them if the weather turns inclement, and acknowledges salutes rendered to the Colors (see Section 1811).

> NOTE: The policy of the Marine Corps is that all sentinels will be armed. Detailed safety instructions, as well as restrictions on the use of the weapon, should be known to all members of the guard, from OD to sentry. Improper or careless use of firearms is an extremely serious matter.

1705 ■ Daily Guard Routine

The daily routine of an interior guard varies somewhat according to the wishes of the commanding officer and the size and missions of the post. But, in the main, guard duty runs as follows:

The normal tour is 24 hours. Anyone detailed for guard duty must be on board and fit for duty at least four hours before commencement of his tour.

Details for guard duty should be published well in advance, by written order, and should specify uniform and equipment which will be needed, together with any other information not covered in standing orders. *Officers detailed for guard duty must be notified in person or by written order, preferably both.* This is the adjutant's responsibility. He also keeps the officer-of-the-day roster, which determines the order in which officers stand watch.

A tour on guard begins with *guard mounting*, when the old (outgoing) and new (incoming) guards are paraded and inspected. Guard mount usually takes place immediately after Morning Colors. After guard mount, old and new officers of the day and sergeants of the guard relieve each other. Thereupon the officers of the day report to the commanding officer, and the new officer of the day assumes duty.

The guard's routine includes execution of colors, posting and relief of sentinels, supervision of working details of prisoners (when a separate brig guard is not maintained), supervision of meal formations, and rendition of honors to the commanding officer, visiting officers, or civilian dignitaries.

Each relief stands watch for four hours before turning over to the next relief. Thus, in a 24-hour tour, each relief stands a total of eight hours on watch—four by day, and four by night.

1706 ■ Challenging and Countersign

"Halt! Who goes there?", the traditional challenge, has been employed by Marines since 1775. As an officer, you should know exactly how to challenge and reply, since a faulty challenge or reply may not only embarrass you, but in combat can cost someone his life.

The *challenge* is used at night or in low visibility to identify anyone approaching a sentinel.

On hearing any suspicious noise, the sentinel brings his weapon to a ready position, and commands, *"Halt! Who goes there?"* The person challenged halts and then identifies himself either by a password, or by some such answer as *"Friend,"* or *"Officer of the Day."* The sentinel replies, *"Advance, Friend, and be recognized."* The person is allowed to approach near enough to the sentinel to be recognized, and is halted again, at which time the sentinel examines him. When satisfied, the sentinel commands, *"Pass, Friend"*; or, if being visited by the officer of the day, reports, *"Post Number . . . secure, Sir."*

It is extremely important, not only as military etiquette, but for your own safety, to reply audibly and promptly when challenged, and to comply exactly with the sentry's orders. The sentry is the man behind the gun. You are *in front* of it.

Challenge and countersign (sometimes called the "password") are used to distinguish between friend and enemy. In the use of this procedure, which takes place only when prescribed by the commanding officer, the person or party approaching a sentry is challenged in the usual way, as described above. Then, after he has advanced the person for recognition, the sentry repeats the secret challenge, an agreed code word to which the person being challenged must respond with the countersign, a second code word which validates the reply. Challenge and countersign change daily and must be kept from the enemy at all costs. For detailed instructions on this important subject, see the *Interior Guard Manual*.

OD AND COMMANDER OF THE GUARD

1707 ■ Officer of the Day and Commander of the Guard

Since officer of the day (OD) and commander of the guard duties will take up much of your energies as a company officer, this section discusses the responsibilities of both jobs. In addition to what you read here, however, you must be thoroughly familiar with the *Interior Guard Manual*, as well as post standing orders which deal with guard duty.

1708 ■ Duties of the Officer of the Day

As officer of the day, you must attend to the following routine duties. More important, however, as the CO's representative, you must be ready to act promptly and sensibly in any contingency not covered by the letter of your orders.

Inspect each relief of the guard, by visiting sentinels at *least once* while that relief is on post. *One inspection must take place between midnight and reveille.* When visiting sentinels, cover the following points:

Verify that the sentinel is on his post, alert, in correct uniform, and correctly armed and equipped.
Question the sentinel on his special orders, and check particularly that he knows:
    The limits and designation of his post
    Location of fire-fighting gear on the post, and how to sound a fire alarm
    Any recent changes in special orders for his post
    The reason why the post is required (his mission as a sentinel)
    Restrictions, if any, on use of his weapon
Verify that he knows verbatim, and understands, the general orders for sentinels. Make him repeat several, and explain them in his own words.

Supervise and coordinate the inspections to be made by your commander of the guard and sergeant of the guard. See that these do not conflict with or duplicate yours.

Take immediate steps, in emergency, to protect life and public property and to preserve order. As soon as the situation permits, report what has happened, and what you are doing about it, to the commanding officer (or to the executive officer, or field officer of the day if your post has one).

Always inform the guard where you can be reached when not in the guardhouse. If possible, leave a telephone extension.

Abstain from alcohol throughout your tour.

Unless otherwise authorized, remain fully clothed at all times. This enables you to turn out immediately in case of fire or other fast-moving emergency. Nothing can get an OD into more trouble than arriving late and drowsy at the scene of trouble.

Inspect galleys and messes, in accordance with local orders, at each meal during your watch. When you do this, look out for:

Cleanliness and sanitary conditions of the galley, messhall, and immediate areas outside (don't forget the garbage house)
Personal cleanliness (clothing, haircuts, fingernails) of cooks and messmen
Quality and sufficiency of meals
Adequacy of food service
Good order and proper uniform of troops being fed

If the brig is under control of the guard, inspect the brig and prisoners. Before you do this, review the *Navy Corrections Manual*. This contains the important rules for brig administration. Brigs and prisoners are "dynamite." Take care that every regulation is complied with. Give each prisoner an opportunity to make complaints. Do not sanction, or hesitate to report to the commanding officer, any violation of required brig procedure.

At guard mount, see that prisoners (other than general court-martial prisoners) whose sentences expire that day are released. Also report to the CO any prisoners confined with no record of charges against them.

Review the guard book before guard mount, and correct any mistakes. In it, log the times when you visited sentries, together with any other information you think proper to place on record. Then attest the correctness of the entire report by signing the guard book, which constitutes your official report.

1709 ■ Relieving as Officer of the Day

After guard mount, old and new officers of the day inspect the guardhouse together and verify the count of prisoners (if under control of the main guard). After that, both report to the commanding officer for relief and posting.

March in, at attention, covered and wearing side-arms (sword or pistol as prescribed by the commanding officer), and halt in front of the commanding officer (old OD on the right). You both salute together. Thereupon the old officer of the day says, "Sir, Captain . . . reports as old officer of the day," and hands the guard book to the CO. The latter reads the guard report, asks any questions which come to mind, and comments as he thinks necessary. Then he informs the old officer of the day, "You are relieved." Thereupon the old OD salutes and withdraws. Then the new officer of the day again salutes, and says, "Sir, Captain . . . reports as new officer of the day." The commanding officer gives the new officer of the day his instructions, whereupon the latter salutes and withdraws. All movements during relief and posting as officer of the day are carried out at attention and in cadence. If an emergency strikes between guard mount and the time when you report to the CO, the senior of the two ODs takes charge of both guards, old and new.

1710 ■ Duties of the Commander of the Guard

As you have seen, the duties of commander of the guard are carried out by the sergeant of the guard if no commander of the guard is detailed. Regardless of whether performed by an officer or NCO, they constitute a useful checklist by which, as officer of the day, you can ensure that your guard is running smoothly.

Inform your officer of the day of any orders that have come to you from anyone other than the OD. Pass on to your own relief all instructions and current information.

See that your guard is properly instructed, and that it performs properly.

Make certain that all inspections (yours and the sergeant of the guard's) are carried out on time and as directed by the OD.

See that sentinels are relieved, Colors executed, the proper bugle calls sounded, bells struck, and guard routine followed.

Ensure that legible copies of general and special orders for each post are mounted both in the guardhouse and under shelter on each post.

Inspect guardhouse and brig thoroughly at least once during your tour.

Inspect each relief of the guard while it is on post. Follow the procedure described in Section 1708. Like the officer of the day, you must make one inspection of sentinels between midnight and reveille.

Parade the guard for inspections as required. In emergency, turn out the guard, sound the appropriate call or alarm, and promptly notify your officer of the day.

If any sentry calls, *"The Guard!"* send help immediately. This is the SOS for a sentinel on post. If he is in serious danger, he may fire his piece three times.

Like the officer of the day, you must keep the guard informed of your whereabouts whenever you leave the guardhouse—if possible, by a telephone number.

Report to the officer of the day if any member of the guard takes sick, quits his post, or has to be relieved for any reason.

It is up to you to keep the guard in prescribed uniforms and equipment.

Make up the details to execute Morning and Evening Colors, and attend Colors to be certain that this ceremony is correctly performed. Ensure that the Colors are properly stowed and are handled only in performance of duty. Finally, report to the OD if a set of Colors is unserviceable (see Section 1817).

Detain any suspicious persons, and report the circumstances to the OD.

You are responsible for prisoners if held by the guard. This means that you must:

See that corporals of incoming and outgoing reliefs verify the prisoners;
Ensure adequate search and medical examination of anyone being confined;
Report all confinements to the OD;
Administer the brig in accordance with the *Brig Manual*.

Write your report in the guard book, and, at the end of your tour, present the guard book to the OD.

1711 ■ Hints for the Officer of the Day

Of all officers, you are the one who can least tolerate any discrepancy, or violation of orders. *Never overlook a dereliction or infraction, however minor.* Be especially alert for:

Unmilitary behavior
Men out of uniform
Traffic offenders
Safety hazards
Unsanitary, unusual, or unsightly conditions
Improper safeguarding of, or undue leniency toward, prisoners
Security of restricted areas

Keep closely posted on the movements and whereabouts of the commanding officer and the executive officer. Try to see the post as it would appear through their eyes, and act accordingly.

Be meticulous in bearing, and conspicuous by your neatness, when on watch. A neat OD has a well turned-out guard. Keep your leather shining. Polish your brightwork and sword. Wear your best uniforms. Set an example for the whole command.

No matter how many times you have stood watch before, review the guard orders as soon as you take over. Changes have a way of sneaking in without warning. "That isn't the way it used to be" is no excuse for a bobble.

Prevent your guard from idling. See that they are instructed in guard orders and routine, and especially in safety precautions. More so-called accidental discharges of firearms take place during guard duty than anywhere else. Sad but true, the overheads of many guardrooms are pock-marked by .45 caliber bullet-holes.

See that reliefs and sentinels are posted in military fashion, by the book.

Keep an eye on the Colors. Such avoidable fumbles as Colors unwittingly hoisted upside down have, on occasion, provided the hapless OD with several days' enforced leisure. Never allow Colors to become fouled or snarled about the pole or halyards.

Visualize every emergency that could happen during your watch. Decide *now* what you will do. What if fire breaks out? A prisoner escapes? A serious automobile accident occurs? Electric power or utilities fail? Disaster occurs in a nearby community? There is a bomb threat or terrorist onslaught? Be forehanded. Know your answer in advance.

Visit the main gate during rush hours. Let yourself be seen, and let the main gate guards know you are on hand to back them up.

Be unfailingly courteous, especially to civilians and visitors. The good name of the post is in your hands when you are on watch.

Avoid gumshoeing. It is one of your main functions to be seen.

Enforce orders to the hilt. If an order is unwise, impractical, or out of date, the best way to get it modified is to enforce it, and to report that you are doing so. Never slough over an order because *you* think it is a "dead letter."

Avoid personal dealings with drunks. Let enlisted members of the guard deal with them while you keep in the background. This will not only save you potential embarrassment but may also save the drunken man from some offense much more serious if done toward an officer. Never allow a drunk to be roughly treated, and, above all, *never detain a supposed drunk without medical examination.* It is easy to confuse seeming intoxication with the symptoms of serious head injury.

Be immediately accessible at all hours. Don't let members of the guard, however well meaning, interpose themselves between you and any sober caller, whether in person or by telephone. You never know who may be calling.

Finally, run your guard the way you know it should be run. You have the responsibility, backed up by almost unlimited authority. If the guard is below standard, you have only yourself to blame.

MILITARY POLICE AND SHORE PATROL

1712 ■ Military Police and Shore Patrol

Whenever enlisted men go ashore on liberty, it is customary to provide military police (MPs), whose job is:

To assist the civil authorities in dealing with men of the Armed Forces;
To maintain discipline and good behavior among Marines and bluejackets ashore, and get them back in good order;
To aid and safeguard liberty men in every possible way.

This duty is described as "military police" when performed by a Marine organization; when a ship or Navy shore station provides such a guard, it is known as "shore patrol." Shore patrol routine and duties are covered in the *Navy Shore Patrol Manual*. In a few places where large military populations are present from all Services, a joint patrol is sometimes organized. This has the title of "Armed Forces Police Detachment."

Where permanent need exists, the military police or shore patrol is organized about a full-time cadre of specially selected officers and enlisted men. A permanent MP detachment is usually administered by a provost marshal and is known as a "provost guard," whereas a corresponding Navy unit is called the "permanent shore patrol" and is headed by a "senior patrol officer," either a Navy or Marine officer.

Marine MPs wear the regulation military police brassard (gold letters, "MP," on a scarlet background), together with belt and prescribed side-arms. Navy shore patrolmen wear a blue "SP" brassard with yellow letters, and generally wear belts and canvas leggings.

Although local conditions vary markedly, MP and SP detachments generally perform the following tasks:

"Town Patrol" of Areas Frequented by Liberty Men. This patrol is made up of noncommissioned or petty officers who work in pairs. When both bluejackets and Marines are on liberty, one patrolman should be a Marine NCO, and the other a Navy petty officer. This permits offenders to be dealt with by members of their own Services.

Maintain close liaison with municipal authorities—primarily, the police, but not overlooking the other emergency services and the prosecuting attorney.

Operate aid stations where liberty men may obtain emergency medical treatment and venereal prophylaxis.

Supervise boat landings and patrol transportation points used by military and naval personnel.

A footnote on dealings with the military police or shore patrol:
Never resist, obstruct, or fail to cooperate with a shore patrolman or an MP, even if he comes from the Army or Air Force. Under joint regulations, MPs and SPs have all-Service authority, with power to enforce any lawful acts or instructions. If you have any complaints, make them through military channels to the proper superior authorities.

1713 ■ Hints on MP and SP Duty

Duty as a military police or patrol officer carries the same kind of problems and responsibilities as guard duty, and, in general, the instructions in Sections 1708 and 1710–1711 (the last paragraph especially) apply, at least in spirit, to MPs and SPs. Even at the expense of repetition, however, the following points should be kept firmly in mind:

When on patrol, make yourself a model officer in every respect—outstanding in dress, irreproachable in bearing and conduct, courteous but firm in execution of your orders. "Fair, firm, and formal" are the "three Fs" of a successful watch officer, MP, or sentry.

Remember that you are a *military* police officer. Don't permit your men to assert police authority over civilians. That is a job for the civilian police. By the same token, keep your men reminded that they are Marines first and foremost, and police only in a secondary and qualified sense. Discourage, and do not tolerate, any symptoms of "law-man" or highway-patrol swagger on the part of any of your Marines.

Never imbibe alcohol while on MP or SP duty. It is a long-standing naval custom that the least evidence that an MP or SP has partaken of alcohol while on duty demands a court-martial. The temptations of military police duty are manifold; the least failing is fatal.

Make yourself and your men conspicuous to liberty men and townspeople alike. This helps to hold down violations and gives assurance to all that the situation is well in hand. Ensure that your men are outstanding in every respect. Bear down on them.

If possible, have medical assistance ready at hand. Should your patrol not include a doctor or corpsman, know where you can get medical aid and how to get it without delay.

Get on easy working terms with the local police. Cooperate sincerely with them and they will do the same with you. *Be unfailingly courteous toward civilians.*

If in a foreign port, obtain a trustworthy interpreter who knows the local customs. Try to select enlisted men who know the language.

Let your enlisted people deal with drunks. Be sure to give a medical examination to any seemingly intoxicated prisoner.

Handle prisoners "by the book." Allow no undue force, "third degree," or abusive behavior toward a prisoner, no matter how he provokes you. Unauthorized treatment of a prisoner is not only unworthy of a Marine officer but a sure road to trouble.

Avoid disorderly public scenes, prolonged disputes, or heated brawls. Get troublemakers, noisemakers, or contrary-minded "ferninsters" back to headquarters, out of the public eye, and deal with them in private.

Know your orders, *Navy Regulations,* and *Uniform Code of Military Justice,* down to the last comma.

Above all, exercise common sense and tact. It is your job to prevent trouble as well as to quell it. When you see a Marine or bluejacket in difficulty, ask yourself, *"How can I help this man?"*

A Marine on duty has no friends.
—Marine Corps proverb

Duty is the great business of a sea officer; all private considerations must give way to it, however painful it may be.
—Horatio Nelson

18

Military Courtesy, Honors, and Ceremonies

MILITARY COURTESY

Military courtesy is the traditional form of politeness in the profession of arms. Though sharing many elements with courtesy in civilian life, military courtesy stems firmly from a traditional code of rules and customs. Just as courtesy in general is said to be "the lubricant of life," so military courtesy helps to ease us along well-worn, tried, and customary paths. Since by its very nature, the life and discipline of the Service is formal, so the form of courtesy used by military men must necessarily be formal, too.

Military courtesy embraces much more than the salute or any other ritual, important as these are. Courtesy is a disciplined attitude of mind. It must be accorded to all ranks and on all occasions. Courtesy to a senior indicates respect for authority, responsibility, and experience. Courtesy toward a junior expresses appreciation and respect for his support and for him as a fellow Marine. Courtesy paid to the Colors and to the National Anthem expresses loyalty to the United States and to the Constitution which we are sworn to uphold and defend, as well as pride and respect for the principles for which those symbols and our country stand.

Military courtesy is a prerequisite to discipline. It promotes the willing obedience and unhesitating cooperation that make a good outfit "click." When ordinary acts of military courtesy are performed grudgingly or omitted, discipline suffers. Discipline and courtesy alike stem from *esprit de corps*. Both are essential factors which help to transform civilians into Marines.

The best fighting outfits can be readily recognized by their standards of military courtesy. The Marine Corps has always stood at the top among the Services by full and willing observance of the twin virtues of soldierly courtesy and discipline.

1801 ■ Conduct Toward Members of Other Services

The minutiae of military courtesy vary little from Service to Service, and from nation to nation. As a Marine (and therefore, as a *professional*), you must learn the meaning and traditions behind the badges, insignia, and titles of the officers and enlisted men of other military Services, both American and foreign.

When you go to duty with another Service or in another country, make it a particular point to know and defer to the customs and traditions of that Service or country. On the other hand, *never forget that you are a Marine*, never feel self-conscious about holding fast to Marine Corps standards of uniform or to the Marine way of doing and saying things. Remember that in every action, great or small, you represent the Corps.

1802 ■ Military Titles, Phraseology, and Address

In Section 715 you will find emphasis on the traditional Marine way of saying things, and at the end of this book a glossary of Marine Corps terms. Know, employ, and enforce the use of those terms. Insignia of rank appear in Figs. 18-1 and 18-2.

Addressing Seniors. When you address an officer senior to yourself, on an official as distinct from a social occasion, it is a long-standing soldierly tradition, not confined to the Marine Corps alone, that you speak in the third person; for example, "Would the Colonel care to sign these papers now?" or "Sir, Lieutenant Wharton reports for duty." This form of address is habitually observed between enlisted men and commissioned officers, and is usual between junior and senior officers when formality is the rule and some disparity of rank exists.

When close association or friendship between individuals would make this mode of address unreasonably stilted, or when the senior indicates that he would prefer that you abandon this formality, you may shift to the direct approach. But never forget that "sir" is an important word in conversation with anyone senior. While you may not be reprimanded on the spot for omission of "sir," that omission is quickly noted and usually remembered.

Speaking to Juniors. To help promote subordination and respect among your juniors, address them by their proper titles *and* their names. Follow the principles laid down in Section 1603, and be wary of over-

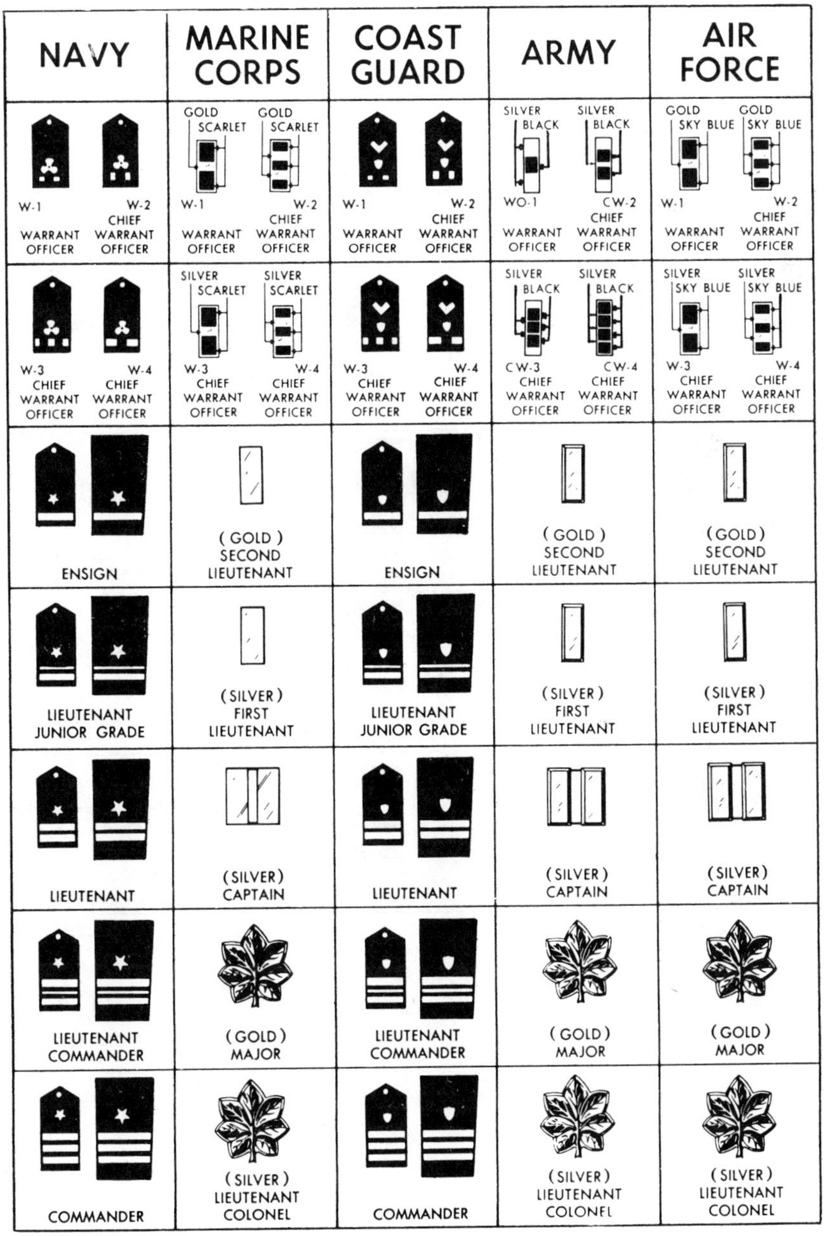

NAVY		MARINE CORPS		COAST GUARD		ARMY		AIR FORCE	
W-1 WARRANT OFFICER	W-2 CHIEF WARRANT OFFICER	GOLD SCARLET W-1 WARRANT OFFICER	GOLD SCARLET W-2 CHIEF WARRANT OFFICER	W-1 WARRANT OFFICER	W-2 CHIEF WARRANT OFFICER	SILVER BLACK WO-1 WARRANT OFFICER	SILVER BLACK CW-2 CHIEF WARRANT OFFICER	GOLD SKY BLUE W-1 WARRANT OFFICER	GOLD SKY BLUE W-2 CHIEF WARRANT OFFICER
W-3 CHIEF WARRANT OFFICER	W-4 CHIEF WARRANT OFFICER	SILVER SCARLET W-3 CHIEF WARRANT OFFICER	SILVER SCARLET W-4 CHIEF WARRANT OFFICER	W-3 CHIEF WARRANT OFFICER	W-4 CHIEF WARRANT OFFICER	SILVER BLACK CW-3 CHIEF WARRANT OFFICER	SILVER BLACK CW-4 CHIEF WARRANT OFFICER	SILVER SKY BLUE W-3 CHIEF WARRANT OFFICER	SILVER SKY BLUE W-4 CHIEF WARRANT OFFICER
ENSIGN		(GOLD) SECOND LIEUTENANT		ENSIGN		(GOLD) SECOND LIEUTENANT		(GOLD) SECOND LIEUTENANT	
LIEUTENANT JUNIOR GRADE		(SILVER) FIRST LIEUTENANT		LIEUTENANT JUNIOR GRADE		(SILVER) FIRST LIEUTENANT		(SILVER) FIRST LIEUTENANT	
LIEUTENANT		(SILVER) CAPTAIN		LIEUTENANT		(SILVER) CAPTAIN		(SILVER) CAPTAIN	
LIEUTENANT COMMANDER		(GOLD) MAJOR		LIEUTENANT COMMANDER		(GOLD) MAJOR		(GOLD) MAJOR	
COMMANDER		(SILVER) LIEUTENANT COLONEL		COMMANDER		(SILVER) LIEUTENANT COLONEL		(SILVER) LIEUTENANT COLONEL	

Fig. 18–1: Commissioned insignia of rank.

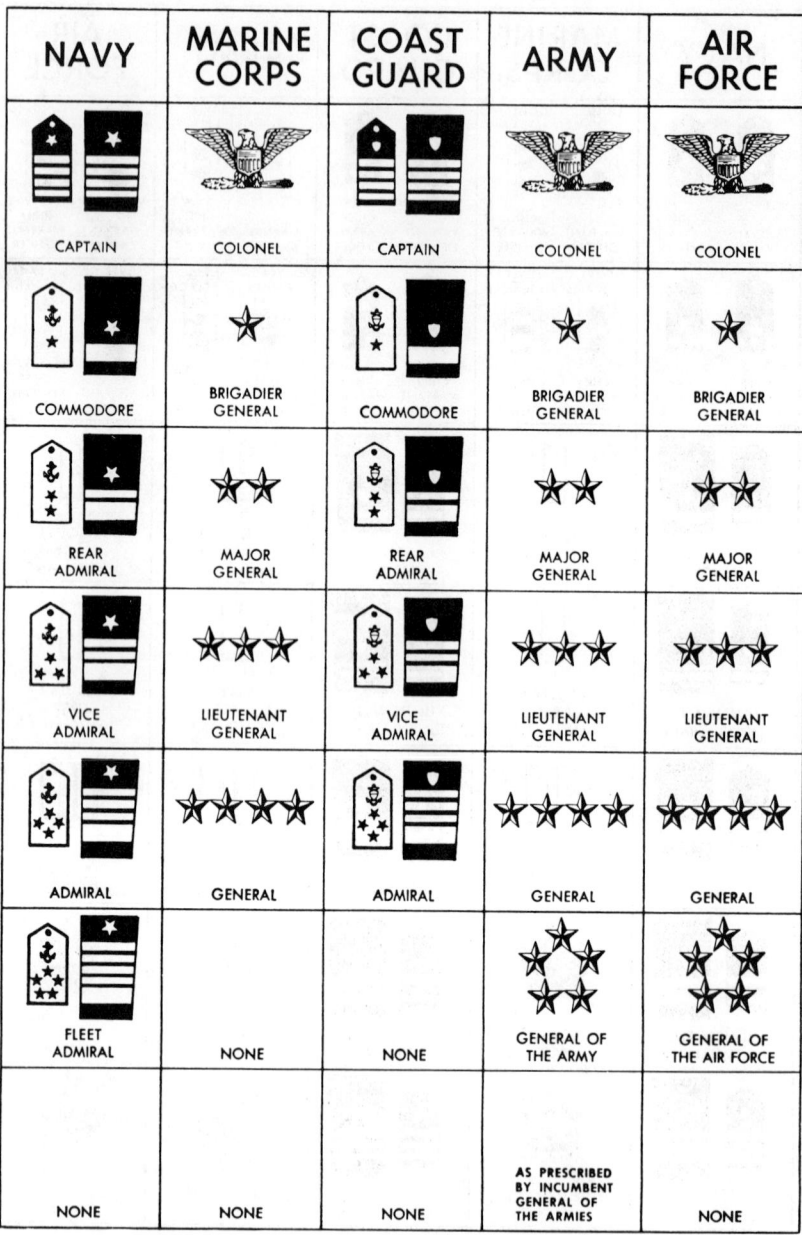

Fig. 18-1: Commissioned insignia of rank.

NAVY	MARINES	ARMY	AIR FORCE	
MASTER CHIEF PETTY OFFICER	SERGEANT MAJOR / MASTER GUNNERY SERGEANT	STAFF SGT. MAJOR / COMMAND SGT MAJOR / SPEC. 9	CHIEF MASTER SERGEANT / CHIEF MASTER SGT OF THE AF	E9
SENIOR CHIEF PETTY OFFICER	FIRST SERGEANT / MASTER SERGEANT	FIRST MASTER SGT. / MASTER SGT. / SPEC. 8	SENIOR MASTER SERGEANT	E8
CHIEF PETTY OFFICER	GUNNERY SERGEANT	SERGEANT FIRST CLASS / SPECIALIST 7	MASTER SERGEANT	E7
P.O. FIRST CLASS	STAFF SERGEANT	STAFF SERGEANT / SPECIALIST 6	TECHNICAL SERGEANT	E6
P.O. SECOND CLASS	SERGEANT	SERGEANT / SPECIALIST 5	STAFF SERGEANT	E5
P.O. THIRD CLASS	CORPORAL	CORPORAL / SPECIALIST 4	SERGEANT	E4
SEAMAN	LANCE CORPORAL	PRIVATE FIRST CLASS	AIRMAN FIRST CLASS	E3
SEAMAN APPRENTICE	PRIVATE FIRST CLASS	PRIVATE	AIRMAN	E2
SEAMAN RECRUIT	PRIVATE	PRIVATE	BASIC AIRMAN	E1

Fig. 18-2: Enlisted insignia of rank.

casual use of first names or nicknames. Formality in speaking to a subordinate is never wrong, whereas informality can be risky and is liable to compromise your position. In particular, never allow casual or even unintentionally disrespectful reference to an absent third person, particularly one senior to you, on the part of one of your juniors.

Shortcuts. It is proper to use shortened titles in conversation or unofficial correspondence. Table 18-1 shows the correct military forms of address on official and unofficial occasions, and when dealing with civilians.

Here are some informalities that usage sanctions:

Medical and dental officers below the rank of commander may be addressed as "Doctor." Your medical officer may be spoken to as "Surgeon."

Any chaplain may be addressed by another officer as "Padre"; Roman Catholic chaplains of whatever rank (and Episcopal chaplains who so prefer) as "Father."

Navy officers of the Supply Corps are occasionally still spoken of as "Paymaster," and may be addressed by other officers of approximately equal or higher ranks as "Pay."

Although no longer prescribed in regulations, custom sanctions a second lieutenant's being addressed or spoken of as "Mr." In the presence of enlisted people, however, it is preferable to speak of him as "lieutenant."

Lieutenant colonels should be addressed as "Colonel."

Generals and admirals, of whatever grade, are spoken to as "General" or "Admiral."

Where the male officer is addressed as "sir," a woman officer may be addressed as "ma'am," or by rank, as "Yes, Major," or "Good morning, Lieutenant." Women warrant officers may be addressed informally as "Miss" (or "Mrs."), as may nurses below the grade of commander.

The first sergeant of a company, battery, or detachment may be addressed privately by officers of the unit as "Top" but never so in the presence of troops. The title "top sergeant," however, is not used in the Marine Corps.

Avoid the unfortunate practice, which has developed in recent years, of referring colloquially to enlisted Marines as "troopers." Basically this is an Army—not a Marine or Navy—term, going back to horse cavalry, or nowadays to the paratroopers; it is inappropriate for Marines (and sounds like the highway patrol). Marines should be referred to collectively as "Marines" or less formally in the traditional Marine usage as "people" (as in the injunction, "You people, square yourselves away.").

Any NCO above corporal may be informally addressed as "sergeant," but it is better to give him his exact rank. Similarly, when an enlisted man speaks of himself or to or of another enlisted man, he should do so by rank—"*Lance Corporal* Daly," not just "Daly."

Drummers and trumpeters (but not bandsmen), regardless of rank, are addressed as "Music," and are spoken of collectively as "field musics."

Navy chief petty officers are habitually spoken to as "Chief."

1803 ■ Pointers on Military Etiquette

This paragraph compiles a miscellany of Marine Corps and Navy customs, courtesies, and points of etiquette, some written, others unwritten —but all important for you to know and observe.

The CO's "Wishes." When your commanding officer says, "I wish," "I desire," "I would like," or similar expressions, these have the force of a direct order and should be complied with on that basis.

Accompanying a Senior. The position of honor for one's senior is on the right. Therefore, in company with a senior, you walk, ride, and sit on the left. When entering a vehicle or a boat, you embark first and take the less desirable places in the middle, or on "jump" or front seats (or forward in a boat); when debarking, the senior leaves first, while juniors follow in order of rank.

When a senior is inspecting, he is followed by the immediate commander of the unit being inspected, who remains on the senior's left, one pace to the rear—*except* that, during inspection of troops in formation, the immediate commander remains on the *right* of the inspecting officer and *precedes* him while inspecting in ranks. For other pointers on inspections, turn to Sections 1612–1616 and Appendix III.

Acknowledging Orders. When a Marine officer or enlisted man receives orders or instructions, he replies, "Aye, aye, sir." This phrase, which descends from the earliest days of the Marine Corps and Navy, is used in both Services. It means: *"I understand the orders I have received, and will carry them out."* **Never permit a subordinate to acknowledge an order by "Very well," "All right," "Yes," or "OK."**

Mounted juniors dismount before addressing or reporting to seniors, except when in the field. Even then, however, dismount if practicable.

When you meet a senior indoors, either in a passageway or on a stairway, give way smartly and promptly. If he is a general or a flag officer, halt and stand at attention with your back to the bulkhead until he has passed. Enlisted men should halt at attention until any passing officer is clear.

When sent for by a senior, juniors must report immediately and in correct uniform. If on the drill field or parade ground, it is customary for juniors to proceed and report at the double.

Table 18-1
Correct forms of address for naval and military personnel.

PERSON ADDRESSED OR INTRODUCED	TO MILITARY PERSONNEL		TO CIVILIANS	
	Introduce as:	Address as:	Introduce as:	Address as:
MARINE, ARMY, OR AIR FORCE OFFICER	Major (or other rank) Smith	same	Major Smith[1]	same
NAVAL OFFICER	Captain Smith	same	Captain Smith[1]	same
NAVY STAFF CORPS OFFICER	Cdr. Smith[2] Chaplain Smith	same same	Cdr. Smith[2] Chaplain Smith	same same
COAST GUARD and COAST AND GEODETIC SURVEY OFFICERS	Same as for same rank in Navy[3]	same	same	same
U.S. PUBLIC HEALTH SERVICE OFFICER (M.D. or D.D.S.)	Dr. Smith[3]	same	Dr. Smith of the Public Health Service	Dr. Smith
U.S. PUBLIC HEALTH SERVICE OFFICER (Sanitary Engr.)	Mr. Smith[3]	same	Mr. Smith of the Public Health Service	Mr. Smith
COMMISSIONED WARRANT OFFICER[4]	Chief Warrant Officer[4] Smith	same[4]	Chief Warrant Officer Smith[4]	same
MIDSHIPMAN OR CADET	Midshipman (or Cadet) Smith	Mr. Smith	Midshipman (or Cadet) Smith	Mr. Smith
WARRANT OFFICER[4]	Warrant Officer Smith[4]	same[4]	Warrant Officer Smith[4]	same[4]

STAFF NCO or CHIEF PETTY OFFICER[5]	Sergeant Major Smith,[5] Master Chief Gunner's Mate Smith	Sergeant Major or Chief	Sergeant Major Smith, Master Chief Gunner's Mate Smith	Sergeant Major or Chief
NONCOMMISSIONED OFFICER OR PETTY OFFICER	Corporal Smith or Gunner's Mate Smith	Corporal Smith or Gunner's Mate Smith	Corporal Smith or Petty Officer Smith	same
PRIVATE OR SEAMAN	Private (or Seaman) Smith	Smith	Private (or Seaman) Smith	Smith

[1] When not in uniform, an officer should be introduced as "of the Navy" or "of the Marine Corps" to distinguish his rank from similar-sounding ranks in the other armed services. Suggested phraseology: "This is Lieutenant Smith of the Marine Corps." Such a form of introduction indicates the officer's rank, service, and how to address him. When status of a woman officer is not clear, add "of the Medical Corps," "of the Medical Service Corps," "of the Navy Nurse Corps," etc.

[2] Add "of the Medical Corps," "of the Civil Engineer Corps," or other corps, when helpful to indicate status of officer. If a senior officer of the Medical or Dental Corps prefers to be addressed as "Doctor," such preference should be honored. Some senior members of the Chaplain's Corps prefer to be addressed by their rank, but it is always correct to address a chaplain of any rank as Chaplain. (See Section 1802.)

[3] In any case where there is reason to believe that the officer's insignia might not be recognized, it is correct to add, "of the Public Health Service," "of the Coast Guard," or "of the Coast and Geodetic Survey."

[4] Male Marine Corps warrant officers appointed in Line occupational fields bear the title of (Chief) Marine Gunner.

[5] All staff NCOs (i.e., those with rank of staff sergeant and higher) are addressed by their particular titles, e.g., "Gunnery Sergeant Hayes," "Master Sergeant Wodarczyk," "Staff Sergeant Basilone," etc.

When a senior enters a room or passes close aboard unorganized groups either indoors or outside, the senior officer or NCO of the group or groups commands "Attention!" All hands come to attention and remain so until the senior has passed. If out of doors and covered, all hands salute.

Uncovering Under Arms. The only exception to the rule that Marines under arms never uncover is at divine service, such as a wedding, when officers may wear swords and still uncover. You do not unsheathe your sword inside a church, however, unless express authority is granted by an appropriate religious functionary.

Permission to Speak to Senior Officers. When one of your enlisted people wishes to speak to the company or detachment commander, he first obtains the first sergeant's permission (see Section 1603). If he desires to speak to an officer of still higher rank or position, he must in turn have his company or detachment commander's permission. When permission has been duly granted, the enlisted Marine reports, "Sir, Private Quick has the first sergeant's (*or* Captain McCawley's) permission to speak to the company commander (*or* to the Battalion Commander)."

Similarly, as a junior officer, you obtain your immediate CO's permission before you seek an official interview with any higher officer.

The reasoning behind this procedure is that the first sergeant or CO can probably solve the problem satisfactorily by himself without the matter having to go higher.

Entering an Office. Enlisted men entering any office should be required to observe the following procedure, regardless of whether an officer is in the office or not: (1) knock; (2) enter and stand at attention immediately inside doorway, uncovered (unless under arms); (3) identify himself by name and rank, and state business.

Although the foregoing sequence is not mandatory for officers, it is a prudent procedure for junior officers to bear in mind, especially when entering the office of one who is appreciably senior. In general, an officer should not "freeze" on entering, however. Having once made his entrance, he should distinctly come to the position of the soldier (to signify his respect for the senior) and then as quickly assume a less formal stance of alert composure.

> NOTE: Never enter the office of a senior while you are smoking, and do not smoke in his office or presence until he invites you to do so.

Delivering Messages. Either an officer or an enlisted messenger with an oral message should follow the customary forms of address. As the originator of an official, oral message, be sure to include "compliments" to any junior officer addressed, and "respects" to any senior. *Examples* (from you, a first lieutenant):

1. "Music, my compliments to Lieutenant Gale, and will he please...."
2. "Runner, present my respects to Captain Biddle, and report that...."

After a message has been delivered, the messenger should report that fact to the officer who sent the message; for example, "Sir, your message has been delivered. Captain Elliott presents his respects. The guard will be paraded at 1000."

On the Telephone. Use moderate and respectful tones and identify yourself and your organization. Be brief. For example, in answering a call: "Company B, 5th Marines, Captain Griffith." *Never* "Hello."

When using a field telephone, always be sure to ring off at the end of a conversation.

Except in emergency, or as staff representative of a more senior commander, a junior officer does not telephone a senior, but presents himself in person. It is the senior's privilege to phone the junior.

Arrival and Departure of the CO. If you are an executive officer or adjutant, never fail to be on hand when the CO leaves or returns to the command, night or day.

On a Ladder or Stairway. When you encounter a senior on a stairway or in a passageway or doorway, you should always give way. In such an instance, greetings are normally exchanged. An enlisted man who is passed by an officer on a stairway should halt and remain at attention until the officer passes.

During Training or Exercises. When drilling or instructing your people on the parade ground or in a training area, and when closely approached by a senior officer, you should come to attention, salute, and inform the senior what you are doing—"Battery F, 15th Marines, sighting and aiming exercises, Sir."

Similarly, on the drill field, never cut across the front of a unit commanded by a senior, thus causing him to halt or mark time, even if you may feel you have the right-of-way.

SALUTES AND SALUTING

1804 ■ The Military Salute

Saluting is a military custom observed by men who follow the profession of arms. It is a matter of pride among Marines, from general to private, to salute willingly, promptly, smartly, and proudly. The good Marine stands out among the other Services by his smart, correct, and cheerful salute, which is as much a hallmark of the Corps as the Globe and Anchor. When you salute or receive a salute, you mark yourself as a Marine who has pride in himself and in his Corps.

As a junior officer you must recognize and teach that the salute is a privilege enjoyed only by military men, and is a mutual acknowledgment of comradeship in the profession of arms.

Origins of Saluting. Over the centuries, men-at-arms have rendered fraternal and respectful greetings to indicate friendliness. In early times, armed men raised their weapons or shifted them to the left hand (while raising the empty right hand) to give proof of amicable intentions. During the Middle Ages, knights in armor, on encountering friendly knights, raised their helmet visors in recognition. If they were in the presence of feudal superiors, the helmet was usually doffed. In every case, the fighting man made a gesture of friendliness—the raising of the empty right hand. This gesture survives as today's hand salute, which is the traditional greeting among soldiers of all nations.

Like the original hand salute and doffing of the cap, the discharge of weapons, presentation of arms, and lowering of the point of the sword were all intended to signify good will. In every case, the one so saluting, in good faith, momentarily rendered himself incapable of using his weapon offensively. The descendants of these earlier gestures are the modern sword salute, present arms, and gun salutes.

Whom to Salute. All military men must salute when they encounter and recognize any person who rates a salute, under circumstances in which the salute is required. An individual with the true soldierly instinct never misses an opportunity to salute a senior. *Those entitled to salutes are:*

1. All commissioned and warrant officers of the Army, Marine Corps, Navy, Air Force, and Coast Guard; of the Reserve components of those Services; and of the National Guard.

2. Officers of friendly foreign powers.

3. In addition, by Service custom, though not by regulation, any high civilian official who is entitled to honors by *Navy Regulations*; and ladies, on occasions when, if in civilian clothes, the hat would be raised or tipped. *Never tip your headgear when in uniform*; substitute a salute.

Officers of the same rank exchange salutes on meeting. The first one to recognize the other initiates the salute.

Enlisted men salute other enlisted men only in formation when rendering reports. Prisoners may not salute or wear the Marine Corps emblem.

Definitions. The following definitions apply to Marine Corps saluting procedure:

Out of doors means "In the open air; or the interior of such buildings as drill halls and gymnasiums when used for drill or exercises of troops; or on the weather decks of a man-of-war; or under roofed structures such as lanais, covered walks, and shelters open at one or both sides to the weather." Synonymous with "on the topside," when used afloat.

Indoors means "The interior of any building ashore, other than a drill hall, gymnasium, or armory."

Between decks means "Any shipboard space below a weather deck, other than officers' country."

Covered and *uncovered* means "When and when not wearing headgear."

Under arms: A Marine is under arms when he has a weapon in his hand, is equipped with side-arms, or when he is wearing equipment pertaining to an arm, such as a sword sling, pistol belt, or cartridge belt. Any Marine wearing an "MP" or "SP" brassard is considered under arms.

Saluting distance means "The maximum distance within which salutes are rendered and exchanged," prescribed as 30 paces. This figure is considered to be one within which recognition of insignia is possible, and approximately that within which friends or acquaintances can recognize and greet each other. The salute should be rendered when six paces from the person (or Color) to be saluted. If the person or Color to be saluted obviously will not approach within this distance, the salute is rendered at the point of nearest approach.

1805 ■ Hand Salutes

Significance. In some Services, the hand salute (Fig. 18–3) has been de-emphasized almost to the vanishing point, based on an erroneous perception that rendition of a salute, rather than being an act of courtesy and soldierly recognition, signifies inferiority and subservience. Nothing could be further from the truth. As in civil life, "on the outside," where you render courtesy to older or more important persons, so, as a junior Marine, you salute first. In returning your salute, the senior in turn salutes you as a brother in arms. Thus, the exchange of salutes is a two-way street.

The manner and enthusiasm with which you render or receive a salute indicate the state of your training, your individual *esprit*, the discipline of your outfit, and your quality as a Marine. *Correct saluting habits characterize a good Marine.*

How to Execute the Hand Salute. Salute at quick time only. If you are at the double and must salute or receive a salute, slow to quick time.

Fig. 18–3: Hand salute.

Stand or walk at attention: head up, chin in, and stomach pulled in. When halted, come to attention distinctly as a preliminary motion to the salute; bring your heels together audibly. Look directly at the person or Color you are saluting. If walking or riding, turn your head smartly toward the person being saluted, and catch his eye. Execute the first movement, holding position until the salute is acknowledged, or you see that it is not going to be, then complete the salute by bringing your hand down smartly. When doing this, keep your fingers extended and joined, your thumb streamlined alongside. In returning a salute, execute the two counts at marching cadence.

In the Marine Corps and Navy it is customary to exchange a greeting with a salute. The junior should always say, "Good morning (or evening), sir," and the senior should unfailingly reply in the same vein—with a smile.

In the Marine Corps and Navy one does not salute when uncovered—that is, when not wearing headgear. The only exception to this rule is that the salute *may* be rendered uncovered when, in a special circumstance, not to salute might cause misunderstanding. For example, when serving with the Army or Air Force (who *do* salute uncovered), you may, if you wish, depart from the naval procedure but beware of contracting the habit.

If carrying a swagger stick during receipt or rendition of a salute, hold the stick snugly under your left arm, large end forward, exactly parallel to the deck, with your left arm in normal position at your side, not swinging.

How Not to Salute. A sloppy, grudging salute, or a childish pretence not to notice anyone to whom a salute is due, indicates unmilitary attitude, lack of pride in self and Corps, and plain ignorance. Evasive, unmilitary saluting betrays the amateur; a sharp, keep salute distinguishes the professional.

Have you ever seen a Marine who amounted to anything who was ragged in his saluting?

Never salute with pipe or cigarette in your right hand or your mouth. If you are chewing tobacco or gum, bring your jaws to rest during the exchange of salutes. As under any other circumstances, it is highly unmilitary to be caught saluting with one hand in your pocket, your blouse unbuttoned, or your cap not squared.

Avoid—and, as an officer, *never tolerate*—trick salutes. The most common aberrations are:

Right wrist bent
Left elbow stuck out at exaggerated, unnatural angle
Palm turned inward, knuckles kept forward
Fingers on right hand bent and flexed inward
Right thumb extended away from fingers
Hips thrust forward, shoulders swayed back

When you find a Marine doing any of these things, no matter how hard he seems to be trying, correct him on the spot, and see that he knows and practices the right way to salute.

1806 ■ **Rifle Salutes**

The rifle salute may be executed from the following positions:

Right or left shoulder arms
Order arms
Trail arms
Present arms

Any individual under arms with rifle salutes by one of the foregoing rifle salutes. The only occasion where a hand salute is executed by a man with a rifle is when the rifle is at "sling arms."

In its four forms, the rifle salute is rendered as follows under the conditions given:

Right or left shoulder arms—when out of doors, at a halt or at a walk.
Order arms—when at a halt, either indoors or out of doors.

Trail arms—when at a walk, indoors or out of doors.

Presenting arms is a special compliment, as a Marine at present arms represents the authority of the nation. The privilege of saluting by presenting arms is reserved for troops in formation and for sentinels on post.

Marines armed with weapons normally carried slung use the hand salute only, and, when so saluting, carry the piece at sling arms, with the left hand grasping the sling to steady the weapon.

1807 ■ Sword Salutes and Manual

You will find the manual of the sword described in Chapter 2, the *Landing Party Manual*, while further information on the sword is given in Sections 1815–1816. Every Marine officer takes pride in being precise, dexterous, and at ease with his sword.

When armed with the sword, you render or return salutes in the following ways:

Fig. 18–4: Sword salutes and manual.

Rendering the Salute
1. *If your sword is sheathed*, and you are not in formation: execute the normal hand salute.
2. *If your sword is drawn and you are halted*, either in or out of formation, execute present sword as prescribed in the manual of the sword. If commanding a formation, which will usually be the case if your sword is drawn, bring your troops to attention before you do so.
3. *If your sword is drawn and you are underway in formation*, execute the sword salute, having first brought your command to attention, if necessary.

When You are Returning a Salute
1. *If your sword is sheathed*, acknowledge by the hand salute.
2. *If your sword is drawn*, acknowledge by the sword salute.

1808 ■ **Individual Saluting Etiquette**

Whether to Salute Once or Twice. After an officer has been saluted initially, if he remains nearby and no conversation takes place, no further salutes are required.

When a junior is spoken to by, or addresses, a senior officer, he salutes initially, and again when the conversation ends or the senior leaves. Throughout the conversation, the junior stands at attention unless otherwise directed by the senior. It should be an instinctive military courtesy on your part, as an officer, to give your subordinates "At ease" or "Carry on" during any extended conversation.

Reporting, Indoors. When you report indoors to an officer senior to you, unless under arms, you uncover, place your cap under your left arm, visor forward, knock, and enter when told to do so. Two paces in front of the senior, halt and report, "Sir, Lieutenant Neville reports to Major Russell." Remain at attention unless told to carry on or be seated. On being dismissed, take one backstep, halt, and then face about and march out. If under arms, remain covered, and salute on reporting, and again on being dismissed. The latter salute is rendered after completion of your backstep.

After entering, do not report until recognized by the officer and until he has completed whatever business he may have in hand.

The foregoing procedure also applies to enlisted men who report to you.

Enlisted Men Not in Formation. When an officer approaches enlisted men who are not in formation, the first to recognize him calls the group to attention as soon as the officer comes within ten paces. Out of doors, if covered (as they should be), all men salute when the officer is within six paces. The salute is held until returned. The men remain at attention until the officer has passed or until he commands, "Carry on," which an officer should be quick to do under informal circumstances.

Profit in this by the example of Major General Lejeune, 13th Commandant, during the Meuse-Argonne Battle in 1918. General Lejeune approached a group of Marines, whom an NCO called to attention. As the men sprang to their feet, General Lejeune checked them, saying "Sit down, men. It is more important for tired men to rest than for the division commander to be saluted."

Overtaking. When you overtake an officer senior in rank proceeding in the same direction, draw abreast on the senior's left, coming to the salute as you do so, and say, "By your leave, Sir." The senior officer acknowledges the salute and replies, "Granted."

When you overtake a Marine junior to you, pass on the right. As you come into view, abreast, salutes are exchanged.

Indoors. Marines not under arms do not salute indoors. When an officer is present, enlisted men uncover in lieu of saluting (see Section 1804). *In an office,* however, men need not cease work when an officer enters unless called to attention. When addressed by an officer, the person so addressed should rise.

In the Mess Hall. Men at meals do not rise when called to attention, but stop eating and keep silent. If spoken to by an officer, an enlisted man gets to his feet and stands at attention. If not under arms, be sure to uncover when you enter a galley, mess hall, or ship's messing compartment.

In Sick Bay. Formal military courtesies are neither rendered nor required in a sick bay. Always uncover when you enter a sick bay or ward.

At the Pay Table. Men not under arms uncover before approaching the pay table, and do not salute. A Marine under arms salutes the paymaster when front and center of the pay table, and again after receiving his pay and before moving off. The paymaster, being occupied, does not acknowledge. For more details on pay call, turn to Section 2007.

In Vehicles. Except when on board public conveyances, such as street cars, buses, and trains, officers in vehicles are saluted as if afoot. The driver of a motor vehicle does not salute if the car is in motion. If the vehicle is not underway, he salutes without rising. Other passengers salute or return salutes as necessary, remaining seated.

When Mounted. Mounted persons salute in the same manner as if on foot, but salutes are not rendered by anyone standing to or leading a horse. A mounted junior always dismounts before addressing a senior who is not mounted; this rule applies to vehicles as well as horses.

During Games. Games are not interrupted at the approach of an officer. Spectators do not rise or salute unless individually addressed by an officer.

In Plain Clothes. Marines in civilian clothes salute officers as usual. Officers wearing plain clothes salute and are saluted, if recognized, as if they were in uniform.

> NOTE: Although nowhere required by regulations, it has become customary on many posts for gate sentries to salute officers' wives as a matter of courtesy when driving through the gate in a vehicle bearing appropriate decal. This is a voluntary act of courtesy and in no sense an expression of subordination.

On Guard
1. *When armed with the rifle,* sentries salute by presenting arms. A sentry walking post halts, faces the officer being saluted, and comes to the present. If then spoken to by the officer (or by any other person), the sentry executes port arms and holds this position throughout the conversation. If speaking with an officer, he does not interrupt the conversation to salute another officer, unless the officer with whom he is speaking likewise salutes; if so, the sentinel presents arms. At the end of the conversation, the sentry presents arms again. During hours of challenging, the first salute or present arms is rendered when the officer has been duly advanced and recognized, as described in Section 1706.

2. *When not armed with the rifle,* a sentry renders hand salutes in the usual way. A sentry armed with submachine-gun, pistol, or (in other Services) the carbine, does not salute during hours for challenging. While challenging, a sentry armed with the pistol remains at raise pistol; one armed with submachine-gun or carbine remains at port arms.

If circumstances are such that payment of compliments interferes with a sentry's performance of duty, he does not salute.

Prisoner guards ("chasers") do not salute except when addressed by an officer. If marching his prisoners, the chaser halts them and takes necessary precautions for their security before rendering the salute. If armed with the rifle, a chaser executes rifle salute, but does not present arms. Prisoners may not salute at any time.

> NOTE: *Never pass between a guard and his prisoners,* and be sure to correct any guard who permits you or any other person to do so.

On Board Ship. Saluting procedure on board a man-of-war is that described in the preceding paragraphs and in Section 2221.

When in Doubt. If you are uncertain as to whether a salute is required, always salute. For a properly trained Marine, there should never be any doubt. Should a doubtful situation arise, however, do not go out of your way to avoid saluting. Having made up your mind to salute, do so properly and smartly. Never give a hesitant, half-hearted salute which suggests only too plainly that you really don't know the score. *Remember, it is better to render five unnecessary salutes than to omit one that you should give.*

1809 ■ Group Saluting Etiquette

Troops in Formation. Troops in formation salute on command only. Officers and NCOs in command of formations render salutes for their respective units. Before rendering a salute, the person in command brings his unit to attention. Individuals armed other than with a rifle (and officers and NCOs whose swords are not drawn) execute the hand salute.

If an officer speaks to an individual in ranks, when the unit is not at attention, the person spoken to comes to attention. At the end of the conversation, he resumes the position of the remainder of the unit.

Troops at Drill and on the March. Troops drilling do not render compliments. The person in command renders salutes for his unit. An officer in a formation is saluted only if he is in command of the entire formation, and he alone returns all salutes. NCOs in charge of detachments or units do not exchange compliments with other units so commanded, except at guard mounting, when the old and new guards do exchange compliments.

Troops marching at ease or route step are called to attention on the approach of a senior entitled to a salute.

Fig. 18–5: Group saluting.

Marine units always begin and end a march at attention. March your unit at attention while within barracks and central areas, and on main roads of your post; and, no matter how tired you are after a day or a night in the field, bring your outfit home with a short, snappy step, pieces aligned, ranks dressed, at regulation cadence, at attention. That is the Marine way.

Groups of Officers. When officers are walking or standing together, or are embarked in a vehicle, *all* render and return salutes as if each were alone.

Formations in Vehicles. Members of formations embarked, *as units*, in military vehicles do not salute individually. The senior person in each vehicle renders and acknowledges salutes. Only the hand salute is employed.

Working Parties. The NCO in charge renders salutes for the entire detail. Individuals come to attention and salute if addressed by an officer, but do not interrupt work at the approach of an officer unless the detail is called to attention.

While Honors Are Being Rendered

1. During ruffles and flourishes by the band or field music, while honors are being rendered, the guard presents arms to the recipient of honors. All persons in the vicinity come to attention and salute, following the motions of the guard (for example, hand salute on present arms; terminate salute on order arms).

2. If ruffles and flourishes are followed by a gun salute, persons in the vicinity but not in formation stand fast at attention until the last gun has fired.

3. On board ship, all hands on the quarterdeck salute while an officer is being piped over the side. If the guard is paraded, follow the motions of the guard in your hand salute.

At Military Funerals

1. The basic rule for saluting at a military funeral is, *salute each time the body-bearers move the coffin, and during volleys and "Taps."* If you are wearing civilian clothes, uncover and hold your headgear over your left breast.

2. During prayers, stand at the pre-1939 position of parade rest without arms, head bowed, as described in the note following Section 727. During the firing of volleys, come to attention and salute.

3. Body-bearers remain covered, both indoors and outdoors, when carrying the coffin. When the remains are lowered into the grave, body-bearers stand at attention, holding the flag waist-high over the grave. The officer in charge of the escort presents this flag to the next of kin, after the ceremony.

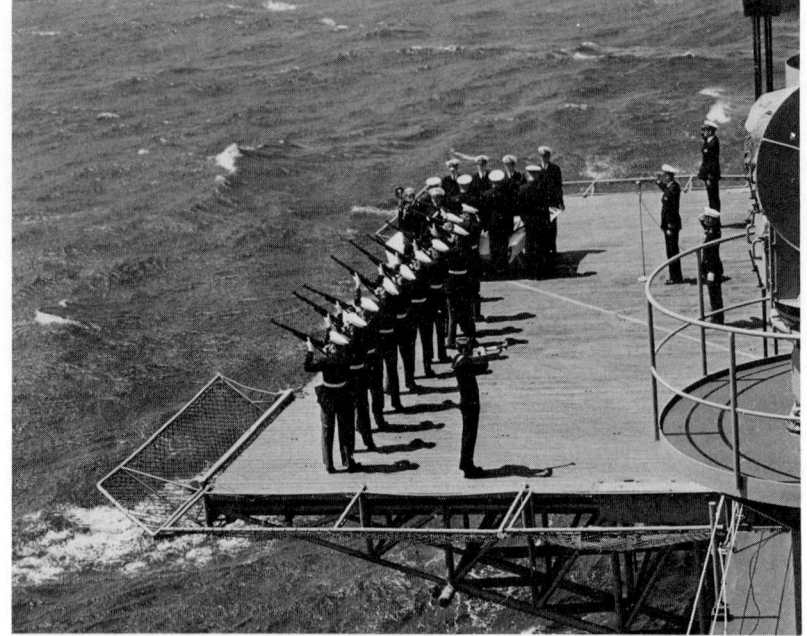

At a military funeral, salute during "Taps," firing of volleys, and when the body is moved.

4. When a military funeral cortege passes, all hands come to attention and salute the remains, using the hand salute if in uniform, and uncovering in the civilian salute, if wearing civilian clothes.

>NOTE: For general information and procedure regarding military funerals, see Section 1813.

1810 ▪ Saluting the National Anthem

When the National Anthem is played, or "To the Color" is sounded, all military personnel come to attention, face toward the music, and salute. You hold your salute until the last note of the music, but remain at attention until "Carry On" is sounded. If the anthem or call is being played incident to a ceremony involving the Colors, face toward the Colors rather than the music.

Troops in Formation. Troops in formation are halted (if on the march) and brought to attention, and the commander salutes, facing in the direction of his unit's original front. If participating in a ceremony that includes rendition of the National Anthem or "To the Color," troops present arms.

Personnel in Vehicles. During the playing of the National Anthem, all vehicles within sight or hearing of the ceremony stop. Passengers do not debark, but remain seated at attention and do not salute. If the passengers comprise a military detail in an official vehicle, the person in charge debarks, faces toward the flag or music and salutes.

Mounted personnel halt and salute without dismounting.

Sentries halt, face in the direction of the flag or music, and render the hand salute or present arms as appropriate (see Section 1808).

In Civilian Clothes. If wearing plain clothes, come to attention, remove your headgear, and hold it over your left breast with your right hand.

Indoors. When the National Anthem is played indoors, you come to attention and face the music. Only men under arms salute.

Foreign National Anthems. Accord the national anthems of friendly foreign powers the same courtesies as our own.

1811 ▪ Courtesy to the Flag

This section confines itself to the courtesies that apply to the National Color (or National Ensign). You will find additional information dealing with flags, colors, and standards in Chapter 7, while Section 1818 in turn covers execution of Morning and Evening Colors, the daily ceremonies which take place when the flag is raised and lowered.

Saluting the Flag. Except at Morning and Evening Colors and on board a man-of-war at anchor, the flag is not saluted when displayed from a mast or flagstaff, nor is any flag saluted unless it is a National Color or Standard as defined in Section 732. When Colors are encased in a protective cover (and said to be "cased"), they are not saluted.

Colors and Standards not cased are saluted when either you or they approach or pass within six paces. Hold your salute until the Colors have passed or been passed by that distance.

In the field or camp, it is customary to display the National Color and unit Battle Color in front of the commanding officer's tent. According to his wishes, this may be done every day, or only on Sundays and national holidays. All hands who approach within saluting distance (six paces) execute a hand or rifle salute as appropriate, holding the salute until six paces beyond. If a Color sentinel is posted, he acknowledges salutes rendered by enlisted men; when officers salute the Colors, he holds his own salute or present arms until the officer has completed his.

Motor Vehicles Passing Colors. When passed by an uncased National Color, all persons embarked in a vehicle remain seated at attention. Vehicles approaching and passing Colors reduce speed; embarked personnel remain seated at attention but do not salute.

Individuals Not in Formation. At the approach of Colors, persons not in formation come to attention, face the Colors, and salute when within saluting distance; if you are passing Colors, continue at attention and salute within saluting distance. Construe this distance literally. Hold your salute and keep your head and eyes turned smartly toward the Colors until they have passed or been passed by six paces. In civilian clothes, render the civilian salute with headgear held over your left breast. If mounted, bring your horse to a walk, and salute without dismounting.

Dipping the Battle or Organizational Color. In military ceremonies Battle and Organizational Colors (see Section 732) are dipped in salute

Dipping the Battle Color takes place only when the National Anthem or its equivalent is played, to honor specified individuals, and during military funerals.

during the playing of the National Anthem, "To the Color," or "Retreat" (in place of the National Anthem), or "Hail to the Chief"; when rendering honors to the organizational commander or individual or higher rank; and, during military funerals only, on each occasion when the funeral escort presents arms.

On these occasions, when passing in review, the Battle Color or Organizational Color (but never the National Color) is dipped when six paces from the individual receiving the salute, and remains dipped until six paces beyond.

Dipping the National Ensign. The National Color or Ensign is never in any circumstances permitted to touch the ground or deck. At sea, however, it is customary for merchantmen to dip their Colors when passing close aboard a man-of-war, and, in reply, the warship runs her Ensign halfway down and then back up again. This is the only time when a National Color or Ensign may be dipped.

1812 ■ Pointers on Saluting

All salutes received by you must be returned unless you are uncovered or unless both hands are fully loaded or occupied. If you are physically unable to return a salute, you should acknowledge it verbally, and should, if possible, excuse yourself to the individual who rendered the salute. If you are uncovered, or in any circumstance when you cannot render a correct salute, you should, if standing still, come to attention; if underway, you should turn your head and eyes smartly (as in "Eyes Right") toward the person or Color being saluted, in a noticeable movement, keeping your arms steady by your side.

When wearing civilian clothes and covered, render and return salutes as if in uniform, except that the civilian salute (headgear held over left breast, in lieu of hand salutes) should be used for:

Salutes to the Colors
Salutes to the National Anthem
Salutes during military funerals

If salutes are to be properly exchanged, both junior and senior must be alert. The junior must spot the approaching senior, and the senior must respond with alacrity. The attitude of seniors toward salutes has a profound effect on the spirit with which any salute is rendered. Enlisted men are discouraged from saluting (and rightly so) if you overlook their courtesy, or seem not to observe them. Such an attitude on your part as a Marine officer is discourteous, and at times downright insulting.

You can do much to foster correct rendering of salutes by inviting them. A pleasant, direct look at a junior as he approaches encourages him to salute with goodwill and generally puts him on his mettle. It is an old trick, when a junior officer seems to need a reminder in his saluting manners, for the senior to salute the junior first, with a solicitous greeting, thus extending a courteous reprimand.

In saluting, *do:*
Begin your salute in ample time (at least six paces away).
Hold your salute until it is returned or acknowledged.
Extend the same military courtesy to female officers as to male officers.
Look squarely at the person or Colors being saluted.
Assume the position of attention.
Have thumb and fingers extended and joined.
Keep hand and wrist in same place, not bent.
Incline forearm at 45 degrees.
Hold upper arm horizontal while hand is at salute.
Bring your heels together audibly.
Place swagger-stick smartly under left arm, cutting away the hand before saluting.

Do not:
Salute with blouse or coat unbuttoned.
Salute with cigarette, pipe, or cigar in mouth.
Have anything in your right hand.
Have your left hand in a pocket.
Salute when in ranks, at games, or part of a working detail.
Salute at crowded gatherings, in public conveyances, or in congested areas, unless addressing or being directly addressed by a senior.
Salute when to do so would physically interfere with performance of an assigned duty.

> NOTE: One of the most unmilitary habits encountered among some Marines, both while saluting and even in ranks, is the ludicrous habit of leaning over backwards (literally) in an effort to stand straight. This swaybacked stance, with stomach and pelvis thrust forward, jaw jutting out, and shoulders too far back, is a caricature of the position of attention. A Marine at attention should stand straight as an arrow, not like a bow.

MILITARY FUNERALS

1813 ■ General Information on Funerals

Navy and Marine Corps funerals are conducted in accordance with *Navy Regulations* and the *Drill and Ceremonies Manual,* both of which you should check carefully if, in any role role other than that of principal, you are to take part in a military funeral. Chapter 23 of this *Guide* contains administrative information on funerals and burials.

Classification. Military funerals are classified as follows:

1. By size of escort (depending on rank of deceased) and type of ceremony, i.e., full honors, simple honors, or modified.

2. By location of military ceremony, i.e., church or chapel service (remains received at church and escorted to graveside); transfer (remains received at station, airport, or cemetery gate and escorted to gravesite); gravesite (remains conveyed to gravesite by civilian undertaker, military participation and ceremony at gravesite only).

Uniforms and Equipment. If the organization providing the funeral escort is authorized blues, uniform should be dress blue A or dress blue/ white A, according to season. Otherwise the uniform should be service dress with large medals instead of ribbons, if blouse is worn.

Body bearers should not wear bayonets or scabbards.

For difficult terrain, mud, or foul weather, units and individuals— such as body bearers, music, and firing party—who must leave paved areas may wear shined boots instead of dress shoes.

Officers of funeral escorts wear mourning band and mourning sword knot, as do pallbearers; noncommissioned officers armed with the sword wear mourning sword knot only (except if acting as pallbearer, when mourning band will also be worn).

When sanctioned by the denomination concerned (as in the case of the Episcopal Church), the officiating clergyman, if so entitled, should wear military ribbons on his vestments.

Dependents' Funerals. Military honors (i.e., firing of volleys and sounding of "Taps") are reserved for deceased military or former military persons. For the funerals of Marine dependents, body-bearers may be assigned and, if desired, the funeral service will be conducted by a Navy chaplain.

Type of ceremony, as classified above, depends on the rank of the deceased and, subject to that consideration, the wishes of the next of kin. Some next of kin may wish only gravesite honors or a reduced escort; some may not wish the firing of volleys. Such wishes are of course governing.

Musical honors, if prescribed by *Navy Regulations,* are rendered during each transfer of remains into, or from, hearse or caisson to church (or vice versa) and from hearse or caisson to gravesite. Next of kin should have an opportunity to select hymns or funeral music to be played by the band, but the Navy Hymn, "Eternal Father, Strong to Save," should always be included.

Rehearsals and Reconnaissance. Unit rehearsals obviously cannot be conducted at church or gravesite, although the various evolutions can be adequately rehearsed on the parade ground. Careful but unobtrusive reconnaissance should, however, be conducted by the adjutant (who acts as officer-in-charge unless otherwise prescribed) and by the escort commander. All Marines assigned to funeral details—especially firing party and body bearers—must have attained the necessary high standards of individual proficiency in their duties for these occasions.

Rules for saluting by those attending military funerals are found in Section 1809.

1814 ■ Funeral Escorts

Officers' Funerals. The basic escort for a deceased officer consists of:

Escort commander (same rank as deceased, if possible)
Staff (colonels and general officers only)
Band
Color guard
Body bearers
Firing party (eight riflemen with NCO-in-charge)
Field music
Personal flag bearer (general officers only)

Troop escort is as follows for the respective officers:

Major General or Senior: Three ceremonial companies (two platoons of three eight-man squads each).

Body-bearers transfer a Marine's remains to the traditional caisson which will carry him to a resting place in Arlington National Cemetery.

Colonel or Senior: Two ceremonial companies composed as above.

Major or Senior: One ceremonial company composed as above (escort commander commands company and has no staff).

Company and Warrant Officers: One ceremonial platoon (three eight-man squads; escort commander serves as platoon leader and has no staff).

Enlisted Marines' Funerals. The funeral escort for a deceased enlisted Marine consists of a noncommissioned escort commander (same rank as deceased, or senior), body bearers, firing party (eight riflemen), field music, and, in the case of gunnery sergeants or above, troop escort consisting of a rifle squad.

Simple Honors Funerals. When next of kin does not desire full honors, the simple-honors funeral escort, for all ranks, consists of an escort commander (not above rank of captain), body bearers, firing party, and field music.

NOTE: For military funerals when personnel are limited, the *Drill and Ceremonies Manual* contains a special procedure requiring only eight enlisted men and an officer or NCO-in-charge.

YOUR SWORD

1815 ■ Rigging Your Sword

Correct wearing of your sword is a special point of professional punctilio. Derived from "Rig It Right," an excellent article in the *Gazette* (June 1961), by Majors T. N. Galbraith and R. N. Good, here is an account of how you should rig and wear your badge as a commissioned officer.

The first step is to get your sword-knot squared away. The way to begin assembly of the knot is to reeve its small end through the eye of the "pommel," slip it back through the two keepers, and hook it to the small metal eye adjacent to the large end. Draw one keeper tight against the pommel, the other over the hook and eye, and you are ready to tie the knot.

Now loop the large end of the knot under the cross guard of the hilt and tie a hitch as shown in Fig. 18–6. If you check this diagram closely, you will see that the knot shown is a clove hitch, not the double half hitch specified in regulations. The fact is, a double half hitch won't hold the knot tight to the cross guard, whereas a clove hitch will.

When the hitch is bent on, draw it taut and, at the same time, work the knot so that the large end doesn't hang below the upper ring mount-

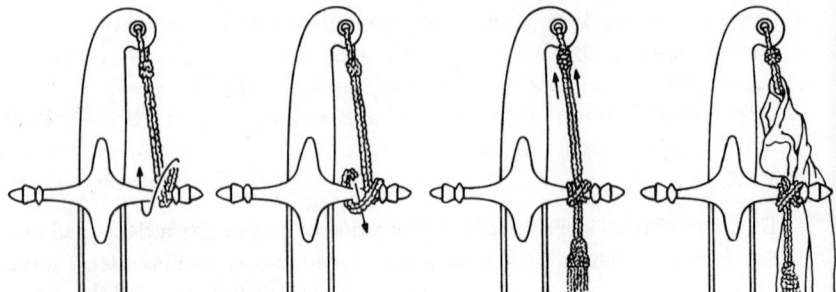

Fig. 18–6: Rigging your sword.

ing on the scabbard. Depending on the length of your particular knot, the portion from the eye of your pommel to the cross guard will possess some degree of looseness. This is all right: the determining factor is the length of the bight hanging free below the cross guard. If the knot hangs down farther than it should, you may find yourself slapped in the face when you present sword.

Mourning Knot. Secure the mourning knot to the leather sword knot between pommel and cross guard by (1) doubling the mourning knot in two; (2) passing its two free ends together around the sword knot and through the middle bend; drawing it taut. Figure 18-6 shows how it should look.

Nomenclature of the Sword and Accessories. Attaching the knot may be the most troublesome part of rigging the sword, but the nomenclature of the sword may also be a source of confusion (see Fig. 18–7). Sword and scabbard are suspended from the *sword sling*. If you are wearing a blouse, the sword sling is attached to your Sam Browne belt by the frog, or, if with cloth belt, to the *shoulder sling*; if you are not wearing a blouse, the sword sling is attached to the *frog* on your belt. In either case, the frog and shoulder sling serve the single purpose of providing a D-ring to which you attach the sword sling.

Attaching Scabbard to Sword Sling. One easy way to attach your scabbard to your sword sling is shown in Fig. 18–7. With the sword sling on and its straps hanging free, attach the sling strap snaps to the scabbard rings. Holding the scabbard by its upper ring, give it a half twist toward the body (clockwise) and hang the upper ring over the sword sling hook.

1816 ■ Manual of the Sword

This is shown in the *Drills and Ceremonies Manual.* In addition to the manual just mentioned, however, two additional positions are sometimes used.

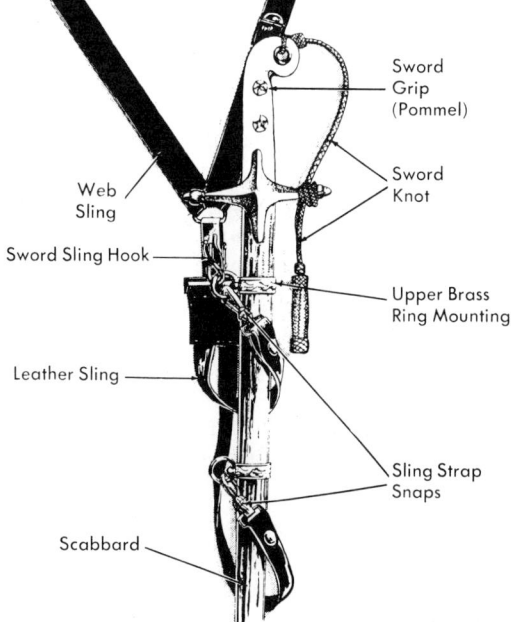

Fig. 18-7: *Nomenclature of the sword, scabbard, and sling.*

Standing at Ease. When it is desired to stand at ease or at rest but with sword drawn (as distinct from "Parade Rest"), thus facilitating quick return to the carry for the purpose of giving commands, the old (naval) position formerly used at rest is both military and convenient. Simply stand with the feet apart, as in "Parade Rest," but with the sword blade carried horizontally without constraint across the front of the body, hilt in the right hand and lower blade in the left.

Carrying Sword When Not in Formation. As your sword is not a fishing pole, a hoe, or a golf club, it should be carried and handled in a military way even when you are not in formation and the sword is unrigged. The proper way to do this is to crook your left arm at right angles across the front of your body, and to place the sword (sheathed in its scabbard) in the crook, *curve of the blade downward* and hilt rearward. The sword will ride easily here as long as you hold your forearm steady, and the appearance will be formal and soldierly.

> NOTE: When underway, with sword drawn, the scabbard will hang and move naturally. Despite jokes to the contrary, it is next to impossible to trip over a scabbard. Few things make you appear more unsure of yourself than clutching at your scabbard while carrying or saluting with your sword.

DISPLAYING THE FLAG

1817 ▪ Displaying the Flag

General. Throughout the Navy and Marine Corps, the National Ensign is displayed from 0800 to sunset (except in ships underway, which fly the Ensign continuously). On shore the flag is flown near post headquarters, or at the headquarters of the senior, when two or more commands are located so close together that separate flags would be inappropriate. Outlying commands or activities display the National Colors in order to make clear their governmental character.

Except when intentionally lowered to half-mast, the flag must be "two-blocked" at all times—that is, it must be hoisted and secured at the *very top* of the staff, or gaff, since any flag not so secured is technically considered to be at half-mast. Display of the flag at half-mast indicates official mourning. On Memorial Day, the flag is half-masted until the completion of the required gun-salute, or until noon, if no salute is fired.

The position of half-mast is midway between the peak (or truck) and the base of the flagstaff, except when the latter has yardarms or is supported by guys, stays, or shrouds, in which case half-mast is halfway between the peak and the yardarm or the point at which guys, stays, or shrouds join the staff.

The church pennant is the only flag ever flown above the Ensign. It is hoisted at the sounding of "Church Call" for divine services on shipboard, and the National Colors are lowered to a position just under the church pennant. When divine services have concluded, the church pennant is hauled down and the National Colors are "two-blocked."

Colors must never be allowed to become fouled. It is an important responsibility of the guard to prevent this. To avoid fouling, they should be raised or lowered from the leeward side of the pole. Should it become necessary to exchange a set of Colors already hoisted, a new set is first run up on a second halyard (this is why flagpoles have two sets of halyards), and the original set is lowered as soon as the new one has been two-blocked.

It is a recognized international *distress signal*, afloat or ashore, sanctioned by law, to fly the National Ensign upside down.

In Battle. It is a very old tradition, although no longer prescribed by *Navy Regulations*, that, on joining action, ships break out National Ensigns at the truck of each mast. The spirit of this tradition should be observed on shore. Any position under attack, at which Colors are normally flown, should keep those Colors flying throughout action, night and day, just as the original Star Spangled Banner flew through the

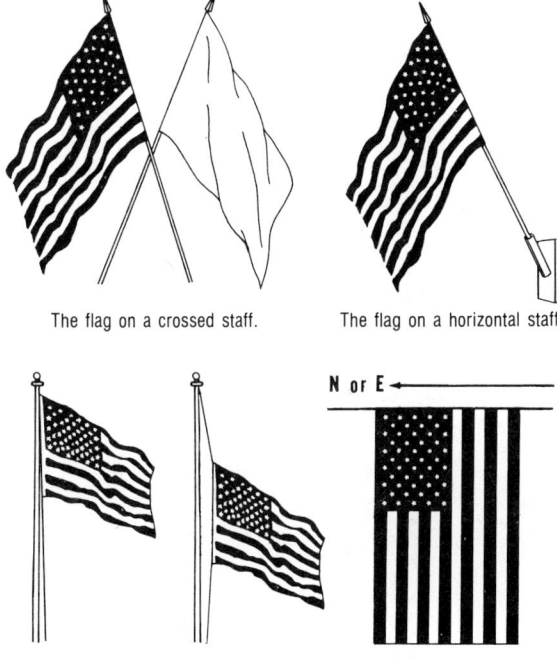

Fig. 18–8: Correct ways in which to display the flag.

night over Fort McHenry at Baltimore (where Marines formed one of the defending units).

Half-Masting the Flag. First, two-block the flag at the truck (top) of the staff, and keep it there until the last note of the National Anthem or "To the Colors"; then lower it to the half-mast position. In lowering the flag from half-mast, hoist is smartly to the truck at the first note of the music, then lower it in the regular manner, as described in Section 1818.

Displayed with Other Flags. The National Colors are always on the right (to your left as you face the displayed flags). If other flags are flown from adjacent poles, the American flag will be the first one raised and the last one lowered.

When displayed from crossed staffs, the National Colors are on the right, and the staff is in front of the staff of the other flag with which it is crossed.

When displayed over a street, the blue field (or "union") of the flag should point north on a street running east-west, and point east on a street running north-south.

When used to drape a coffin, the flag should be placed so that the union would cover the head and left shoulder of the body within.

Foreign Flags. Except in cases of official ceremonies, the carrying of foreign flags by members of the U. S. Armed Forces is not authorized. An example of an official ceremony would be the arrival or departure of a foreign head of state. Rulings as to whether given events may be considered official ceremonies should be obtained from Marine Corps Headquarters.

1818 ■ Morning and Evening Colors

Colors are the most important ceremonies of the working day, and must be conducted with precision and ceremony. Executing Colors is the responsibility of the guard of the day, and should be personally supervised

Morning Colors are executed on the final stroke of eight bells.

by the commander of the guard (see Chapter 17). Honors to be rendered by individuals and formations are described in Section 1810.

Raising the Flag (Morning Colors). The *color guard*, a noncommissioned officer and two privates, forms at the guardhouse, with the NCO (carrying the folded Colors) in the center. The color guard marches to the flagstaff, halts, and bends on the flag to the halyards. The halyards are manned by the two privates, and the NCO holds the flag until it is hauled free of his grasp. He must see that the Colors never touch the ground. At precisely 0800 the signal to execute Colors is given from the guardhouse by the corporal of the relief on watch. The field music then makes eight bells, and, after the last stroke, the music begins, and the flag is hoisted smartly. When the flag is clear, the NCO comes to hand salute. As soon as the flag is two-blocked, the privates manning halyards likewise come to hand salute and hold this position throughout the National Anthem or "To the Color," after which the halyards are triced. In saluting during Colors, members of the Color Detail should avoid looking up at the Colors, and should salute in the normal manner and stance.

The *guard of the day and band*, or field music, parade facing the flagpole. At Morning Colors, following the last stroke of eight bells, attention is sounded by bugle, followed in turn by the National Anthem (if a band is present) or "To the Color" (by field music). The guard is brought to present arms on the call to attention. If foreign forces are present, the band renders prescribed honors to foreign ensigns after playing the U. S. National Anthem. Hand salutes and present arms end on the last musical note, after which "Carry On" is sounded.

In the absence of a band, "To the Color" is sounded by field music. If no music is present, the signals for attention, hand salute, and carry on must be given by whistle, which is most undesirable. *Even if your outfit does not rate or include a music, you should make every effort to obtain a bugle and train a nonrated man to sound the calls required for Colors.* This is where initiative, enterprise, and spirit of "make-do" can show.

Lowering the Flag (Evening Colors). Evening Colors is executed by the same guard details as Morning Colors, and the ceremony is virtually a reverse performance of the latter. The flag is lowered precisely at sunset, the exact daily time of which should be kept in a table in the guardhouse. Beginning with the first note of the music, the flag is slowly lowered, in time with the music, so that it will be in the hands of the NCO of the color guard as the last note sounds. In the absence of a band, "Retreat" is sounded by field music.

After being lowered, the flag is folded in the shape of a cocked hat. The correct procedure for folding a set of Colors may be found in Fig. 18–9 and should be followed.

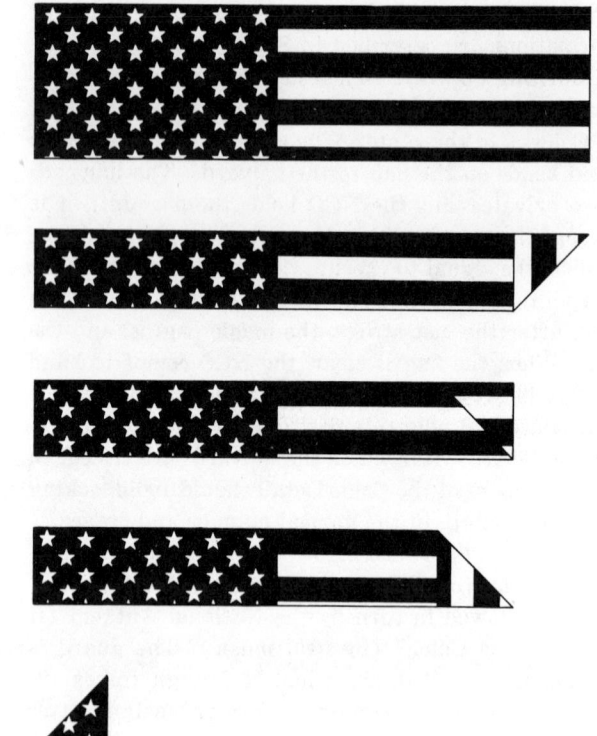

Fig. 18–10: Folding the flag.

Standing lights (such as street lights, aircraft obstruction lights, etc.) throughout the post should not be turned on until after the last note of Evening Colors.

1819 ■ Display of Personal Flags or Pennants

At Commands Ashore. The personal flag or pennant of a general or flag officer is displayed, day and night, in the headquarters area (usually from a staff on the headquarters building). When an officer entitled to a personal flag makes an official visit or inspection at some other activity of his command, his flag is hauled down and shifted to the activity that he is visiting. If this latter activity is in turn commanded by a flag or general officer, the inspecting officer's personal flag displaces that of the local commander.

When a foreign ensign or personal flag is displayed ashore during an official visit by, or gun salute to, a foreign officer or civil official, it is broken at the normal point of display of the local commander's flag or pennant, and the latter is in turn shifted to some other point within the command.

If the points of display of two or more personal flags are so close together that it would be inappropriate to fly them in competition, so to speak, the senior officer's personal flag is displayed alone. Similarly, if two or more civil officials who rate personal flags are present officially at the same time, only the flag of the senior is broken.

It is a Marine Corps custom that, on conclusion of a tour in command, a general officer may retain his personal flag.

On Vehicles. Any officer entitled to a personal flag or pennant may display this forward on a vehicle in which he is riding officially. Alternatively, he may mount plates, forward and aft, bearing the number of stars appropriate to his rank. Marine Corps and Army generals have scarlet plates (but the arrangement of stars on Marine general officers' flags and plates corresponds to that of flag officers of the Navy rather than for Army generals). Navy and Air Force flag or general officers have blue plates. A personal flag and a set of such plates are never displayed at the same time from the same car. When the officer who rates the flag or plates is not in the car, the flag should be furled and cased, and the plates cased. This is a point on which drivers should be carefully schooled.

Aboard Ship or in Boats. The rules for display of personal flags and pennants afloat are complex and precise, and must be carefully followed. They are given in *Navy Regulations.* If going to sea, you should learn these rules, since, as officer of the deck in port, you will need to know them. Consult *The Watch Officer's Guide.*

HONORS, OFFICIAL VISITS, AND CALLS

1820 ■ Honors, Official Visits and Calls

As a junior officer, your first contacts with honors, official visits, and official calls will probably occur on board ship or when you find yourself detailed to command a guard of honor. Like all military etiquette, the subject demands precise attention and compliance with every rule and ground-rule. Since the Marine Corps prides itself on being the most military of the Services, make a point to know and observe all the ins and outs of honors and official visits.

The following definitions may be helpful:

Official visit: A formal visit of courtesy which requires special honors and ceremonies.

Official call: An official but informal visit of courtesy which does not require honors or ceremony. Note the distinction between official calls, discussed in this chapter, and personal calls, which are covered in Sections 2407–2409.

Guard of the day: For rendering honors, *Navy Regulations* provide that the "guard of the day" (normally not part of the interior guard, except on board ship) shall be not less than one rifle squad.

Full guard: Not less than one rifle platoon.

Guard of honor: Any guard, not part of the interior guard, which is paraded ashore for rendition of honors. When the interior guard turns out in compliment to an individual, it is spoken of as *the guard*, not as a guard of honor.

Compliment of the guard: This honor consists of an interior guard turning out and presenting arms, as a compliment to visiting officers or civilian dignitaries.

Shipboard compliments: In addition to honors by a guard, shipboard compliments may include any or all of the following elements, which are dispensed with ashore:

Manning the rail on weather decks by the ship's company
Piping alongside and over the side
Sideboys

Since Marines in a ship's company are fully occupied with other elements of rendering honors to visitors, these shipboard compliments are performed only by Navy officers and enlisted men.

Honors. Arrangements for rendering honors ashore are usually coordinated by the commanding general's aide. On a post or unit not commanded by a general officer, arrangements are made by the adjutant.

Ashore and afloat, we render the same salutes, honors, and ceremonies, as practicable, at Marine Corps posts and stations, and in naval ships and stations. Wherever Marines are present, they provide the honor guard. Detailed pointers and procedures on honors are in Section 1822.

It goes without saying that troops paraded as honor guards must be the *best*. See to it that your guard is correct, snappy, and immaculate—a reflection of the Marine Corps at its smartest.

Official Visits and Calls. Official visits and official calls, as we discuss them here, are paid only by officers in command, and are distinct from the personal calls described in Chapter 24.

Generally speaking, official visits are more often paid by commanders afloat, whereas on shore official calls are substituted.

On taking over a command, you must make an official call on the senior to whom you have reported for duty. This call is at his headquarters, and is not the domestic "visit of courtesy" mentioned in Section 2408, which is likewise required.

In addition, unless the senior indicates otherwise, the following other official calls are required:

1. A call by the commander of an arriving unit on his immediate superior, if present; and on the senior Marine or Navy officer present.

2. A call by commanding officers on an immediate superior in the chain of command upon arrival of the latter.

3. A call by an officer who has been senior officer present, upon his successor.

4. A call by the commander of a unit or ship arriving at a Marine Corps post or naval station, upon the commander of such activity; except that when the arriving commander is senior, the local commander makes the call.

5. Calls on high civil officials (state and territorial governors, and U. S. diplomatic and consular officials) as prescribed in *Navy Regulations*.

6. When in the vicinity of a command ashore belonging to another U. S. Armed Service or to a friendly foreign power, the senior Marine officer present in command arranges with the other commander concerned for an exchange of official visits or calls as appropriate. Check *Navy Regulations* for calling procedure on foreign officials.

When you leave your command for an official visit, or return therefrom, you receive the honors prescribed for such a visit, except that your own organization remains in uniform of the day and omits gun salutes.

Official calls or visits must be paid expeditiously as they become due, or on the first working day thereafter. They must be returned within 24 hours or on the first subsequent working day.

Circumstances permitting, generals and flag officers return in person official visits or calls by officers of the grade of colonel or higher. The chief of staff or deputy commander returns official calls or visits by officers below colonel.

Officers below general or flag rank return all calls and visits in person.

High foreign officials (other than chiefs of state) return in person visits or calls by a general or an admiral. Otherwise they return such visits by a suitable representative.

Before making or returning an official visit or call, check both *Navy Regulations* and *Marine Corps Uniform Regulations* for the proper uniform. If unable to get access to these publications, or if you cannot find your answer, *you will never be far wrong wearing undress blue or white*, according to season and climate.

1821 ■ Official Visits and Calls Ashore

Before you make or receive an official call or visit ashore, be sure to have all arrangements taken care of well in advance. This requires liaison between the maker and the recipient of the call, which is up to the aide or the adjutant, as the case may be. Coordinate the following details especially:

Time and exact place call is to be paid
Uniform
Entrance to post, station, and headquarters which caller will use
Transportation
Honor guard, if required (see Section 1822)
Use of calling cards
Refreshments to be served, if any
Specific units, places, or installations to be visited
Arrangements to break or haul down personal flags, or fire gun salutes

1822 ■ Procedure and Pointers on Rendering Honors

Preparations. The general's aide (or the adjutant) virtually always has ample advance notice to prepare for an official visit that requires honors, and to notify those concerned. Ashore, the casual or surprise official visit is rare.

A well-trained, smartly turned-out saluting battery should be considered an essential part of the honor guard.

When detailed to command an honor guard, you should immediately visit the aide (or report to the adjutant), and obtain all possible information. Check with the bandmaster on the timing of ruffles and flourishes, and with the NCO in charge of the saluting battery. If the saluting guns are remote from the honor guard's parade, see that absolutely foolproof communications are established between the parade and the saluting battery. It is elementary, but vital, that the saluting battery knows how many guns are to be fired; also that a standby piece (if available) and spare rounds are in instant readiness to fire in the event of hangfire or malfunctions of a saluting gun.

NOTE: If for any reason your saluting battery doesn't have a stopwatch for timing the rounds (or if the watch stops permanently), the NCO in charge may obtain the correct interval between rounds by repeating, "If I wasn't a gunner, I wouldn't be here—Fire One (two, three, etc.)." This is a trick that dates back to early days in the sailing Navy.

Make a personal reconnaissance of the honor guard's parade, and know exactly where the guard and band and the recipients of honors are to be posted. Determine what markers and guidons are required, and who will supply and locate them. If possible, give your platoon leaders and leading NCOs an opportunity to look over the ground. If you are going to provide an escort of honor, verify the route of march.

As to the guard itself, spare no effort to make it the finest in the Marine Corps. See that every man is immaculately turned out, that the guard is perfectly sized, and that the entire formation, if possible, is adequately rehearsed. If troops for the guard come from units other than your own, have no hesitation in returning substandard men to their units, reporting that you have done this (and why) to the adjutant.

Procedure for Honors on an Official Visit Ashore. The following general procedure can serve as a guide for rendering honors ashore. (Remember, however, that almost every post has its ground rules and standing operating procedures, and be sure to consult these when preparing to render honors.)

1. Well before the time of arrival of the visiting official, complete the preparations discussed above, parade your guard and band, have the Color detail standing by to break or haul down personal flags involved, and have the saluting battery manned and ready. If possible, especially on a large post, have communications which can apprise you, up to the last moment, of the visitor's movements and approach (field radio equipment is ideal).

2. When the recipient of honors arrives and debarks from boat, train, car, plane, or helicopter, "Attention" is sounded by bugle. The local

commander (or whoever is receiving him) greets the dignitary and conducts him to his post front and center of the guard.

3. When the official takes post, the commander of the honor guard brings the guard to present arms. All hands in the vicinity but not in formation come to hand salute, following the motions of the guard.

4. When the guard has been presented, and the commander has executed his salute, the band sounds off with ruffles, flourishes, and other musical honors. The personal flag or National Color, as specified for the individual in question, is broken on the first note of the music.

5. The guard is brought to the order after the last note of the music, or when the commander of the guard has exchanged salutes with the official, if there is no music. If a gun salute is rendered, the first gun is fired immediately after the last note of the music, and the guard remains at the present throughout the salute, as do all hands in the official party, who hold salutes throughout the gun salute. Persons not in the official party but in the immediate vicinity, remain at attention during the gun salute. If the National Color, or a foreign flag or ensign, is to be displayed during the gun salute only, it is broken on the first gun, and hauled down on the last.

U. S. Marine Honor Guard being inspected by the Captain-General Royal Marines (Prince Philip of Edinburgh) during state visit to Washington.

6. On completion of the musical honors or the gun salute, if fired, the honor guard commander brings his guard to order arms, executes present sword to the person being honored, and reports, "Sir, the honor guard is formed." If the personage desires neither to inspect the guard nor that it pass in review, the honor guard remains at attention. For procedure to be followed for marching the guard in review or for its inspection by the personage, see *Drill and Ceremonies Manual*.

Honors on Departure from an Official Visit Ashore. In general, departure honors reverse those given on arrival.

1. Departure honors commence when the visiting official has completed his personal leave-taking from the senior officer present. Either the latter officer himself or the aide will signal this to the commander of the guard.

2. Gun salutes, if rendered on departure, must begin before the individual actually leaves (that is, while he is within earshot). His personal flag, or any national ensign displayed, is hauled down on the last gun.

3. An honor guard is not normally inspected on departure. Hold the guard on parade until the official is out of sight.

Pointers on Honors, Ashore and Afloat. Your "bible" for honors incident to official visits and calls is *Navy Regulations*. The table of honors and ceremonies given here (Table 18–2) is a compilation of the information in those articles, assembled for quick reference. You must in addition be meticulously familiar with the *Drills and Ceremonies Manual*, which gives detailed instructions for rendering all types of honors.

In addition to guards paraded in receiving the President, or any foreign sovereign or chief of state, or member of a reigning royal family, all officers not required elsewhere form on the left of the honor guard, in dress uniform with swords. Troops not otherwise occupied should form on parades adjacent to the guest's route of inspection, and also line the route. On board ship, men not manning the rail fall in at quarters.

Afloat, all ships present, other than the one receiving the President or sovereign, man the rails and fire the required national salute on the arrival and departure of the distinguished guest.

The officer of the day or officer of the deck attends the arrival and departure of any distinguished visitor, whether the visit is official or not.

Afloat, side-honors only (i.e., side-boys only, no guard or band) are rendered when a general officer, flag officer, or another commanding officer comes on board without his flag or pennant flying. If he so requests, full honors may be rendered on departure, but otherwise side-honors only are again the rule. While side-honors are being rendered, all hands on deck and in view of the gangway stand at attention, facing the gangway; they salute as the officer reaches the top of the accommo-

Table 18-2:
Table of honors for official visits.

Rank	Uniform	Gun salute Arrival	Gun salute Departure	Ruffles and flourishes	Music	Guard	Side boys	Crew	Within what limits	Flag What	Flag Which truck	Flag During
President	Full dress	21	21	4	National anthem*	Full	8	Man rail	....	President's	Main	Visit
President or sovereign of a foreign country	do.	21	21	4	Foreign national anthem	do.	8	do.	....	Foreign ensign	do.	do.
Member of reigning royal family	do.	21	21	4	do.	do.	8	do.	....	do.	do.	Salute
Ex-President or President-elect	do.	21	21	4	Admiral's march	do.	8	Quarters	....	National	do.	do.
Secretary of State when acting as special foreign representative of the President	Full dress	19		4	National anthem	Full	8	Quarters	....	Secretary's	Main	Visit
Vice-President	do.		19	4	Admiral's march	do.	8	do.	....	Vice-President's	do.	do.
Speaker of the House of Representatives	do.		19	4	do.	do.	8	....	....	National	Fore	Salute
Governor of a state of the United States	do.		19	4	do.	do.	8	....	Area under his jurisdiction	do.	do.	do.
Chief Justice of the United States	do.		19	4	do.	do.	8	....	....	do.	do.	do.
Ambassador, High Commissioner, or special diplomatic representative whose credentials give him authority equal to or greater than an Ambassador	do.		19	4	National anthem	do.	8	....	Nation or nations to which accredited	do.	do.	do.
Secretary of Defense	do.	19	19	4	Honors march#	do.	8	Quarters	....	Secretary's	Main	Visit
Deputy Secretary of Defense	do.	19	19	4	do.	do.	8	do.	....	Dep'y Sec.'s	do.	do.
Prime minister or other cabinet officer of a foreign country	Dress		19	4	Admiral's march	do.	8	....	....	Foreign ensign	Fore	Salute
Cabinet officer other than Secretary of Defense	do.		19	4	Honors march#	do.	8	Quarters	....	National	do.	do.
Secretary of the Navy	Full dress	19	19	4	do.	do.	8	do.	....	Secretary's	Main	Visit
Secretary of the Army or Air Force	Dress	19	19	4	do.	do.	8	....	....	National	Fore	Salute
President pro tempore of the Senate	do.		19	4	do.	do.	8	....	....	do.	do.	do.
Director of Defense Research and Engineering	Dress	19	19	4	Honors march#	Full	8	Quarters	....	National	Fore	Salute
Assistant Secretary of Defense	Dress	17	17	4	do.	Full	8	Quarters	....	Ass't Sect'y's	Main	Visit
Under Secretary and Assistant Secretaries of the Navy	do.	17	17	4	do.	do.	8	do.	....	Under or Ass't Sec.'s	Main	Visit
Under or Assistant Secretary of the Army or the Air Force	do.	17	17	4	do.	do.	8	....	....	National	Fore	Salute

450

Governor General or Governor of territory, commonwealth, or possession of the U.S. or area under the administration of the U.S.	do.	17	4	Admiral's march	do.	8	Area under his jurisdiction	do.	do.
Committee of Congress	do.				do.	8		do.	do.
Envoy Extraordinary and Minister Plenipotentiary	Dress	15	3	Admiral's march	Full	8	Nation to which accredited	National	Fore Salute
Minister resident	do.	13	2	do.	do.	6	do.	do.	do.
Chargé d'affaires	do.	11	1	do.	do.	6	do.	do.	do.
Career Minister or Counselor of Embassy or Legation	do.		1	do.	do.	6	do.		
Consul General or Consul or Vice Consul when in charge of Consulate General	do.	11	1	do.	do.	6	District to which assigned	National	Fore Salute
First Secretary of Embassy or Legation	Of the day with sword				Of the day	4	Nation to which accredited		
Consul or Vice Consul when in charge of Consulate	do.	7			do.	4	District to which assigned	National	Fore Salute
Mayor of an incorporated city	do.				do.	4	Within limits of mayoralty		
Second or Third Secretary of Embassy or Legation	do.					2	Nation to which accredited		
Vice Consul when only representative of the U.S. and not in charge of Consulate or Consulate General	do.	5			Of the day	2	District to which assigned	National	Fore Salute
Consular agent when only representative of the U.S.	do.				do.	2	do.		

Military and Naval Officers, United States and Foreign[11]

Chairman of the JCS	Dress	19	4	Admiral's march[11]	Full	8	Quarters		
Chief of Staff, U.S. Army	do.	19	4	General's march	do.	8	do.		
Chief of Naval Operations	do.	19	4	Admiral's march	do.	8	do.		
Chief of Staff, U.S. Air Force	do.	19	4	General's march	do.	8	do.		
Commandant of the Marine Corps	do.	19	4	Admiral's march	do.	8	do.		
Fleet Admiral or General of the Army or the Air Force	Dress	19	4	Admiral's march[11]	Full	8	Quarters		
Admiral or General	do.	17	4	do.	do.	8			
Naval or other military governor, commissioned as such by the President, within area under his jurisdiction	do.	17	4	do.	do.	8		For United States officers, personal flag at the main during the salute	
Vice Admiral or Lieutenant General	do.	15	3	do.	do.	8		For officers of foreign nations,	
Rear Admiral or Major General	do.	13	2	do.	do.	6			

451

452

Commodore or Brigadier General	do.	11	do.	6
Captain, Commander, Colonel or Lieutenant Colonel	Undress		Of the day	4
Other commissioned officers	Of the day with sword		do.	2
Official not herein provided for			Honors as prescribed by the senior officer present; such honors normally shall be those accorded the foreign official, when visiting officially a ship of his own nation, but a gun salute, if prescribed, shall not exceed 19 guns.	
Foreign officer of the armed forces, diplomatic or consular representative in country to which accredited, or other distinguished foreign official			Honors for an official or officer of the United States of the same grade, except, that equivalent honors shall be rendered to foreign officers who occupy a position comparable to Chairman JCS, CNO, Chief of Staff Army, Chief of Staff Air Force, or CMC.	the foreign ensign at the fore during the salute

NOTES

1. All other ships present man rail and fire national salute at official reception or departure of President.
2. For President of United States, president of a foreign republic, foreign sovereign or member of reigning royal family, officers assembled on quarterdeck in full dress, crew man rail, and other officers unemployed formed forward of guard, men not occupied fall in at quarters.
3. For others for whom full dress uniform is prescribed designated officers assembled on quarterdeck and formed forward of guard.
4. If a flag or commanding officer comes on board without flag or pennant flying, only side honors shall be given unless he should request full honors on departure. All persons on the quarterdeck shall stand at attention by command without bugle.
5. No officer in civilian clothes shall be saluted with guns or have a guard paraded in his honor.
6. When side honors only are rendered to a flag or commanding officer, officers and men on deck and in view from the gangway shall stand at attention facing the gangway, and salute as the officer appears over the side and shall remain at the salute until the end of the pipe.
7. The officer of the deck shall attend at the gangway on the arrival or departure of any commissioned officer or distinguished visitor.
8. All honors except attendance at gangway by the officer of the deck, except as social courtesy may demand, shall be dispensed with:
   a. When officers are in plain clothes.
   b. From sunset to 0800 (except that for foreign officers, side shall be piped during daylight).
   c. During meal hours of the crew for officers of U.S. Navy or Marine Corps.
   d. When exercising at general drills or when undergoing Navy Yard overhaul, for officers of USN and USMC.
   e. For ships with less than 180 men in the seaman branch, for officers of USN, USMC, USCG, USA, and USAF, except when advance notice of an official visit has been received.
9. The guard and band shall not be paraded on Sunday for officers of USN, USMC, USCG, USA, or USAF officers.
10. All sentries on the upper deck or in view from outside shall salute all commissioned officers passing them close aboard, in boats or otherwise.
11. Admirals and Marine Corps Generals receive the "Admiral's March"; Generals (Army/Air Force) receive the "General's March."
* "Hail to the Chief" may be used in lieu of National Anthem on either arrival or departure. When specified by the President, "Hail to the Chief" may be used while the President and his immediate party move to or from their places while all other stand fast.
# Honors march is a 32-bar medley in the trio of "The Stars and Stripes Forever."

dation ladder or brow, and remain at the hand salute until the end of the boatswain's pipe, following the motions of the side-boys.

Honors are dispensed with under the following circumstances:

When the visiting officer is in plain clothes, or visits the post unofficially;
Between sunset and 0800 (except that foreign officers may be rendered honors at any time during daylight) ;
During meal hours for the troops (except in the case of foreign officers) ;
When a ship is engaged in maneuvers, general drills, or undergoing overhaul in a navy yard, or is in action;
When a unit or post ashore is carrying on tactical exercises or emergency drills;
On Sundays or national holidays (except in the case of foreign officers) ;
For ships with a complement of less than 180 men in the seaman branch.

Honors in the Field. Despite the rigors of field service, Marine units make every effort to render appropriate honors when so serving. The spirit, if not the letter, of the preceding paragraphs must be faithfully observed. It is the distinguishing mark of really professional troops that in face of handicaps and obvious obstacles to smartness, they nevertheless remain smart and military and do the best they can with what they have. Marines in the field may well remember what was said of England's Brigade of Guards: "They die with their boots clean."

For rendition of honors in the field, the most important points are that:

Men be in clean, homogeneous uniforms;
Equipment (especially weapons) be in first-class, evident serviceability;
Individuals be smart, keen, and clean;
The place for rendering honors be not subject to enemy shelling or observation;
Military readiness or combat operations be not interrupted.

POINTERS FOR AIDES-DE-CAMP

1823 ■Duties and Relationships

Duty as an aide-de-camp (usually short-titled "aide") is one of the most exacting details which a young officer can receive. If you are so assigned, you may take it as a compliment to your military and personal character—a compliment that you must do your best to live up to.

As an aide, you are always on duty, and this duty is always personal and confidential, and always official. Your duties are only such as your general personally directs. On the other hand, if you are to succeed, you must quickly learn to anticipate your general's desires and needs, and take care of them *without being told*. No matter what the circumstances, your first thoughts should be for your chief's safety, reputation, convenience, and pleasure. Any duty he asks of you should therefore be promptly performed.

Intelligence, tact, loyalty, absolute discretion, and military smartness are the most important characteristics of a good aide, with sensible frankness not far behind. Although, by direction, an aide must often serve as an extra pair of eyes and ears for his chief, he must avoid becoming a tale-bearer and should, whenever consistent with loyalty and fairness toward his chief, do his utmost to protect other officers from having chance indiscretions reach the attention of higher authorities.

Aside from your chief, the two most important persons with whom you routinely deal are his chief of staff and his wife. Establishment of a cooperative, deferential, helpful relationship with the chief of staff—while being careful never to betray any confidence of the general—is essential. Toward his wife you must be unfailingly helpful, tactful, and mannered.

To sum up, virtually all *arrangements* that concern the general end up as your ultimate responsibilities. Within this general premise, your job breaks down functionally into (1) scheduling, (2) paperwork, (3) protocol, and (4) personal needs.

Keeping your general on schedule is of overriding importance; it is also one of your most difficult tasks. Remembering that "punctuality is the politeness of kings," you must stay on top of his itinerary or other program for each hour and minute.

At briefings, ensure that the general has good background familiarity with the subject. On visits, see that he is prepared for and familiar with the people he can be expected to meet, and, where appropriate, with the missions and general situation of units concerned.

Your main job in the realm of paperwork is to keep track of all papers going in or out, setting them into proper priority, depending on deadlines, actions required, and inherent importance.

The responsibilities of protocol and personal needs are covered in subsequent paragraphs.

1824 ▪ Travel Arrangements

Before leaving:
Prepare an itinerary, giving hours and modes of arrival and departure at destination and all intermediate stops, and furnish a copy to the chief of staff and any other interested parties.

Obtain a program for each official stop, to include schedule of events, uniforms required, times, and other necessary information. This program must, of course, have been coordinated by the host activity and be in complete agreement with what is expected thereat.

Inform your chief of the uniforms he will require throughout the trip, and on what occasions.

Obtain your chief's orders and transportation-requests (if required), and see that transportation is arranged.

Issue instructions for forwarding mail or messages.

See that all baggage for the official party is suitably tagged and identified.

Determine what, if any, papers or files the general will require on the trip and arrange for their handling and stowage, especially that of any classified matter.

Check communications arrangements to be sure the general can be reached rapidly no matter where he is.

During travel:

If traveling commercially, keep air and railroad timetables handy, and, no matter how you travel, know hours scheduled for arrival and departure. Know places and times for connections.

Prescribe uniforms for aircrew and stewards, and ensure that other members of the party are informed as to correct uniform during travel, on arrival, and for scheduled events.

When on board government aircraft, be sure the pilot sends a message ahead, stating the composition of the party, ETA, and transportation required on arrival. If traveling commercially, send such a message yourself. Be sure that the host activity is informed if any ladies are in the party, or of changes in schedule.

Take care of all tickets, baggage-checks, baggage-handling, and transportation. In this capacity, your first responsibility is to take care of the general's gear and keep track of it at all times. This particularly includes his official and classified papers.

Obtain copies of daily papers published at the principal places en route, and see that they reach your chief.

Keep track of time-zone changes and the dateline. Remind the general to set his watch.

Take with you: station-lists or rosters of officers at activities to be visited; a copy of the "Blue Book"; official and personal stationery and postage stamps as required; carbon paper, notebook, pen and sharp pencils; a supply of the general's visiting cards; ample cash and a supply of personal checks on the general's bank; cleaning gear; liquor and refreshments as may be required; spare insignia and ribbons; and the general's personal flag and vehicle-plates (if visiting an activity where such are possibly not available).

Keep a running record by name, rank or title, and address, of all persons to whom "bread and butter" notes or letters should be sent; if you have time, rough out such notes before memory fades.

After return:

Write or prepare, for the general, official and personal letters of appreciation to all who extended special courtesies.

Prepare for the general's signature his itinerary and travel claim, being careful not to omit miscellaneous expenses which may properly be claimed.

Obtain and deliver all personal mail held for your chief.

1825 ■ Duties of an Aide in Garrison

Since you are expected to be the social arbiter and expert on the staff, you should know and possess both *Service Etiquette* (Oretha D. Swartz, published by the U. S. Naval Institute) and *Naval Customs, Traditions, and Usage*, Lovette (also a Naval Institute publication). These sheet-anchor books can be relied on for tested and correct advice in virtually any situation involving Service social usage or protocol. In addition, you should keep an up-to-date *Combined Lineal List*, being careful to keep track of all promotions, retirements, changes of status, and yearly promotion zones.

Be punctilious in neatness and correctness of dress and uniform, and help the general to be so, too.

Courtesy and thoughtfulness are indispensable in an aide. You should never be too busy to be courteous to all comers. Especially avoid a nose-in-the-air attitude toward brother officers.

Stay as much as possible within call of your chief.

Make a daily appointment schedule for your chief (and, if he desires, save it for record). Keep track of his engagements and commitments, and calls to be returned (see Section 1821), and see that he is reminded as necessary.

Whenever anyone calls on the general officially or semiofficially, meet him on arrival and accompany him to his vehicle on departure.

It is not only proper but in order for you to invite your chief's attention to anything that may be amiss as to his uniform or dress, and also to remind him of any social amenities or courtesies that may have been overlooked. It is up to you to know your chief's shortcomings, and to protect him against them.

Supervise the performance of drivers, orderlies, secretaries, stewards, and all others who serve the general. Keep them on their toes personally and professionally, and weld them into a team. See that the general's office and outer office are attractive and efficient.

Assist your general and his wife in preparations for all social functions to be given by them. Supervise the issuance of all invitations, make sure that dates and times are correct and that the desired uniform or costume is correctly specified, and keep track of RSVPs. (It is a helpful practice for those attending formal dinner parties, if you can find the time, to put the guestlist on the back of each invitation.)

On social occasions, keep close by your chief, seeing that he, his lady, and senior persons with whom he may be talking are supplied with refreshments. Take post in the receiving line next to your chief, on the approach side. You need not shake hands except with guests you know. The most important thing is to get each name correctly and announce it clearly and distinctly to your chief, even in the cases of people he knows well.

Make the acquaintance of aides assigned to other flag or general officers in the immediate area. By close coordination and mutual support, you may be able to prevent many slip-ups through such helpful relationships.

1826 ■ Duties of an Aide in the Field

Your duties as an aide to a general officer in the field are quite different from those in garrison, although the spirit in which they are performed and the basic relationships remain unchanged.

Subject to his wishes, you must accompany your chief wherever he goes. In any case you must always keep his personal situation-map, and other maps or status boards, absolutely up to the minute. Pay particular attention to locations of front lines, of installations be may wish to visit, and, above all, of unit command posts. The last information is important not only to the general and his driver, but also to you, as the general may often use you to convey personal messages from him to regimental and other commanders.

Be alert as to the military situation and be ready to obtain any information the general wishes, either from staff sections and subordinate headquarters or, if necessary, by personal reconnaissance.

See that your chief's personal wants are cared for, and that he always has everything he needs or desires. Have arrangements been made for laundry? For keeping his weapon and gear in shape? For his foxhole?

Introduce visiting officers, official visitors, correspondents, and other persons having business with the general.

Arrange and control all transportation for the general.

Supervise the general's drivers, orderlies, cooks, and stewards as you would in garrison, but be sure these people are reminded that, in the field, they are combat Marines, too, and must be prepared to defend the general and his area in the event of surprise attack or enemy penetration of the command post area.

Supervise and act as caterer for the general's mess. Be sure that any fatigued, wet, or cold officer or enlisted man who sees the general (especially people from front-line units) always gets a cup of hot coffee or, if appropriate, a drink. Have plenty of coffee for drivers and runners, day or night.

Supervise the security arrangements for the general's area.

Work closely with the Headquarters Commandant in such arrangements as digging a suitable head, erection of tentage, digging in tents, camouflage of the area, water supply, electricity, and facilities.

Above all, do everything in your power to protect and defend your general, and to shelter him from unnecessary strain and fatigue.

CEREMONIES

Ceremonial duties are written deep into our history as a Corps. Marines have always striven to excel in this field, and we have good reason to be proud of our record. We should continue so to strive and succeed. Every officer taking part in a ceremony—especially when, as is often the case, little time is available for practice—should realize how broadly revealing of wider professionalism our parade-ground performance can be. Precision drill, immaculately turned-out troops, disciplined marching, and fine bearing—all these furnish, for the public to see, evidence of Marine Corps alertness, determination to put out only our best, and pride in Corps and selves. It is no coincidence that among the units and corps famous for ceremonial prowess and spit-and-polish are also to be found some of the world's most redoubtable fighting formations.

1827 ■ Types of Ceremonies

The Marine Corps and Navy have eight military ceremonies which may be performed on shore. These ceremonies are in the form prescribed by the *Drill and Ceremonies Manual*, and may only be modified when the nature of the ground, or exceptional circumstances, require that change be made.

The title, and a brief description or discussion of each ceremony, follows:

A *review* is a ceremony at which a command, or several commands, parade for inspection by, and in honor of, a senior officer; or in honor of a visitor or a civilian dignitary. In a review, the individual being honored passes on foot or in a vehicle throughout the formation, which is then marched past him.

Presentation of decorations is the ceremony at which decorations are presented. This ceremony follows, in part, that prescribed for a review; it is noteworthy in that, regardless of rank, the individuals who have been decorated receive the review side by side with the reviewing officer. In modified form, this ceremony can be adapted for such occasions as presentation of commissions or enlisted warrants, commendations, and so forth.

A *parade* is the ceremony at which the commanding officer of a battalion or larger unit forms and drills the entire command, and then marches them in review. The battalion parade is the most common form of periodic ceremony, and, under normal garrison conditions, is usually performed each Saturday morning. Together with guard mounting, described below, the parade is probably the most important ceremony for you to know by heart. "Memorize every comma in it!" Captain Lewis ("Chesty") Puller used to enjoin his Basic School lieutenants.

Escort of the National Color is known less formally as "Marching on (or off) the Colors." That is, when the Colors are to take part in a ceremony, be presented to a unit, or turned over to some institution or person for safekeeping, they are ceremonially received and escorted from their place of safekeeping (usually the CO's headquarters), and are similarly returned, by a picked escort. The ceremony for this occasion corresponds somewhat to portions of the famous British ceremony *Trooping the Color*, and is derived from that.

Escort of Honor is the ceremonial escorting of a senior officer or other dignitary during an official visit, or on arrival or departure.

Military funerals are covered in the *Drill and Ceremonies Manual*, as well as in Sections 1809, 1813–1814, and 2325 of this *Guide*. The ceremonial forms followed in military funerals are among the oldest in the profession of arms; some parts, such as the firing of volleys (originally to frighten evil spirits) can be traced to pagan times.

Inspections, as you have seen from Section 1612, run to all types. The ceremonial inspection of troops in ranks has as its object the general military appearance and condition of individual uniforms and equipment within a command. Officers headed for sea duty should note that personnel inspection on board ship follows considerably different lines, and frequently varies from ship to ship. Be sure you know your own ship's ground rules and inspection procedure.

Guard Mounting is the ceremony whereby a guard is organized from guard details, is inspected before assuming the guard, and then relieves an outgoing or "old" guard. This is a very old ceremony, portions of which antedate the Revolutionary War and go back to the British Army. Guard mounts may be *formal* or *informal*, according to weather, size of guard, availability of music, or local conditions. Most commands in garrison perform at least two formal guard mounts weekly; the ceremony is relatively tricky, and is one which you, as a junior officer, must learn thoroughly. Ability to run a formal guard mount is the mark of a well-drilled lieutenant.

Morning and Evening Colors, although sometimes regarded as parts of the daily routine, should be considered ceremonies, if only out of respect to the significance of the daily raising and lowering of the National Ensign. You will find the execution of Colors discussed in Section 1818 of this *Guide,* as well as by the *Drill and Ceremonies Manual.*

In addition to the foregoing eight ceremonies of general character, we employ specific ceremonial forms on the occasion of *change of command, relief of the sergeant major,* and for celebration of the *Marine Corps Birthday.* (Information on the observance of the birthday of the Corps may be found in Chapter 24 of this *Guide.*)

Another form of ceremony not covered in any official regulations is the *tattoo* (sometimes called "searchlight tattoo"). A tattoo is an evening parade conducted under flood- or searchlights, embellished with traditional, historic, or display drills and special musical features, usually climaxed by lowering of the Colors, playing of "Taps" and sometimes a traditional evening hymn. Evening parades at 8th and Eye, though not so entitled, are in fact a form of tattoo.

1828 ▪ Precedence of Forces in Parades or Ceremonies

To avoid conflicts at parades or ceremonies, the places of honor are allocated in order of Service seniority, and, since you may readily find yourself at the head of a Marine detachment in a parade or ceremony, you should know your own place, and those of other components relative to your own. As prescribed in law (*Federal Register,* volume 14, 19 August 1949, page 2503), the precedence of U. S. forces in parades or ceremonies is as follows (reading from the head to rear of column, or from right to left in line):

1. U. S. Corps of Cadets (U. S. Military Academy)
2. Midshipmen, U. S. Naval Academy
3. Cadets, U. S. Air Force Academy
4. Cadets, U. S. Coast Guard Academy

5. United States Army
6. **United States Marines**
7. United States Navy
8. United States Air Force
9. United States Coast Guard
10. Army National Guard of the United States
11. Organized Reserve Corps, U. S. Army
12. **Marine Corps Reserve**
13. Naval Reserve
14. Air National Guard of the United States
15. Air Force Reserve
16. Coast Guard Reserve
17. Other training organizations of the Army, **Marine Corps**, Navy, Air Force, and Coast Guard, in that order.

When the Coast Guard is serving as part of the Navy, in time of war or emergency, the precedence of Coast Guard units and personnel shifts to position immediately after Navy units and personnel.

Bear in mind, as a Marine, that although the Air Force is one of the three larger Services, it is nevertheless junior in ceremonial precedence to the Marine Corps. Never accede to erroneous assignment of fourth, or junior, place to Marines, following Air Force units, as is sometimes carelessly done on the basis of size.

Since the place of honor is the head of column, or right of the line, foreign units should be assigned that post of honor in any ceremony or procession. Where several foreign units of mixed nationality are present, they should be placed in alphabetical order, ahead of any U. S. forces, if the ceremony is conducted by U. S. forces or on American soil.

A review renders honor to a civilian dignitary or senior officer.

The official who organizes and coordinates a street parade or procession is entitled the *grand marshal* or sometimes, the *marshal*. If your unit is misplaced, he is the official who should rectify the mistake.

1829 ■ General Appearance of Troops and Units

The Marine Corps has long enjoyed a worldwide reputation for smart appearance and soldierly performance of every task. This reputation has been enhanced by continually demonstrating to the American public that our execution of peacetime functions is excelled only by our performance in battle.

During peacetime the reputation of the Corps is maintained to a considerable degree by creating favorable, highly military impressions in parades, ceremonies, and other functions. It is therefore a responsibility of all officers, and especially commanding officers, that marching units in the public eye fully meet the standards by which the Marine Corps is measured. Men in key positions must have perfect posture; troop leaders must excel in command presence; uniforms and equipment must be in outstanding condition and appearance. All such public appearances should be preceded by ample drill and specific rehearsal as needed.

1830 ■ Pointers on Ceremonies

Know your parade ground. If possible, not only make a personal reconnaissance of the parade ground or area where a ceremony is to be held, but conduct a rehearsal on the ground. At a minimum, be sure your leading NCOs and unit guides know the layout of the ground, and how the field is to be marked.

Markers. Dress guidons (see Section 732) mark the boundaries and the reviewing point for a parade ground. In addition, it is sometimes customary to place small metal discs on the ground to mark the posts of unit guides and other key personnel. The adjutant places markers and guidons, but every officer and NCO must know the system and layout of markers. In addition, guides and leaders should know the line-up of "landmarks" adjacent to and visible from the parade ground, so as to be able to march in exactly straight lines and column without wavering or falling off to right or left. Guides and leaders should keep their heads up and their lines of sight directly to the front and well out, so as to be able to "navigate" on guidons, markers, and landmarks.

Officers Center is an evolution that should be gone over until all concerned are *letter perfect*. Every individual participant is on display, and this evolution comes at the high point of the parade. Properly executed, "Officers Center" should seem to be the movement of a single man. Manuals of sword and guidon count here as at no other time.

Marines are traditionally known for their smart, soldierly appearance in parades and ceremonies.

Photographers. Photographers, both official and otherwise, can do more to detract from the formality and solemnity of a military ceremony than anyone else. Keep them under strict control, preferably in a suitably located, enclosed vantage point, from which they can get good pictures but not mar the occasion by capering about at will.

Cadence. The regulation cadence is 120 steps per minute, and this is the "tempo" at which a military band plays marches. That is, the bass drummer hits his drum 120 times per minute, with a heavier downbeat or thump on the first and succeeding alternate beats. For parades or ceremonies, it makes for smarter appearance to have a short, snappy step, and, if possible, a slightly accelerated cadence. Thus cadence should never fall below 120. When the band is not playing, individual foot movements during a ceremony, notably those by the adjutant when taking post, are traditionally executed at markedly accelerated cadence, with short steps.

Use of P. A. Systems and Amplifiers. In general, except for the largest ceremonies and under special conditions, it is most unmilitary to employ a public-address system for commands or other purposes incident to military ceremonies. Regimental and battalion commanders and adjutants should pride themselves on their *voice of command* and should, if necessary, practice to strengthen and increase its carrying power.

Stepping off in Time with the Music. Units must step off on the left foot, as is well known, and must accomplish this on command of the leader, *and* on the first beat of the music—a combination which often defeats inexperienced junior leaders, and one which you as a Marine must be prepared to lick.

One method of achieving this result—which requires briefing your men and some rehearsing, but is well worth it—is to give your preliminary command just in advance of the music, and have all hands drilled to step off automatically on the first note of the music, without any command of execution from you; in other words, to *let the first note of the music be the command of execution.* This is particularly effective on parade, after the commands have been given to pass in review.

For guard mounting, and for many other ceremonies where units march onto their parades to music, troops must be brought to right shoulder arms at the first note of Adjutant's Call, and marched off at the first note of march music. This, too, requires coordination by leader and unit. A recommended sequence for these evolutions—"by the numbers"—has been published by Master Sergeant G. P. Finn in *The Marine Corps Gazette*, and is summarized as follows:

1. Bear in mind that "Adjutant's Call" is a 16-beat call, and the "first note of march music" will therefore be count 17.

2. The signal for the first note of "Adjutant's Call" is given by the drum major, who brings down his baton, and can thus be seen by all hands.

3. Give your commands in time with "Adjutant's Call," on successive beats as shown in this diagram, in which numbers correspond to beats in the call:

1	2	3	4
RIGHT	SHOUL-	DER	ARMS
5	6	7	8
	(troops execute the movement)		
9	10	11	12
	(pause)		
13	14	15	16
FOR-	WARD	(pause)	MARCH

4. Rehearse this a few times with a field music, and, in the old Marine phrase, "You've got it made."

Command of Mixed Detachments of Seamen and Marines. When a mixed (or composite) detachment of seamen and Marines is formed for a parade, the Marines occupy their post of seniority and honor at the head of column or on right of line, *but* the senior line officer present, of the Marine Corps or the Navy, according to date of rank, commands the entire detachment. This rule does not apply when Navy and Marines form separate detachments. It usually occurs when a ship parades a landing force as a unit (of which the Marines form part).

Close Order Drill. The object of drill is to teach troops by exercise to obey orders, and to do so in the correct way. For this reason, slovenly drill is harmful. Close order drill is one foundation of discipline and *esprit de corps.* Well-executed, confident, precise ceremonial close order drill is therefore the foundation of success in ceremonies.

Uniform for Inspections, Parades, and Ceremonies. Where possible, undress or dress uniforms (i.e., blues or white-blue-whites with ribbons or medals) should be prescribed for inspections, parades, and ceremonies. Additionally, swords should be worn on such occasions in preference to pistols and belts. If blues are not authorized for the command, large medals may be prescribed on ceremonial occasions for wear with the service blouse.

Music Played During "Sound Off." At a review or parade, when a foreign visitor or officer of another Service is being honored, the march played during "Sound Off" should if possible be one traditional to his country or branch of the Service. At ceremonies conducted by Marine artillery units, "The Caisson Song" is normally played during "Sound Off." For a parade on the occasion of a Marine's retirement, it is a pleasant and appropriate courtesy to ascertain whether there is any particular march he would like to have played on "Sound Off." When several individuals are being so honored, the senior, of course, gets his choice. At the very end of a retirement ceremony, the band should play "Auld Lang Syne."

Law, order, duty and restraint, obedience, discipline . . .
—*Rudyard Kipling,* M'Andrew's Hymn

19

Notes on Military Justice

This chapter contains a general description of the system of military law, both judicial and nonjudicial, now in force throughout the U. S. Armed Services. Since this system is elaborate, technical, legalistic, and at times inflexible, remember that what you read here are at best only notes on the high spots. There can be no substitute for thorough knowledge of the *Manual for Courts-Martial* and the *Manual of the Judge Advocate General of the Navy*. But these alone will often not suffice.

Any substantive question, especially one regarding the disciplinary process at any stage, should be referred to your staff judge advocate or other qualified legal officer because the answers to many legal questions can be found only in cases and he should be familiar with them.*

Military law governs individual conduct and performance of duty in the Naval Services. It also provides means for enforcing the rules; that is, the nonjudicial powers of commanding officers, and the courts-martial system. The prime essential, however, is always to ensure justice in every case.

As a Marine officer, you must know military law. It is part of the tradition of Marine Corps discipline that legal proceedings are conducted with expedition, firmness, and expertness. On board ship with the Navy,

---

*Military Law, by Commander E. M. Byrne, JAGC, U. S. Navy (U. S. Naval Institute, Annapolis, Md. 21402) is an authoritative basic text on military law, which covers virtually all topics a Marine Corps company officer should be familiar with.

Marine officers are called on to perform various legal functions; and here, as ashore, you must set an example by your competence and knowledge.

1901 ■ Sources of Military Law

The written sources of military law include the Constitution of the United States, the Uniform Code of Military Justice (known as "UCMJ"), and other acts of Congress. The Uniform Code of Military Justice leaves the administration of that Act up to the President and the military secretaries. Pursuant to this authority, the President, by executive order, promulgated the *Manual for Courts-Martial, 1969* (cited in this chapter as *"MCM"*), and the Secretary of the Navy published the *Manual of the Judge Advocate General of the Navy* (cited as *"JAG Manual"*). These two publications, plus *Navy Regulations* and certain general orders, constitute the sources of written military law as it applies to the Navy and Marine Corps. You must be thoroughly familiar with these publications, not to mention the *Marine Corps Manual* and pertinent general orders.

Other sources of military law include: decisions of the courts (including U. S. Court of Military Appeals and Navy Department Court of Military Review); directives of the President; directives of the Secretary of Defense and Secretary of the Navy; opinions of the Attorney General and the Judge Advocate General of the Navy; court-martial reports; and customs and usage of the Service.

1902 ■ Civil and Military Law

In addition to being subject to the federal and state laws that bind all citizens of the United States, members of the Armed Forces are subject to a third body of law and a separate jurisprudence which govern the Armed Forces. This body of law includes the statutes and regulations setting forth the rights, liabilities, powers, and duties of officers and men in the military services. Thus, members of the Armed Forces may be brought before civil or military tribunals and are generally answerable to both bodies of law. Breaches of the peace and other minor offenses by Service personnel which violate both civilian and military law will often be tried by court-martial, although this does not exclude exercise of civil jurisdiction as well. When an offense violates state, Federal, and military law at the same time—for example, a serious crime such as murder—the authority which first obtains control over the offender may try him. For, just as civil courts may not interfere with military courts (other than by writ of *habeas corpus*), neither do military authorities have power to interfere with civil courts.

A member of the Marine Corps accused of an offense against civil authority may, upon proper request, be delivered to the civil authority for trial. Regulations promulgated by the Secretary of the Navy covering this are found in the *JAG Manual.*

> NOTE: Under the frequently unclear and controversial provisions of the so-called O'Callahan case, the Supreme court has ruled that certain crimes by military personnel are "non–Service-connected" and must be tried in civil courts rather than before military tribunals.

In certain foreign countries, the United States has *"status-of-forces" agreements*, which among other things prescribe conditions under which U. S. military personnel may be delivered to local authorities for trial in local courts (or, alternatively, tried by U. S. military courts). These agreements vary from country to country.

1903 ▪ Uniform Code of Military Justice ("The Code")

On May 5, 1950, the Uniform Code of Military Justice (hereafter cited as the Code) was approved by President Truman. The agencies which, under the Code, administer military justice are shown in Fig. 19–1.

Instructions and Publication. Certain articles of the Code must be carefully explained to every enlisted man when he begins active duty, then again after six months, and also when he ships over, and a complete text of the Code must be available to every person on active duty in the Armed Forces of the United States.

At frequent intervals the "punitive articles" (those dealing mainly with offenses and punishments) must be published to troops, to the crew of a naval vessel, and to the personnel of shore stations. This is known —in the old Navy phrase—as "reading the Rocks and Shoals."

Jurisdiction. All persons in the Armed Forces are subject to the Code. Reciprocal jurisdiction between Services is provided, but the exercise of jurisdiction is in accordance with regulations prescribed by the President and is resorted to only when an accused cannot be brought before a court-martial of his own Service. For the purpose of these regulations, the Navy and Marine Corps are considered to be "the Naval Services."

Rights of the Accused. In addition to the constitutional rights which are enjoyed by all American citizens (excepting those expressly or by implication inapplicable to the Armed Forces), the most important rights of an accused person under the Code are as follows: the right to be warned before interrogation if suspected of an offense; the right to a preliminary investigation before trial for an offense; the right to challenge members of the court, both for cause and peremptorily; the right

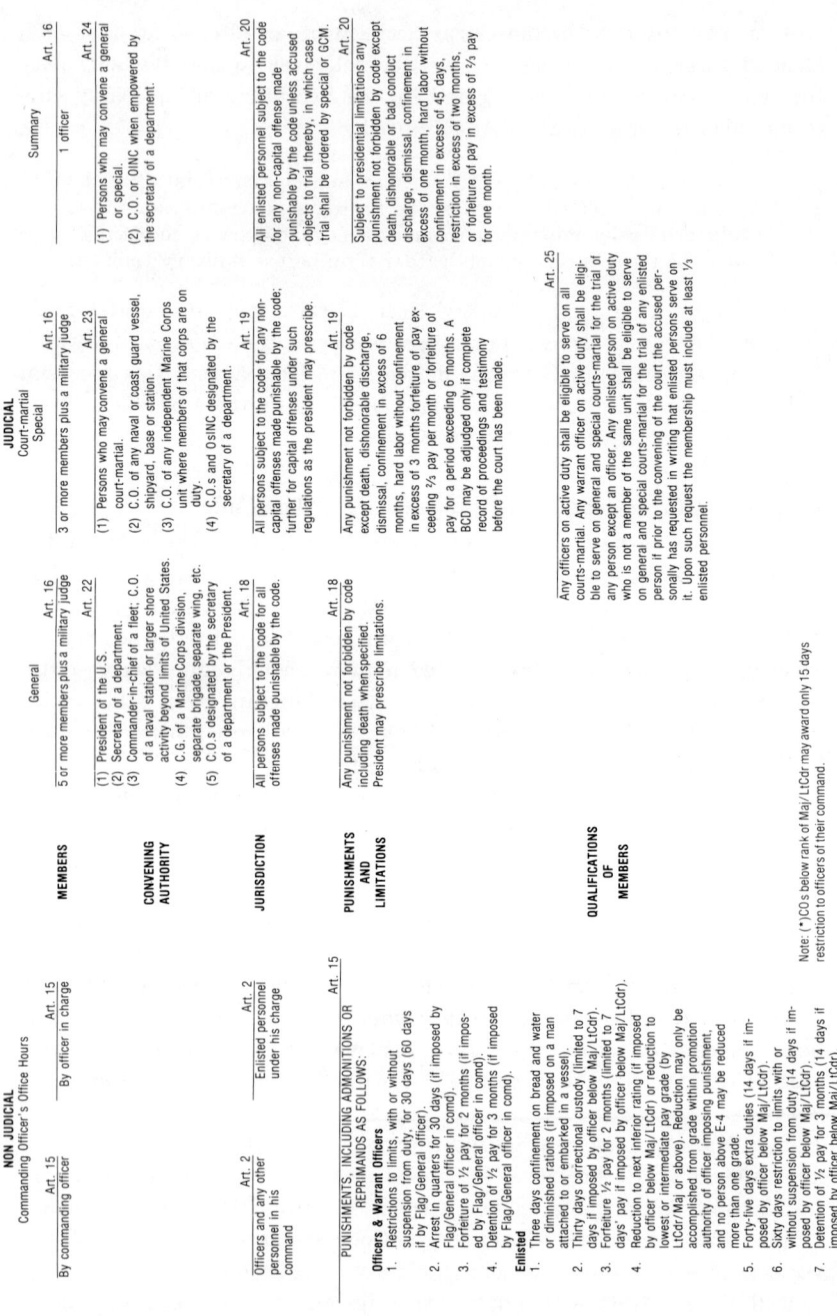

Fig. 19–1: *Administration of military justice under UCMJ.*

to have the findings and sentence of the court made known to him as soon as determined; the right, if convicted, to testify under oath or to make an unsworn statement to the court as to matters in extenuation and mitigation; the right to forward a brief of matters which should be considered in review of the case; and the right to counsel at specified stages of the foregoing proceedings.

Review and Appeals. The Code establishes elaborate machinery and channels for review and appeal of courts-martial. In addition to review by various officers in the chain of command, courts-martial may be scrutinized by *Courts of Military Review* (composed of three or more officers or civilian lawyers qualified to practice before Federal courts or before the highest court of a state); by the *Judge Advocate General of the Navy*; and by the *Court of Military Appeals.* This last tribunal, in Washington, is composed of three civilian judges, appointed for 15 years, with the same qualifications as other Federal judges.

Approval. Sentences of death and those involving a flag or general officer must be approved by the President. Sentences dismissing an officer, cadet, or midshipman must be approved by the Secretary of the Navy. Sentences to dishonorable or bad-conduct discharge, or confinement of one year or more, are not executed until affirmed by a board of review (or the Court of Military Appeals, if the case comes before the Court).

OFFICERS PERFORMING LEGAL DUTIES

The *Staff Judge Advocate* is the senior Marine officer lawyer, certified in accordance with the UCMJ; he performs the staff legal duties of a command.

A *Judge Advocate* is a Marine officer lawyer certified in accordance with the UCMJ, to perform duties as trial and/or defense counsel. In addition, he is authorized to review trial records of summary and non-BCD special courts-martial.

A *Military Judge* is a judge appointed by the Judge Advocate General of the Navy, who serves on general and special courts-martial in the same capacity as that of a civilian judge. If the accused requests and the military judge consents, a military judge may sit as a one-officer court-martial to determine the issue of guilt or innocence and adjudge sentence if found guilty.

A *Military Magistrate* is an officer of field grade with no other legal or provost duties, who reviews and confirms all decisions to confine individuals prior to trial or other action of higher authority. Not later than 72 hours after a Marine is ordered into pretrial confinement, he must be brought before the cognizant military magistrate for hearing.

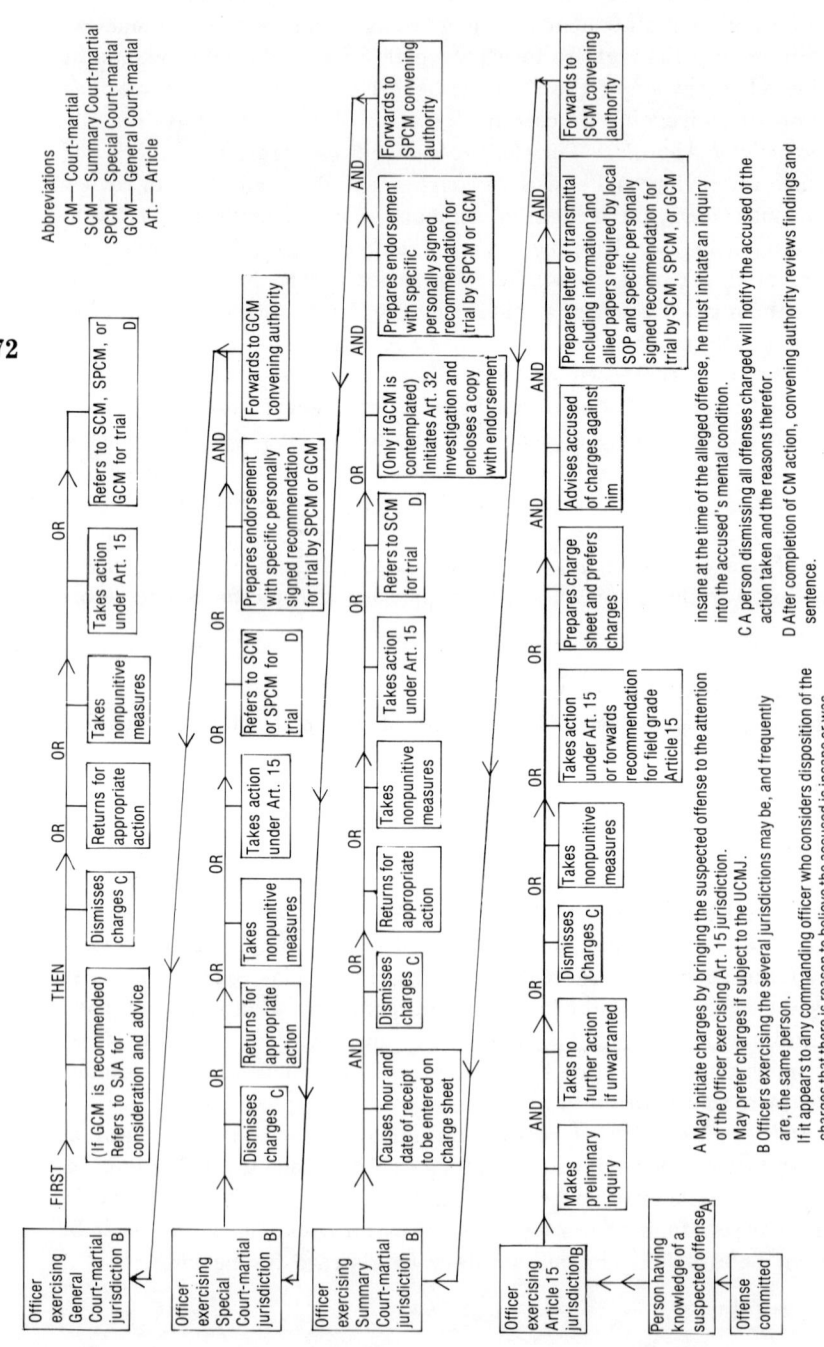

Fig. 19-2: *Disposition of a case under UCMJ.*

A *Legal Assistance Officer* is an officer designated by the commander to give legal advice to members of the command on personal legal problems involving civilian law generally.

A *Legal Officer* is an officer (nonlawyer) designated by a commanding officer to perform legal duties, of purely military nature, within the command. This officer does not render legal assistance (see above), but can answer questions regarding the UCMJ.

1904 ■ Common Offenses and the Small Unit

The Punitive Articles. Articles 77–134 of the Code ("the Rocks and Shoals") divide punishable offenses into three general groups. *First*, crimes common to both civil and military law, such as murder, rape, arson, burglary, larceny, sodomy, and frauds against the United States; *second*, purely military offenses, arising out of military duties and having no counterpart in civilian life, such as desertion, willful disobedience of lawful orders of superior officers and noncommissioned officers, misbehavior before the enemy, and sleeping on watch; and *third*, a general group of offenses based on two articles which do not specify any particular acts of misconduct but cover a variety of transgressions harmful to the Service in general terms. Article 133 applies only to officers and midshipmen. It makes punishable "conduct unbecoming an officer and a gentleman."

Article 134 applies to all persons who are subject to military law. Offenses punishable under this article include: disorders and neglects prejudicial to good order and discipline; conduct tending to bring discredit upon the armed forces; crimes and offenses covered by Federal laws other than the Uniform Code of Military Justice. This general article ensures that there will be no failure of justice simply because an offense is not specifically mentioned in an article of the Code.

Among the foregoing punitive articles, those most violated are as follows:

83: Fraudulent enlistment
85: Desertion
86: Unauthorized absence
87: Missing ship or unit movement
92: Disobedience of orders
93: Maltreatment
107: Knowingly signing false return, record, etc.
108: Unlawful disposition of Government property
111: Drunken or reckless driving
113: Drunk or asleep on watch, or quitting post without proper relief

121-3: Theft or misappropriation
128: Assault
134: General article (conduct to prejudice of good order and discipline; scandalous conduct)

Offenses in the Small Unit. As a company officer, you should familiarize yourself with the most commonly encountered offenses, mainly order violations, which arise within the platoon or company/battery. These are:

Disrespect toward superior commissioned officer (Art. 89)
Assaulting or disobeying a superior commissioned officer (Art. 90)
Insubordinate conduct toward warrant officer, noncommissioned or petty officer (Art. 91)
Failure to obey orders or regulations (Art. 92)
Larceny and wrongful appropriation (Art. 121)
Assaults (Art. 128)
Drug offenses (Art. 92 or 134)
Unauthorized absence (Art. 86)
Drunk or asleep on watch (Art. 113)

In order to deal effectively with the above offenses, you should familiarize yourself with the elements that go into each, together with the possible defenses against such charges. Otherwise, you cannot effectively use the Code as a tool for maintaining effective discipline.

1905 ■ Investigations, Warnings, and Evidence

There are two kinds of *punitive investigations*: preliminary inquiries (preliminary to an Article 15 hearing, see below); and so-called Article 32 investigations, which are preliminary to a general court-martial. Since the latter must be performed by a field officer, they lie beyond the purview of this *Guide*.

Preliminary inquiries are a common occurrence within units, and should thus be understood by all junior officers. Typically, within the Marine division, a company commander who receives a report of misconduct directs that a preliminary inquiry be conducted by an officer or staff NCO of his command. The purpose is to provide the CO with sufficient information so that he can intelligently dispose of the case. Depending on his wishes, the inquiry may be oral or written.

What you are looking for in a preliminary inquiry boils down to three elements: (1) Has any offense chargeable under UCMJ been committed? (2) Who committed it? (3) What is the gravity of the offense in light of the circumstances?

Your job is not to perfect a case or "hang" an accused but to collect all evidence, favorable or unfavorable, to enable your commander to dispose of the matter.

A preliminary inquiry is inherently informal. It is up to you to go out and get the information. Likely places to start include the logbook of the OD; military police "blotter"; civilian police; hospitals and dispensaries; judges advocate; and witnesses otherwise identified. What you learn should be distilled into findings of fact, together (if requested by the CO) with any opinions or recommendations arising out of the inquiry.

> NOTE: The Naval Investigative Service (NIS) investigates felonies and other serious offenses, while some major commands have Marine CID personnel to investigate serious or complicated offenses.

Warnings. Because both the Constitution and Article 31, UCMJ, protect a Marine from being forced to incriminate himself, every accused or suspect must be fully warned of certain rights, or evidence obtained cannot be used against him. The elements of such a warning include:

Nature of offense with which he is or may be charged;
That he has an absolute right to remain silent;
That any statement made may be used against him in any subsequent trial or proceeding;
That he has the right to consult a lawyer, and have counsel present during all questioning, and may seek counsel's advice before answering any question;
That he may obtain a civilian lawyer at his own expense;
That if he cannot afford or does not desire civilian counsel, he may have a military lawyer at no cost;
That he may discontinue an interrogation at any time at his own option.

Evidence. Without attempting to summarize the laws of evidence, which are precise and complex, it is enough to say that even junior officers should be familiar with them for two reasons: (1) no evidence that is obtained in any manner contrary to law can be used against an offender; and (2) much evidence is originally uncovered, either at first instance (for example, by an OD) or during preliminary inquiry, by junior nonlawyer line officers—i.e., you. Thus the admissibility (which is to say, the usability) of evidence often depends on correct decisions at the outset, based on your knowledge of the rules.

Searches, which frequently turn up evidence, are of two kinds: (1) the limited search of an individual and his immediate area, incident to a lawful apprehension upon probable cause; and (2) searches authorized by a commanding officer on probable cause within his command (i.e., a barracks). The laws of search, which are part of those of evidence, are also precise and complex, but you should be acquainted with them. Many an otherwise well-founded case has failed because an officer has conducted an overbroad or otherwise improper search, which in turn denies admissibility of evidence so obtained.

NONJUDICIAL PUNISHMENT

1906 ■ Nonjudicial Punishment (CO's Office Hours)

Legally speaking, a "commanding officer" is one who is authorized to award summary (nonjudicial) punishment, and to award or recommend courts-martial at Office Hours (called "Captain's Mast" afloat). For the purposes of Article 15, MCM, a commanding officer is the CO of a company/battery, or higher unit. If, instead of being officially designated "commanding officer," the officer exercising command has the title "officer-in-charge" (OIC), he may only inflict nonjudicial punishment within diminished limits (see Fig. 19–1). Under the law, distinctions are made between officers in command, with progressively increased limitations on their powers, as follows. Flag and general officers in command, and officers having general court-martial jurisdiction have the greatest scope of nonjudicial punishment. Among commanding officers not in the foregoing class, those of or above the rank of major/lieutenant commander have considerably increased authority over that possessed by COs of company grade.

In addition, nonjudicial punishment may be imposed by any officer possessing general court-martial jurisdiction, or any officer senior in the chain of command to one otherwise so authorized.

Nonjudicial punishment under Article 15, UCMJ, is intended to take care of offenses too serious to be dealt with by a mere rebuke, but not serious enough to warrant court-martial. In this connection, it should be borne in mind that, even though COs have nonjudicial punitive powers virtually tantamount to those of a summary court-martial, punishment under Article 15 should not be used as a substitute for court-martial in serious cases, such as major military offenses or crimes that are felonies in civilian law.

Preliminary Report and Investigation. The customary procedure for putting a man on report is as follows:

An officer submits a report against an enlisted man directly to the executive officer or adjutant of the command concerned. When a *non-*

commissioned officer makes the report, the offender is brought before an officer of the command by the NCO making the charge. If the officer decides that the facts warrant action, a written report is submitted to the executive officer or the adjutant, giving the name of the offender, the offense charged, the name of the NCO making the charge, and any witnesses.

The executive officer or adjutant makes, or causes to be made by the offender's company commander, the provost-marshal, or other responsible person, a thorough investigation of the charges. (See Section 1905 for details on the conduct of the preliminary inquiry, or investigation into an offense.) For company-level proceedings see Fig. 19–3.

At company level, each morning, the first sergeant informs the commanding officer of men placed on report during the preceding day. At battalion level, this is done by the executive officer or adjutant.

Officer offenses, when they occur, are by custom the province of battalion-commanders or higher. They are dealt with by special reports, and handled separately and privately.

Unit Punishment Book (UPB). Every unit whose commander has Article 15 powers must keep a Unit Punishment Book, which is simply a record of each case considered at Office Hours. The UPB also records each individual's acknowledgement that he has been apprised of his rights under Articles 15 and 31, and his waiver of right to trial by court-martial. The first sergeant or sergeant major takes care of this prior to Office Hours and obtains the man's initials in the appropriate spaces in the UPB. At this time, the accused is also told that, although he has no right to legal representation at Office Hours, he may obtain a personal representative to speak in his behalf; he may also call witnesses and cross-examine witnesses against him.

The UPB is an important administrative record, which is liable to inspection at any time, incident to a case, or by higher authority or the IG. A sloppy or improperly kept UPB can get you into trouble.

1907 ■ Office Hours Procedure

Office Hours, as we have seen, is the Marine Corps equivalent of Mast. Like Mast, Office Hours can be, and frequently is, devoted to nondisciplinary matters such as praise, special requests, and the like. Here, however, we are concerned only with the legal and disciplinary aspects of Office Hours. But bear in mind that Office Hours is not merely an administrative procedure, but also a ceremony intended to dramatize praise and admonition. Like any ceremony, it should be dignified, disciplined, especially set apart in the daily routine, and carefully planned (see Section 1611).

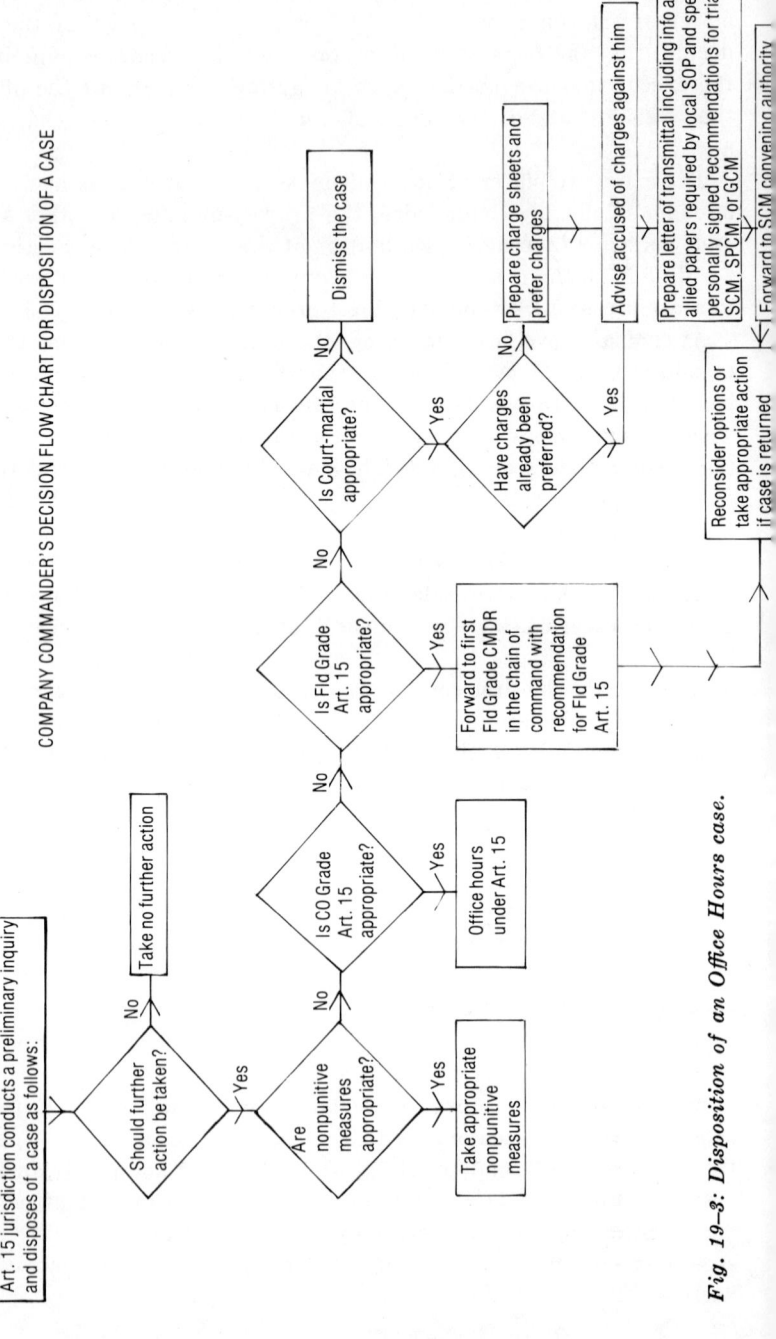

Fig. 19–3: Disposition of an Office Hours case.

Office Hours should be:

Held at a set time, in a set place, usually the office of the commanding officer.

Attended by immediate commanding officers and first sergeants (or platoon-sergeants if within a company) of men required to appear, whether for praise, reproof, or request.

Supervised by the adjutant and sergeant major if at battalion level, otherwise by the company first sergeant.

Held in full, immaculate uniform of the day.

Every officer attending Office Hours should prebrief himself on the cases that concern him. If one of your men is up, take a careful look at his service record book, and talk with his squad leader and platoon sergeant. Assure yourself that he is in tip-top condition as to uniform, cleanliness, and military demeanor. If you yourself hold Office Hours, be sure to review the service records and individual cases before you call in the men concerned. This does not mean that you should prejudge in any sense of the word. However, it does ensure that you focus your thoughts on the man and on his case.

Here is a typical Office Hours procedure:

1. Ten minutes before the scheduled time, the sergeant major (first sergeant) assembles all men who are to appear, together with any enlisted witnesses and the respective first (or platoon) sergeants, who bring the men's service record books (if these are not already in the hands of the sergeant major). At the same time, immediate commanding officers and any officer witnesses report to the adjutant, who conducts the officers into the commanding officer's office, where they are then seated.

2. At the appointed time the adjutant (or company executive officer) stands on the left of the commanding officer with the service record books; these should be opened, tabbed appropriately for ready reference. First (platoon) sergeants stand in a group to one side. The sergeant major (first sergeant) conducts in the first man and reads aloud the charge or report against him, while the adjutant (executive officer) places his service record book before the CO. The man stands uncovered and at attention throughout, one pace in front of the commanding officer's desk.

3. After the charges have been read, the CO must satisfy himself that the man understands his rights under Articles 15 and 31, UCMJ, and, as a matter of prudence, should again warn him as follows: "Private ———————, you do not have to make any statement regarding the offense of which you are accused or suspected. I must warn you that

under Article 31, UCMJ, any statement made by you may be used as evidence against you in a trial by court-martial. Also that, if you so desire, you have the right to trial by court-martial rather than accept nonjudicial punishment here at Office Hours. What have you to say?"

This gives the accused a chance to tell his side of the case if he wishes to do so. Witnesses may be called—usually the reporting officer and witnesses to the offense. The accused must not be compelled to make a statement; he does not have to admit guilt or produce evidence.

4. After all explanations have been heard and the commanding officer has considered the report of preliminary investigation, the CO has four courses of action (see Figs. 19-2 and 19-3):

He may dismiss the accused, either accepting the explanation or giving a warning.

He may award nonjudicial punishment.

He may order the accused to be tried by special or summary court-martial (or he may recommend such trial, if he himself is not authorized to convene these courts).

For a very serious offense, he may order or recommend that an investigation be conducted under Article 32, UCMJ, to determine whether or not the accused should be tried by general court-martial. An Article 32 or pretrial investigation must be conducted before a case can be referred to a general court-martial.

5. At the conclusion of the hearing, the sergeant major (first sergeant) commands, "1. About. 2. FACE. 3. Forward. 4. MARCH." On the command "MARCH," the man marches out of the office, and the process is repeated in the next case.

When meritorious cases (such as presentation of Good Conduct Medals, promotions, or special commendations) are involved, disciplinary cases should be paraded in the rear of the CO's office, to watch the proceedings. They are then marched out and individually brought in again as described above.

1908 ■ Appeal from Nonjudicial Punishment

When nonjudicial punishment is imposed at Office Hours, the man must be informed of his right to appeal his punishment to the immediate superior in command if he feels (1) the punishment is unjust; or (2) it is disproportionate to the offense. This appeal must be in writing and must be submitted within 15 days. At the point when an accused indicates a desire to appeal, and while appeal is in progress, he may not be required to undergo any part of his punishment involving restraint of liberty or extra duties. The immediate superior in command (or the officer who imposed punishment) may at any point suspend probationally any part of the unexecuted punishment, or remit, mitigate, or set it aside.

1909 ■ Limits of Nonjudicial Punishment

At Office Hours, under Article 15, UCMJ, the commanding officer, in addition to, or in lieu of, admonition or reprimand, may impose one, or certain combinations, of the sentences which are given below:

Upon officers and warrant officers:

1. Restriction to limits, with or without suspension from duty, for 60 days (limited to 30 days if imposed by an officer other than a general or flag officer in command or one exercising general court-martial jurisdiction).

2. Arrest in quarters for 30 days (if imposed by a general or flag officer in command or an officer exercising general court-martial jurisdiction).

3. Forfeiture of one-half pay for 2 months (if imposed by a general or flag officer in command or an officer exercising general court-martial jurisdiction).

4. Detention of one-half pay for 3 months (if imposed by a general or flag officer in command or an officer exercising general court-martial jurisdiction).

Upon enlisted persons:

1. Confinement on bread and water or diminished rations for 3 days, if imposed upon a person attached to or embarked in a vessel.

2. Correctional custody for 30 days (limited to 7 days if imposed by an officer below the grade of major or lieutenant commander).

3. Forfeiture of one-half pay for 2 months (limited to 7 days' pay if imposed by an officer below the grade of major or lieutenant commander).

4. Reduction to the next inferior rating (if imposed by an officer in a grade below major/lieutenant commander). Reduction to the lowest or any intermediate pay grade, if imposed by an officer in the grade of major (lieutenant commander) or above. In all cases, a reduction may be accomplished only if the rating from which demoted is within the promotion authority of the officer imposing punishment, or a subordinate. However, an enlisted man in a pay grade above E-4 may not be reduced more than one pay grade, except that in time of war or national emergency declared by Congress, a two-pay-grade reduction is permissible, if the Secretary of the Navy authorizes such reductions.

5. Extra duties for 45 days (limited to 14 days if imposed by an officer below the grade of major or lieutenant commander).

6. Restriction to limits, with or without suspension from duty, for 60 days (limited to 14 days if imposed by an officer below the grade of major or lieutenant commander).

7. Detention of one-half pay for 3 months (limited to detention of 14 days' pay if imposed by an officer below the grade of major or lieutenant commander).

The foregoing provisions are in addition to or in lieu of admonition or reprimand. In addition, certain combinations of the above punishments are permitted, involving deprivation of liberty and forfeiture or detention of pay.

While the law permits reduction of two grades for NCOs and petty officers above pay grade E–4 (corporal), Navy Department regulations provide that no such person can be reduced more than one grade. And, even more important as far as Marines are concerned, *no staff NCO may be reduced nonjudicially at Office Hours under Article 15, UCMJ, except by the Commandant himself, as he is the only one who has the authority to promote a staff NCO.*

On the point as to who may impose commanding officer's punishment aboard ship, only the captain has such power *over members of the ship's company* (the Marine detachment included), even though the title of the CO, Marine detachment, is also that of commanding officer. On the other hand, the disciplinary authority of the commanding officer of an embarked, separate organization of Marines (a floating battalion, for example) remains unaffected whether afloat or not, insofar as members of his command are concerned (see Section 2214).

Office Hours punishment is not considered as a conviction insofar as the offender's record is concerned. Remember, also, that under no circumstances may an offender awarded extra duty be placed on guard to work it off. And do not forget to keep the unit record of nonjudicial punishment as required by the Code and Departmental regulations.

The Commandant of the Marine Corps favors resort to nonjudicial punishment whenever justice and discipline may be attained by its use. Proceedings are brief and direct, and punishment is prompt, which in turn is far more conducive to discipline than punishment long deferred. The accused's rights are protected, as he has a right to appeal to the next higher authority in the chain of command.

One last word: When you hold Office Hours, do so with the greatest respect for each man's individuality. Not only must the punishment fit the crime, as we have already said; it must fit the man.

NAVAL COURTS-MARTIAL

1910 ■ Summary Courts-Martial

Summary courts-martial may be convened by any person who may convene a general or special court-martial, the CO of a ship, naval station, or Marine Corps post, and the commanding officer or officer-in-charge of any other command, when empowered by the Secretary of the Navy. In general, the level at which a summary court may be ordered is that of the battalion or squadron.

> NOTE: Commanding officers have virtually the identical nonjudicial punitive authority as that which a summary court-martial possesses judicially. It may therefore be assumed (and this was one of the intents of Congress in enacting the law) that there will be far less recourse to the summary court now that Article 15 is adequate. Even so, remember that a summary court is not restricted in the same sense as nonjudicial punishment when it comes to adjudging reduction in rank. This could be a decisive consideration in certain cases.

A summary court-martial consists of one commissioned officer whenever practicable not below the rank of captain, USMC, or equivalent; when practicable, this officer should have at least six years' commissioned service. When only one officer is attached to a command, the commanding officer himself shall be the summary court-martial, in which case no convening order is required. The summary court-martial officer is not sworn, performing his duty under overall sanction of his oath of office.

The reporter—if one is required (see *MCM* and *JAG Manual*—may be any person who is competent to keep the record of the case.

Witnesses testify under oath. Examination is conducted by the summary court-martial officer. Rules of evidence are binding.

Since a summary court may award punishment, court decisions require that an accused be given counsel. If he doesn't or can't afford a civilian lawyer, a qualified judge-advocate will be appointed.

Limits of Punishment. The provisions dealing with summary court punishments are somewhat complex, as certain combinations of punishments are permitted. In general, in any case of doubt on this point, seek the advice of your staff judge advocate.

Objection to Trial by the Accused. An accused in the Marine Corps or Navy must signify his willingness to be tried by summary court-martial by signing a statement to that effect. If he objects, he may be ordered to trial by a general or special court-martial, as appropriate.

Record. The record of a summary court-martial is written on a standard summary court-martial form, customarily printed or typed. Detailed instructions and information as to preparation of the record of trial are found in the *Manual for Courts-Martial.* The accused is informed of the findings (and sentence if convicted) in the course of trial. An entry reflecting trial by SCM is also made in the man's service record book. Action taken on the case subsequent to that by the convening authority must be communicated to the convening authority and to the accused's commanding officer for implementation, including notation in the accused's service record book.

1911 ■ Special Courts-Martial

Special courts-martial may be convened by any of the officers shown in Fig. 19-1, and specifically, insofar as the Marine Corps is concerned, by any officer with general court-martial authority; any general officer in command; commanding officer of any battalion or squadron; director, Marine Corps District; commanding officer of any Marine brigade, regiment, detached battalion, or corresponding unit; commanding officer of any aircraft group, separate squadron, station, base, Marine barracks, or auxiliary airfield; commanding officer of any independent Marine Corps unit or organization where members of the Corps are on duty; any inspector-instructor; and by any other CO or OinC when so designated by the Secretary of the Navy. A special court-martial may try officers or enlisted persons for any offenses (except capital) which the convening authority deems appropriate.

A special court-martial is composed of a military judge, of not less than three members, who may be commissioned officers, warrant officers, or (if the accused is an enlisted person and so requests) enlisted men of any of the Services (including Coast Guard, Coast and Geodetic Survey, and Public Health Service, when actively serving under the Navy Department), but such enlisted men should be of the same Service as the accused.

> NOTE: An accused has the option in both general and special courts-martial to request trial by military judge alone, in which case the judge alone hears arguments, adjudges guilt, and hands down sentence. In this case, there are no other members of the court.

When a full special court-martial sits, the function of the members (sometimes called "the panel") is to determine guilt or innocence and adjudge sentence. The senior member of the court, usually a captain or higher—even though he no longer presides (which is the judge's duty) —is called the president.

The trial counsel conducts the prosecutor's case; the defense counsel acts as defense attorney for the accused. In the Naval Services, a reporter transcribes the testimony and keeps the record under guidance of the trial counsel. The orderly acts as guard and messenger, and, if need be, escorts the accused.

If an accused enlisted man requests in writing that enlisted members be included in the special court-martial trying his case, at least one-third of the members must be enlisted, unless that many cannot be obtained. Enlisted members cannot be from the same unit as the accused. When enlisted members cannot be obtained, the trial may still be held, but convening authority must give the reasons in writing. No member of a court-martial should be junior to the accused, and warrant officers or enlisted men may not under any circumstances sit as members for the trial of an officer.

> NOTE: While the law provides for a variant form of special court without military judge and with diminished powers, it is the policy of the Marine Corps that all Marine special courts shall have military judges and full powers. For that reason, this *Guide* omits other reference to the lesser type of court.

1912 ■ General Courts-Martial

General. The highest naval court, the general court-martial, may be convened by the President; the Secretary of the Navy; the Commandant of the Marine Corps; the commanding generals of the Fleet Marine Forces; the commanding general of any corps, division, aircraft wing, or brigade; the commander-in-chief of a fleet; the commanding officer of a naval station or Marine Corps post or large shore activity beyond the continental limits of the United States; any general officer or his immediate successor in command of a unit or activity of the Marine Corps, and such commanding officers as may be authorized by the President and the Secretary of the Navy. General courts may try anyone who is subject to the Code, and award any punishment authorized by law.

Composition. A general court-martial is composed of not less than five officers (one-third enlisted members, *if* an enlisted accused so requests). A military judge serves with each general court-martial. The president must if practicable be at least a captain or equivalent, but Navy Department policy requires that he must be a senior officer. For the trial of a Marine or a Navy staff corps officer, at least one-third of the members of the court should be of the same corps as the accused officer. Unless unavoidable, all members should be senior to the accused. When a regular officer is tried, it is customary for a majority of the members also to be regulars.

Investigation of Charges. Charges may not be referred to a general court-martial unless they have been formally investigated. This investigation may be ordered by the immediate CO of the accused or by the CO of a higher unit. The officer conducting the investigation should be of field grade and should possess legal training and experience. His job is neither to build up nor to whitewash a case, but simply to ascertain the facts thoroughly and impartially under Article 32, UCMJ.

Proceedings of a general court-martial are conducted with special military formality. The military judge presides, while trial counsel prosecutes and defense counsel defends.

Although special courts have jurisdiction to try officers, by custom of the Marine Corps, officer cases are reserved for general court martial. All cases involving officers must be reviewed by the Secretary of the Navy. Those of flag or general officers must be reviewed by the President.

Table 19-1
Court-Martial Punishments

■ Type of Punishment	■ General Court-Martial	■ Special Court-Martial	■ Summary Court-Martial
Admonition	Yes	Yes	Yes
BCD (Enlisted Only)	Yes	Yes	No
Confinement	Yes	Yes (not in excess of 6 months)	Yes (only enlisted below 5th pay grade; not in excess of 1 month)
Bread and Water*	Yes	Yes	Yes
Death	Yes		
Detention of Pay (Enlisted only)	Yes	Yes	Yes
DD (Warrant Officer and Enlisted Only)	Yes	No	No
Dismissal (Officers Only)	Yes	No	No
Fines	Yes	Yes	Yes
Forfeiture	Yes	Yes (not in excess of ⅔ pay per month for 6 months)	Yes (not in excess of ⅔ of 1 month's pay)
Hard Labor (without confinement—enlisted only)	Yes (not in excess of 3 months)		Yes (not in excess of 45 days) (only enlisted below 5th pay grade)

■ Type of Punishment	■ General Court-Martial	■ Special Court-Martial	■ Summary Court-Martial
Life Imprisonment	Yes	No	No
Loss of Rank, Promotion, Seniority	Yes (Seniority only)	No	No
Suspension from Rank, Command Duty	No	No	No
Reduction of Officer	No	No	No
Reduction to Lowest Enlisted Grade	Yes	Yes	Yes (only enlisted below 5th pay grade)
Reprimand	Yes	Yes	Yes
Restriction to limits	Yes (not in excess of 2 months)	Yes (not in excess of 2 months)	Yes (not in excess of 2 months)

*Subject to various administrative limits and as provided by Art. 125, MCM.

1913 ■ Duty as a Court Member

Second lieutenants rarely if ever serve as court members, but first lieutenants and captains often do. Despite the importance of this duty, no special preparation is required. Your one big responsibility as a member is to be there, smartly turned out with a fresh haircut, in the prescribed uniform, prepared to look and be alert. From the time court convenes until it finally adjourns, no matter how pressing may be your regular duties, your duty as a member is primary and comes first.

As trial progesses, the military judge will explain all applicable points of law. Your job is to listen to the evidence adduced by both sides, determine the facts fairly and impartially, and then apply the law, on which the judge will have instructed you, to the facts as you see them. Every member has an equal vote regardless of rank. You may not divulge your deliberations.

1914 ■ Role of Counsel

Counsel on both sides before Marine Corps courts-martial must be qualified judges-advocate and possess legal qualifications appropriate to the level of court.

Trial counsel prosecutes cases for the government, but he is more than a prosecutor: he has a responsibility to help the court develop the truth and safeguard the rights of the accused. You will often have dealings with trial counsel in preparation for cases in which you may have conducted an inquiry or investigation or may be called as a witness.

Defense counsel conducts the defense, to which every accused is entitled. It is the duty of defense counsel:

To undertake the defense regardless of personal opinion as to the guilt of the accused;

To disclose to the accused any interest you may have in connection with the case, or any ground of possible disqualification, and any other matter which might influence the accused in the selection of counsel;

To represent the accused with undivided fidelity; and

Not to divulge his secrets or confidence.

1915 ■ Arrest, Restriction, and Suspension

When charged with an offense, anyone subject to the Code may be ordered into restriction, arrest, or confinement; this is administrative, not a punishment. Or, pending investigation of charges, an accused may be restricted to a specified area at specified times. Administrative restriction differs from Office Hours punishment called "restriction" in that administrative restriction is usually but not always imposed to facilitate investigation. A person under administrative restriction engages in all usual military duties, whereas a person under arrest is not on duty. Arrest is a moral restraint; confinement entails physical restraint.

Arrest or confinement should be no more rigorous than required to ensure appearance by the accused; it should be imposed, normally, only just before trial. When an officer is under arrest, he must remain within the limits assigned; he cannot officially visit his commanding officer or other superior officer unless sent for, or on approval of a written request for audience.

An officer under arrest should not ordinarily be deprived of the use of any part of the ship or post to which he had access before arrest. But, on board ship, if suspended from duty, he may not visit the ship's bridge or quarterdeck, except in case of danger to the ship.

Persons in confinement are in custody of the officer of the day or of the brig officer. They may not be subjected to cruel or unusual punishment. They must be visited periodically, to ascertain their condition and to care for their needs. See and comply with *The Corrections Manual*.

> NOTE: In reading the above, you should distinguish between pretrial restraints, which are discussed, and the officer punishment, awarded by a commanding officer, of restriction to limits, suspension from duty, or arrest in quarters—all generically known in the Naval Services as "hack." An officer in such status is spoken of as being "under hack." (See Section 1909.)

Physical Restraints. As an officer of the day you will frequently have the decision of arresting, apprehending, or confining enlisted people, and, on certain occasions, of applying such physical restraints as irons or straitjackets, whose use is carefully restricted by Navy Regulations and other instructions.

On probable cause to believe that an offense has been committed, any of the following may apprehend (arrest) any enlisted man: (1) officers; (2) warrant officers; (3) NCOs; (4) enlisted men on duty as military police. As OD, you may apprehend and also—subject to ultimate affirmation by a military magistrate—confine. Orders for confinement (which you may receive from a commanding officer) may be oral or written, direct or conveyed through a staff officer.

You (as an OD) are required by UCMJ to accept any prisoner brought in to you by a commissioned officer with a signed, written report of an offense. Not later than your relief as officer of the day, you must report to the commanding officer the full details of any such confinements.

Instruments of restraint (handcuffs, etc.) may never be used for punishment. They are authorized only for safe custody and no longer than is strictly required: (1) to prevent escape during transfer; (2) on medical grounds certified by the medical officer; (3) by order of the CO or OinC, to prevent a man from injuring himself.

COURTS OF INQUIRY AND INVESTIGATIONS

1916 ■ Courts of Inquiry and Investigations

General. Courts of Inquiry and Investigations (sometimes called *"JAG Manual* investigations") are primarily fact-finding bodies (unless specifically directed by the convening authority to express opinions or to make recommendations).

They thus perform no real judicial function, and are in no sense the trial of an issue or of an accused person. They are convened to inform the convening authority of the facts involved. The court of inquiry is the senior, and thus deals with more serious matters.

Because of probable legal sequels, it is advisable to convene a court of inquiry to investigate the following:

Loss of life under peculiar or doubtful circumstances, or from accident; a serious fire; loss, stranding, or serious casualty to a ship of the Navy; major loss or damage to Government property.

In case of loss of life, a medical officer should be a member of the court of inquiry or investigation. The investigating body must determine, if possible, whether death was caused through the intent, fault, negligence, or inefficiency of any person in the Naval Services. However, no opinion will be expressed—as used to be required—regarding the misconduct and line-of-duty status of an individual in the report of investigation of his death, or any endorsement thereon. Unless the body is missing, the investigative body performs the duties of inquest.

Types of Fact-Finding Bodies. There are three types of *JAG Manual* investigations: Court of Inquiry, Board of Investigation, and single-officer individual investigations.

Court of Inquiry. The court of inquiry is the formal fact-finding body of the Naval Services. It may be convened by anyone authorized to convene a general court-martial or by anyone else designated by the Secretary of the Navy. A court of inquiry consists of three or more officers and counsel for the court. Any person whose conduct is subject to inquiry shall be designated as a party to the inquiry. Any person who has a direct interest in the subject of inquiry may request to be designated as a party. Any person so designated must be given due notice of that fact, and he enjoys the right—among other rights (see the *JAG Manual*)—to be present, to be represented by counsel, to cross-examine witnesses, and to introduce evidence. Witnesses may be summoned to be examined upon oath before courts of inquiry just as for courts-martial. Courts of inquiry return findings of fact, and do not express opinions or make recommendations unless required to do so by the convening authority.

The proceedings of a court of inquiry are similar to those prescribed for a general court-martial.

Board of Investigation. A board of investigation consists of two or more officers, one of whom should be of field grade. By Marine Corps custom, such boards may only be convened by battalion or more senior commanders. Proceedings may be either formal or informal; the board has authority to examine witnesses on oath, to receive depositions and other documents, and to make such findings, together with such opinions and recommendations, as may be directed by the precept. Do not confuse reports of such proceedings (or of individual investigations) with certain reports required by *Navy Regulations* or with reports of investigations conducted by the Inspector General.

Investigations. An investigation, consisting of an informal proceeding conducted by a single officer, may be ordered by any commander with Article 15 powers. Save that it is nonpunitive and thus outside the disciplinary process, it is not unlike the preliminary inquiry into an offense, described earlier in this chapter. The conduct of an investigation may well be one of your earliest legal assignments on your own, so it behooves you to be able to handle it in competent fashion.

Lost, Damaged, or Destroyed Government Property. A common form of investigation often falling to junior officers is an investigation—still occasionally referred to by its traditional name as a "board of survey"— into the loss, damage, or destruction of Government property. When such an investigation falls to your lot, it is imperative that you first consult *Marine Corps Supply Manual, Vol. I,* which supplements the *JAG Manual* with much special information on boards of survey.

> NOTE: An important part of any investigation dealing with death, injury, or individual performance of a Marine (e.g., a serious traffic accident involving a Government vehicle) is to determine whether it took place in line of duty or involved individual misconduct. Such findings have wide repercussions in subsequent handling of claims against the Government, Veterans' Administration determinations, and so on. For this reason, the responsibility of an investigation or board of investigation is heavier than may at first seem to be the case.

1917 ▪ Administrative Discharge Boards

These boards, convened by officers with general court-martial jurisdiction, hear cases of individuals whose separation from the Corps by administrative discharge (as distinct from a punitive discharge) has been recommended. Since the Commandant specifically requires that, for obvious reasons, such boards and their staff shall be composed of experienced officers, you would hardly be expected to serve on one; but you may well—as, for example, the respondent's platoon leader—have to appear before or deal with such a board.

In general terms, the board, like an investigation, seeks to determine facts and, based on these, it recommends either (1) that a Marine be retained in the Corps; or (2) that he be given an administrative discharge of character and type recommended by the board.

---

*Law is a regulation in accord with reason,
issued by a lawful superior, for the common good.*
—*Thomas Aquinas,* Summa Theologica

20

Housekeeping

In military parlance, the term "housekeeping" connotes the humdrum but necessary stewardship of administration and maintenance that a unit requires for day-to-day existence. "Housekeeping" thus embraces such matters as police and maintenance, mess management, supply and property, clothing, equipment, transportation, pay, post exchange, and special services. Afloat, housekeeping focuses especially on cleanliness and upkeep.

Like a home with a slovenly housekeeper, the organization where housekeeping is below par finds itself perpetually beset with irksome disorders and nagging minor problems. A prerequisite to tactical efficiency is housekeeping so streamlined that the unit can pursue its military missions unhampered by the distracting demands of inefficient administration.

Good housekeeping characterizes a smart, professional command, and is expected of every Marine.

2001 ■ Police and Maintenance

Police and maintenance are the janitorial side of administration. Except on large posts, both functions come under a single officer. "Police" has to do with tidiness and good order; "maintenance" means upkeep.

Every unit, afloat or ashore, has a *police sergeant*, a noncommissioned officer who supervises cleaning details, trash collection, minor repair, and upkeep. Despite his title, the police sergeant may be any rank from corporal up. He should be selected by virtue of cost-consciousness, powers of observation, forceful character, ability to work independently, ingenuity, and tinkering bent. The police sergeant is the key man in your unit's housekeeping set-up.

The police sergeant's workshop is known as "the police shed." This is anything from a storeroom to a separate building that houses tools, scrap materials, cleaning gear, paint, and salvaged items, which an energetic police sergeant will habitually recover wherever found adrift. As can be realized, the police shed, properly administered, may resemble a miser's lair.

The labor force for police details comes from varying sources. Except in small organizations, the police sergeant has one or more assistants, ordinarily jackleg artificers known collectively as "the police gang." The police gang is supplemented by working details of prisoners from the brig, and by men awarded extra duties. Much of the effectiveness of extra duties as a disciplinary measure depends on the personality and executive abilities of the police sergeant. If the supply of malefactors is inadequate, the first sergeant supplies working parties from men available. Needless to say, close liaison should be maintained between the first sergeant, the police sergeant, and the gunnery sergeant.

When police and maintenance efforts bulk sufficiently, a unit or station *maintenance officer* (see Section 802) is detailed. His duties, on enlarged scale, are much the same as those of the police sergeant. Whenever the commanding officer conducts an inspection, he is accompanied, among others, by the maintenance officer (if there be one) and the police sergeant.

2002 ■ Subsistence and Mess Management

"An army travels on its belly," wrote Napoleon, and so does the Marine Corps. There is no more direct way to the heart of a Marine than through his stomach. This being the case, every officer must know the rules and arts of mess management, which are basically set forth in the *Marine Corps Manual*, and set forth in greater detail in *Marine Corps Supply Manual*, and in the *Marine Corps Subsistence Manual*.

A word is in order at this point to describe how messes are organized. By the term "mess" we mean the General Mess, where enlisted members of the command are fed, rather than any of the various types of officers' mess. Marine Corps messes today mainly operate on a cafeteria system, but a few "family-style" messes remain, in which food is brought to the tables by messmen.

The General Mess and its management represent one of the most important responsibilities of command. The commanding officer must ensure without fail that his men are served meals which, in the traditional guardbook phrase, are ". . . well served and well prepared, of good quality, and sufficient in quantity."

On large posts with several separate messes, and in Marine divisions and air wings, a *consolidated mess system* is employed. This simply means that all messes are centralized for operations, under a single *mess officer*; that a central butcher shop is maintained; and that central storage is provided for dry stores and other mess supplies. In addition to the mess officer, who is essentially an operator, the *food services officer*, a specialist in mess management, supervises training of mess personnel, advises the CO and supply and commissary officers on mess matters, and systematically inspects all messes. On small posts, there is no food services officer. For that matter, on many small posts, especially those satellited on the Navy, Marines eat in the Navy mess.

Both the food services officer and post mess officer are specialists, but *unit mess officers* are not—and that is where you come in. Every unit with its own mess has a mess officer, usually a lieutenant. This officer in turn has a noncommissioned assistant, the *mess sergeant*. The quality and standing of any given mess usually reflect the energy, imagination, and capability of the mess officer and mess sergeant, working as a team.

You may expect with all certainty to be a unit mess officer at some time or other in your career, and the experience will be invaluable in preparing you for command. Thus you should know how a typical mess is organized and how it operates.

The mess sergeant is the leading NCO of the mess. He must be a capable executive, a good cook, and an efficient culinary planner.

The chief cook is senior cook in the galley force and supervises the cooks in preparation of food.

The chief messman, usually an NCO, is in charge of messmen. He is responsible for the cleanliness, sanitary conditions, and neatness of the galley, messhall, and adjacent outside space. The chief messman is a key man. Although most tables of organization do not contain a chief messman billet, you should nevertheless try to find the right man—a forceful, meticulous NCO—and detail him permanently. A slipshod, crummy messhall manned by idle, unclean messmen can usually be traced back to an inefficient chief messman.

The storeroom keeper assists the mess sergeant by keeping the galley stores and provisions.

Cooks are divided into watches, regulated by the chief cook. Each watch should be headed by a rated cook, known as the "cook on watch." Depending on the size of the galley and galley force, the cook on watch may be assisted by other cooks or by "strikers," as apprentice cooks are known. Messmen who show a bent for cooking are assigned as strikers.

*Messme*n are the "hewers of wood and drawers of water" for the mess. Messmen serve food, wash dishes, wallop pots, police the messhall and galley, and function as the mess sergeant's labor force. On posts where civilian messmen are not authorized, messmen are detailed monthly from the nonrated men of the command, in accordance with the *Marine Corps Manual*. The normal assignment of messmen is one for every 20 members of the command in "family-style" messes, and one for every 25–30 in cafeteria-style. An NCO should never serve as a messman (except as chief messman).

2003 ▪ The Unit Mess Officer

The duties of the unit mess officer are as prescribed by the commanding officer. If you are detailed as unit mess officer, immediately look up appropriate references in the *Marine Corps Manual, Marine Corps Supply Manual,* and the *Marine Corps Subsistence Manual*. Despite some variations from unit to unit, all unit mess officers should:

Chow in the field in Korea. Every Marine unit must be prepared to feed itself efficiently under field conditions.

Make frequent spot checks and inspections of the galley and messhall, paying particular attention to:

Personal cleanliness of cooks and messmen (clean, regulation clothing; clean hands and fingernails; obvious general health);
Cleanliness and good order of cooks' and messmen's quarters, including condition of weapons and individual equipment;
Contents of garbage cans (to eliminate waste and to spot badly prepared food);
Good order and sanitary condition of storerooms and reefers (look especially for signs of spoilage or evidence of rodents or insects);
Cleanliness of messhalls (properly washed dishes and utensils; immaculate decks and table tops; condiments covered);
Sanitary garbage stowage and disposal.

Attend *at least* one meal daily. Attend breakfast at least once a week. See that food is served hot and appetizingly, and that there is enough for all. Enforce wearing of prescribed uniform by troops being fed.

Subject to unit policies, provide separate messing spaces in the General Mess for officers (when subsisting in the mess); for staff NCOs; for sergeants and corporals; and for nonrated men. Special tables for staff NCOs and sergeants and corporals—screened off, if possible—are most important and should be provided whenever physically possible.

Stand by the mess and galley area for all inspections by the commanding officer. Have your mess sergeant and chief messman with you.

Be prepared at all times (especially if in an FMF mess) to take the field and serve rations under field conditions.

See that no Marine ever goes hungry or misses a meal through conflict of duties. This means that special servings, both group and individual, may frequently be required for men on watch, drivers, travelers, and the like. Never brush off such men with a sandwich. Indoctrinate your mess force that any Marine who has unavoidably missed a meal *must* receive hot chow and hot coffee, day or night. *A proper galley has hot coffee available to all comers, all the time.*

2004 ■ Clothing Your Marines

Instructions for wearing the uniform, and specifications of all articles of uniform, may be found in *Marine Corps Uniform Regulations*. Procurement, issue, and inspection of clothing—with which we are concerned at this time—are covered by *Individual Clothing Regulations*.

It is your responsibility, as a Marine officer, that every Marine under your command always be properly uniformed and always possess the required regulation clothing correctly marked.

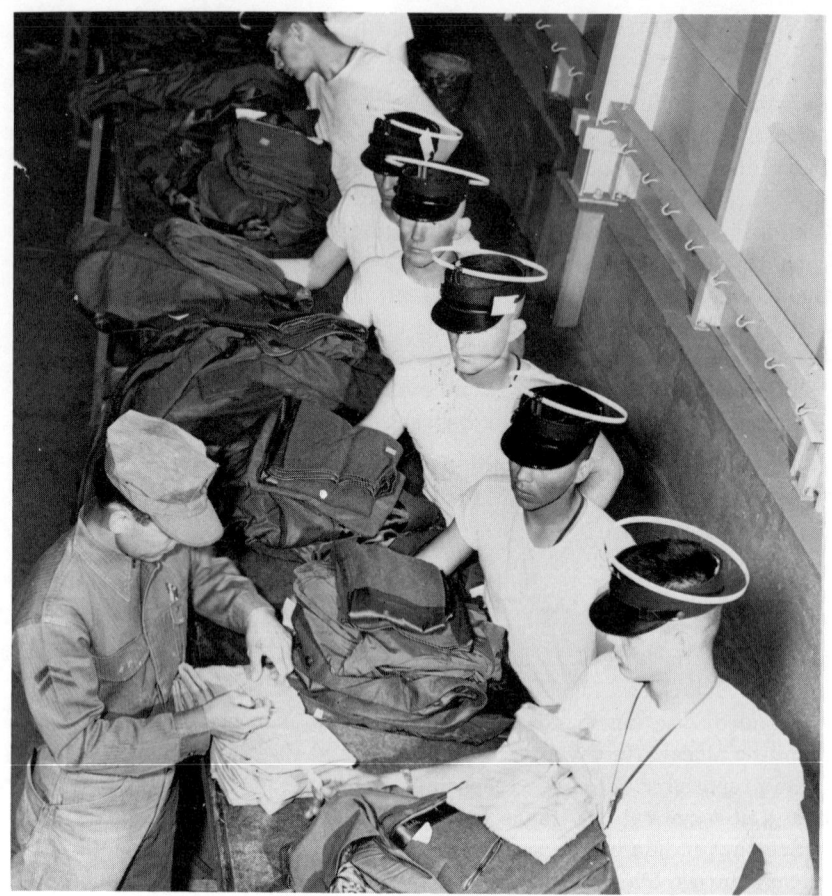

Initial clothing issue is an important moment in a Marine's career.

Clothing issues to enlisted men must be witnessed by a senior noncommissioned officer to ensure that the clothing fits and is in proper condition when issued. As *witnessing officer*, he should make each man try on all outer clothing and shoes. If a good fit seems impossible from stock sizes, you may recommend alterations at Government expense; if the CO approves, the alterations are then made. Before issue, altered clothing is again inspected. *No other alterations of uniform are permitted.*

Every unit has periodic *clothing inspections* for all hands. When inspecting, ensure that:

Each man has the required quantities of clothing.

Clothing is marked as required by *Uniform Regulations*; that it bears correct rank insignia; and that it is in the hands of the man whose name appears thereon (it is an offense to possess clothing belonging to another Marine without the CO's written permission). Clothing is serviceable, or that steps are being taken to replace unserviceable items.

Whenever a Marine is transferred or joins, his platoon leader should give him an individual clothing inspection.

Special clothing, such as cooks' and messmen's uniforms, cold-weather gear, flight gear, and chemical warfare clothing, is organizational property and issued to individuals on receipt, like equipment. All other clothing is supplied to enlisted Marines on the basis of a *monetary clothing allowance* credited to the individual's pay record, against which charges are debited as clothing is issued. If a man loses or mutilates clothing, he must then replace it by cash purchase. At stated intervals, each Marine receives a *maintenance allowance*, a pay-record credit to permit him to replace worn-out items. If a Marine needs clothing but hasn't enough money on the books, the issue is made, and checkage is entered against future pay. This system makes it impossible for a Marine to have an excuse for not having required and serviceable uniforms. It is your responsibility to make this system work.

2005 ■ Supervision of Uniforms

Here are ways to keep your men properly uniformed:

Carry out frequent, systematic clothing inspections with careful follow-up of deficiencies. This means keeping written individual records.

Never allow men to go on liberty unless fully and properly uniformed or dressed in appropriate civilian clothing.

Place full-length mirrors in passageways and exits of barracks, and in headquarters entrances, so that men will cultivate the habit of self-inspection.

Rotate the command through various uniform combinations on successive days or at troop-inspections.

Obtain cleaning, washing, and pressing equipment (irons and ironing boards) for squadrooms and barracks. Recreation funds may be expended for such items.

Require inspection of individuals by an officer or NCO before they go on liberty. Require gate sentries to turn back men in improper uniform or, if wearing civilian clothing, otherwise inappropriately attired.

Limit wearing of utility clothing in every way possible. Make spot checks to ensure that men wear dungarees only when authorized to do so.

Inculcate officers and NCOs with their responsibility to enforce proper wearing of the uniform by Marines at all times. You have no excuse for disregarding a breach of uniform regulations with the famous last words: "He isn't one of *my* men. Let his own outfit catch up with him." When you encounter a man out of uniform, make him correct it on the spot. If the violation is conspicuous, report him to his organization commander.

2006 ■ Individual Equipment

In addition to uniforms, every Marine is issued a weapon and individual equipment (known as "782 Equipment"). It is one of your first responsibilities to see that the arms and equipment of your men are on hand, ready, and serviceable.

On entrance into the Marine Corps, each recruit is issued his rifle. It is up to him to keep that rifle (or its successors) in top condition, since he will have to depend upon it in combat. Even when temporarily armed with other organizational weapons, the Marine is responsible for his rifle's safekeeping and maintenance. Like Oscar Wilde's *Picture of Dorian Gray*, a Marine's rifle is a mirror of its owner; the rifles of a platoon or a detachment are the mirror of the platoon leader or detachment commander.

In addition to his weapon, a Marine's individual equipment comprises other items that he requires in order to fight and survive in the field. These articles are issued on memorandum receipt by his organization, and are recovered or exchanged as necessary. The major items of individual equipment are: pack, bayonet, mess gear, cartridge belt, canteen and cup, poncho, first-aid packet, shelter half, steel helmet, helmet liner and cover, and entrenching tool.

Your responsibility for your men's equipment is similar to your responsibility for their clothing. It is a matter of constant supervision and inspection. Remember that each item of equipment is U. S. Government property, bought and paid for by the taxpayers—including you and the Marine who carries it. Field service and combat consume equipment, and this is to be expected. What is not to be expected, and will not be tolerated in the Marine Corps, is carelessness and negligence toward the weapons and equipment on which Marines' lives depend.

2007 ■ Pay

Detailed information on pay is contained in Chapter 13. From a housekeeping point of view, our interest lies in the command responsibilities involved in paying the troops. These responsibilities are simple.

The most important responsibility of a commanding officer insofar as pay is concerned is to see that the men are paid on time and correctly. When a Marine paymaster is present, this will look out for itself. For outlying detachments, or for units dependent on other Services or on visiting disbursing officers, the problem sometimes requires close supervision.

In FMF units, especially in the field, platoon-leaders often are called on to distribute paychecks to their men. If conditions permit, you should do this on a single occasion with pay line by rank, pay table, and the platoon sergeant as witness. You and the platoon sergeant should be seated, the table should be covered with a blanket, and men being paid should be in clean uniform and uncovered. As their names are called, they step front and center, stand at attention while being paid, and move off at attention. Silence should be enforced in the pay line. In addition, you will of course have to pay men who are sick or cannot make pay call because of duties, prior to returning the pay list and undelivered checks to the disbursing officer for ultimate payment.

Officers have the duty of advising their men in money matters and of educating them to save and to take advantage of the opportunities which the Government provides.

Every Marine, within his means, should participate in the U. S. Savings Bond Program (see *Navy Comptroller Manual*). For men who cannot afford a bond each month, the "Bond-a-Quarter" system should be explained.

2008 ■ Property

Handling and accounting for public property consumes much of the energy of the Marine Corps. The golden rules on property are found in *Marine Corps Supply Manual*. You, and every officer in the Corps, should know these rules and definitions, the most important of which are summarized in the following paragraphs:

Anyone who possesses Government property or who commands those who possess it, whether it is in use or in storage, has *responsibility* for that property, whether he has signed a receipt for it or not. Responsibiliy—in the supply sense—means the obligation of anyone who is required to have personal possession of, or supervision over, public property, to ensure that it is procured, used, and disposed of only as authorized. When you have public property in your custody, you assume, as a public trust, responsibility that this property will be utilized only as authorized by law or regulations.

The CO of any post or unit, however, has *command responsibility* over all the public property of the command. It is his job to ensure that all such property is safeguarded, maintained, and accounted for. An officer has *accountability* (and is known as "an *accountable officer*") when specifically detailed to duty involving pecuniary responsibility for Government funds and property. An *accountable officer*—as distinguished from a *responsible officer*—must keep formal records and stock accounts subject to audit by higher authority.

As you can see, virtually every Marine officer has some type of *responsibility* for Government property, whereas relatively few officers are *accountable*. It is unlikely that you will become an *accountable officer* unless you specialize in supply; you may well be a *responsible officer* tomorrow.

2009 ■ A Responsible Officer's Duties

As a *responsible officer* you have certain basic obligations with regard to public property.

You are personally (and pecuniarily) responsible for all nonexpendable property issued to you. You are also responsible that all nonconsumable but expendable property issued to you be used only for the purposes authorized. Some such items, like 782 gear (see Section 2006), which, owing to their nature, may often be in short supply and are always pilferable, may require control by individual memorandum receipt.

You must inspect your property frequently to ensure serviceability, safekeeping, and proper use. At least once a year you must take physical inventory and adjust discrepancies with the accountable officer.

You must possess in serviceable condition all equipment shown on your unit's *allowance list*. The allowance list for a unit is prepared by the supply officer and contains all items allowed for the organization by Tables of Equipment (T/E) and Tables of Allowance (T/A) and any supporting allowance items except repair parts.

You must keep records reflecting the status of equipment and property for which you are responsible.

It is desirable that you designate in writing one or more representatives, officer or enlisted, who are authorized to receipt for property in your name. If an officer is such a representative, he is known as the "property officer." If an NCO, he is known as "the property sergeant."

Turn in to the appropriate supply agency any property grossly in excess of authorized allowances or not needed for fulfillment of your missions. *But remember* the old saying, "A good property sergeant is never caught short."

When relieved by another officer, you must conduct a joint inventory and adjust any discrepancies with the *accountable officer*. Your relief must then sign for all nonexpendable property carried on the equipment custody records. If you are relieving, and circumstances prevent joint inventory and immediate signature, you are nevertheless responsible for all property on hand. Before you sign any equipment custody records, however, be sure to make an immediate inventory of all property.

Ensure that your officers and enlisted men are instructed in care, use, and maintenance of Government property, and that all hands are totally cost-conscious. The men you select for safekeeping property—your property sergeant, in particular—must be chosen with great care. Do not entrust keys of storerooms or chests to enlisted men or civilians without constant officer supervision.

2010 ■ Expenditures of Property

Even with the most careful stewardship, property wears out and supplies are expended. The Marine Corps recognizes this; our supply system permits expenditure of material, and there are procedures for fixing responsibility for unusual or improper loss or damage to Government property.

Nonconsumable expendables and consumables are expended on issue by the supply officer. Thus, no formal accountability exists for these items. Nevertheless you must ensure that there is sufficient control over such material to guarantee proper end use as well as ordinary economy.

If nonexpendable property is unavoidably lost or destroyed, this fact should immediately be brought to the attention of the supply officer, who may drop the property, with the approval of the commanding officer, through a special adjustment to the account.

If culpability or negligence be the suspected cause of loss of a nonexpendable item, an investigation should be held to determine the exact circumstances surrounding the loss. If appointed a member of such a board, immediately look to the *Marine Corps Supply Manual*.

Checkage of individual pay is a means of recovering the value of lost, damaged, or destroyed property from anyone who acknowledges responsibility therefor. An individual cannot be compelled to checkage of his pay, but he can be subjected to disciplinary action, which most Marines are anxious to avoid (see *Marine Corps Supply Manual*).

2011 ■ Hints on Property

In addition to the advice in *Marine Corps Supply Manual*, a few hints on the management of your property are in order:

Remember the maxim quoted earlier—"A good property sergeant is never caught short." It is always best to keep *a little ahead* on the items in your property account.

Make friends with your *accountable officer*. Keep him candidly informed on the state of your property account. Though you may have shortages, he in turn may have overages. Let him share your problems.

Keep in touch with salvage and reclamation activities. It is often possible to adjust an awkward debit balance through assists from reclamation.

Look and plan ahead. Nothing is worse than getting caught short because of lack of ordinary foresight.

Be cost-conscious. The money available to your unit is not limitless, and supplies cost your unit money.

Follow through. Your responsibility doesn't end with the placing of requirements for material with the supply officer. It ends only when the items in question are physically either in the hands of the end user or in your storeroom. As the Duke of Wellington wrote in 1810:

> It is very necessary to attend to all this detail and to trace a biscuit from Lisbon into a man's mouth on the frontier and to provide for its removal from place to place by land or by water, or no military operations can be carried out.

Find an observant, aggressive property sergeant who is quick to assert title to ownerless public property adrift.

Take pains to determine and enforce responsibility for Government property among your subordinates. When one of your men loses or damages property, institute checkage against him. This not only reimburses the Government for the loss and thus clears your books, but it also reminds all hands that property is to be respected and cared for.

Keep a neat, uncluttered storeroom. Allow no "grab-bag" accumulations in dark corners. Inspect your storeroom frequently—and unannounced.

Finally, remember the old saw which supposedly covers every known category of Government property: *"If it's small enough to pick up, turn it in; if you can't move it, paint it."*

2012 ■ Exchanges

The missions and activities of Marine Corps Exchanges are described in Section 808.

Exchanges are managed by an officer, designated in writing by the commanding officer as the *exchange officer*. The exchange officer may have other officer assistants if the exchange is large or has several branches. Subordinate to the exchange officer is an enlisted superintendent, the *steward*. The steward is a key man, who should be selected for acumen, integrity, and conservative habits.

To assist and advise the CO and the exchange officer is the job of the *exchange council* and the *exchange enlisted committee*. The council consists of not less than three officers (including the exchange officer). Assisted by the exchange enlisted committee, the council advises the commanding officer concerning the operations of the exchange, takes inventories, and performs such other duties as prescribed by the CO or the *Exchange Manual* (see page 458). In addition to assisting the council, the exchange enlisted committee considers suggestions or complaints from enlisted men, and funnels them to the council.

The "bible" on post exchanges is the *Marine Corps Exchange Manual*. If you are assigned as exchange officer or as member of an exchange council, reach for this volume.

2013 ■ Special Services and Recreation

"Special services" embrace a number of nonmilitary welfare activities within the Marine Corps, chief among which is unit recreation. *Marine Corps Special Services Manual* compiles all information on special services. Remember, however, that no matter what machinery may be set up for welfare purposes, nothing can supersede or diminish the commanding officer's paramount responsibility to lead, care for, counsel, and educate his command.

Special services include: recreation; athletics; education (nonmilitary programs intended to raise the general educational level of the Corps); information; and "personal affairs." To implement these programs, every command has a *special services officer*, and (often as additional duty), education, athletic, and personal affairs officers.

The mainstay of post or unit recreation is the *recreation fund*, which provides for the recreation, amusement, and welfare of all hands. The recreation fund gets its money from exchange profits and from the Central Marine Corps Recreation Fund. Since recreation funds are "nonappropriated funds" (that is, not provided from Government appropriations), some latitude exists in spending them for items not covered by official grants but nevertheless desirable for the welfare or morale of the command. *Marine Corps Special Services Manual* lists expenditures that may be made from a recreation fund, such as: athletic equipment; athletic and marksmanship prizes; dances, picnics, and beer parties for the unit as a whole; musical instruments; washing machines and electric irons for barracks; and television and radio sets. The foregoing items, remember, are only examples. Before making a purchase from recreation funds, consult the *Manual*, not only to ensure that your project is authorized, but also to make certain it does not fall among the prohibited transactions also covered.

The recreation fund is administered by the *recreation council*, which consists of not fewer than three officers, one of whom is the special services officer. On small posts, the exchange officer may also serve as custodian of the recreation fund. In this case, the exchange council "doubles in brass" as recreation council. The recreation council audits the fund, inventories and controls all recreation property, and ensures compliance with established recreation policy.

The *custodian of the recreation fund* takes charge of recreation funds and property, and supervises recreation activities. He is usually the special services officer.

2014 ■ Transportation

Every unit or station that has vehicles includes a *motor transport officer*, who, under the S-4, is responsible for the upkeep and operation of the unit's transportation. The motor transport force is made up of drivers and mechanics. Motor transport operations are supervised by a noncommissioned *dispatcher*, who assigns vehicles to particular runs and who keeps the unit's transportation operating in accordance with policy and regulations. The dispatcher is the man whom you call when you need transportation, and he is the "front man," so to speak, of the motor transportation organization.

When dealing with drivers and transportation, keep these pointers in mind:

Cars and drivers are for *official* business; use them accordingly. However, commanding officers, senior staff officers, and aides-de-camp frequently have cars and drivers assigned for use as required. Misuse of Government transportation is a serious matter.

If you are the senior officer in an official vehicle, you are responsible for its safe operation and proper employment. If the driver breaks rules, you are responsible. Never, except in emergency, order a driver to transgress a regulation or safety precaution. If you do, be ready to explain.

You may not drive a Marine or Navy vehicle without a Government Motor Vehicle Driver's License. In addition, some commands require a unit driver's license. Other commands prohibit any officer from driving official vehicles except in emergency.

When duty requires a driver to make a run during meal hours, it is *your* responsibility to see that arrangements are made to feed him (not just a cup of coffee and a sandwich, but a proper hot meal). Always ask your driver if he has been fed, and see that he is taken care of. If you are not satisfied, call the dispatcher or, if necessary, the motor transport officer himself. In extreme cases, you may have to arrange directly with the mess sergeant. *Regardless of how you do it, see that your driver is fed.*

Avoid keeping vehicles waiting. Send them back to the motor pool, with instructions to return at a specified time (or call up when you need a return trip). When using an official car in connection with an official social function, avoid having your driver wait outside at your pleasure. Be meticulously punctual; keep to schedule.

Play fair with the dispatcher, and he will play fair with you.

Require drivers to be military and correct in manner and uniform. Because of the nature of their work, motor transport men sometimes tend to wear utility clothing in preference to uniform of the day. This tendency must be sternly checked.

If you are concerned with motor transport operation, "safety" and "preventive maintenance" are key words.

Let all things be done decently and in order.
—I Corinthians, 14:40

21

Life in the Field

Life in "the tented field" or under canvas takes up a good share of the normal Marine Corps career, and what you make of it is largely up to you. An experienced soldier can be comfortable, clean, and well fed under conditions that seem miserable to the recruit. This chapter gives a few facts and tips which will make field service at least a livable, if not invariably a pleasant, experience for you and for the Marines under you.

2101 ■ Make Camp

The steps in making a camp are much like the steps involved in reconnaissance, selection, and occupation of a tactical position. In fact, for a tactical camp, the principles are identical.

Reconnoiter for a good campsite. Do this in person, if possible; if not, at least make a thorough map reconnaissance. Look for a site that is tactically sound (or one which supports your mission, if not tactical); one that is sheltered from the weather, well drained, easy of access, moderately level, and convenient to a good source of water.

Organize your advance detail. It is the advance detail's job to lay out the camp, build as much of the camp as time and manpower permit, and guide the main body into camp. The advance detail includes: artificers for camp construction, engineer equipment operators, mess personnel, hospital corpsmen, communicators, motor transport men, and a small labor force. Depending on the size of the camp, you may equip your advance detail with no more than a set of pioneer tools, or you may need surveying instruments, heavy engineer equipment, earth-moving gear, and demolitions. Be sure, above all, that the advance detail contains energetic, resourceful noncommissioned officers, led if possible by an experienced gunnery sergeant and including at least one corpsman.

Occupy the campsite. First, make camp for the advance detail. Make every step count toward the finished camp. Start from the beginning with a well-defined, orderly, and symmetrical camp plan in mind (see Fig. 21-1). Give sanitation first priority as you approach each task. *Field Manual 21-10, Military Sanitation,* is an excellent reference. Establish water points for drinking and for washing; dig heads immediately, and mark them plainly; locate your galley on the side of the camp away from the heads and close to the water point; dig a garbage pit. Be careful to show respect for graves or shrines in or near the campsite.

Establish communications and open your message center. Develop and mark access roads and the camp road net. Destroy as little timber and vegetation as you can, for the sake of shelter and for camouflage. From the outset, establish a regular camp routine. Make full use of your field music if you have one. Remember, the bugle was used to control armies 30 centuries before the public address system and radio were invented. If one can be obtained, get a ship's bell to strike the hours and sound the alarm. As the camp nears completion and occupancy, generate campetitive spirit as to who has the best tent area, best-looking motor pool, etc.

Be prepared for emergencies. Never neglest tactical security; establish a camp guard; promulgate a fire bill; know your sources of outside help; keep your fuel dump, ammunition, and weapons secure against fire

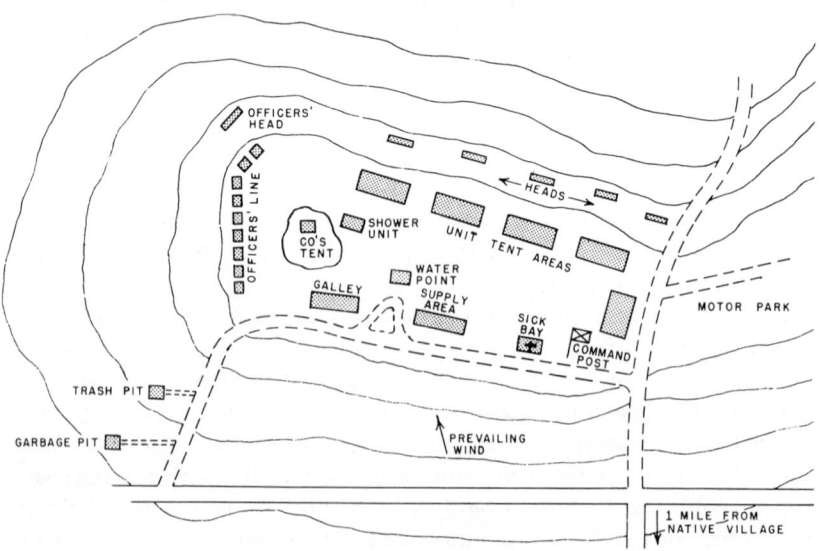

Fig. 21-1: Typical non-tactical camp layout.

and theft; establish liaison with your military and/or civilian neighbors. To supplement camp illumination, spot vehicles with headlights bearing on essential areas (sick bay, communications shack, ammunition stowage, gates, guard tent, and so forth) in case generators fail.

Finally, as you make camp, work your men hard, but take good care of them. Give them an opportunity to bathe, and, if possible, to swim each day. Feed well, and waste no time in getting an effective galley and mess into operation. Get all hands a beer ration, and ensure availability of exchange supplies.

P.S.—Don't forget a good tall flagpole for the Colors!

2102 ■ Living Under Canvas

"Under canvas" covers everything from life under a shelter half up to a squad tent. Even so, the fundamentals are the same.

To be comfortable under canvas, you must:

Don't forget the flagpole when you build a camp.

Keep dry. This means erecting your tent properly, ditching it to prevent being flooded out, making sure your tent is leakproof, and not allowing branches or heavy brush to touch the surface of the tent. It only takes a little rain to kill a night's sleep and make you miserable. In humid conditions, put a blanket between your sleeping bag and air mattress, to collect condensation. *Keep a dry supply of toilet paper.*

Sleep soft. Make your bed before dark. Select a level spot, free from rocks or roots. If you have an air mattress, inflate it; if not, hollow out a hip-hole, or form-fitting depression about three inches deep, where you expect your hips to be when you lie down (this is most important for comfortable sleeping on the ground).

Cheat the insects. Dust your bedding with insecticide. Rig a mosquito net carefully before dark, and take good care of your net to prevent holes or rips. A mosquito net with a hole in it merely serves as a mosquito trap, with you as live bait. If possible, get the hospital corpsman to spray down your tent area, and use an insecticide "bomb" if you can get one. If nothing else is available, rub all exposed areas of skin with insect repellent.

Keep clean. Wash and shave, using your helmet as basin and bucket. You can rig a washstand by resting the helmet inside three tent-pegs, or sticks in triangular formation driven upright into the ground. Keep a small steel mirror with your toilet articles. If you are going to be in camp longer than overnight, get hold of an expeditionary can to provide water storage for your tent.

Keep warm. If the weather is cold, winterize your tent by throwing dirt around the bottom edges and packing it to keep drafts from blowing under. Any kind of insulating layer between ground and sleeping bag will help you to sleep warm. Use your poncho, a blanket, even your parka. You can pile layer after layer on top without doing much good, whereas insulation between you and the ground will keep you snug. If you are in a semipermanent setup, with canvas cots, a few layers of newspaper or magazines make a fine underpinning for a warm bunk. If you camp in one site for a while, fluff your sleeping bag each day; this prevents matting underneath and consequent loss of insulation.

Make a deck. If you can find scrap boards, make a deck for your tent. Even if you have only one piece of plank, use this beside your bunk as a footing to keep your feet out of the dirt and off the ground when you dress or undress. If you have footlocker or seabag, be sure to rest it on trestles or duckboards so as to permit circulation of air underneath and avoid the damp. When you get a tent-deck, beware of snakes nesting under it (or for that matter crawling into warm places in your tent at night).

Make a light. Always have a flashlight handy and in working order. If you can, obtain a gasoline pressure-lantern such as comes in the standard Marine Corps illuminating chest. Possibly you may have a pressure lantern of your own. If so, keep a few spare mantles handy. *Never forget a dry supply of matches.*

Keep safe. If in an area subject to enemy attack or air raid, dig a foxhole immediately. In the rainy season (*if* drainage permits), dig your foxhole inside your tent or under a shelter half. Arrange your gear inside the tent so that your weapons, ammunition, and equipment avoid direct contact with the ground and stay free from dust, mud, or rain. Learn the location of guy-lines so you won't trip over them at night or when diving for your fighting hole. On any tents with metal stakes, put tin cans atop the stakes to prevent people from injuring themselves in the dark.

2103 ■ Sanitation

The fundamentals of field sanitation, both for individuals and units, are:

Personal cleanliness
Disposal of waste (trash, garbage, excreta, and urine)
Control of insects and pests
Disease-free food service and water supply

In the Spanish-American War, the U. S. Army sustained more than four deaths from disease for every man killed in action. Approximately 50 percent of the Army in Cuba became casualties to yellow fever, malaria, and enteric disease. A substantial fraction of these appalling nonbattle casualties could have been averted by sanitary techniques known even then—*plus* effective leadership. By comparison—same island, same war—the U. S. Marine battalion (Huntington's) which served in Cuba throughout the war, suffered only 2 percent sanitary casualties and had no nonbattle deaths at all. *Today, there is no medical excuse for casualties resulting from bad sanitation. Poor sanitary conditions, and the inevitable resultant casualties, are a direct reflection of poor leadership, weak discipline, and inadequate supervision.*

2104 ■ Individual Hygiene

Keep as clean as you can. As a minimum, bathe feet, hands, and private parts. If possible, change your underwear and socks after bathing.

Change your underwear and socks at least twice a week. If you cannot wash your clothing, crumple it, shake it out, and hang it outside in the sun, turned inside out, for at least two hours.

Get out of wet clothing as quickly as you can. This is particularly important in cold-weather operations.

Drink only water that has been purified. *Regard every water source as contaminated until you know otherwise.* Drink plenty of water but do not drink large amounts at one time. In warm weather, increase your intake of salt, if necessary by salt tablets from your sick bay. Use your own canteen or cup only.

Inspect your body frequently for lice. If you cannot wash, dust with insect powder.

Carry toilet paper and keep it dry. Inside your helmet-liner is a good place. *Relieve yourself only at prescribed heads and urinals.*

Have a short, regulation haircut, and keep your fingernails short and clean. Brush teeth at least once daily.

Avoid direct contact with damp ground, especially when hot or perspiring. Likewise avoid drafts.

Keep your mess gear scrupulously clean. Wash mess gear in hot, soapy water, and rinse it in boiling water, before and after each meal. Be sure no garbage, grease, or food particles adhere to mess gear between meals.

Take care of your feet. This means properly fitted shoes and socks, daily washing, toenails trimmed straight across, use of shower clogs when in garrison, and correct treatment of blisters and corns.

Use your mosquito net and keep it in good condition. Scrupulously observe any malarial prophylactic regimen that the surgeon may prescribe.

As an officer, set an example of personal cleanliness and sanitary discipline to your men. Wash and shave daily. Observe all the precautions just described, and require that your men do likewise.

2105 ■ Camp and Unit Sanitation

Lay out and build camps and bivouacs around the sanitary plan. Make sanitation convenient for all hands.

Control your water supply. Purify drinking water by accepted techniques. Establish and plainly mark separate water points for washing, cooking, and human consumption. Locate washing points downstream from points used for human consumption.

Give particular care to location, maintenance, and cleanliness of heads and urinals. Follow prescribed designs in construction. *Flyproof every head.* Locate heads and urinals conveniently to living areas but away from the galley, messhall, and water supply. Never locate a head uphill on a drainage line leading down toward a water source. When a head goes out of use, it must be filled and covered, and then marked with a sign, "HEAD CLOSED (Date)."

Sanitary conditions in the galley are as important as in the head. Galley, garbage house, and, if possible, the mess itself must be flyproof. Garbage must be kept covered and must be removed to garbage pits at least once daily. Empty food containers should either be burned or buried, both for camouflage and to prevent flies from swarming. For any but the most temporary bivouac, a grease trap must be built for disposal of waste grease and greasy water. Most important of all, the galley area must include facilities so that each man's mess gear (plus all pots and food containers) can be thoroughly washed down in boiling soapy water, and then rinsed twice in boiling clear water.

Wage ceaseless war on insects and pests. The principal enemies are: flies (enteric diseases); mosquitoes (malaria and yellow fever); lice, ticks, and mites (typhus and its cousins); cockroaches; and all rodents.

The simplest way to control flies is to cut off their nourishment by screening heads, galleys, and messes. This keeps flies away from food and thus from poisoning you. Watch for young flies, particularly *inside* a screened structure. The presence of small young flies means that flies are breeding nearby, usually inside.

Drainage of standing water, coupled with oiling of stagnant pools and low, swampy spots, kills mosquito larvae as they breed.

Clearing and burning tall grass, brush, and undergrowth helps to reduce the danger from ticks and mites. Human cleanliness and rodent control stop lice.

Cockroach control is a stubborn problem and requires constant effort. Cleanliness, discovery and attack of hiding places, and insecticides are the prime essentials (but insecticides alone are a mere palliative).

All insect control problems can be reduced in a large camp area if aerial dusting with insecticides is carried out immediately before occupation of the site.

2106 ■ Feeding in the Field

The essentials of mess management in garrison have already been covered, in Section 2002, and the sanitary aspects of your field mess and galley have just been discussed in Section 2105. This section supplements both the foregoing with a few hints on feeding in the field.

The simplest type of field rations has the components packaged for individual or small group consumption without being prepared by a galley. Typical of these are the "C" rations, which, in various stages of improvement, constituted the staple of World War II, the Korean War, and Vietnam. These rations are packaged to provide balanced days of subsistence, with components for breakfast, dinner, and supper, together with certain other necessities in an "accessory packet," which includes

A model camp erected by Fleet Marine Force units at Culebra.

such items as toilet paper, cigarettes, chewing gum, heat tablets for cooking, powdered milk and coffee, and can-opener. Although field rations can be eaten without heating, their edibility increases considerably with heating, which may be carried out either in the can or in mess gear. If you heat rations in the can, you may immerse the can in hot water, or warm it over an open flame. In the latter case *open the can before heating* or it will explode with dangerous violence. Aside from an open fire, the usual methods of heating rations are: heat tablets, sterno (solidified alcohol), or a gasoline stove of the Coleman or Primus type.

A stove of this kind is an excellent investment for any officer as part of his kit for field service.

If you find yourself in a situation which permits a little "housekeeping," you will find that field rations—although well-seasoned for the average palate—can be improved by such gastronomic aids as garlic salt, curry powder, paprika, and bouillon cubes.

The next step up the subsistence scale from individual rations is a "B" ration, which is a garrison ration in which fresh components requiring refrigeration have been replaced by equivalent canned items. The "B" ration, of course, is prepared and served on a unit basis through the mess, and thus requires a galley.

The essential features and requisites of a galley in the field are:

1. Shelter from the elements, and, if necessary, from enemy fire and observation.
2. Access for ration vehicles or carrying parties.
3. Water supply for cooking and washing, plus a pot-walloping area nearby.
4. Ration storage safe from human, rodent, and insect incursions.
5. Fuel storage for stoves and lanterns. (*Note:* Establish a preventive maintenance area for stoves in conjunction with the refueling point.)
6. Refrigeration and electric power are the ultimate refinements of a galley in the field. Although not necessities, both are prized luxuries toward which every mess officer and mess sergeant bend their efforts and acquisitive talents.

As a final thought, remember that important as is good food in garrison, it is ten times more so in the field. It is just as true of Marines today as it was in Shakespeare's time:

> Give them three great meals of beef, and iron and steel. They will eat like wolves and fight like devils.

2107 ■ Hints for Tropical Service

Disease and climate are enemies that all soldiers in the tropics must fight. Both enemies can be conquered by common sense, plus heightened attention to the precautions which you would take anywhere.

The most serious tropical diseases are insect-borne. These are malaria, dengue fever, yellow fever, elephantiasis, and the various dysenteries. In addition, humid climates beget fungus complaints on a scale unknown elsewhere. Control of all these (except fungus) is a function of insect control and sanitation, plus rigid observance of prophylactic measures prescribed by your surgeon. Fungus in the tropics, as anywhere else, thrives on heat, dampness, insufficient ventilation, and negligence at its inception.

Tropical climates may be dry and hot, or hot and humid. Of these, the dry climate is of course the more healthful.

In dry, hot regions, such as Arabia or the Persian Gulf, protection against the direct rays of the sun is paramount. Always remain covered. Do not overexpose large skin areas to sunburn. If available, wear light-colored uniform, such as khaki, which reflects heat, rather than dark-colored, heat-absorbent clothing. Use Polaroid sunglasses or you may be blinded by glare. Beware of heat exhaustion; and keep your salt balance up by use of salt tablets. Do not overheat if you can help it.

Bathe often when in the tropics.

In hot and humid climates, special attention to sanitation and personal hygiene is called for. Here, too, dehydration and heat exhaustion must be guarded against. Dampness constitutes an enemy both to you and to your equipment, which will quickly deteriorate under the ravages of rust, rot, and mildew. Clean and oil your weapon and its magazines more frequently. Bathe as often as you can, with special attention to your feet and body areas where perspiration naturally collects. (Soap, incidentally, is often a good fungicide.) If you have been operating in paddy or swamp, get your feet dry and powdered whenever you can, change socks (wool, *not* cotton, the kind you can rinse, wring out, and, if necessary, wear damp).

The essential items of clothing are shirt, trousers, and shoes—just those. If necessary, or if you prefer, you can do without socks and drawers or undershirts. A mosquito headnet and heavy cotton gloves, tight enough to prevent bites at night, can be godsends. Hang your shirttail out, not tucked in, to allow body ventilation. Under no circumstances continue the self-defeating garrison practice of shining field boots. (The original concept of the World War II field shoe was expressly to allow open pores in the leather for passage of air and essential ventilation of feet—a practical fact which was quickly forgotten.)

Travel light. Hang gear from your shoulders or waist, rather than your back. A poncho and items slung from your belt are better than a pack. But a heavy knife, such as the Marine Corps combat (or "Kabar") knife, is essential.

Avoid excessive use of strong drink. Expose yourself as little as possible to biting insects, and try to hold down the biting area of your body they can get access to. After nightfall, try to keep under cover or, if available, behind screening. Mosquitoes have killed more men in the jungle than armed enemies. If possible, sleep and keep your gear off the ground. Another good reason for keeping off the ground, in many parts of the world, is snakes. However, remember that snakes, carnivorous animals, even fire ants and other such fierce insects, are more concerned with avoiding than attacking you.

2108 ■ Cold Weather Hints

Survival and military efficiency in extreme cold weather constitute transcendent tests of officer and NCO leadership. Severe cold renders the most simple operations difficult, prolonged, and clumsy, and enervates, paralyzes, and numbs the individual. In winter warfare, compared to the tropics, disease plays a minor role, while climate is the major problem. Cold is largely overcome by intelligent use of cold-weather clothing designed for military operations. It is your job, as a leader, not only to see that your men *know* how to wear cold-weather clothing, but (much more difficult) that they take the pains to do so. Frostbite, when not fatal, is crippling; the paraplegic amputees from the early Korean winter campaigns were virtually all the result of severe frostbite (and, behind that, incompetent leadership). *Frostbite is not only avoidable, but, if it occurs, is reprehensible and may well warrant disciplinary action.*

The rule for cold-weather survival is the word "C-O-L-D":

  C: Keep yourself and clothing *C*lean.
  O: Avoid *O*verheating.
  L: Wear clothing and equipment *L*oosely.
  D: Keep *D*ry.

520 Other problems of cold-weather operations are: danger from fires due to overheated stoves; carbon monoxide poisoning in improperly ventilated shelters; and temporary blindness resulting from snow or ice glare. All these can be prevented by leadership and supervision.

Before you go into the field in arctic or subarctic climates, the Marine Corps will probably send you through cold-weather indoctrination at the Mountain Warfare Training Center at Pickel Meadows, Bridgeport, California. If unable to obtain such training, be sure to get the effective manual on cold-weather operations and survival.

There is nothing magical or even superhuman about successful warfare in extreme cold. It is a technique that can be learned, and, for your own sake and your men's, you must learn it.

2109 ■ Break Camp

The main points to be remembered in vacating a camp site are as follows:

 a. *Always leave the site in better condition than you found it.* Good camping grounds are rare; think of the next unit. Every individual and every unit, from squad up, must police its area prior to vacating.

 b. *The site must be left completely sanitary.* Heads and garbage pits must be filled and properly marked. Trash must be collected and disposed of.

 c. *Gear adrift must be collected and salvaged.* There is no worse reflection on an organization than a miscellaneous accumulation of jettisoned or forgotten property and equipment on a former campsite. Nothing useful should be left behind. No documents of any sort (especially if classified) should be lying about.

 d. *Fires must be extinguished.*

 e. *Officers and NCOs must inspect their areas prior to vacating.* In most instances, a rear party under officer command should be detailed for final police, check for property left behind, and detailed inspection of all areas. The officer should be required to report to the commanding officer on the condition of the campsite when finally vacated.

It is unworthy of the Marine Corps and an inexcusable mark of disorganization, low morale, and defective leadership for a unit to leave a slatternly camp area behind.

2110 ■ Care of Field Gear

Your helmet, next to your weapon, is your most important piece of combat equipment. It may seem practically indestructible, but even helmets can be abused.

Don't use it for cooking or expose it to flame. This weakens or destroys its temper and thus reduces its protective strength.

If you employ the helmet as an intrenching tool or for any kind of digging, you will scratch off its dull, camouflage coating.

Helmets are nonmagnetic so that the steel won't interfere with your compass. However, if the helmet is badly scraped, scratched, or dented, this may remagnetize it.

Protect the liner from being cracked or deformed by rough use.

Don't sit on the helmet or you may break the chin strap loops.

Canteens should be thoroughly dried out after use, or else they will collect corrosion and/or unpleasant taste. If you have one of the old-type canteen covers with felt liner, hang onto it; this feature is designed so that, when the felt is dampened, it keeps the canteen nicely cooled. Once a canteen is dried out, screw on the cap so that the cork liner won't shrink, dry up, or dry out.

Mess Gear. The most important rule regarding mess gear is to keep it clean and sanitary. Careful drying will avert corrosion.

2111 ■ Useful Articles for Field Service

We all remember the White Knight in *Through the Looking Glass*, who took the field burdened with everything from plum cake to a birdcage. From time to time, you will encounter soldiers with that propensity. The jetsam of every beach and battlefield testifies how much useless gear we tend to carry around.

Bear that thought in mind before you add another gadget to your personal kit.

Remember also that the day may come when you have to carry all your gear by yourself; never accumulate more than you can handle on your own. As General R. S. Ewell remarked, "The path to glory cannot be followed with much baggage."

With the foregoing caution in mind, look through the list of useful items which follows. As you do so, remember that the nature of service ahead may well eliminate some articles and dictate others not listed. In other words, the following list of items for field service is *a guide only*. No checklist can possibly be a substitute for common sense.

USEFUL ARTICLES FOR FIELD SERVICE

Wrist watch (luminous dial)
Jackknife (should have corkscrew)
Miniature can-opener (get from C-ration can, and carry with your dog-tags)
*Pliers
*Medium screwdriver
*Wire coat-hanger(s)
*Double socket
Flashlight, waterproof
*Stove, individual, gasoline (Coleman or Primus)
*Lantern, pressure, gasoline
Polaroid glasses
Tennis shoes (wear at night, so you can remove field boots but still turn out in a hurry)
Ballpoint pen
Small bottle bismuth and paregoric
Insect and foot powder
*Shower clogs
Housewife
Steel mirror
Multicolor automatic pencil
Pocket notebook
Air mattress
Cane (with alpenstock-type metal tip, if available)
Pocket flask, metal
Plastic bags (those used to pack radio batteries can be cumshawed from battalion or company radiomen; they are particularly useful for keeping things dry, such as socks, map, pistol, etc.).
Sponge (practically weightless supplement to towel or substitute for washcloth in warm climate with limited laundry facilities; carry in helmet liner).

*For static situation or nontactical camp only.

As you lay in your stock of such items, try to ascertain which may be available by issue (air mattress or flashlight, for example). Avoid purchasing what you can be sure of being able to draw. Especially avoid buying nonregulation combat equipment or weapons, such as fancy pistols or knives. The Government spends millions a year to make certain that the combat gear issued to you is the finest and most reliable. You merely make a fool of yourself and set a poor example to your men if you are not content with regulation equipment and weapons.

P.S.—When you go into the field, take a paperback edition of some good book with you. You will be amazed how often it can chase monotony and provide escape and relaxation when most needed.

Henceforth, in fields of battle,
the tents shall be our home.
—Old Hymn

22

Service Afloat

Sea duty is the oldest and original duty of Marines, dating from the Athenian fleets of the fifth century B.C., and carrying on through Roman times when separate legions of *milites classiarii* ("soldiers of the Fleet") were assigned to duty afloat. In the seventeenth century, when the British and Dutch organized the first modern corps of Marines, it was for duty as ships' detachments, and it was for this same purpose that U. S. Marines were first raised. And the anchor in our Marine Corps emblem symbolizes today that the Marine is first and foremost a maritime soldier whose natural medium is the sea.

Since World War II and the Korean War, and even more today, the great permanent expansion of the Marine Corps has unfortunately been accompanied by a corresponding reduction in the number of ships of the Navy carrying Marine detachments, with the result that sea duty is a much rarer tour for the young Marine officer than in the past. Thus when you receive orders to report on board the USS *Tuscarora* for duty, you are not only embarking upon a tour that previous experience has little prepared you for, but you may not even be able to turn for advice to a contemporary in your present outfit. Nonetheless, sea duty is one of the most rewarding tours that can come your way. (Some old-timers will say you aren't a real Marine until you have served in a ship's detachment.) It is an opportunity and a professional privilege.

2201 ■ Sea-Duty Training and Indoctrination

Because the Navy forms its opinions of the Marine Corps to a large extent from observation of seagoing Marines, and because ships' detachments represent the Corps "in every clime and place," officers and men detailed for sea duty are carefully selected. Enlisted men going to sea are first sent to the Infantry Training School (ITS) at Camp Pendleton, then to the Sea School at San Diego Recruit Depot. An officer going to sea is ordered for a few days' temporary additional duty (TAD) orientation at Sea School. In addition to the foregoing, on receiving orders to sea, write or phone the officer-in-charge, Sea School, and ask him to send you any indoctrinatory material that may help you get the most out of your limited time at the school.

In addition to Sea School, Ships' Detachment Supply Officers (SDSOs) are stationed respectively at San Diego and at Marine Barracks, Portsmouth, Virginia. If you have time, you should visit the office that supports your detachment, for essential information on fiscal and supply matters peculiar to service afloat.

By way of background and reference reading, you should be familiar with Marine Corps Orders on assignment criteria for security forces and on the personnel reliability program, as well as the *Navy Corrections Manual* (a SecNav Instruction). The nominal cost of a current *Bluejacket's Manual* (U. S. Naval Institute, Annapolis) and *Watch Officer's Guide* (U. S. Naval Institute) will be repaid many times. In addition, obtain a copy of the Marine Corps Educational Center's pamphlet ECP 1-17, *Service Afloat*, as well as of Admiral Lovette's invaluable *Naval Customs, Traditions, and Usage* (U. S. Naval Institute). If you have a file of *Marine Corps Gazettes* handy, turn to the September 1961 issue for an excellent article on sea duty by Major W. M. Cryan, and, much earlier, to June 1927 for a similar article—essentially

Sea duty is the oldest and original duty of Marines. Seagoing Marines have played traditional parts as prize crews, riflemen, and landing forces throughout U. S. history.

still valid and containing much excellent advice—by Captain L. C. Shepherd, Jr. Since your duties will deal with gunnery and gunfire support, make every effort to obtain any appropriate naval gunnery schooling available; and also get fully up-to-date on Naval Justice, since Marines are traditionally expected to be legal experts.

THE SHIP AND SHIPBOARD LIFE

> *The words Marine and Mariner differ by one small letter only; but no two races of men, I had well nigh said no two animals, differ from one another more completely.*
> —Captain Basil Hall, RN, 1832

2202 ■ Reporting Aboard

Your orders, in case of sea duty, will direct you to report to a ship for duty and, in peacetime, will ordinarily state the port in which you are to report. When this is impossible for security reasons, you will be directed to report to some shore command, such as a naval operating base, a naval shipyard, or a naval district. This headquarters will further direct you where and when to join your ship. Arrive in the specified port at least as early as the night before you are expected to report.

Ships in port periodically send boats to the regular fleet landings. Plan to have your baggage and yourself at the officers' landing by 0730. Better still, if you know your ship is in port, go to the landing and take the next boat; that way you will run no risk of missing ship.

The uniform for Marine Officers reporting for sea duty is undress blue or, in warm climes or places, undress white (or for some ships, summer service A).

As you reach the top of the gangway or accommodation ladder, come to a halt, face aft, and salute the Colors. Then face and salute the officer of the deck (OOD), saying, "I request permission to come aboard, sir. I am Lieutenant Holcomb reporting aboard for duty."

The OOD will return your salute, and will probably shake hands and welcome you aboard. He will ask for a copy of your orders for the log. The OOD's messenger or a sideboy will escort you below to the ship's secretary, or, in smaller ships, to the executive officer. You will find the other activities to which you must report much more conveniently located than ashore.

Within 48 hours, make a reporting call on the Captain in his cabin; the executive officer will arrange the time, if you ask. Ascertain that the Captain is in. Give his orderly your card and ask him to report to the Captain, "Lieutenant Holcomb would like to pay his respects to the

Every battleship, large aircraft carrier, and cruiser has its detachment of Marines. Sea duty is a Marine mission which dates from earliest antiquity. Here the Commander in Chief inspects the guard on the USS Des Moines.

Captain." The uniform will be prescribed by the executive officer. Remove your cap before entering the cabin. Be alert for the Captain's dismissal; the call will usually last not longer than ten minutes.

> NOTE: If your ship is in home waters the Captain may desire that you call on him ashore at home rather than in his cabin.

After reporting to the executive officer, next report to the commanding officer of the Marine detachment. He knows the ropes and will advise you as to subsequent moves, which will ordinarily follow in this sequence:

Report to your head of department (usually the weapons officer).

Report to the first lieutenant, the "housekeeper" of the ship, who is in charge of assignment of rooms. You will mess in either the wardroom or junior officers' mess (the "steerage") if your ship has one.

Report next to the mess treasurer and pay your mess entrance fee, (known as "the mess share") (see Section 2412).

Go to the disbursing officer and present your original orders with two certified copies, including all endorsements. The disbursing officer will take up your accounts and pay you any money you have coming. In some cases, he can pay you travel mileage; in any case, he will help you claim reimbursement.

2203 ■ Ship's Organization

One of your first tasks should be to study the ship's organization. Each ship prepares its own organization book, which varies somewhat from those of other ships, but all have certain essentials in common (see Fig. 22–1).

The Captain of a naval vessel, the senior line officer assigned to the ship's company, has full command and responsibility for his ship, and exercises authority and precedence over all persons serving in his ship. He is also charged with the supervision of all persons temporarily embarked in the ship (see Section 2214). His authority, responsibility, and duties are described in *Navy Regulations*, which every officer going to sea should study.

The Executive Officer is the executive arm of the commanding officer. As such, he is the Captain's direct representative and is responsible for the prompt and efficient execution of his orders. The executive officer works through the heads of departments, who assist him in organization, administration, operation, and fighting the ship. In addition to his general responsibilities just mentioned, the "exec" directly oversees such administrative functions as morale, welfare, berthing, training, personnel administration, religious, and legal matters.

Departments and Activities. Under the executive officer, the tasks of the ship are divided among the departments and activities and then further subdivided into divisions.

Most ships have the following departments:

Weapons	Medical
Deck	Supply
Operations	Engineering
Navigation	

The Weapons Department includes all the ship's weapons and the people required to fight and maintain them. In combatant ships the Marine detachment forms part of this department.

The Deck Department, headed by the first lieutenant, has charge of all deck seamanship and ground tackle, rigging and superstructure and exterior of the hull, and operation and upkeep of all boats. In ships not primarily concerned with offense, the deck department also includes air

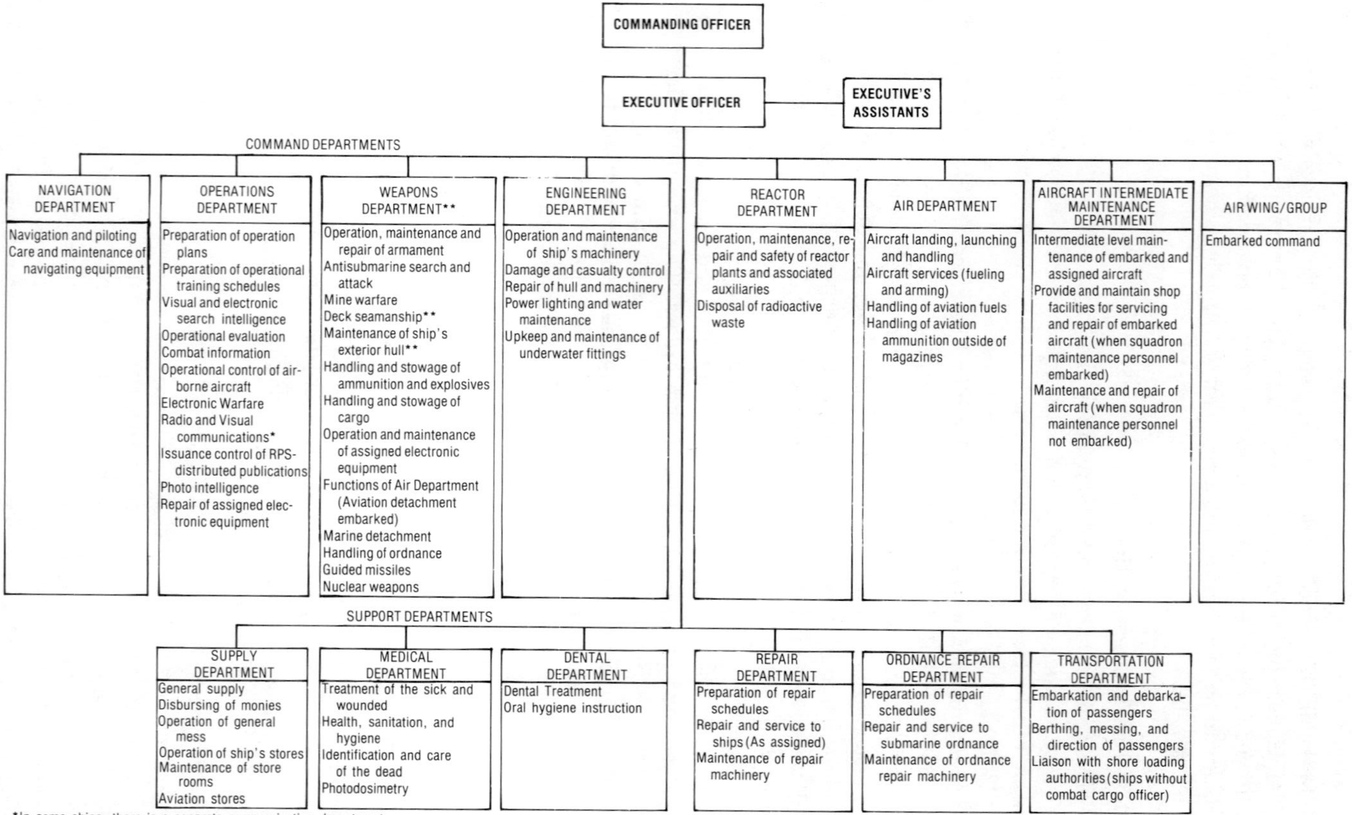

Typical Organization of a Large Ship

COMMANDING OFFICER

EXECUTIVE OFFICER

EXECUTIVE'S ASSISTANTS

COMMAND DEPARTMENTS

NAVIGATION DEPARTMENT
- Navigation and piloting
- Care and maintenance of navigating equipment

OPERATIONS DEPARTMENT
- Preparation of operation plans
- Preparation of operational training schedules
- Visual and electronic search, intelligence
- Operational evaluation
- Combat information
- Operational control of airborne aircraft
- Electronic Warfare
- Radio and Visual communications*
- Issuance control of RPS-distributed publications
- Photo intelligence
- Repair of assigned electronic equipment

WEAPONS DEPARTMENT**
- Operation, maintenance and repair of armament
- Antisubmarine search and attack
- Mine warfare
- Deck seamanship**
- Maintenance of ship's exterior hull**
- Handling and stowage of ammunition and explosives
- Handling and stowage of cargo
- Operation and maintenance of assigned electronic equipment
- Functions of Air Department (Aviation detachment embarked)
- Marine detachment
- Handling of ordnance
- Guided missiles
- Nuclear weapons

ENGINEERING DEPARTMENT
- Operation and maintenance of ship's machinery
- Damage and casualty control
- Repair of hull and machinery
- Power lighting and water maintenance
- Upkeep and maintenance of underwater fittings

REACTOR DEPARTMENT
- Operation, maintenance, repair and safety of reactor plants and associated auxiliaries
- Disposal of radioactive waste

AIR DEPARTMENT
- Aircraft landing, launching and handling
- Aircraft services (fueling and arming)
- Handling of aviation fuels
- Handling of aviation ammunition outside of magazines

AIRCRAFT INTERMEDIATE MAINTENANCE DEPARTMENT
- Intermediate level maintenance of embarked and assigned aircraft
- Provide and maintain shop facilities for servicing and repair of embarked aircraft (when squadron maintenance personnel embarked)
- Maintenance and repair of aircraft (when squadron maintenance personnel not embarked)

AIR WING/GROUP
- Embarked command

SUPPORT DEPARTMENTS

SUPPLY DEPARTMENT
- General supply
- Disbursing of monies
- Operation of general mess
- Operation of ship's stores
- Maintenance of store rooms
- Aviation stores

MEDICAL DEPARTMENT
- Treatment of the sick and wounded
- Health, sanitation, and hygiene
- Identification and care of the dead
- Photodosimetry

DENTAL DEPARTMENT
- Dental Treatment
- Oral hygiene instruction

REPAIR DEPARTMENT
- Preparation of repair schedules
- Repair and service to ships (As assigned)
- Maintenance of repair machinery

ORDNANCE REPAIR DEPARTMENT
- Preparation of repair schedules
- Repair and service to submarine ordnance
- Maintenance of ordnance repair machinery

TRANSPORTATION DEPARTMENT
- Embarkation and debarkation of passengers
- Berthing, messing, and direction of passengers
- Liaison with shore loading authorities (ships without combat cargo officer)

*In some ships, there is a separate communication department.
**In carriers and cruisers, there are both weapons and deck departments.

Fig. 22-1: Typical organization of a large ship.

operations and gunnery/weapons, and the aviation officer and gunnery officer then assist the first lieutenant. On board both carriers and cruisers, the deck and gunnery departments are separate.

The Operations Department is responsible for collection, evaluation, and dissemination of combat and operational information required to enable the ship to perform her assigned missions and tasks. The two principal divisions in this department are communications and CIC (combat information center).

The Navigation Department, headed by the navigator, is responsible for the safe navigation and piloting of the ship. The navigator will usually be senior to all watch and division officers.

The Engineering Department operates and maintains all propulsion and auxiliary machinery and electrical equipment, and is responsible for the repair or control of emergency or battle damage. On board nuclear-powered ships, there is also a separate *Reactor Department*, which operates and maintains the nuclear plant.

The Supply Department is charged with all logistic functions for the ship (procurement, stowage, issue, and accounting of supplies and equipment) plus disbursing and operation of the general mess (crew's mess) and ship's store. Marine Corps clothing is requisitioned from Marine supply sources. Responsibility for Marine Corps equipment also rests with the Commanding Officer of the Marine detachment.

The Medical Department has its normal responsibilities for the ship's sanitation and hygiene, and for the health of all hands.

Other Departments. According to function and mission, certain ships will also have *Air Departments* (aircraft carriers, seaplane tenders, helicopter assault ships, etc.), *Dental Departments* (when a dental officer is assigned), *Repair Departments* (repair ships and tenders), and *Military Departments* (in transports manned by civilians and permanently assigned to the Military Sealift Command, or MSC).

2204 ■ Ship's Regulations

Usually in one binder with the Ship's Organization, Ship's Regulations (which are the Captain's standing orders to all hands) contain indispensable information as to the administration of the ship and of the department and division to which you are assigned. The ship's secretary or the aide to the executive officer will issue them. Study them carefully; these regulations will answer many questions and save much embarrassment.

Study the ship's plans, a copy of which can be borrowed from the first lieutenant. Supplement them by a tour of the ship. If no guide is provided for such a tour by new officers, ask your roommate or some friendly officer to show you around. Visit the bridge, forecastle, combat information center, crew's quarters, galley, central station, plotting rooms, all directors, one mount or turret in each battery, handling rooms, ship control stations, the engine room, and one fireroom.

The Plan of the Day is an important document, issued daily by the executive officer, giving the next day's schedule of routine work or operations and any variations or unusual additions. The Plan of the Day promulgates the orders of the day, drills and training, duty and liberty sections, working parties, and movies or recreational events.

The Boat Schedule is promulgated when in port and not lying alongside a dock or pier. Obtain a copy and keep it with you, especially when going ashore.

2205 ■ The Quarterdeck

The quarterdeck is a portion of the ship's main deck (or occasionally a prescribed area on another deck) set aside by the Captain for official and ceremonial functions. Certain parts of the quarterdeck (usually the starboard side) are reserved for the Captain or for an admiral if embarked. The remainder is reserved for the ship's officers. The detachment parade of the Marines is either on or immediately adjoining the quarterdeck. The rules, traditions, and etiquette of the quarterdeck are among the most venerable in the Service, and their strict observance by all hands is the mark of good sea manners.

Never appear on the quarterdeck unless in the uniform of the day, except when crossing to enter or leave a boat, or when otherwise required by duty.

Do not be seen on the quarterdeck with hands in pockets, or uncovered, and do not remain on the quarterdeck for any length of time if in civilian clothes.

Salute the quarterdeck every time you set foot on it, even from other parts of the ship.

Do not smoke on the quarterdeck until after Evening Colors.

Do not skylark or engage in recreational sports on the quarterdeck unless specifically so authorized by the Captain, and then, of course, only after working hours.

Remain clear of those portions of the quarterdeck reserved for Captain or Admiral.

2206 ■ The Wardroom

The wardroom is your home—and your club. Here you meet and get to know your fellow officers; it's up to you to make them shipmates.

In Navy messes, tradition is important, and seniority is well recognized. The executive officer sits at the head of the senior table; other officers are seated down the table in strict order of seniority, and on down the other junior tables to the last, which is called "the Fourth Ward."

Unless you are on duty under arms, remove your cap when entering the mess. Never unsheathe your sword in a mess (see Section 2418). Save your quarrels for elsewhere.

This does not mean that the silence of a library need be maintained. A noisy mess is often the sign of a happy mess. Between meals, you gather in the wardroom for a moment of relaxation, discussion of problems, a game of acey-deucey or cribbage, or a quiet cup of coffee. It is also often a place for fun, on a Mess Night, when wives and sweethearts are entertained, or on the occasion of a ship's party. It can be all these things, or it can be just a place to eat. It depends upon you and upon the other members of the mess.

Officers' messes (see Sections 2410–2414) are organized as business concerns, with a mess fund to which you contribute your share on joining. Monthly assessments are made, from which costs of food, periodicals, decorations, and other essentials and conveniences are paid. This fund is administered by a mess treasurer. In addition, on board some ships a junior officer—it could be you—is designated as "mess caterer" and put in charge of menu planning, detailed supervision of meal service, and so on. A good way to get this job is to complain about any of these matters. Avoid doing so unless you have better ideas.

Enlisted cooks, stewards, and stewardsmen man the wardroom, "steerage," pantries, and officers' galleys, and take care of officers' rooms. The senior wardroom steward (or "number one boy," in old Navy parlance) supervises the other stewards.

The senior Navy line officer (usually the executive officer) is mess president. Be in the mess before mealtime so that you can take your seat when the "Exec" seats himself. Etiquette once required that an officer remain seated (unless excused) until the mess president rose. While this custom is not observed in all ships today, this does not excuse you from showing ordinary politeness and deference.

Wardroom country is "out of bounds" to enlisted men except when on duty or in special circumstances. Do not use your stateroom as an office. See that enlisted men have little need to enter wardroom country. When they do, require that they uncover (unless under arms), keep quiet, and refrain from profane language.

2207 ■ Wardroom Etiquette

Be punctual for meals. If unavoidably delayed, express your regrets to the senior member.

If necessary to leave before the senior member has risen, ask to be excused.

Do not lounge about the wardroom out of uniform. Some latitude may be allowed during "all hands" evolutions which require working uniform, but be certain that the Captain approves such departure from the basic rule.

Introduce guests to as many wardroom officers as possible, and *always* to the senior member and those at your own table. Entertain only such guests as your messmates and their families will be happy to meet.

Each guest is considered a guest of the wardroom. Be friendly and sociable with all guests.

Except on mess or party nights, officers' guests should leave the ship by four bells of the first watch (2200).

Only officers on the sick list may have meals in their rooms. This does not preclude your having a cup of tea or coffee in your cabin when you are working there.

Do not loaf about the wardroom during working hours. If you have nothing else to do, catch up on your professional reading and study the Ship's Organization and Regulations.

Do not be boisterous in the wardroom; be considerate of your messmates.

When a visiting officer enters, introduce yourself, extend the courtesies of the ship, and try to help him in any way you can.

Observe mess rules—not to talk shop at meals, not to talk religion or politics, not to play the radio or phonograph during meals—or whatever they may be.

Be just and pleasant in your dealing with stewardsmen; make complaints to the mess treasurer.

Do not abuse the privilege of the watch boy by sending him on long errands; he is there to serve *all* officers.

Gambling, drinking, and possession of liquor on board ship, except for medical purposes, are serious offenses.

Pay your bills promptly. Within 24 hours of reporting, pay the mess treasurer your mess bill and mess share in advance.

When necessary, admit ignorance. Experienced officers respect a frank admission and detest bluff. But endeavor to find out what you did not know.

2208 ■ Shaking Down

Your stateroom will be small, or, if large, crowded. Junior officers are usually doubled- or quadrupled-up. Some ships have junior officer bunk rooms containing 6 to 12 bunks.

Space will be cramped. You will probably have an upper bunk, comfortable but not luxurious, a share in a desk with drawers for stowing clothing, a chest of drawers or part of one, and some hanging space. Some rooms have air ports, but most are force-ventilated or air-conditioned.

Your stewardsman (or "room boy") will supply you with bed linen, blankets, and towels; he will keep your bed and cabin as clean and neat as you require, and will pick up laundry. Deal with him tactfully but firmly.

As the junior Marine, you can expect considerable good-natured running. Get to know your roommates, and, if possible, to like them. In cramped quarters in a ship, it requires a nice adjustment to live in harmony with a number of other positive personalities.

As just suggested, Marines at sea traditionally get a certain amount of teasing harassment from Navy shipmates. Remember it is in fun, don't let it get under your skin, and don't hesitate to slip in your own digs as targets present themselves.

Make a definite effort to get along. It's a matter of give-and-take; be sure you give more than you take.

2209 ■ Calls

The exchange of calls is thoroughly discussed in Sections 1820 and 2407–2409 of this *Guide*. Here, we shall consider only the ways in which calls by officers serving in a ship differ from those ashore.

Your call on the Captain is made in his cabin as previously described in Section 2202. In wartime, social calls on officers at their homes ashore are generally discontinued. In peacetime, the new officer and his wife should call on the Captain and his family at home ashore as soon as convenient; either uniform or civilian attire is appropriate for this call.

Other social calls should be made ashore on the executive officer, your head of department, the other heads of departments, and the commanding officer of Marines. Bachelor junior officers are not required to call on all the ship's married officers in their homes, but this is a pleasant thing to do. If married, you may expect all married officers (including the Captain) and their wives to call on you as soon as you are settled. Such calls must be promptly returned. There is no custom to prevent you from calling upon anyone you like and wish to know better.

Calling hours are 1700 to 1900. A call is considered made if you leave cards, when the family are not at home; but try to time your calls so that you do find people at home.

THE MARINE DETACHMENT

Every battleship (when in commission), large aircraft carrier, and cruiser (together with certain amphibious ships), has its detachment of Marines. In addition, each major amphibious ship carries Marine combat cargo or embarkation officers, and, in the case of certain ships on which Marine aircraft are to be based, Marine aviation officers. Moreover, to meet special security requirements, nuclear submarine tenders now carry Marine guards, too.

Marines on board a man-of-war are a distinct and integral part of the ship's company. On board a combatant ship, the Marine detachment is one of the gunnery divisions, usually in the antiaircraft battery. Shipboard Marines do the following jobs:

1. Provide ship's landing force for limited operations ashore;
2. Form provisional battalions of Marines, together with other ships' detachments, for landing operations by the Fleet;
3. Man guns and fire control instruments on board ship;
4. Provide internal security (guard and orderly) services for the ship.

In addition, Marine officers frequently serve as the ship's gunnery air spotters.

Marine detachments on board battleships and large carriers include, on the average, two officers and about 65 enlisted Marines. On board cruisers the detachment has two officers and approximately 40 enlisted Marines. In addition, each admiral's headquarters has a special detachment called a "flag allowance." Marine detachments on board amphibious command ships (AGCs) consist entirely of communication personnel and do amphibious communication duties only.

The Marine detachment is always a separate division. Marines participate in such all-hands evolutions as taking on ammunition, and provisioning ship. In port, when moored to a pier, Marine sentries guard the brows and dock as required. The detachment commander performs a dual role. He is, in effect, head of department with respect to internal administration and security duties of the Marines; but since the Marine detachment is also a division, the Marine officer is a deck or gunnery division officer as well.

2210 ■ The Detachment Commander

To be detailed as CO of a ship's Marine detachment is one of the finest opportunities for independent command that is open to a company officer. At sea, you are truly on your own, and in general, so long as you discharge the responsibilities imposed on the Marine detachment, you will be left alone to get results in your own fashion both by Marine Corps Headquarters and by the naval authorities. No officer can ask any more. Moreover, the intimate experience that you gain as to Navy ways, operating procedures, and capabilities and limitations will prove invaluable to you throughout the rest of your career, and all the more so as you advance in rank. When you become a senior officer, you will find that many of your shipmates from junior officer days with the Fleet are now senior officers in the Navy, and you will have relative ease in transacting Marine Corps business with them as an old friend, which they might hesitate to conduct so readily with a stranger. *From the viewpoint of your career, sea duty is good duty, and should be assiduously sought.*

The commanding officer of Marines is responsible to the Captain for the efficiency, shipboard administration, and total performance of the detachment. He is responsible to the Commandant of the Marine Corps for the detachment's personnel administration, its training in general military subjects, and the Marine Corps property in his charge. He is responsible to his head of department for the gunnery training of the Marines and bluejackets (if any) under his supervision, and for upkeep and operation of ship's equipment, supplies, and spaces assigned to the Marine detachment.

Marine officers may be assigned as officer or junior officer of the deck in port. Underway, junior Marine officers may serve as junior officer of the deck; specially qualified Marine officers may take the deck underway, although this is unusual. Generally speaking, if you have any choice in the matter, you should, if only as a matter of professional pride, try to qualify as a watchstander underway. This will gain you the respect of all Navy officers on board, and will put you on a footing of complete equality as a shipmate who pulls his weight and a true man-of-warsman.

In addition to the foregoing duties you may find yourself assigned as shore patrol officer (see Chapter 17) or as a naval gunfire air spotter or as a shore bombardment liaison officer.

2211 ■ The Junior Marine

The junior Marine officer's most important responsibility is to be prepared to take command of the detachment on short notice. The essential first step in this process is to gain the confidence both of the detachment commander and of the ship's officers.

Shipboard duties of the junior Marine, other than those normal in any small Marine unit, may, in addition to watchstanding on deck, bridge, and battle station, comprise any or all of the following: supervision of the care, cleaning, and upkeep of the battery and fire-control gear entrusted to the Marines; trial counsel for special courts-martial; brig officer; damage-control party officer.

Keeping the *Watch, Quarter, and Station Bill* is part of the junior Marine's duties. This is a chart that lists every man in the detachment by name, rank, bunk and locker assignment, cleaning station, watch and liberty section, and battle station. This bill is arranged in standard tabular form for each division (including the Marines) and, when up-to-date, gives a graphic picture of what every man in the detachment does, and when.

Within the Marine detachment—subject, of course, to the policies of the detachment commander—routine duties of the junior Marine should include the following:

Make at least one daily inspection of sentries and orderlies (supplement this by a nocturnal inspection whenever you come off a night watch).

Take an informal turn around the Marine compartment at least once daily for cleanliness, good order, and upkeep.

When prisoners are confined, inspect brig for cleanliness and good order, hear prisoner complaints, and survey brig log.

Read and initial the detachment guard book before its presentation to the CO.

Arrange with the First Sergeant to check all outgoing routine correspondence and returns prior to their presentation to the CO.

Act as detachment training and educational officer; conduct instruction in the NCO school, and supervise all correspondence courses and study by Marines.

Stand by all Marine cleaning stations and batteries as they are inspected.

> NOTE: An old-time seagoing Marine practice is to require a second lieutenant to make out, with the First Sergeant's guidance, all required reports and returns for at least a one-month cycle. This provides excellent training in basic administration, helps him to learn the detachment, and teaches him to respect the First Sergeant.

GUARD DUTY AFLOAT

2212 ■ Shipboard Guard

Marines perform all guard duty on board ships possessing Marine detachments. This is the oldest mission of Marines.

The same regulations and routine prevail for guard duty afloat—subject to shipboard conditions—as obtain on shore. The ship's Captain has the same responsibilities for the interior guard as the commanding officer ashore. The *officer of the deck* (see Sections 2215–2220) has the same general relationship toward interior guards as the officer of the day ashore. For shipboard guard duty and security, the *commanding officer of Marines* functions much like the provost marshal on a shore station, and, for this purpose, is a "head of department," directly responsible to the Captain. General and special orders for the guard are promulgated by the commanding officer of Marines, but must be approved and countersigned by the Captain.

A ship's detachment is often called "the ship's guard." One of the guard's important duties is to render honors as prescribed in *Navy Regulations*. For this purpose, the Marine detachment maintains a *"guard of the day,"* from which routine guard details are provided. The guard of the day, which functions directly under the officer of the deck, remains ready between Morning and Evening Colors to turn out immediately, in prescribed uniform, on the quarterdeck to render honors. The guard of the day is paraded by the sergeant of the guard, *not* by the detachment officers. The *"full guard,"* so called, consists of the entire complement of Marines, less men detailed to other duties, and is commanded by an officer. Ordinarily, the band turns out on occasions when the full guard is required.

Aside from honors, the guard of the day usually parades for an informal guard mount, and for Morning and Evening Colors when the ship is in port (but not underway). Informal guard mounting may be combined with Morning Quarters, when the entire Marine Detachment parades. Formations for Colors take place on the ship's fantail, in sight of the flagstaff. Guard mounting, quarters, and other routine formations of the Marine detachment take place at the detachment parade, which is located on or as near the quarterdeck as the ship's topside arrangement permits.

Marine detachments aboard carriers average 2 officers and 65 enlisted Marines. At sea, you are on your own, and the CO of a ship's detachment has fine opportunities for independent command.

2213 ■ Shipboard Posts Maintained by Marines

The security posts maintained for the safety and good order of the ship correspond to similar posts on shore, although some general orders (such as that pertaining to challenging) are relaxed or modified. Certain shipboard guard duties are peculiar to sea duty, however:

Special Weapons Security. On board carriers, submarine tenders (or any other ship carrying nuclear weapons), the Marine guard mans special security posts to safeguard the weapons in question.

Life-Buoy Watch. When required, he is posted in the vicinity of the ship's life buoy(s) when underway. The life-buoy sentry keeps a bright lookout for any person falling overboard, or alert for the cry "Man overboard!"; he releases the life buoy, and keeps the man in the water in sight.

Communications Orderly. This orderly delivers radio messages (dispatches) or signals to officers concerned. This duty demands intelligence, knowledge of the ship, and knowledge by name, face, stateroom, and duty, of all officers on board.

Orderlies. Flag officers and the Captain of the ship are authorized by *Navy Regulations* to have personal orderlies. In addition, if the detachment's commitments permit, it is customary to assign an orderly to the executive officer, as well as to any embarked senior Navy or Marine officer who is not a member of the ship's company. Shipboard orderlies

The Captain's orderly must be sharp and intelligent.

not only perform the usual personal services associated with orderly duty, but act as messengers and as receptionists for the officer to whom assigned.

The qualities required for orderly duty are high, and only the best Marines in the detachment should be detailed as orderlies. In selecting and training orderlies, look for intelligence, alertness, discretion, tact, neatness, military bearing, and knowledge of the ship and the ship's company. The closest contact that most naval officers have with the Marine detachment is through its orderlies. Put your best in the show-window.

Brig Watch. The CO of the Marine detachment is responsible to the Captain for the operation of the ship's brig. The sergeant of the guard performs duties as brig warden, and the corporal of the guard acts as turnkey. In large ships, instead of coming under the guard of the day, the brig may be operated as a separate section with the First Sergeant as brig warden, and a permanently detailed turnkey and assistant turnkey. Although the Captain is ultimately responsible for enforcement of standards prescribed in *The Corrections Manual,* all Marine officers and men performing duties in connection with the brig should be thoroughly conversant with this publication.

2214 ■ Guard Duty in Embarked Marine Corps Units

When a Marine Corps unit *not* part of the ship's company goes to sea, the guard duty required is a combination of that maintained ashore and that described above for a Marine detachment.

Bear in mind two principles of command relations: *first,* the Captain of a ship has paramount authority and responsibility for safety, good order, and discipline over everyone embarked in his ship, whether or not under his military command; *second,* the commanding officer of troops embarked in a ship retains his military command authority and responsibility for his officers and men, subject only to the overriding authority and responsibility of the Captain. The CO of Troops is the senior troop commander embarked in the ship, and is ordinarily specifically designated as commanding officer of troops by higher troop authority.

Consistent with the foregoing principles, the Captain of the ship can call upon the embarked troop units to establish a guard to assist in maintaining the security and safety of the ship by manning such posts as life-buoy, control of circulation of passengers, and communication orderlies, and by performing any other necessary guard duties. For internal order, security, and control of embarked units, the troop commander (Commanding Officer of Troops), with the concurrence of the Captain, may establish any posts he considers necessary. In general, he organizes his guard like an interior guard ashore, providing an officer of the day as his direct representative for supervising the troop guard and carrying out troop orders, ship's regulations, and special instructions of the ship's Captain.

All orders to troops embarked in a ship, including instructions for the troop guard, are transmitted through the Commanding Officer of Troops.

For further information on guard duty and transport routine, consult *Landing Force Manual 18, Troop Life and Training Aboard Ship*, as well as the ship's regulations of the transport, and standing operating procedures.

OFFICER OF THE DECK

2215 ■ Marines as Watch Officers

Navy Regulations give the commanding officer of a ship authority to "assign to duty in charge of a watch, or to stand a day's duty, any commissioned or warrant officer who is subject to his authority and who is, in the opinion of the commanding officer, qualified for such duty."

Specifically, Marine officers below major may be assigned duty as officer of the deck (OOD) in port. Marine officers on the junior watch list may stand junior watch at sea. Some commanding officers of ships consider certain Marine officers "qualified for such duty," for the latter —including the author of this work—have been duly qualified and have served as officers of the deck underway.

2216 ■ Status, Authority, and Responsibility of the Officer of the Deck

When on watch as officer of the deck, an officer is the direct representative of the Captain; every person in the ship who is subject to the orders of the commanding officer, except those noted below, is subordinate to the officer of the deck and must carry out his lawful orders. A young officer should, however, exercise this authority with tact and understanding. While on watch, the OOD is responsible for the safety of the ship and for execution of the ship's routine.

The *executive officer* has authority to direct the OOD in matters concerning the general duties and safety of the ship. When the Captain is not on deck, the executive officer may tell the OOD how to proceed in emergency, or may even "take the deck" himself.

> NOTE: In large ships, there are usually detailed a Command Duty Officer, CDO (ordinarily a commander), and an assistant Command Duty Officer, ACDO (usually a lieutenant commander). When such officers are detailed, reports by the OOD to the Captain or executive officer are submitted via this chain.

Command authority may be delegated by the Captain to another officer for a specified watch, which authority in general extends to the same limits as that described above for the executive officer; but such authority does not extend to the executive officer.

2217 ■ Junior Officer of the Deck (JOOD)

Any junior Marine officer may be assigned as junior officer of the deck, in port or underway. It is customary *not* to assign the commanding officer of the Marine detachment to a junior watch list, in port or underway. Any other Marine officer serving in a ship should be prepared and qualified to stand watch on deck, either in port or at sea.

2218 ■ Governing Publications for Deck Watchstanders

Familiarize yourself with the following publications:

Navy Regulations	*Type Doctrine*
General Signal Book	*Type Tactical Bulletins*
General Tactical Instructions	*Watch Officer's Guide*
Current Fleet Tactical Orders	*Naval Shiphandling*

The *Watch Officer's Guide* and *Naval Shiphandling* are not only invaluable for watchstanding in port, but essential for learning to take the deck underway, either as JOOD or OOD.

The classified publications listed above pertain to tactics and tactical organization; you may obtain them for study from the ship's communication officer. You must thoroughly know these instructions before you can perform your duties on deck properly.

2219 ■ Hints for the Officer of the Deck in Port

When coming on watch, plan the action to take in an emergency; what would you do for man overboard, your ship dragging anchor, another ship dragging down on you, sighting a mine or suspicious object in the water close aboard, etc.?

When taking over the deck, inform yourself as to the following:

Berth number, anchor bearings	Head of Department (or Command) duty?
Anchor, scope of chain	
Depth of water	Boats in water, location, fueled?
Swivel and hawse (if moored)	Duty boats, relief boats?
Lines and connections (at dock)	Boat officers?
Weather	Boats, special trips
State of tide	Absentee and PAL lists
Gyro in use	Duty lists
Boilers and auxiliaries	Orders for the Day not executed

Ships present
S.O.P.A.
Guard and Medical Guard
Where are?
  Admiral
  Captain
  Executive Officer
  Heads of Department
Who is senior line officer on board?

Smartness of ship:
  Canvas, bedding, clothes, bunting, lights, Irish pennants, etc.
Orders, special, to execute
Liberty sections, time commences and expires, number of men ashore
Watch relieved?
Ship's plan of the day

Reports to Commanding Officer. When OOD in port you must make these reports to the Captain:

The hours 0800, 1200, 2000 (requesting permission to strike the bells)
Weather changes, shift of wind, barometric change, squalls, rain, snow, etc.
Movements of all men-of-war and large merchant vessels
Arrival and departure of own ship's planes
Serious injury to personnel or material
Flag officers or commanding officers coming or going, in boats or on the dock and headed for the ship
Dragging anchor
Winding of chronometers (with 1200 report)
Changes of condition of readiness, condition of watertight integrity of ship
Aircraft warning signals
Any emergency, or unusual phenomenon
Relief of the OOD

Reports to Executive Officer. In addition to the foregoing, these are reports you must make to the executive officer (or command or head of department duty officer, when the executive officer is ashore):

Special boat trips, departure and return of special details, etc.
Execution of special orders
Compliance with dispatch requests or orders from other ships or commands
Receipts and transfers of drafts or individual men; detachment or reporting aboard of officers
Special details, inspections, ship's exercises
Any departure from routine

Smartness of Ship. The ship is only as smart as you are on your watch. The following items, if frequently checked, will keep the ship looking smart:

The Corporal of the Guard supervises the guard of the day aboard ship.

Uniform of the day must be worn by all hands topside when not performing work of a nature that ship's regulations authorize to be done in working clothes.

Uniforms of the watch must be clean and correct.

Clothes must not be hung up to dry in exposed parts of the ship instead of in drying rooms; trice up and pipe down clotheslines in accordance with ship's regulations.

Keep swab racks out of sight and clear of passageways.

Keep halyards taut; halyards taut at 0800 may be slack an hour later, owing to weather changes.

Inspect, and have your J.O. frequently inspect, for lines over the side, dangling lines, canvas straps, and Irish pennants.

Inspect boat crews' uniforms each time they leave ship; require clean, neat uniforms, properly worn. Allow no lounging in boats underway, no sitting on rails aft, or other unmilitary conduct. Require coxswains to salute when officer passengers embark or debark, and when acknowledging orders.

Have your Junior OOD make frequent inspections of the topside and boats in the water.

Keep OOD booth neat, clean, with gear such as binoculars, spyglasses, notebooks, logs, and so forth, stowed. Return mess gear to pantry immediately after use.

Be alert to reply to dipping Colors from merchantmen (see Section 1811) and passing honors from warships.

Smartness of the Quarterdeck. The OOD is responsible for the general appearance and smartness of the quarterdeck. To maintain a clean, taut quarterdeck:

Keep unauthorized men and idlers clear.

Keep quarterdeck free of stores, motion-picture cans, and other loose gear.

When stores are handled, spread tarpaulins.

Prevent spilling oil, food, paint or other matter that makes spots.

Have any spots or dirt cleaned immediately.

Make frequent sweepdowns (use your sideboys).

Allow no skylarking or whistling on or near quarterdeck.

Accident Prevention. Exercise particular care in bad weather; also when men are working aloft, over the side, or in confined spaces; and when inflammables or explosives or any other dangerous materials are being handled. Men working over the side should wear life jackets and have preventers tended by another man with no other duty. Life rings and life buoys must be ready for instantaneous use.

At anchor, put over a drift lead, take bearings and plot them, and take every precaution to detect and prevent dragging anchor.

Keep a watch on boats and aircraft within visual range of the ship; send aid promptly if needed.

Boats. Boats are one of your main responsibilities and can cause you considerable trouble while on watch; take these precautions:

Never shove off a boat prior to the scheduled time for a trip.

Ensure that boats are properly manned and equipped.

Know the rated capacity of a boat. *Never* allow it to be overloaded. If the weather is foul, or brewing up, underload your boat appreciably, and require that passengers and crew wear life jackets.

Smoking is never permitted in a Navy or Marine boat.

Before sending away a boat, be sure it is properly equipped. Check especially for life jackets, fire extinguisher, filled water-breaker, adequate fuel, boat compass, and operative running lights. Know, and be sure the coxswain knows, the "Rules of the Road."

Navy Regulations prohibit a Marine officer from taking command of or responsibility for a Navy boat. This does not apply, however, to boats specifically assigned to Marine posts or units.

See that safety appliances are installed and in order; be sure crews understand their use and observe prescribed safety precautions.

Keep informed of all boats and other craft which come alongside or shove off from the ship; keep gangways clear, allowing boats alongside minimum time.

See that boats are handled smartly, and that boat crews know and observe the rules of the road, as well as the regulations for honors and ceremonies.

See that boats are properly secured at booms when not in use.

During fog or foul weather, do not hesitate to send boat officers with hazardous or distant trips.

Boat crews must always be kept warm and well fed. See that they are supplied with necessary foul-weather clothing, also that they have hot meals no matter when they return or leave the ship. There is no excuse for a hungry boat's crew.

In most ships the OOD can authorize special boat trips. But be sure that enough boats remain available for scheduled trips. When you authorize a special trip, such as to the repair ship or the hospital ship, pass the word so that all departments may take advantage of it. Special trips that will keep a boat away from the ship more than an hour should be approved by the executive officer.

Send guard-mail trips at scheduled times. Resist the temptation to have guard-mail boats make side trips. Upon entering port, make a guard-mail trip immediately to the division flagship, or to the force flagship or senior ship of your task force if your division flagship be absent.

Marines do not man boats on board ship. On Marine Corps stations, where required, Navy-type boats are assigned, and are manned and maintained by Marine crews, usually under supervision of a boat officer, dock officer, or similarly entitled staff officer. It is a point of pride that, when the Marine Corps operates a Navy boat, we not only observe every sea-going rule and precaution, but do so with man-of-war smartness. A Marine boat ought to outshine every Navy boat in sight, both in appearance and in smart and seamanlike handling.

Be sure a correct copy of the boat and guard-mail schedule is posted in the OOD's booth.

Inspection of Provisions. Unless a supply officer is on board, you, as OOD, inspect for *quantity* provisions delivered on board by a contractor. See that a medical officer, or, in his absence, a hospital corpsman, inspects fresh provisions as to *quality.*

Security. The security of the ship is your most important responsibility. Be zealous and uncompromising to ensure that:

Prescribed security measures are taken.

Men on watch are alert.

Required security reports are received.

Required inspections are made.

Only authorized persons, articles, and materials are admitted into the ship.

Miscellaneous. Write your log and sign it before leaving the quarterdeck or bridge; events are then fresh in mind. Take care that it is neat and complete. Check the details on the left-hand page, for quartermasters sometimes do make mistakes.

Select the circuits for use in passing words on the public-address system. Do not cut in circuits not concerned with the news being disseminated. Require boatswain's mates to adhere exactly to prescribed phraseology.

Keep a close check on your Junior OOD. He can get you into more trouble than any other person in your watch.

Either you or your Junior OOD must inspect all liberty men going ashore. They must be entitled to liberty and be in correct, regulation uniform or, if so authorized, in acceptable civilian clothes. Men in nonregulation uniforms must never be permitted to leave the ship. In embarking liberty parties, embark NCOs and petty officers ahead of nonrated men.

Cultivate the habit of keeping an eye aloft. Watch for signals in the air; for bunting, commission pennant or colors fouled; for absentee pennants incorrectly displayed; for improper smoke; and for Irish pennants.

Avoid unfairness in calling on the respective divisions for working parties; see that such details go to the divisions directly affected. As a Marine—on watch or off—remember that Marines should participate only in working parties directly concerned with the detachment, or in all-hands evolutions (such as ammunitioning, provisioning, and so forth). Otherwise, Marines are *not* part of the general labor pool for ship's work.

Be particular about your own uniform. Officers on watch wear uniform of the day, with gloves and spyglass. Side-arms may be prescribed. Require sailors of your watch to be in neat, regulation uniforms, with shoes well shined, hair cut, and clean shaven.

> NOTE: Marine OODs should always wear Sam Browne belt when on watch (unless pistol and web belt are prescribed). If ship is in the tropics and Dress Blue C is prescribed, wear the belt proper with shoulder strap detached. In either case, unless pistol is required, wear sword-sling with eyelets over the hook.

When leaving the ship, all persons must report to the OOD or his representative, giving their authority. Attend the gangway for everyone coming or going, or have your representative do so.

2220 ■ Officer of the Deck Underway

Since the number of Marine officers who will stand watch underway is limited, and since the subject is one that calls for detailed discussion, space precludes its treatment here. Should you be called on to stand Junior OOD watch underway, or should you aspire ultimately to take the top watch, as some Marines have done in the past, consult *The Watch Officer's Guide*, published by the U. S. Naval Institute. This compact manual is "the bible" for Navy watchstanders.

COURTESY, ETIQUETTE, AND HONORS

2221 ■ Shipboard Courtesy and Etiquette

On many occasions during your career, you will find yourself serving or embarked in naval vessels, on a Navy staff, as part of the ship's Marine detachment, or as a passenger. Because of this, and because, as a Marine officer, you are a member of the Naval Services, you must comply meticulously with the courtesies and customs practiced on board men-of-war.

Ladders and Gangways. The starboard accommodation ladder is reserved for officers; if there are two starboard ladders, one ladder may be designated for flag and general officers. The port ladder is for enlisted men. When a ship is alongside a dock, the officers' gangway usually leads to the quarterdeck; the enlisted men's gangway is forward or aft, as the case may be.

Coming on Board and Leaving a Man-of-War. As a Marine officer, whether on duty or as a visitor, you will always be welcome on board a Navy ship. In fact, you should seize every opportunity to visit each type and class of ship so that you may increase your fund of seagoing knowledge. The Navy is rightly flattered by such visits, and will do everything possible to make your stay instructive as well as socially pleasant. Always pay your respects to the Marine officer when visiting a warship that has a Marine detachment.

If the ship is docked, there will be little complication. If the ship is anchored in the stream, obtain permission from the ship's senior officer present at the landing (or from the boat coxswain, if no officer is on hand) to go out in one of the ship's boats. If no boats are at the landing, ask the shore patrol representative when the next boat is due. If you come off in a motor launch loaded with enlisted passengers, you will probably be taken to the port ladder.

Observe boat etiquette. Defer to seniors in the boat, and introduce yourself.

On reaching the quarterdeck, either from the gangway or an accommodation ladder, halt, face aft (or toward the National Ensign), and salute the Colors. Immediately afterward, render a second, distinct salute to the officer of the deck or junior officer of the deck, and say, "Sir, I request permission to come on board." When it is time to leave the ship, render the same courtesies in reverse order, saying, "Sir, I request permission to leave the ship."

> NOTE: If you are a member of the ship's company, or embarked in any capacity, you report to the OOD as follows:
> *On coming aboard,* "I report my return on board, sir."
> *On leaving the ship,* "I have permission to leave the ship, sir."

Boat Etiquette. When boarding a small boat, juniors embark first and sit forward, leaving the sternsheets for seniors, who embark last. The most senior officer in the boat sits farthest aft, at the centerline, or elsewhere as he wishes. When debarking, officers do so in order of rank. Officers or enlisted men in the boat rise and salute when a senior officer boards or debarks.

When another boat passes close aboard with a senior officer embarked and in view, or when a senior officer passes close aboard on shore, the senior officer and the coxswain in each boat render hand salutes. Seated officers do not rise to salute; coxswains rise unless to do so would be dangerous or impracticable. When a boat is crowded, juniors rise and yield seats to seniors. If there are not enough seats, take the next boat.

Marine officers (although line officers) and officers of the Navy staff corps, when senior in a boat, receive and return salutes and are otherwise accorded the deference due individuals of their seniority, *but* the senior Navy line officer or petty officer in the boat, regardless of how junior he may be, is in charge of the boat and is responsible for its navigation and for the safety of personnel and material embarked. This is provided by *Navy Regulations,* and should be remembered by you as a Marine officer, in case you are inadvertently directed to act as a boat officer.

During Colors, boats underway within sight or hearing lie to, or, if necessary, proceed at slowest safe speed. The coxswain (or boat officer, if embarked) stands and salutes unless dangerous to do so. Other persons embarked remain in place and do not salute.

Shipboard Amenities and Saluting. While in general the amenities and rules for saluting set forth in Chapter 18 apply on board ship, the following special points should be observed.

Fig. 22-2: When embarking, junior officers board first.

Men at work, at games, or at meals, are not required to rise when an officer other than the Captain, or a flag or general officer, or an officer senior to the Captain passes, unless attention is called, or when passageway must be cleared. It is customary for all officers to uncover when entering a sick bay or space in which food is being prepared or served; when a senior officer does so, this indicates that he does not desire men present to be brought to attention.

Juniors give way to seniors in ship's passageways, and particularly when going up and down ladders.

"*Gangway!*" is a command given by anyone who sees an officer or civilian dignitary approaching a gangway, ladder, or passage that is blocked. Never use "Gangway!" except for an officer or senior civilian. For others, "Coming through!" is appropriate. The senior officer, NCO, or petty officer present must clear passage after "Gangway!" has been given.

The ship's Captain, any officer senior to the Captain, all flag and general officers, the executive officer, and inspecting officers are saluted at every meeting except in officers' country, heads, and messing compartments.

Fig. 22–3: When disembarking, senior officers are first to leave the boat.

On the first meeting of the day, salute each officer senior to you; thereafter, salutes are dispensed with except when an officer is directly addressed by a subordinate, or in the cases of senior officers listed above.

Sentries at gangways salute all officers coming on board or leaving the ship. Sentries posted on the topside also salute officers passing close aboard in boats.

When passing honors are being exchanged between men-of-war, or when ruffles and flourishes are sounded on the quarterdeck, all personnel on weather decks, not in formation, come to attention and salute.

Navy (*but not Marine*) formations on board ship are dismissed with the command, "Leave your quarters," whereupon all members of the formation salute, the officer in charge returns the salute, and all hands fall out.

Ship's sentries posted on the dock, when a ship is moored alongside, carry out normal saluting procedures for sentinels ashore.

When Colors go in port and the ship is in port, all hands on weather decks or on the pier (if ship is berthed alongside) face aft and salute.

2222 ■ Display of Personal Flags and Pennants Afloat

On Board Ship. A flag officer or unit commander afloat displays his personal flag or command pennant from his flagship, but not from more than one ship at a time. If two flag officers are embarked in the same ship, only the senior's flag flies. When a civil official who rates a personal flag is embarked for passage in a naval vessel, his flag is flown, but if an officer rating a personal flag or command pennant is also on board, both his flag and that of the civilian are displayed.

In Boats. An officer in command, or chief of staff acting for him, when embarked officially in a naval boat, flies his personal flag or command pennant from the bow. Officers who rate neither display a commission pennant. Officers who rate personal flag or command pennant may display a miniature of such flag or pennant from the vicinity of the coxswain's station when embarked on other than official occasions. Civilian officials display their flags, if any, from the bow, if embarked in a naval boat.

2223 ■ Official Visits and Calls On Board Ship

Official Visits. Insofar as practicable, the same honors and ceremonies are rendered for an official visit afloat as for one ashore. In addition, however, on board ship the compliments mentioned in Section 1820 are added, such as manning the rail and piping and tending the side.

If *Navy Regulations* call for a gun salute on departure of the visitor, the salute is fired when he is clear of the side (to avoid blast), and his flag or pennant (if he rates one) is hauled down with the last gun of the salute.

Official Calls On Board Naval Vessels. The procedure for receiving official callers on board U. S. Navy ships is more formal than ashore. According to the rank of the visitor, and the occasion, the side may be

Fig. 22-4: The coxswain and the senior officer in the boat render hand salutes.

piped; sideboys, guard, and band are paraded; and certain officers attend the side. If you or your ship is on the receiving end of the call, you can find the details (and should check these carefully in advance) in such publications as *The Watch Officer's Guide* or *The Naval Officer's Guide*, both of which will be readily available on board, as well as in *Navy Regulations*. For a table of honors to be rendered for the various military and civil officials of the United States and foreign countries, see Chapter 18.

2224 ■ Housekeeping Afloat

Much advice and information already presented in Chapter 20 applies to Marine detachments on board ship. As you have seen, a man-of-war's Marine detachment is also a deck and gunnery division, and the commanding officer of Marines thus has additional duty as "division officer," a status roughly analogous to that of a company commander ashore. The Marines man and maintain certain guns and fire-control instruments, and have their own part of the ship (where the detachment lives, works, and maintains its headquarters—styled, sailor-fashion, as "the Marine office").

The term, "housekeeping," however, has no exact equivalent afloat. "Ship's work," or " keeping the ship," includes some housekeeping functions, but does not encompass all shoreside connotations of the phrase. In examining "housekeeping" for Marines at sea, it will be helpful to consider some of the similarities and the differences.

Police. "Police," in the Navy sense, means the ship's police force. Under the Chief Police Petty Officer (or "Chief Master-at-Arms," to use the old Navy title), each division, other than the Marines, details a police petty officer ("master-at-arms" or "jimmy-legs") whose job is to act as a kind of military policeman in enforcing good order and ship's regulations. It is a popular misconception that Marines perform this function and act as "the Navy's police force" on board ship. Nothing could be further from the truth.

Within the Marine detachment, however, the term "police" has its normal meaning, as in Section 2001, and the detachment police sergeant performs the duties usually associated with that title anywhere in the Corps.

Subsistence and Mess Management. Like "police" (in the Navy sense), subsistence and mess management are functions of the ship's executive officer. The supply officer and his assistants perform the duties that ashore would fall to the post mess officer as described in Section 2003. In most ships, a central enlisted mess is operated on a cafeteria basis. Among enlisted men, only chief petty officers have a separate mess; the Marine first sergeant and gunnery sergeant are, for messing purposes, considered to rate as chief petty officers.

Messmen are detailed from privates and Pfcs of the Marine detachment on the basis provided by *Navy Regulations*. Never detail an NCO as a messman. If the detachment includes a rated cook, he should be assigned to duty in the ship's galley. *But never lose sight of him as a Marine.*

> NOTE: In recognition of Marines' continuing and important security functions, many ships do not call on the Marine guard for messmen. If you have any latitude in this, bend every effort to prevent detail of a Marine for mess duty.

Clothing and Small Stores. Marine Corps clothing for the detachment is obtained via Sea Duty Supply Officers at San Diego and Norfolk.

Marines may likewise avail themselves—for cash purchases—of the ship's "Clothing and Small Stores," where many items of regulation (Navy) clothing may be bought at considerable saving over shoreside prices—such articles as handkerchiefs, socks, underwear, and so forth. Obviously, the Marine officers and NCOs must prevent Marines from improper wearing of Navy articles of uniform in lieu of prescribed Marine items of similar type.

Pay. Pay day for the crew (including Marines) is held on the 15th and 30th of each month. When Marines are paid, the first sergeant and clerk assist, and the junior Marine officer may be called on for his share of duty among the other junior officers as a pay call witnessing officer, just as he would on shore. Whenever the paymaster picks up or pays in cash, the detachment provides an armed escort.

Ship's Store. The ship's store is the seagoing equivalent of the post exchange ashore. The ship's store is operated by the supply officer. It stocks stationery, candy, toilet articles, insignia, and the usual selection of post exchange supplies, including, in most ships, "pogey bait" and soda-fountain ("gedunk") delicacies such as are found ashore.

Ship's Welfare and Recreation. On board ship, Welfare and Recreation embraces the gamut of activities associated with Special Services on the beach (see Section 2013). Both as individuals and as a division, Marines take part in the ship's athletic and recreation programs. A ship's *special services officer* administers the funds provided for these purposes. Be sure the Marine detachment gets its share.

2225 ■ Shipboard Cleanliness and Upkeep

As its "part of the ship," the Marine detachment usually has a berthing compartment, a storeroom, an office, several guns, a gunnery control station, and adjacent topside deck space. The cleanliness and upkeep of these spaces and structures are the responsibility of the Marine detachment's CO in his capacity as a division officer. Each division performs its own minor repairs; more extensive repair, when needed, is the job of the shipfitters and "repair gangs," based on "work requests" submitted by the division officer (detachment commander).

As ashore, your detachment will have a police sergeant. Instead of his police shed, though, he will have a *gear locker*. In lieu of a police gang—career men in their field—the police sergeant will levy on every Marine in the detachment to keep the ship's "Marine country" a model space, spic and span, an example to the bluejackets.

In your responsibility for structural upkeep and cleanliness of the ship, you should assign certain "cleaning stations" to your NCOs, and, if you are in command, to your junior officer(s).

> NOTE: On board some ships, the immediate approaches to the Captain's cabin (or flag quarters if an admiral is embarked) are assigned as a Marine cleaning station, and should be a source of special pride and solicitude by the detachment. Marines get this duty because the Captain *knows* the job will be well done.

2226 ■ Navy Property Accounting

Much of the equipment and supplies in the hands of the Marine detachment will be Navy, rather than Marine Corps property, and thus must be handled and accounted for under the Navy property system. In this system, all items are divided into functional accounting categories, or "titles." Those with which you will be concerned are:

Equipage (Title B). Title B covers nonexpendable equipment on board ship. Anything classed as "Title B" must be carefully safeguarded, inventoried, and covered by receipts and records. The ship's supply officer keeps records on Title B items on *Equipage Stock Card and Custody Records*, which serve the same purpose as consolidated memorandum receipts ashore. The Marine detachment CO signs for his Title B gear on this form and keeps a duplicate in his own files. In turn, responsibility for Title B items is subdivided within the detachment by means of similar forms. As equipage is received or expended, the respective custody cards must be kept up to date.

Title B gear is inventoried at least annually, and also whenever the ship's supply officer or other responsible custodian changes. By Navy custom, routine inventory usually takes place during the third quarter of the fiscal year. An incoming custodian must complete his Title B inventory within 20 days of assuming actual custody; shortages or unserviceable items must then be reported and recommended for survey.

Shipboard control of equipage is a prime responsibility, which the Navy takes very seriously. Just as on the beach, ability to control valuable property, both in fact and on the record, is rightly considered a significant index of your military efficiency.

Operating Expenses of Ships in Commission (Title C). Although a multitude of maintenance and running expenses (even crew's pay and rations) is chargeable to Title C, the Title C items that concern the Marine detachment consist mainly of expendable supplies issued for cleaning, repair, and administration. Anything classed under Title C is expended when issued. Examples are: stationery, toilet paper, chipping hammers, swabs, and paint. Ordinarily, the ship's first lieutenant and gunnery officer establish allowance lists for each division, covering Title C material, which is then drawn once a week, usually on Monday morning. Then it is up to the division concerned (the Marine detachment, in our case) to see that the supplies are not carelessly expended or wasted. If for some unforeseen reason, the routine apportionment of consumables runs out, the division officer can ordinarily get more by pleading his case in a special written request to his head of department. Needless to say, this situation should be avoided.

HINTS FOR SEAGOING MARINE OFFICERS

In addition to Marine duties, seagoing Marine officers perform the usual duties of a division officer. Don't be too proud to learn from the Navy. Make it your goal to be as salty (in a military way) as any of your messmates. Never lean on the lifelines; always know which is the lee side; learn what the different boatswains' calls mean. Make yourself into a true seagoing man-of-warsman.

Know which NCO is responsible for each cleaning station; know who your gun-captains and key members of gun-crews are.

Keep a sharp eye on the corners, out-of-way areas, and tops of angle irons and beams, under lockers, and behind and under gear. The center of the deck takes care of itself.

Give special attention to gaskets, knife-edges and dogs on hatches, scuttles, ports, and doorways. The condition of these items is a direct measure of the watertight integrity of the ship.

Timely, energetic scrubbing prevents wasting paint to cover up dirt. Paint not only costs money but greatly increases danger from fire.

Know your whole ship; know every fitting, rivet, and detail in your own spaces.

Take a turn around your battery daily with the gunnery sergeant and with the battery gunner's mate. Visit all security spaces daily, with the gunnery sergeant and sergeant of the guard.

Ship's detachments are justly noted for the splendid appearance of rifles. By applying several coats of a preparation of olive oil and orange shellac to previously scraped rifle stocks and bayonet grips, the grain of the wood can be brought out in a beautiful glosslike finish. This work should preferably be done by a single individual, such as the detachment armorer, who can master the technique and mix, and thus get very best results.

Enlisted dress buttons, issued with a very thin layer of gold plate, can be stripped down for polishing by working over with women's nail-polish remover. This gives a button that can be brought to a fine, high polish with jeweler's rouge, Brasso, or any good-quality brightwork polish compound.

Avoid oversupervising your NCOs. This is more difficult on board ship, not only because of the proximity in which officers and men work and live, but also because the Navy expects junior officers to concern themselves with many details that, in the Marine Corps, are entrusted to noncommissioned officers. Do not spoil NCOs—and also pick up bad supervisory habits—by breathing down their necks at every turn. Tell your people the "what"; leave the "how" to them.

In most ships, custom requires division officers to be in their parts of the ship by 0800 each working day. Your presence shows interest in the work in progress. When anything important is going on in your part of the ship, be there.

If on board a carrier, be exceedingly careful on the flight deck and be sure your Marines are well schooled as to safety precautions. Sad but true, while pilots are the most likely to sustain injury about the flight deck, the second most likely are the Marines.

Be scrupulously neat and clean. A self-respecting officer never appears unshaven before his shipmates and his men. If you find one of your men unshaven, make him shave immediately.

Salute admiral, captain, executive officer, and any other officer of the rank of commander or above, whenever you meet them about the ship, except when standing watch on the bridge. Salute whenever making reports. Salute all officers senior to you when you meet them for the first time during the day, and give them a cheerful, "Good morning, sir."

Observe the local rule for permission to leave the ship. In some ships, you must obtain permission not only from the Marine officer but from the head of department also; in other ships permission is assumed to be granted after working hours, if your work is done. A few ships require one Marine officer on board at all times.

Learn about RHIP ("Rank has its privileges"). Tread safely until experience, higher rank, and increased responsibility bring you greater privileges. Reserve jokes and wisecracks for your contemporaries. Toward seniors, maintain an attitude of respectful friendliness.

Try to get as much accomplished during normal working hours as you can, so that your men get the benefit of their free time. But work as long as a job remains to be done, working hours or not.

Most Navy personnel at all levels have a high regard for Marines, which usually lasts as long as the Marines continue to function at the level of professional competence which the Navy has a right to expect of them. Relationships between Navy and Marines on board ship are therefore generally healthy and rewarding. If you cannot adjust yourself to cooperating and working with those whose methods admittedly differ from our own, you have no business on sea duty.

Ensure that you (and if possible new Marines coming on board) have a full seabag before reporting. Items of uniform are hard to come by at sea in the middle of the Indian Ocean. (In particular, you should have not less than three dress white cap-covers, which soil easily on board ship but must be kept white and spotless.)

If your sword-drill is weak or rusty, you should brush up intensively prior to coming to sea.

Take every opportunity to get your detachment ashore for training, even if no more than close-order drill on a Navy Yard pier. Should your ship be at a base that includes a Marine Barracks, try to work out arrangements to have your detachment take part in weekly parades or reviews, along with the barracks troops. If at bases where the terrain permits, such as Guantanamo Bay, Guam, Panama, or Roosevelt Roads, get permission to land the Marines for a day of hiking and field training. Be alert for opportunities for competitive rifle and pistol shooting with units ashore, with military units in friendly foreign countries, or with other ships. Remember that the IG will visit your detachment once a year and will inspect training, physical fitness, weapons qualifications, close-order-drill, and so on. All the above requirements, and more, must be concurrent with shipboard duties. You cannot afford to coast.

Finally, when on sea duty never neglect to call upon the local Marine commanding officer ashore, or overlook the help you can receive from brothers in arms who wear the Globe and Anchor.

GLOSSARY FOR SEAGOING MARINES

Here you will find a glossary of commonly used shipboard terms and phrases that every Marine should know. By using and understanding these, you will prove to Navy shipmates that Marines can be, and are, just as salty as any sailor.

Abaft: behind or farther aft: astern or toward the stern.
Abeam: at right angles to the centerline of and outside a ship.
Aboard: on or in a vessel. *Close aboard* means near a ship.
Absentee pennant: special pennant flown to indicate absence of commanding officer, admiral, his chief of staff, or officer whose flag is flying (division, squadron, or flotilla commander).
Accommodation ladder: a portable flight of steps down a ship's side.
Adrift: loose from moorings, or out of place.
Aft: in, near, or toward the stern of a vessel.
Afternoon watch: the 1200–1600 watch.
Ahead: forward of the bow.
Ahoy: term used to hail a boat or a ship, as "Boat ahoy!"
All night in: having no night watches.
Aloft: above the ship's uppermost solid structure; overhead or high above.
Alongside: by the side of a ship or pier.
Amidships (or midships): in middle portion of ship, along the line of the keel.
Anchor ball: black shape hoisted in forepart of a ship to show that ship is anchored in a fairway.
Anchor buoy: a small buoy secured by a light line to anchor to indicate position of anchor on bottom.
Anchor detail: group of men who handle ground tackle when the ship is anchoring or getting underway.
Anchor's aweigh: said of an anchor when just clear of the bottom.
Anchor watch: detail of men standing by at night as a readiness precaution while ship is in port.
Astern: toward the stern; an object or vessel that is abaft another vessel or object.
ASW: antisubmarine warfare.
Athwart, athwartships: at right angles to the fore and aft or centerline of a ship. Pronounced "A'thort."
Avast: a command to cease or desist from whatever is being done.
Aweigh: position of an anchor just clear of the bottom.
Backstay: a stay supporting a mast from aft.
Backwash: water thrown aft by turning of ship's propeller.
Ballast: heavy weight in the hold of a vessel to maintain proper stability, trim, or draft. A ship is *in ballast* when she carries no cargo, only ballast.
Barbette: heavily armored cylinder within which turret rotates: extends from upper part of a turret down to the lowest armored deck.

Barge: craft used to haul material, as a coal barge; a power boat used by flag officers, as admiral's barge.
Batten: long strip of steel or wood that wedges the edge of a tarpaulin against the hatch.
Batten down: to cover and fasten down: to close off a hatch or watertight door.
Battle lantern: battery-powered portable electric lights for emergency use.
Battle lights: dim red lights that furnish sufficient light for personnel during darken-ship period.
Beacon: conspicuous mark or structure used to guide ships.
Beam: width; breadth; greatest athwartships width of a vessel.
Bear: to lie in a certain direction from the observer.
Bear a hand: speed up the action; lend a helping hand.
Bearing: direction of an object, expressed in degrees either as *relative* or *true* bearing.
Belay: to cancel an order; to stop; to firmly secure a line.
Below: short for *below decks*; below the main deck.
Bend on: to secure one thing to another, as bend a flag onto a halyard.
Berth: space assigned a vessel for anchoring or mooring.
Bight: middle part of a line as distinguished from the end and the standing part; a single complete turn of line; bend in a river or coastline.
Bilge: lower part of vessel where waste water and seepage collect.
Billet: allotted sleeping space; a man's position in the ship's organization.
Binnacle: large stand used to house a magnetic compass and its fittings.
Bitt: strong iron post on ship's deck for working or fastening lines; almost invariably in pairs.
Bitter end: the free end of a line, wire or chain.
Blinker: lamp or set of lamps, triggered to a telegraph key; used for sending flashing light message.
Block: an item of deck gear made of one or more grooved sheaves, a frame (casing or shell), supporting hooks, eyes or straps; may be metal or wood.
Bluejacket: a Navy enlisted man below the grade of CPO.
Boats: small open or decked-over craft propelled by oars, sails, or some type of engine. This term also applies to larger vessels built to navigate rivers and inland waters; calling a *ship* a *boat* is not good Navy talk.
Boat boom: a boom to which boats secure. It is swung out from the

side when the ship is anchored or moored.

Boat chock: a strong deck fitting that supports one end of a boat that is resting on deck.

Boat fall: rigging used to hoist or lower a ship's boats.

Boat gripe: lashing used at sea to secure against the strongback a boat hanging from the davits and away from the ship's side.

Boathook: wooden staff with metal hook and prod at one end; used to fend off or hold on.

Boat painter: line attached to the stem ringbolt of a boat; used for securing it. Also a short piece of rope secured in the bow of a boat; used for towing or making fast. Not to be confused with the sea-painter, which is a much longer line.

Boat sling: rope or chain sling used for hoisting or lowering larger-size boats with a single davit or crane.

Boat station: allotted place of each person when boat is being lowered.

Boatswain: warrant officer in charge of deck work. Pronounced "bosun."

Boatswain's call: See "boatswain's pipe."

Boatswain's chair: line-secured board on which a man sits as he works aloft or over the side.

Boatswain's locker: compartment where deck gear is stowed.

Boatswain's pipe: small, shrill silver whistle used by boatswain's mate to pass a call or pipe the side. Never say "boatswain's whistle."

Bollard: wooden or iron post on a pier or wharf to which mooring lines are secured.

Boom: projecting spar or pole that provides an outreach for extending the foot of sails, or for mooring boats, handling cargo, and so on. Rigged horizontally or nearly so.

Boot topping: surface of the outside plating of ship or boat's side between light and load lines.

Bow: forward section of a vessel.

Bowline: one of the most used knots; used to make a temporary eye in the end of a line.

Boxing the compass: naming all the compass points and quarter points in their proper order, from north, east, south, through west.

Break: to unfurl a flag with a quick motion. In ship construction, an abrupt change in the fore-and-aft contour of a ship's main deck.

Breaker: a small container for stowing drinking water carried by boats or rafts; a wave that breaks into form against the shore.

Breast line: a mooring line running at right angles from the ship's fore-and-aft line.
Bridge: raised platform from which ship is steered, navigated, and conned; usually located in forward part of the ship.
Brightwork: metalwork that is kept polished rather than painted.
Broad command pennant: personal command pennant of an officer, not a flag officer, commanding a major unit of ships or aircraft.
Broad on the starboard beam or port beam: bearing 090° or 270° relative to the bow of the ship.
Broad on the starboard or port bow: bearing 045° or 315° relative to the bow of the ship.
Broadside: simultaneous firing of all main-battery guns on one side of a warship.
Broadside to: at right angles to the fore-and-aft line of a ship.
Brow: large gangplank leading from a ship to a pier, wharf, or float; usually equipped with rollers on the bottom and hand rails on the side.
Bulwark: raised plating or woodwork running along the side of a vessel above the weather deck. Helps keep decks dry and prevents men and gear from being swept overboard.
Bumboat: small civilian boats used in port to sell merchandise.
Bunker: storage space for fuel.
Buoy: floating marker anchored by a line to the bottom, which by shape and color conveys navigational information; may be lighted or unlighted. Pronounced "boo-ee."
Burgee: swallow-tailed flag.
Burgee command pennant: personal command pennant of an officer not a flag officer, commanding a division of minor war vessels or major subdivision of an aircraft wing.
Cabin: captain's living quarters; covered compartment of a boat.
Call: series of notes played on boatswain's pipe indicating commands.
Calking, caulking: burring or driving up the edges of steel plates along riveted seams to make them watertight; forcing a quantity of sealing material into the seams of a deck or ship's side to make them watertight. Pronounced "kawking."
Camel: large fender float used for keeping vessel off wharf, pier, or quay; usually consists of one or more heavy timbers.
Can buoy: cylindrical, flat-topped metal buoy.
Capstan or capstan head: that part of vertical shaft windlass around which a working line is passed.
Cardinal point: one of the four principal points of the compass—north, east, south, and west.

Cargo net: heavy, square, rope net used for slinging cargo.

Cargo whip: rope or chain used with a boom which is used for handling cargo. One end has a heavy hook; the other end is rove through a block and taken to the winch. Also called *cargo hoist*.

Carry away: to break loose, tear loose, or wash overboard.

Carry on: an order to resume work or duties.

Cast: act of heaving the lead into the sea to determine depth of water; to direct the ship's bow in one direction or another when getting underway.

Cast loose: to let go a line or lines.

Cast off: to throw off; to let go; to unfurl.

Catapult: shipboard mechanism for launching aircraft or drones.

Catwalk: elevated walkway between bridges; commonly found on tankers. Also called *fore and aft bridge, connecting bridge,* and *monkey bridge*.

Centerline: imaginary line running from ship's bow to stern.

Chafing gear: guard of canvas or rope around spars, hawsers, chocks, or rigging to prevent chafing.

Chain locker: compartment in which chain cable is stowed.

Chains: platform or a general area on either side of forward part of a ship where leadsman stands as he takes soundings.

Chain stopper: short length of chain fitted with a pelican hook and secured to an eyebolt on the forecastle; used for quickly letting go the anchor or for securing the anchor in stowed position.

Charley Noble: galley smokepipe.

Chart: nautical map used as an aid to navigation.

Charthouse or chartroom: compartment on or near the bridge for handling and stowage of navigational equipment.

Check: to slack off slowly; to stop a vessel's way gradually by a line fastened to some fixed object or to an anchor on the bottom; to ease off a line a little, especially with a view to reducing the tension; to stop or regulate the motion, as of a cable when it is running out too fast.

Chipping hammer: small hammer with a sharp peen and face set at right angles to each other; used for chipping and scaling metal surfaces.

Chock: steel deck member, either oval or U-shaped, through which mooring lines are passed. Usually paired off with bitts.

Chockablock: completely full; full to the top.

Chronometer: an especially accurate timepiece, set to Greenwich time; used for navigation.

Clamp down: going over a deck with damp swabs; a lesser form of swabbing down.

Cleat: a small deck fitting of metal with horns; used for securing lines; also called *belaying cleat*. Short piece of wood nailed to brow or gangplank to give surer footing.

Clinometer: bridge and engine-room instrument that indicates amount of a ship's roll or degree of list.

Close aboard: nearby.

Close up: a flag or pennant is close up when it is all the way up on its halyard (equals "Two-blocked").

Clothes stop: small cotton lanyard used for fastening clothes to a line after washing them, or for securing clothes that are rolled up.

Clove hitch: a knot much used for fastening a line to a spar or stanchion. A clove hitch will not slip.

Coaming: raised framework around deck or bulkhead openings and cockpits of open boats to prevent entry of water.

Cockpit: well or sunken space in a boat for the use of the boat crew or passengers; the pilot's compartment in an airplane.

Coil: laying down line in circular turns, usually one turn atop the other.

Collision mat: a mat used to temporarily close a hole in a ship's hull below the waterline.

Commission pennant: long, thin, seven-star pennant flown by a ship to indicate that the ship is commissioned in the U. S. Navy.

Companionway: set of steps or ladders leading from one deck to another.

Compartment: space enclosed by bulkheads, deck, and overhead, corresponds to a room in a building.

Condenser: device for converting exhaust steam from engines into water for reuse in the boilers.

Conn: to direct the helmsman as to movement of helm, especially when navigating in narrow channels on heavy traffic.

Conning tower: heavily armored structure just forward of and slightly below the bridge, for conning the ship in battle. Found on larger warships.

Convoy: a number of merchant ships sailing under the escort of warships and patrol craft.

Counter: part of the ship's side at the stern.

Country: the general area occupied by living quarters, such as officers' country, wardroom country, CPO country.

Course: direction steered by a ship.

Coxcombing: a type of fancy ropework used around tiller handles, boat hooks, stanchions.

Coxswain: enlisted man in charge of a boat; usually acts as helmsman. Pronounced "coksun."

CPO: chief petty officer.
Cradle: a stowage rest for a ship's boat.
Crossing the line: crossing the earth's equator.
Crosstree: superstructure member at top of a low mast or between two such masts; runs athwartships.
Crow's nest: lookout's stand high on a mast or crosstrees.
Cut of the jib: general appearance of a vessel or of a person.
Cutwater: forward edge of the stem at and below the waterline.
Damage control: measures necessary to keep ship afloat, fighting, and in operating condition.
Davit: shipboard crane that can be swung out over the side; used for hoisting and lowering boats and weights. Often found in pairs. Pronounced "day-vit."
Davy Jones's locker: the bottom of the sea.
Day's duty: tour of duty on shipboard lasting 24 hours.
Dead ahead: directly ahead of the ship's bow, bearing 000° relative.
Dead in the water: said of an underway ship that is making neither headway nor sternway.
Deadlight (ventilating deadlight): an arrangement of baffles to permit air while preventing the passage of light. Usually seen on Navy ships as a circular device that fits into ports.
Dead reckoning: navigator's estimate of ship's position from the course steered and the distance run.
Deck gang: men of the ship's deck and gunnery department; all the deck force.
Deckhand: seaman of the deck department.
Deckhouse: structure built on an upper or weather deck; it does not extend over the full breadth of the ship. Deckhouses are typical of smaller vessels.
Deck seamanship: branch of seamanship embracing the practical side, from the simplest rudiments of marlinespike seamanship up to navigation; includes small-boat handling, ground tackle, steering, heaving the lead, signaling, etc.
Deep: the distance in fathoms between two successive marks on a lead line, as "By the deep, four."
Deeps: in a lead line, the fathoms which are not marked on the line.
Deep six: a term meaning to dispose of by throwing over the side.
Degaussing gear: electrical gear which sets up neutralizing magnetic fields to protect the ship from magnetic-action mines. Pronounced "de-gow'sing."
Depth charge: explosive charge used against submarines.
Derelict: abandoned vessel at sea, still afloat.

Dinghy: small, handy boat, 16 to 20 feet in length, propelled either by oars or sail.
Dip: lowering a flag part way in salute or in answer, and hoisting it again. A flag is "at the dip" when it is flown at about two-thirds the height of the halyards.
Director: electro-mechanical device for directing and controlling gunfire.
Displacement: weight of water displaced by a ship.
Ditty bag, ditty box: small container used by sailors for stowage of personal articles or toilet articles.
Division: in an organization of ship or plane groups, the unit between sections and squadrons. In shipboard organization, a number of men and officers grouped together for command purposes.
Dock: artificial basin for ships, fitted with gates to keep in or shut out water; water area between piers.
Dog: small, bent metal fitting used to secure watertight doors, hatch covers, scuttles, etc.
Dogwatch: one of the two-hour watches from 1600 to 2000.
Doldrums: areas on both sides of the equator where light and variable breezes blow.
Dolphin: cluster of piles for mooring.
Dory: small, double-ended, flat-bottomed pulling boat, used chiefly by fishermen.
Double-bottoms: watertight subdivisions of ship, next to the keel and between outer and inner bottoms.
Double up: to increase the number of ship-to-pier-to-ship turns of a mooring line.
Downhaul: line or wire that pulls something downward.
Draft: depth of water from the surface to the ship's keel: a detail of men.
Draft marks: numeral figures on either side of the stem and sternpost, used to indicate the amount of the ship's draft.
Dressing ship: to display the national ensign at all mastheads and the flagstaff; *full dressing* further requires a rainbow of flags bow to stern over the mastheads.
Drift lead: sounding lead and line dropped over side of a ship to detect dragging of the vessel.
Dunnage: loose material placed in holds for cargo to rest on, or jammed between the cargo to wedge it.
Ease her: a command to reduce the amount of rudder or helm.
Ease off: to ease a line; slacken it when taut.
Ebb tide: tide falling or flowing out.

Engine-order telegraph: signaling gear for transmitting speed and direction orders from bridge to engine room.

Ensign: colors, national flag. Also, junior commissioned officer in the Navy. Pronounced "en'sin."

Even keel: floating level; no list.

Eyes: foremost part of weatherdeck in the bow of the ship.

Fairway: in inland waters, an open channel or midchannel.

Fake down: Coiling down a line in long flat bights so that each run of line overlaps the one underneath and makes the line clear for running.

Fall: entire length of rope in a tackle; the end secured to the block is called the *standing part*; the opposite part, the *hauling part*. Also, the line used to lower and hoist a boat.

False keel: thin covering secured to lower side of main keel of ships; affords more protection.

Fancywork: intricate, symmetrical rope work used for decorative purposes.

Fantail: main deck section in the after part of a flush-deck ship.

Fast: snugly secured; said of a line when it is fastened securely.

Fathom: a six-foot unit of length.

Fender: canvas, wood, rope gear or old rubber tire used over the side to protect a ship from chafing when alongside a pier or another ship.

Fend off: to push away; pushing away from a pier or another ship when coming alongside, to prevent damage or chafing.

Fid: a wooden marlinspike used in separating the strands for splicing.

Fire control: shipboard system of directing and controlling gunfire or torpedo fire.

Fire control tower: may be either a separate structure or a part of the conning tower containing fire control equipment; typical of major warships.

Fire main: system of pipes that furnish water to fireplugs.

First lieutenant: officer in charge of cleanliness and general upkeep of a ship or shore station. This is a duty, not a rank.

First watch: the 2000–2400 watch.

Fix: determination of a ship's position by using one or more navigational methods.

Flag bag: container for stowage of signal flags and pennants; rigged with different slots to take the flags' snaps and rings.

Flag officer: an officer of the rank of commodore or above; so called because he is entitled to fly his personal flag, which, by stars, indicates his rank.

Flagstaff: small vertical spar at the stern on which the ensign is hoisted when the ship is moored or anchored.
Flank speed: a certain prescribed speed increase over standard speed; faster than full speed, but less than emergency full speed.
Flare: outward and upward curving sweep of a ship's bow; outward curve of the side from waterline to deck level. Also, a blaze to illuminate or attract attention.
Flat-top: slang for aircraft carrier.
Fleet: organization of ships and aircraft under one commander; normally includes all types of ships and aircraft necessary for major operations. Also to draw the blocks of a tackle apart.
Flemish: to coil line flat on deck in a clockwise direction, each fake outside the other, all laid snugly side by side; begins in the middle and works outward.
Flight deck: deck on aircraft carrier.
Flood tide: tide rising or flowing toward land.
Flotsam: floating wreckage. See "jetsam."
Fluke: flat end of an anchor which bites into the ground.
Flush deck: continuous upper deck extending from side to side and from bow to stern.
Flying bridge: a bridge extending out from the control tower.
Fore and aft: running in the direction of the keel.
Forecastle: upper deck in the forward part of the ship. Pronounced "foke'-sul"; abbreviated fo'c'sle.
Forecastle deck: partial deck over the main deck at the bow.
Foremast: on a two-masted ship, the first mast abaft the bow.
Forenoon watch: the 0800–1200 watch.
Forestay: a stay supporting a mast from forward.
Forward: toward the bow; opposite of *aft*.
Foul: jammed; not clear for running.
Foul anchor: anchor with its cable twisted around it.
Founder: to sink.
Foxtail: small hand brush.
Frame: ribs of a vessel; numbered from forward to aft, they serve as reference points.
Freeboard: height of a ship's sides from waterline to main deck.
Full speed: a prescribed speed that is greater than standard speed but less than flank speed.
Funnel: ship's smokestack; stack.
Furl: fathering up and securing a sail or awning; opposite of *spread*.
Gaff: small spar abaft the mainmast from which the national ensign is flown when the ship is underway.

Gale: a wind between a strong breeze and a storm; wind force 28 to 55 knots.

Gangplank: See "brow."

Gangway: opening in the bulwarks or the rail of the ship to give entrance; an order to stand aside and get out of the way.

Gear: general term for lines, ropes, blocks, fenders, etc.; personal effects.

General alarm: sound signals used for general quarters and other emergencies.

General quarters: battle stations for all hands.

Gig: one of the ship's boats designated for commanding officer's use.

Gimbals: a pair of rings, one within the other, with axes at right angles to each other; supports the compass and keeps it horizontal despite the ship's motion.

Glass: barometer or quartermaster's spyglass.

Glasses: binoculars.

Go adrift: to break loose.

Grab-rope: a rope secured above a boat boom or gangplank; used to steady oneself.

Granny knot: a knot similar to square knot; does not hold under strain.

Grapnel: small anchor with several arms; used for dragging for lost objects or for anchoring skiffs or dories.

Gratings: wooden or iron openwork covers for hatches, sunken decks, etc.

Gripes: metal fastenings for securing a boat in its cradle; canvas bands fitted with thimbles in their ends and passed from the davit heads over and under the boat for securing for sea.

Grommet: ring of rope formed by a single strand laid three times around; a metal ring set in canvas, cloth, or plastic.

Ground tackle: term referring to all anchor gear.

Gun mount: a gun structure with one to four guns; may be open or enclosed in a steel shield. Enclosed mounts are not as heavily armored as *turrets* and carry no gun larger than 5-inch.

Gunwale: upper edge or rail of a ship or boat's side. Pronounced "gunnel."

Guy: a line used to steady and support a spar or boom in a horizontal or inclined position. See "stay."

Gyrocompass: compass used to determine true directions by means of gyroscopes.

Gyrocompass repeaters: compass cards electrically connected to gyrocompass and repeating the same readings.

Gyropilot: automatic steering device connected to the repeater of a gyrocompass; designed to hold a ship on its course without a helmsman. Also called *automatic steerer, iron mike, iron quartermaster*.

Hail: to address a nearby boat or ship. Also a ship or man is said to *hail from* such and such a home port or home town.

Halfdeck: partial deck below the main deck and above the lowest complete deck.

Half hitch: usually seen as two half hitches; a knot used much for the same purposes as a clove hitch.

Half-mast: position of the ensign when hoisted halfway; usually done in respect to a deceased person.

Halyard or halliard: line used for hoisting flags or sails.

Hand lead: a lead weighing from 7 to 14 pounds, secured to a line and used for measuring the depth of water or for obtaining a sample of the bottom. Pronounced "led."

Hand rope: See "grab-rope."

Handsomely: to ease off a line gradually; to execute something deliberately and carefully, but not necessarily slowly.

Handy billy: small, portable, powerdriven water pump.

Hangfire: gun charge that does not fire immediately upon closing the firing key, but some time later.

Hatch: an opening in the ship's deck, for communication or for handling stores and cargo.

Hawsepipes and hawseholes: the steel castings in the bow through which anchor cables run are hawsepipes; the openings are hawseholes.

Hawser: heavy line, 5 inches or more in circumference, used for heavy work such as towing or mooring.

Headroom: clearance between decks.

Headway: forward motion of a ship.

Heave: to throw or toss; to pull on a line.

Heave away: an order to start heaving on a capstan or windlass, or to pull on a line.

Heave in: an order to haul in a line or the anchor cable.

Heave 'round: to revolve the drum of a winch or windlass so as to pull in a line or anchor cable.

Heave short: an order to heave in on anchor chain until the ship is riding nearly over her anchor.

Heave to: to bring the ship's head into the wind or sea and hold her there by the use of engines and rudder.

Heave the lead: to employ the lead line.

Heaving line: a small line with a weight on one end; weighted end is thrown to another ship or to a pier so that a larger line may be passed.

Heel: to list over.

Helm: the helm proper is the *tiller*, but the term is often used to mean the rudder and the gear for turning it.

Helmsman: the man at the wheel; the man who steers the ship.

High line: line running between ships that are replenishing.

High-line transfer: method of sending men, supplies, etc., from one ship to another while underway.

Hoist: display of signal flags on halyard. The end of a flag or pennant to which the halyard is secured. Also, to raise a piece of cargo.

Hold: space below decks for storage of ballast, cargo, etc.

Holiday: an imperfection or vacant space in an orderly arrangement; spots in painting or cleaning left unfinished.

Hove taut: pulled tight.

Hug: to keep close. A vessel might *hug* the shore.

Hulk: a worn-out and stripped vessel unable to move under her own power.

Hull: framework of a vessel, together with all her decks, deckhouses, inside plating, or planking, but exclusive of masts, rigging, guns, and all superstructure items.

Hull down: said of a distant vessel when only her stack-tops and mast are visible above the horizon.

Idler: member of ship's company who does not stand night watches.

Inboard: toward the ship's centerline.

Inshore: toward land.

Intercardinal points: the four points midway between the cardinal points of the compass: northeast, southeast, southwest, northwest.

Interior communication (I.C.): telephone or communication systems inside a ship.

Irish pennant: unseamanlike, dangling loose end of a line or piece of bunting.

Island: superstructure on an aircraft carrier; contains conning tower, navigation bridge, etc.

Jack: flag similar to the union of the national ensign; flown at the jackstaff when in port; plug for connecting an electrical appliance to a power or phone line.

Jackbox: fitting on a bulkhead into which telephone or power lines are plugged.

Jack-o'-the-dust: enlisted man serving as assistant to the ship's cooks.

Jackstaff: small vertical spar at the bow of a ship from which the jack is flown.

Jacob's ladder: light ladder made of rope or chain with metal or wooden rungs; used over the side and aloft.
Jetsam: goods, cast overboard to lighten a ship in distress. See "flotsam."
Jettison: to throw goods overboard.
Jetty: breakwater built to protect a harbor entrance or river mouth.
Jury rig: makeshift rig of mast and sail or of other gear, as jury anchor, jury rudder; any makeshift device.
Kapok: water-resistent fiber stuff packed into life jackets to make them buoyant.
Keel: backbone of a ship, running from stem to sternpost at the bottom.
Keelhaul: to reprimand severely.
Keelson: timber or steel fabrications bolted on top of a keel to strengthen it.
King post: short mast supporting a boom.
Knife edge: smooth, polished edge of the coaming against which the rubber gaskets of watertight doors and scuttles press when closed; furnishes better watertight integrity.
Knock off: to cease what is being done; to stop work.
Knot: one nautical mile (6080.2 feet) per hour. (Never say "knots per hour." This would be the same as saying "miles per hour per hour.") Also, a tie or fastening formed with rope.
Landfall: first sighting of land at the end of a sea voyage.
Landing party: force of infantry from ship's crew detailed for emergency, riot, or parade duty ashore.
Landlubber: seaman's term for one who has never been to sea.
Lanyard: a line made fast to an article for securing it; for example, a *knife lanyard, bucket lanyard.*
Lash: to tie or secure by turns of line.
Lay: the direction of the twist of strands of a rope.
Lead or sounding lead: weight used for soundings; that is, for measuring the depth of the water.
Lead line: line secured to the lead used for soundings.
Leadsman: seaman detailed to heave the sounding lead.
Lee: direction away from the wind.
Lee helmsman: assistant or relief helmsman.
Leeward: in a lee direction. Pronounced "lu'ard."
Leeway: drift of a vessel to leeward.
Lie off: order to a boat to await word from the OOD to come alongside.
Lie to: said of a vessel when underway with no way on.

Life buoy or life ring: a ring or U-shaped buoy of cork or metal to support a person in the water.
Lifejacket or life preserver: a belt or jacket of buoyant or inflatable material; worn to keep a person afloat.
Lifeline: line secured along the deck to lay hold of in heavy weather; line thrown on board a wreck by a rescue crew; knotted line secured to the span of lifeboat davits for the use of the crew when hoisting and lowering.
Lifelines: lines or metal pipes stretched fore and aft along the weather decks to furnish shipboard personnel safety against falling or being washed overboard.
Liferaft: float constructed either with a metallic tube covered with cork and canvas, or made of balsa wood or other suitable material.
Lighter: small vessel used for loading and unloading ships anchored in harbor.
Lightship: small ship equipped with a distinctive light and anchored near an obstruction to navigation or in shallow water to warn shipping.
Line: seagoing term for rope; the equator.
List: inclination or heeling over of a ship to one side.
Locker: small metal or wooden stowage space; either a chest or closet.
Log: instrument for measuring a ship's speed through the water. Also, a short term for *logbook*.
Logbook: a book containing the official record of a ship's activities and of other pertinent or required data.
Log room: engineers' shipboard record room.
Lookout: seaman assigned duties involving watching and reporting to the OOD any objects of interest; the lookouts are "the eyes of the ship."
Loran (*lo*ng *ra*nge *n*avigation): a navigational system that fixes the position of a ship by measuring the difference in the time of reception of two synchronized radio signals.
Lubber's line: line marked on inner surface of compass bowl to indicate direction of ship's bow.
Lucky bag: locker for stowage of personal gear found adrift.
Magazine: compartment used for stowage of ammunition and explosives.
Main battery: the largest-caliber guns carried by a warship.
Main deck: highest complete deck extending from stem to stern and from side to side.
Mainmast: second mast from bow of a ship that has two or more masts. If a ship has but one mast, that mast is considered the mainmast.

Man-of-war: fighting ship; warship.

Manrope: side rope to a ladder used as a handrail; rope used as a safety line anywhere on deck; rope hanging down on the side of a ship to assist in ascending the ship's side.

Mark: call used in comparing watches, compass readings, or bearings; fathoms in a lead line that are marked. Also, a model or type of a piece of equipment, as Mark XIV torpedo.

Marlinspike: pointed iron instrument used in splicing line or wire.

Marry: placing two lines together, as in hoisting a boat; to sew together temporarily the ends of two lines for rendering through a block.

Meal flag: Echo Flag, which is hoisted from port yardarm of a Navy ship at anchor when crew is at mess, sometimes called "the bean rag."

Mess: to eat; group of men eating together.

Messenger: light line used for hauling over a heavier rope or cable; for example, the messenger is sent over from the ship to the pier by the heaving line and then used to pull the heavy mooring lines across. Also, an enlisted man who runs errands for the OOD.

Messman (or slang, messcook): enlisted man who performs duties in mess hall or galley.

Midshipman: A cadet in training at the U. S. Naval Academy or a Naval ROTC Unit; a naval or Marine officer aspirant.

Midwatch: the 0000–0400 watch.

Misfire: powder charge that fails to fire when the firing key has been closed.

Monkey fist: a knot, with or without a weight enclosed, worked in the end of a heaving line to form a heavy ball to facilitate throwing the line.

Mooring: securing a ship to a pier, buoy, or another ship; or anchoring with two anchors.

Mooring buoy: a large, well-anchored buoy to which one or more ships moor.

Mooring line: one of the lines used for mooring a ship to a pier, wharf, or another ship.

Morning watch: 0400–0800 watch.

Motor launch: large, sturdily-built powerboat used for liberty parties and heavy workloads.

Motor whaleboat: a 26-foot powerboat pointed at both ends.

Nautical mile: 6,080.2 feet, or about a sixth longer than a land mile.

Navy yard: traditional term for naval shipyard.

Nest: two or more vessels moored alongside one another; boat stowage in which one boat nests inside another.

Nothing to the right (left): order to helmsman not to let the ship go to the right (left) of the designated course.

Nun buoy: cone-shaped buoy used to mark channels; it is anchored on the right side, entering from seaward, and is painted red.

Oakum: a calking material made of old, tarred, hemp rope fiber.

Officer of the deck (OOD): the officer on watch in charge of the ship, equivalent to the Officer of the Day ashore.

Officer of the watch: See "watch officer."

Oiler: a tanker—a vessel especially designed to carry fuel oil.

Oil king: petty officer in charge of fuel oil storage.

Oilskins: waterproof clothing.

On the bow: bearing of an object ahead somewhere within 45° to either side of the bow.

On the quarter: bearing of an object somewhere astern of the ship, 45° to either side of the stern.

OOD: officer of the deck.

Orlop: partial deck below the lower deck; also the lowest deck in a ship having four or more decks.

Outboard: toward the side of the vessel, or outside the vessel entirely.

Overhang: projection of ship's bow or stern beyond the stem or sternpost.

Overhaul: to separate the blocks of a tackle; to overtake a vessel; to clear or repair anything for use.

Pad eye: metal eye permanently secured to deck or bulkhead.

Painter: a line in the bow of a boat for towing or making fast.

Paravanes: torpedo-shaped devices towed on either side of ship's bow to deflect and cut moored mines adrift.

Part: to break, snap, or carry away a line.

Pass a line: to carry or send a line to or around an object, or to reeve through and make fast.

Pass the word: to repeat an order or information to all hands.

Passageway: corridor or hallway on a ship.

Patent log: device for measuring ship's speed through the water. See "taffrail log."

Paulin: see "tarpaulin."

Pay out: to increase the scope of anchor cable; to ease off or slack a line.

Peacoat: a short, heavy blue coat worn by enlisted men below the grade of CPO.

Peak: topmost end of the gaff; from this point the ensign is flown while the ship is underway.

Pelican hook: hinged hook held in place by a ring; when the ring is knocked off, the hook swings open.

Pelorus: navigational instrument used in taking bearings; consists of two sight vanes mounted on a hoop revolving about a dumb compass or a gyro repeater.

Pennant: three-sided flag, swallow-tail flag, or four-sided flag that tapers toward the end.

Pier: a harbor structure projecting out into the water with sufficient depth alongside to accommodate vessels.

Pigstick: a small spar that projects above top of mainmast; commission pennants are usually mounted on this.

Pile: pointed spar driven into the bottom and projecting above the surface of water; when driven at the corners of a pier or wharf, they are termed *fender piles*.

Pilot: an expert who comes aboard ships in harbors or dangerous waters to advise the captains as to how the ship should be conned; also a man at the controls of an aircraft.

Pipe down: an order to keep silent; also used to dismiss the crew from an evolution.

Pipe the side: ceremony at the gangway in which sideboys are drawn up and the boatswain's pipe is blown when a high-ranking officer or distinguished visitor comes aboard.

Pitch: the heaving and plunging vertical motion of a vessel at sea.

Pivot point: point in a ship about which she turns.

Plan of the day: schedule of day's routine and events ordered by executive officer; published daily on board ship or at a shore activity.

Plank owner: a person who served on board ship from its commissioning.

Platform deck: partial deck below lowest complete deck; called first, second, etc., from the top where there is more than one.

Plimsoll mark: a mark on the side of merchant ships to indicate allowed loading depths.

Pointer: member of gun crew who controls vertical elevation of a gun in aiming at a target; that is, he positions the gun up and down. See "trainer."

Pollywog: person who has never crossed the Line (the equator).

Poop deck: partial deck at the stern over the main deck.

Port: left side of ship facing forward; a harbor; an opening in the ship's side, such as a *cargo port*. The usual opening in the ship's side for light and air is also a *port*. The glass set in a brass frame that fits against it is called a *port light*.

Preventer: line used for additional safety and to prevent loss of gear under heavy strain or in case of accident.

Protective deck: deck fitted with heaviest protective plating.

Pudding: bulky rope fender attached to a strongback, or to a boat or tug's stem or gunwales.

Punt: rectangular, flat-bottomed boat usually used for painting and other work around waterline of a ship.

Quadrantal correctors or spheres: two iron balls secured at either side of the binnacle; help compensate for ship's magnetic effect on compass. Also known as "the Navigator's balls."

Quarter: that part of ship's side within 45 degrees on either side of the stern.

Quarterdeck: that part of the main (or other) deck reserved for honors and ceremonies and as the station of the OOD in port.

Quay: a wharf; a landing place for receiving and discharging cargo. Pronounced "key."

Radar picket: ship stationed at a distance from the main force for the purpose of picking up by radar the approach of an enemy.

Radio direction finder (RDF): apparatus for taking bearings on the source of radio transmissions.

Rail: top pipe of the lifeline pipes that extend along various outboard sections of weather decks; uppermost edge of a bulwark.

Rail loading: loading a davit- or crane-supported boat while it is swung out and even with the deck.

Rake: angle of a vessel's masts and stacks from the vertical.

Rakish: having a rake to the masts; smart, speedy appearance.

Rate: enlisted rank in the Navy. A rate identifies a man by pay grade or level of advancement, within a *rating*, a rate reflects levels of aptitude, training, experience, knowledge, skill, and responsibility. See "rating."

Rat guard: a sheet metal disk constructed in conical form with a hole in the center and slit from the center to the edge. It is installed over the mooring lines to prevent rats from boarding ship from the shore over the mooring lines.

Rated man: Any petty officer, or noncommissioned officer of Marines, as distinct from seaman and privates.

Rating: seagoing equivalent of an MOS. Men in pay grades E-1, E-2, and E-3 are not considered as possessing ratings.

Ratline: short length of small stuff running horizontally across shrouds; used for a step.

Ready room: compartment on aircraft carriers in which pilots assemble for flight orders.

Reefer: refrigerator vessel for carrying chilled or frozen foodstuffs.

Reeve: to pass the end of a rope through any lead, such as a sheave or fairlead.

Relieving (the watch, the duty, etc.): to take over the duty and responsibilities, as when one sentry or lookout relieves another. Those who relieve are *reliefs*.

Request mast: mast held by captain or executive officer to hear special requests for leave, liberty, etc.

Rig: general description of a ship's upper works; to set up, fit out, or put together.

Rig ship for visitors: word passed as a warning to all hands to have ship and their persons in neat order for expected visitors.

Rigging: general term for all ropes, chains, and gear used for supporting and operating masts, yards, booms, gaffs, and sails. Rigging is of two kinds: *standing rigging*, or lines that support but ordinarily do not move; and *running rigging*, or lines that move to operate equipment.

Riser: a pipe running vertically between decks with branch connections or off-shoots.

Roll: the side-to-side motion of a ship at sea.

Rope: general term for cordage over one inch in diameter. If smaller, it is known as cord, twine, or string; if finer still, as thread or double yarn. It is constructed by twisting fibers or metal wire. The size is designated by the diameter (for wire rope) or by the circumference (for fiber rope). The length is given in fathoms or feet.

Ropeyarn sunday: a time for repairing clothing and other personal gear. (Usually Wednesday afternoon at sea.)

Rudder: a flat, vertical, mobile structure at the stern of a vessel; used to control vessel's heading.

Ruffles: roll of the drum used in rendering honors.

Running lights: lights shown by ship or plane when underway between sunset and sunrise.

Sally ship: evolution in which the crew runs from side to side together to cause the ship to list, used to help free a grounded ship.

Salvage: to save a ship or cargo from danger; to recover a ship or cargo from disaster and wreckage.

Samson post: in small craft, a single bitt amidships or in the bow.

Scope: length of anchor cable out.

Scow: large, open, flat-bottomed boat for transporting sand, gravel, mud, etc.

Screw: the propeller; the rotating, bladed device that propels a vessel through the water.

Scupper: opening in side of ship to carry off water from waterways.

Scuttle: small opening through hatch, deck, or bulkhead to provide access; similar hole in side or bottom of ship; cover for such an opening; to sink a ship intentionally by boring holes in the bottom or by opening seacocks.

Sea chest: sailor's trunk; intake between ship's side and sea valve or seacock.

Seacock: valve in a pipe connected to the sea; a vessel may be flooded by opening the seacocks.

Sea ladder: rope ladder, usually with wooden steps, for use over the side. Also known as a *Jacob's ladder*.

Sea lawyer: enlisted man who likes to argue; usually one who thinks he can twist regulations and standing orders around to favor his personal inclinations.

Sea marker: dye for brightly coloring the water to facilitate search and rescue.

Sea painter: a long line running from well forward on the ship and secured by a toggle over the inboard gunwale in the bow of a boat.

Seaworthy: capable of putting to sea and meeting usual sea conditions.

Second deck: complete deck below the main deck.

Section: a unit of a division or watch.

Secure for sea: extra prescribed lashings on all movable objects.

Seize: to bind with small rope.

Semaphore: code indicated by the position of the arms; hand flags are used to increase readability.

Serving: additional protection over parceling, consisting of continuous round turns of small stuff.

Set: direction of the leeway of a ship or of a tide or current.

Set taut: an order to take in the slack and take a strain on running gear before heaving it in.

Set the course: to give the helmsman the desired course to be steered.

Set the watch: the order to station the first watch.

Shackle: U-shaped piece of iron or steel with eyes in the ends through which a bolt passes to close the U.

Shaft alley: spaces within a ship surrounding the propeller shaft.

Shakedown cruise: cruise of newly commissioned ship to test and adjust all machinery and equipment and to train the crew as a working unit.

Sheer: longitudinal upward curve of a deck; amount by which the deck at the bow is higher than the deck at the stern. Also, a sudden change of course.

Sheer off: to turn suddenly away.

Shellback: man who has crossed the equator and been initiated.

Shell room: compartment for stowage of projectiles.

Shift colors: to shift the ensign from the gaff to the flagstaff upon mooring or anchoring; from the flagstaff to the gaff upon getting underway.
Shift the rudder: an order to swing the rudder an equal distance in the opposite direction.
Ship: a general term for large ocean-going craft or vessels; to enlist *(ship in)* or reenlist, as to *ship over.*
Ship's company: all the officers and men serving in, and attached to, a ship; *all hands.*
Shipshape: neat, orderly.
Shore patrol: Petty and noncommissioned officers detailed to maintain discipline, to aid local police in handling naval personnel on liberty or leave, and to assist naval personnel in difficulties ashore.
Shore up: to prop up.
Short stay: when anchor chain has been hauled in until amount of chain out is only slightly greater than depth of water and ship is riding almost directly over the anchor.
Shroud: side stay of hemp or wire running from masthead to rail to give athwartship support to the mast.
Sideboys: nonrated men manning the side when visiting officers or distinguished visitors come aboard.
Side lights: red and green running lights carried on port and starboard sides respectively.
Single up: to reduce the number of mooring lines out to a pier preparatory to sailing; that is, to leave only one easily cast-off line in each place where mooring lines were doubled up for greater security.
Skids: beams fitted over decks for stowage of heavy boats.
Slack: the part of a line hanging loose; to ease off; state of the tide when there is no horizontal motion. See "stand."
Slings: fitting for hoisting a boat or other heavy lift by crane or boom; consist of a metal ring with four pendants. Two of these pendants are for athwartships steadying lines, the other two shackle to chain bridles permanently bolted to the keel of the boat.
Slip: to let go by unshackling, as an anchor cable; space between two piers; waste motion of a propeller.
Small craft: generally, all vessels less than small-ship size.
Small stuff: small cordage designated by the number of threads (nine-thread, twelve-thread, etc.) or by special names, such as marline, ratline stuff, etc.
Smart: snappy, seamanlike.
Snipes: slang for members of the engineering department.

Snub: to check suddenly.
Sonar: (*s*ound *n*avigation *a*nd *r*anging): device for locating objects under water by emitting vibrations similar to sound and measuring the time taken for these vibrations to bounce back from anything in their path.
Sound: to measure depth of water by means of a lead line. Also, to measure the depth of liquids in oil tanks, voids, blisters, and other compartment or tanks.
Sound-powered phone: shipboard telephone powered by voice alone.
Spanner: tool for coupling hoses.
Spar: steel or wood pole serving as a mast, boom, gaff, pile, etc.
Spar buoy: long, thin, wooden spar used to mark channels.
Speed cone: cone-shaped, bright-yellow signal used when steaming in formation to indicate engine speeds.
Spitkit: derisive term for small, unseaworthy vessel.
Splice: to join two lines by tucking the strands of each into the other.
Splinter screen: protective plating around a gun mount.
Spring: mooring line leading at an angle of about 45° off centerline of vessel; to turn a vessel with a line.
Squall: sudden gust of wind.
Square away: to get things settled down or in order.
Squilgee: drier for wooden decks made of a flat piece of wood with a rubber blade and a long wooden handle. Pronounced "squeegee."
Stack: ship's smoke pipe. See "funnel."
Stack cover: canvas secured over the top of a stack when it is not in use.
Stadimeter: instrument for measuring distance from an object.
Stage: platform rigged over ship's side for painting or repair work.
Stanchion: wood or metal upright used as a support.
Stand by: preparatory order meaning "Get ready," or "Prepare to."
Standard speed: speed set as basic speed by officer in command of a unit.
Starboard: right side of a ship looking forward.
Stateroom: officer's shipboard bedroom.
Station keeping: the art of keeping a ship in her proper position in a formation of ships.
Stay: piece of rigging, either wire or fiber, used to give fore-and-aft support to a mast. See "guy."
Steady: order to steersman to hold ship on course.
Steerage way: slowest speed at which a ship can be steered.
Stem: upright post or bar at most forward part of the bow of a ship or boat. It may be a casting, forging welding, or made of wood.

Stern: after part of a ship.
Stern fast: line used to secure a boat's stern.
Stern sheets: space in a boat abaft the after thwart.
Sternway: backward movement of a ship.
Stopper: short length of rope or chain firmly secured at one end; used in securing or checking a running line.
Stove: broken in; crushed in.
Stow: to put gear in its proper place.
Strip ship: to prepare ship for battle action by getting rid of any unnecessary gear.
Strongback: spar lashed to a pair of boat davits; acts as a spreader for the davits and provides a brace for more secure stowage of lifeboat at sea.
Superstructure: all equipment and fittings, except armament, extending above hull.
Superstructure deck: partial deck higher than the main, forecastle, and poop deck, and not extending to the ship's sides.
Swab: a rope or yarn mop.
Swamp: to sink by filling with water.
Sweepers: men who use brooms in cleaning ship when "clean sweep down" is ordered.
Swing ship: moving the ship through the compass points to check the magnetic compass on different headings and make up a deviation table.
Swivel: metal link with an eye at one end, fitted to revolve freely and thus keep turns out of a chain.
Tackle: arrangement of ropes and blocks to give mechanical advantage; to purchase, that is, a rig of lines and pulleys to increase available hauling force. Pronounced "take-el."
Taffrail: a rail at the stern of a ship.
Take a turn: to pass a turn around a cleat, bitts, or bollard and hold on.
Talker: man who handles a sound-powered phone during drills or combat.
Tarpaulin: heavy canvas used as protective covering.
Taut: with no slack. Also, strict as to discipline.
Tend: to man; direction and cable leads when ship is anchored.
Tender: an auxiliary vessel that supplies and repairs ships or aircraft.
Thwart: crosspiece used as a seat for a boat. Pronounced "thort."
Tide: the vertical rise and fall of the sea caused by gravitational effect of sun and moon.

Tiller: short handle of metal or wood used to turn a boat's rudder.
Toggle: wooden or metal pin in a becket for rapid release.
Tompion: plug placed in muzzle of gun to keep dampness and foreign objects out. Pronounced "tompkin."
Top: platform at top of mast; to *top a boom* is to lift up its end.
Topside or topsides: above decks.
Tow: to pull through the water; vessels so towed. The usual towing vessels in Navy talk are *tugs*, not *towboats*.
Track: path of a vessel or aircraft.
Trades: generally steady winds of the tropics that blow toward the equator, NE in the northern hemisphere, and SE in the southern.
Train: to traverse a gun horizontally onto a target.
Trainer: gun crew member who controls horizontal movement of gun in aiming it at a target.
Transom: athwartship piece bolted to sternpost; planking across stern of square-sterned boat.
Trice up: to hitch up or hook up, such as trice up a shipboard bunk.
Trick: period of time a helmsman is at the wheel, as "to take a trick at the wheel."
Trim: angle to the horizontal at which a ship rides; that is, how level the ship sits in the water; shipshape.
Truck: flat, circular piece secured at top of mast or at top of flagstaff and jackstaff. Also, uppermost part of a mast.
Turn to: an order to begin work.
Turnbuckle: metal appliance with a thread and screw capable of being set taut or slacked and used for setting up standing rigging.
Turret: heavily armored housing containing a grouping of main battery guns. It extends downward through decks and includes ammunition handling rooms and hoists. See "gun mount."
Two-blocked: when two blocks of a tackle have been drawn as closely together as possible. The official term is now *Close-up*.
Unbend: to cast adrift or to untie.
Uncover: to remove headgear.
Underway: a ship is underway when not at anchor, made fast to the shore, or aground. She need not be actually moving; she is underway as long as she lies free in the water.
Union: inner upper corner of a flag. See "jack."
Unship: to remove from place; to take apart.
Up anchor: the order to weigh anchor and get underway.
Up and down: perpendicular (pertaining to the anchor cable).
Up end: to stand an object on one of its ends.
Upper deck: partial deck amidships above main deck.

Uptake: enclosed truck connecting the boiler(s) to the stack.
Veer: to slack off, to let anchor cable, line, or chain run out by its own weight. Also, when the wind changes direction clockwise or to the right, it is said to veer.
Very well: reply of a senior (or officer) to a junior (or enlisted man) to indicate that information given is understood, or that permission is granted.
Vessel: inclusive term for *ships, small craft,* and *boats.*
Wake: the track left in the water behind a ship.
Wardroom: officers' mess and lounge on board a ship.
Watch: a period of duty, usually of four hours' duration.
Watch and watch: alternating four hours on watch with four hours off watch. Most off-watch periods are of eight to twelve hours' duration.
Watchcap: knitted wool cap worn by enlisted men below CPO in cool or cold weather; canvas cover placed over a stack when not in use.
Watch officer: an officer regularly assigned to duty in charge of a watch or of a portion thereof; for example, the OOD, or the engineering officer of the watch.
Watch, quarter, and station bill: a large chart showing every man's location in the ship's organization and his station in the various shipboard drills.
Water breaker: drinking-water cask or container carried in boats.
Waterline: point to which ship sinks in water; line painted on hull showing point to which ship sinks in water when properly trimmed.
Watertight integrity: system of keeping ship afloat by maintaining watertightness.
Waterway: gutter at side of ship's deck to carry water to scuppers.
Weather: exposed to wind and rain: to the windward, as "to face the weather," or "to weather a storm."
Weather cloth: canvas spread for protection from wind and weather.
Weather deck: portion of main, forecastle, poop, and upper deck exposed to weather.
Weather eye: to keep a weather eye is to be on the alert.
Weigh: to lift the anchor off the bottom.
Well deck: a low weather deck.
Whaleboat: double-ended lifeboat, pulled by oars and/or fitted with sails; when equipped with an engine it is called a *motor whaleboat.*
Wharf: harbor structure alongside which vessels moor. A *wharf* generally is built along the water's edge; a *pier* extends well out into the harbor.
Wheelhouse: pilothouse; the topside compartment where on most

ships the OOD, helmsman, quartermaster of the watch, etc., stand their watches.

Where away?: answering call requesting location of object sighted by lookout.

Whipping: keeping the ends of a rope from unlaying, by wrapping with turns of twine and tucking the ends.

Winch: hoisting engine secured to the deck; used to haul lines by turns around a horizontally driven drum.

Windlass: anchor engine used for heaving in the anchor.

Wind scoop: metal scoop fitted into a port to direct air into the ship for ventilation.

Windward: into the wind; toward the direction from which wind is blowing; opposite of *leeward*.

Wire rope: rope made of wire strands, as distinguished from *fiber rope*; sometimes called a *cable*, in error.

With the sun: in clockwise direction; the proper direction in which to coil a line; right-handed.

Work a ship: to handle ship by means of engines and other gear; for example, to work a ship into a slip using engines, rudder, and lines to docks.

Yard: spar attached at the middle to a mast and running athwartships; used as a support for signal halyards or signal lights; also a place used for shipbuilding and as a repair depot, as Philadelphia Naval Shipyard.

Yardarm: either side of a yard.

Yardarm blinker: signal lights mounted above the end of a yardarm and flashed on and off to send messages.

Yaw: zigzagging motion of a vessel as it is carried off its heading by strong overtaking seas. This motion swings the ship back and forth across the intended course.

NAVY RATING ABBREVIATIONS

The Navy's system of ratings (i.e., petty officer ranks and specialties) with its many different abbreviations may at first appear confusing but is something seagoing Marines must understand. The list below gives the basic abbreviations and their titles. Rank is shown in the following way by suffixes to the basic rating:

Suffix	*Denotes rank as*	*Example*
CM	Master chief PO (E-9)	GMCM
CS	Senior CPO (E-8)	HMCS

Suffix	Denotes rank as	Example
C	Chief, or CPO (E-7)	BMC
1	1st Class PO (E-6)	YN1
2	2nd Class PO (E-5)	DT2
3	3d Class PO (E-4)	QM3

NOTE: Abbreviations for E-3, E-2, and E-1 carry no grade suffix as above.

Abbr	Title
AA	Airman Apprentice
AB	Aviation Boatswain's Mate
AC	Air Controlman
AD	Aviation Machinist's Mate
AE	Aviation Electrician's Mate
AFCM	Master Chief Aircraft Maintenanceman
AG	Aerographer's Mate
AK	Aviation Storekeeper
AM	Aviation Structural Mechanic
AN	Airman
AO	Aviation Ordnanceman
AQ	Aviation Fire Control Technician
AR	Airman Recruit
AS	Aviation Support Equipment Technician
AT	Aviation Electronics Technician
AVCM	Master Chief Avionics Technician
AX	Aviation ASW Technician
AZ	Aviation Maintenance Administrationman
BM	Boatswain's Mate
BR	Boilermaker
BT	Boilerman
BU	Builder
CA	Construction Apprentice
CE	Construction Electrician
CM	Construction Mechanic
CN	Constructionman
CR	Construction Recruit
CS	Commissaryman
CT	Communications Technician
CYN	Communications Yeoman
DA	Dental Technician Apprentice
DC	Damage Controlman
DK	Disbursing Clerk

Abbr	Title
DM	Illustrator Draftsman
DN	Dental Technician (E-3)
DP	Data Processing Technician
DR	Dental Technician Recruit
DS	Data Systems Technician
DT	Dental Technician
EA	Engineering Aid
EM	Electrician's Mate
EN	Engineman
EO	Equipment Operator
EQCM	Master Chief Equipmentman
ET	Electronics Technician
FA	Fireman Apprentice
FN	Fireman
FR	Fireman Recruit
FT	Fire Control Technician
GM	Gunner's Mate
HA	Hospital Corpsman Apprentice
HM	Hospital Corpsman
HN	Hospital Corpsman (E-3)
HR	Hospital Corpsman Recruit
IC	Interior Communications Electrician
IM	Instrumentman
JO	Journalist
LI	Lithographer
MA	Machine Accountant
ML	Molder
MM	Machinist's Mate
MN	Mineman
MR	Machinery Repairman
MT	Missile Technician
MU	Musician
OM	Opticalman
PC	Postal Clerk
PH	Photographer's Mate
PICM	Master Chief Precision Instrumentman
PM	Patternmaker
PN	Personnelman
PR	Aircrew Survival Equipmentman
PT	Photographic Intelligenceman
QM	Quartermaster

Abbr	Title
RD	Radarman
RM	Radioman
SA	Seaman Apprentice
SD	Steward
SF	Shipfitter
SH	Ship's Serviceman
SK	Storekeeper
SM	Signalman
SN	Seaman
SPCM	Master Chief Steam Propulsionman
SR	Seaman Recruit
ST	Sonar Technician
SW	Steelworker
TA	Steward Apprentice
TD	Tradesman
TM	Torpedoman's Mate
TN	Stewardsman
TR	Steward Recruit
UT	Utilitiesman
YN	Yeoman

... that no persons be appointed to offices, or inlisted into said battalions, but such as are good seamen, or so acquainted with maritime affairs as to be able to serve to advantage by sea.
—Resolution of the Continental Congress to raise Marines, 10 November 1775

A ship without Marines is like a coat without buttons.
—David G. Farragut

23

Personal Affairs

Your first responsibilities as a Marine officer are to Corps and country. Hardly second, however, is your responsibility to your family, and your responsibility to organize your affairs so that they can continue undisturbed through all the ups and downs and sudden turnings in a Service career.

Sudden death is only one contingency you must anticipate. What if you are captured? Prematurely retired? Ordered overseas where your family cannot follow?

Reflect on these possibilities. Put your house in order. *Keep* it in order.

> NOTE: Because personal affairs are personal, the advice which follows, though based on much Service experience, is given with diffidence. You can take it or leave it.

YOUR ESTATE

2301 ■ A Balanced Estate

Under the selection system of promotion, most officers retire between 45 and 60, and all must retire by 62. Thus a Service career is shorter than that in any other profession. Today's laws have considerably lessened assurance of adequate retirement income even for the physically retired. If you are over 50 or physically handicapped after retirement, your prospects for employment are poor. For all these reasons you must lose no time in laying the foundations of a balanced estate: an estate which reflects well-planned objectives; which is built on a prudent insurance program, wise investments, and property ownership; and which affords shockproof protection against the unexpected.

2302 ■ Survivor Benefits

Current laws, which took effect in 1957 and have since been amended and extended, provide a greatly improved structure of benefits for eligible survivors of all officers and men who die on active service, or after separation from active service if, in the latter case, death results from a condition incurred or aggravated on active service. These benefits are described in this chapter as follows: Federal Government insurance (Section 2303); Social Security (Section 2304); Survivor Benefit Plan (SBP) (Section 2306); death benefits (including back pay, death gratuity, dependency and indemnity compensation, pension for non–service-connected death, compensation for unused leave), Section 2327; other benefits (Section 2328).

2303 ■ Life Insurance

From the moment you take out life insurance, you create a cash estate of the amount of that policy, an estate whose proceeds are not taxable under the inheritance laws of most states. Life insurance provides an estate while you are in a low-income bracket, and protects the future of your wife and family during your younger years against the occupational hazards of your profession. Finally, life insurance gives a modest return on your investment and can help maintain your standard of living after retirement.

On a straight life policy you pay premiums over a lifetime. It is the most widely used type of insurance because it provides lifetime protection at less cost than any other permanent insurance.

Term insurance, which has no cash value and only gives temporary protection over a certain period, is the cheapest of all. On it, you pay premiums for a set term—usually from one to fifteen years—and its protection ends at the end of that period. Term insurance also goes up (in most cases) at the end of each five years, and becomes virtually prohibitive at age 70.

Limited payment policies provide lifetime protection and contain cash values just as straight life policies do. It is more expensive than straight life, however, because it is designed to give you a paid-up policy at the end of a certain number of years or at a certain age. After that, you are still fully protected, but you no longer have premiums to pay.

Endowment policies are essentially a form of insured savings. This type of policy provides for the payment of its face value to you at a future date elected by you. If you die before that date, the face value goes to your beneficiary. Premiums on endowment policies are higher than on any other insurance, since the emphasis is on savings rather than protection.

Your Life-Insurance Program. A sound life-insurance program varies with income, age and number of your children, your own age, and your probable number of years remaining upon the active list. Periodically, you must overhaul your program, considering carefully the number of years your children will remain dependent; their educational requirements; your outside income; your wife's employment capabilities; your income after retirement; and any experience qualifying you for civil employment.

Consider the needs of your family five, ten, twenty years from now, and your probable income. Think about retirement income, education, and cash for the down payment on a house. Contrary to some opinion, not all endowment policies are bad; neither are short-term policies. Short-term policies cost more in the long run, but you buy them while receiving full pay; when you retire, they are paid up.

But first *by all means join the Navy Mutual Aid Association.* Its death benefit is now (1977) $16,500, and the Association gives more than helpful assistance to beneficiaries of its members. Details on Navy Mutual Aid are given below and in Section 2305.

Only second to Navy Mutual Aid, you should immediately enroll in the *Group Insurance Plan of the Marine Corps Association,* which pays an active-duty death benefit up to $30,000. Details on the MCA Group Benefits Plan are given below.

Try to complete the greater part of your insurance program by the time you have served 10 years as an officer, so that a 20-year plan can be paid out before retirement. As soon as a child is born, take out an education policy for it, either an endowment policy on the life of the child, or on your own life; if on the child, arrange for a policy that guarantees continuation of premiums should you die. Consider an education policy on your own life, either endowment or retirement-income plan. If you do not need the money for children's education, let the policy mature and receive it either as a lump-sum payment or as retirement income at the age of 60 or 65.

At least every five years, review your program. You may well find that changes in income or employment, and a different family situation, indicate modifications. Look through a sound guide on insurance.

Government Life Insurance. There are three types of U. S. Government life insurance, one currently open to all hands on active duty, and two others of earlier vintage, which are still in force for those who hold, and for most but not all of those who have held, policies established under their provisions.

Servicemen's Group Life Insurance (SGLI) provides up to $20,000 term life insurance in addition to any other Government insurance carried, for all persons on active duty from 24 May 1974. Your pay is checked $3.40 per month for this coverage, and you are covered for the full amount until and unless you cancel all or part of your coverage, or until (as the law permits), on separation or retirement, you convert SGLI into VGLI (see below), with the ultimate option of further conversion into permanent protection, *without medical examination*, with one of many fine insurance companies. Death claims are handled by Marine Corps Headquarters and by the commercial company which is the prime insurer.

NOTE: Under no circumstances should you cancel any part of this marvelous insurance protection; nor, however, should you cancel any other insurance you may carry. Instead, you should consider SGLI as extra protection above and beyond your permanent insurance plan.

In addition to SGLI, just described, which is current, the other Government life insurance schemes are these:

1. *Veterans' Group Life Insurance* (VGLI), which took effect on 1 August 1974, is a five-year nonrenewable term policy that has no cash, loan, paid-up, or extended values. VGLI automatically covers Marines who, after the above date, are separated or retired (or reservists released from active duty over 30 days). VGLI takes effect at the end of the 120-day free coverage under SGLI following separation or retirement as above, only if payment for at least the first month of the required premium has been made before the end of the 120-day SGLI free period. Depending on age, monthly VGLI premiums for $20,000 coverage (1977) range from $3.40 to $6.80, and must be paid directly to: Office of Servicemen's Group Life Insurance (OSGLI), 212 Washington Street, Newark, N.J. 07102, which will answer any questions on SGLI/VGLI.

At the end of its five-year term, VGLI may be converted to an individual insurance policy with an eligible company *without medical examination*, as noted above.

2. *United States Government Life Insurance* (USGLI) was established during World War I and continued until World War II.

3. *National Service Life Insurance* (NSLI) was established during World War II and continued until the Korean War.

Neither of the latter two programs is open, except to those previously insured thereunder. If, however, you had a policy, either USGLI or NSLI, which was in force on or before 25 April 1951, which you have allowed to lapse, or surrendered for cash while on active duty before

1 January 1957, you may replace or reinstate this policy at any time while you remain on active duty or within 120 days after active duty terminates. This is a most valuable privilege and should be taken advantage of. As long as such Government insurance can be had, it constitutes by far the cheapest and the best.

In addition to the foregoing, postservice nonparticipating insurance for service-connected disability is still available for regulars and reservists who are disabled while on continuous active duty. It is also open to reservists disabled under certain conditions when on active duty for training.

If eligible on separation from the service, be sure to take advantage of this opportunity. *You cannot get a better insurance bargain*, unless it be SGLI, described above.

On Government insurance matters, refer also to the *Handbook for Retired Marines*, published by Headquarters, U. S. Marine Corps, and frequently updated.

NMA and MCA Insurance. *Navy Mutual Aid (NMA)*. The Navy Mutual Aid Association, Navy Department, Washington, D.C. 20370, is a semiofficial, nonprofit life insurance group for officers, midshipmen, and cadets of the Navy, Marine Corps, and Coast Guard. The death benefit is $16,500 today, with the possibility of further increases in terminal dividend if favorable actuarial trends continue as at present. Rates are minimum. On receipt of notice of your death from the Navy Department, the Association instantly wires $1,000 to your beneficiary, and the rest of the benefit will be paid in accordance with the desires of the beneficiary. NMA will also render help to your surviving dependents in settlement of all other claims. Perhops the most important service of this kind performed by NMA is the assistance provided in securing service-connected compensation for widows, orphans, and dependent parents of deceased members of the Association (see Section 2327). In the event compensation is disallowed initially (as sometimes happens), the Association provides, without charge, competent legal representation before the Veterans Administration Board of Appeals, in order to obtain the best possible settlement.

> MORAL: Every Marine officer should be a member of Navy Mutual Aid. *Attend to this now.*

Marine Corps Association Group Benefit Program (MCAGBP) incorporates plans designed and operated by Marines. It is open to any member of the Marine Corps Association (MCA) on active duty or in the Class II Reserve (see Chapter 9), if under 65. Depending on the plan chosen, and on the age of the insured, benefits may go as high as

$40,000. On reaching either age 60 or 65, you may convert to any permanent life insurance or endowment policy then being issued by the insuring company, an old-line commercial firm. Rates are minimum. One of the attractive features of this plan is that, unlike most group insurance schemes, you may, after separation or retirement, continue your low-cost protection until age 65 so long as you retain MCA membership (another good reason to take the *Gazette*).

To inquire, or better, join, address:

Administrator, MCAGBP
PO Box 812
Red Bank, N.J. 07701

596 **Notes on Life Insurance.** If you hold it, don't let Government insurance lapse. In Service or out, it is the *best* insurance you can get.

Don't take out insurance haphazardly. Follow a program.

Keep your beneficiaries up-to-date. Name contingent beneficiaries. Remember to include the phrase ". . . or to the survivor or survivors thereof," which is all-inclusive. If a beneficiary dies, make a prompt change in beneficiary. Consider the effect if both beneficiary and you should die in the same accident.

Review settlement arrangements with your insurance agent or broker periodically to insure that they are adapted to your present circumstances.

Will your wife have funds immediately after your death? Navy Mutual Aid is splendid for this purpose.

Pay premiums by allotment. Regulations permit indefinite allotments for insurance premiums, which continue after retirement. Payment by allotment prevents lapse of policies.

Don't overload yourself with insurance against remote dangers, but be sure your policies protect against all expected military hazards. Many insurance companies have restrictions as to war, flying hazards as pilot or crew of a military aircraft, and so on.

Do not place all your insurance with any one company. Protection is enhanced by diversification among several good companies.

If you have your policy made payable to your estate, payment of proceeds will be delayed until completion of administration. This could be costly and reduce your estate, while delaying benefits to your dependents.

Although insurance is something you should attend to promptly, be wary before you start signing. Avoid fly-by-night companies and insurance agents who hover about newly commissioned lieutenants. Deal with sound, well-known companies, and select your agent with discrimination. Seek advice from your unit insurance officer.

2304 ■ Social Security

Contributory Social Security coverage is extended to all hands in uniform. Your contribution is made through an automatic checkage of a percentage of your basic pay. In addition to your military retirement pay and any disability compensation paid by the Veterans Administration, Social Security provides monthly income for:

You, on reaching age 65 (or age 62, if you apply to receive the smaller payments due at that time);
Your wife, if you die and your minor children remain in her care;
You, your wife, and children, if you should be totally disabled;
Your wife or widow, if not entitled earlier, on attaining age 60;
Your children under 18, or older if incapable of self-support, after your death or while you are disabled;
Your dependent parents.

The payments for a family group may go as high as the legal maximum even though you have only paid into the Social Security program through taxation of your basic pay for a few years. The amount of your Social Security benefits is determined by your "average monthly wage" during the years you were contributing. The exact amount differs in almost every case and must be worked out. If you are eligible for any or all of the Social Security benefits just mentioned, you must apply for them, since benefits are not paid automatically. You must file an application and it must be in the hands of the Social Security Administration before they can pay you. File immediately when you become eligible, since back payments are limited by law. The local post office can furnish you with the address of the nearest Social Security district office. You should get in touch with them on attaining age 65, or when your wife reaches 60, or at any time if disabled before reaching age 65. When you die, your next of kin should check with the Social Security office to see if there is an entitlement to survivor's insurance.

Regular or Reserve Marines may benefit from Social Security wage credits if the following conditions are met:

You have had at least 90 days' active service (unless discharged for disability).
If you were discharged, conditions must have been other than dishonorable.
If no other periodic benefits based on the same period of military service are being paid by any other Federal agency (retired pay does not count), *except* the Veterans' Administration.

An application must be filed with the Social Security Administration (or with a U. S. Foreign Service officer, if outside the United States) to start payments.

You need not have wage credits for military service added to your record currently. These credits are recorded when a claim is made for retirement or survivor's insurance payments.

Your Social Security number is important for both you and your dependents to know. It can be found in your officer's qualification jacket (OQJ) and must, of course, accompany claims or inquiries. Moreover, it is your basic number for military administrative purposes and is used by the Internal Revenue Service in connection with all your tax returns and related records. You should memorize this number and, as a precaution, record it not only in your notebook, but also in your safe-deposit box and with any emergency papers you keep, such as insurance policies, and so forth.

To assist in computing where you stand under Social Security, the Social Security Administration encourages every insured individual—you—to request a Statement of Wages every three years from the Social Security Administration, Baltimore, Maryland. In this way you can determine whether their records are complete and you are getting credit for all earnings on which Social Security tax has been paid.

You can get full information on these and other matters of interest by applying to the nearest Social Security Administration office, and usually from your unit personal affairs officer. Before retirement, investigate your Social Security rights and credits, and be sure your wife is acquainted with her rights under this law.

2305 ■ Insurance Death Claims

Government Insurance death claims should be submitted as follows:

On Active Duty. Beneficiary will be mailed forms by the Veterans Administration. The VA is notified by Marine Corps Headquarters; no further proof of death is required. Beneficiary must fill out the form and return it.

Retired or Separated from Service. Beneficiary should apply to the nearest VA Regional Office for necessary forms, or see the legal assistance or personal affairs officer at the nearest Navy or Marine Corps station; and if these are inaccessible, the beneficiary should seek assistance from his nearest state or other service organization. Proof of death must be furnished; beneficiary should get certified copies of public death record, coroner's report, death certificate of attending physician, or death certificate of naval hospital. If one of these is not available, beneficiary should obtain an affidavit from persons who viewed the body and knew the deceased when living.

Veterans Administration central and regional offices will be helpful, but because of a backlog of claims, there may be appreciable delay before payment. Navy Mutual Aid, if you are a member, can keep claims moving.

Beneficiary should hold your Government policy until claim is paid; *do not* send it with claim.

On commercial policies, death claims should be submitted as follows:

Beneficiary should consult local representatives of each company, or write direct to the home office. Take the following steps:

Give insured's name.

Give insurance policy numbers.

Request necessary forms to make a death claim.

Return *by certified mail* the accomplished forms, with return receipt requested.

Send a certified copy of death certificate, affidavit of death as described above, or, if death occurred at sea or abroad, a certified copy of the official notification of death.

You yourself should list the commercial insurance companies which insure your life in the Record of Emergency Data (NAVMC 10526) in your qualification jacket; Marine Corps Headquarters will notify the companies in case of death. Most companies accept such notification as proof of death.

Some companies require submission of policy before paying claim.

Your insurance agent will assist your beneficiary with this paperwork.

Navy Mutual Aid. Deaths occurring on active duty or at a naval hospital are reported to Navy Mutual Aid via official channels. In case of death occurring elsewhere or under circumstances wherein an official death message may not have been sent to the Navy Department, the next of kin should notify Navy Mutual Aid via the most rapid means. As previously noted, $1,000 will be immediately transmitted to the beneficiary, and the rest of the benefit will be paid in accordance with the desires of the beneficiary. A letter containing details of what must be done, enclosing all forms to be signed, is sent to your beneficiary at once. Navy Mutual Aid also notifies civilian insurance companies with which you may be insured of your death and the address of your next of kin. The paymaster holding your accounts is notified of your death and the name and address of next of kin, for the purpose of expediting payment of arrears of pay, unused leave compensation, and death gratuity, if eligible.

> NOTE: In general it is unwise for your dependents to put claims for Government benefits in the hands of private attorneys, as this may simply cause unnecessary delay and will certainly entail added expense.

2306 ■ Survivor Benefit Plan (SBP)

This plan provides survivor income up to 55 percent of your retired pay to your surviving spouse and dependent children.

In the past, surviving members of a retired Marine's family often found themselves with little or no income after the retiree's death. The SBP fills that gap; until its enactment, retired pay ended with your death, unless you had elected to take part in the old Retired Servicemen's Family Protection Plan (RSFPP), known in turn originally as the Contingency Option Act (COA).

You will be automatically enrolled in the SBP with maximum coverage when you retire, if you have a spouse or dependent child at retirement time, unless you specifically elect a lesser coverage, or decline participation before the day you become entitled to retired pay.

If you have no spouse or dependent child when you retire, you may either join the plan then by naming someone else as beneficiary, or begin participation later if you acquire a spouse or child after retirement.

The cost of SBP will be checked from your retired pay.

Since the Government pays a substantial part of the SBP costs, your loss of retired pay for participation is considerably lower than if you had purchased the same commercial coverage at retirement time.

When retired pay goes up with the Consumer Price Index (CPI), withholding will increase accordingly.

SBP survivor benefits are based on your retired pay at time of death, or escalated base amount, not that initially received or elected when you began. Also, SBP payments after your death will increase as and whenever retired pay goes up, based on the CPI.

The decision to elect or not to elect the SBP is a big one. It is not a substitute for life insurance (it is taxable as an annuity, you have no equity in the plan, and you cannot cash it in or borrow against it). Whether it is best for you depends on your personal situation. Basically, if you have a long life-expectancy on retirement, are well-fixed, with adequate life insurance and a solid estate, the plan has the disadvantages that you will probably receive reduced retired pay for many years, your widow may remarry (at which time payments cease unless remarriage is after age 60) or die soon (at which time payments cease), and she would receive little benefit. On the other hand, if you are in such poor health that you cannot obtain additional insurance, it could be an excellent means of augmenting insurance and other survivor benefits, at a relatively small cost, and should be carefully pondered.

2307 ■ Other Kinds of Insurance

Automobile Insurance. Your car can cause you much grief if not properly insured.

Auto insurance is available to cover liability for bodily injury, property damage, medical payments, collision or upset, fire and lightning, and transportation, theft, windstorm, earthquake, explosion, hail, or water damage. Liability awards for bodily injury and property damage are very high and tend to rise with no end in sight. Your insurance agent can recommend how much coverage you should carry in each category. Collision or upset coverage is also very expensive; you should therefore take out a "deductible" policy—$100 deductible for each accident, as nearly every accident now costs that much or more. This protects you against heavy damage to, or total loss of, your car.

Many states and virtually all Marine posts require public liability and property damage insurance before you can be issued a license for your car.

Personal Property Insurance. Personal property (clothes, jewelry, silverware, furniture, and so on) should be covered against fire, theft, transportation. Inexpensive "floater" policies for officers are written by many companies. One of the best and least expensive underwriters is United Services Automobile Association, San Antonio, Texas, an association of officers of all Services who mutually insure each other against automobile liabilities and loss incurred to personal effects. Reflect upon this fact: *Government liability for effects destroyed or damaged while in Government custody is limited to $15,000, regardless of how much more you may lose.*

Fire Insurance and Personal Liability Insurance. Take out fire insurance on any house or other real property of your own. Another valuable coverage is personal liability insurance, which protects you against claims for injuries by persons visiting your home, or by servants or workmen, or for damage done by pets, children, wife or self (usually including damage arising out of sports), and damage done to the property of others by such accidents as falling trees, or fire originating on your property. This insurance is inexpensive, but invaluable when trouble comes.

2308 ■ **An Insurance Checklist**

Over and above Social Security coverage, do you have Navy Mutual Aid insurance? Do you have enough commercial life insurance, in addition, to provide an adequate cash estate? Is there an educational policy for each child?

Is your car insured for bodily injury, property damage, and the contingencies listed in Section 2307? Do you carry a personal property "floater" policy? Is your home insured against fire or other disaster?

Do you have the name and address of your insurance agent(s)? Does your beneficiary?

2309 ■ **Real Estate**

While you are young, with small income and few obligations, it is probably better to rent quarters for your family. As you get older and have children, you may agree with many officers that it is advantageous to own your own home. Marines have some advantage in this, as there are a few localities where they may be ordered to duty over and over again. For example, a ground officer would serve most of his stateside duty in the vicinity of Camp Lejeune, Washington-Quantico, and San Diego or Pendleton; an aviator would have maximum service in the vicinity of El Toro, Washington-Quantico, and Cherry Point.

Some officers find it financially advantageous to buy a house where they have duty and are not assigned quarters, then sell or rent when ordered to other shore duty, or else leave the family in their own home while on sea or expeditionary service. While absentee landlord is certainly a difficult role, it is worthwhile to have a home available when you return, and you can approach retirement with something besides canceled checks and rent receipts. When retirement comes, you have an asset that will permit you to buy a house wherever you decide to settle, if the city where you already own a house does not suit you. And if you die on active duty, your family will have a home, or an income from the real estate you leave. Moreover, the Federal Housing Authority (FHA) will in most cases be able to grant you a low-interest long-term mortgage, which will facilitate purchase or construction of the home you want, or purchase of a second home, provided that your first FHA-sponsored home had to be disposed of because of military orders.

Consider carefully the terms of ownership of any real property you buy, before the deed is prepared. Consult the legal assistance officer. Joint ownership or transfer of property to your spouse by deed may offer material advantage to your estate. Leave with your valuable papers a list of your real estate holdings, giving: description and location of all holdings; location of deeds, mortgages, or other papers; original cost, depreciated cost, estimated present value, and present ownership status.

> NOTE: If you do rent housing, you should insist on a "military clause" in the lease. This clause generally states that the tenant may terminate the lease subject to payment of a certain sum, and allows the tenant to end the lease on 30 days' written notice to the landlord, for any one of several reasons, such as permanent change of station, release from active duty, etc. Your legal assistance office can give you desired wording and other details.

2310 ■ Control of Property

An individual may use or control his estate himself or through an agent acting for him under power of attorney. Remote control of one kind or another is often necessary during a service career.

Joint Ownership. To facilitate use of property, and to provide for its disposition on death, an individual may arrange for most property to be held in joint tenancy (with his wife or other beneficiary) with right of survivorship, thus enabling his joint tenant to use and control the property jointly during his lifetime, and after his death to obtain full title as survivor. Property held jointly cannot be disposed of by will if your joint tenant survives you, but it is wise to include provision for its disposal should your joint tenant die before you do.

The advantages of joint tenancy are less expense, less inconvenience, and less time required to dispose of property after a death. But there are also disadvantages. The provisions of the Soldiers' and Sailors' Civil Relief Act exempting military personnel from state or municipal taxation do not apply to your wife's interest in property.

Real estate is not the only property that may be held in joint tenancy. Joint bank accounts and joint ownership of securities, with right of survivorship, have some advantages.

Note that joint ownership of Government savings bonds, if held in safekeeping with the Treasury Department, does not necessarily ensure flexibility. Such a bond cannot be withdrawn by a joint owner unless the purchaser has registered with the safekeeping agency a sample of the co-owner's signature, along with written authority to withdraw the bond.

Automobiles. Joint ownership of the family car also has advantages. Serious loss may result if your wife or another family member drives your individually owned car after your death. The best plan is to hold the title to the car in joint tenancy. The certificate of title and the insurance policy should bear the names of the *joint* owners.

While joint ownership of automobiles is desirable, the wife's interest in this personal property is taxable. Payment of taxes in a state where you live temporarily may be avoided under the Soldiers' and Sailors' Civil Relief Act of 1940, if the car is registered in your name, in your own state of legal residence. A power of attorney to your wife, however, will enable her to transfer title, secure registration, sell, or buy a car during your lifetime.

If, as sometimes happens, you or your wife should have to dispose of your car in a direct sale to some other person, the following guidelines are generally used in figuring fair depreciation on automobiles for resale as used cars. During the first year of ownership, a car loses 25 per cent of its initial purchase price; for each succeeding year, it depreciates 20 per cent of its current value. Taking the case of a new car which cost $4,500, the following would be its sale prices:

During first year after purchase	$3,375
During second year after purchase	$2,700
During third year after purchase	$2,160
During fourth year after purchase	$1,730

Naturally, the foregoing scale is flexible and always depends on the condition of the car, existing market, and so on.

Joint Bank Accounts. If suddenly ordered to expeditionary service, or upon his sudden death, an officer who carries his bank account in his own name only may deprive his dependents temporarily of access to funds at a time when they are most needed. Investigate the advantages of joint accounts—at least during times when you are separated from your family.

2311 ■ Investments

An investment program which includes both Government savings bonds and sound stocks should be one of your undeviating objectives. Beginning with your first paycheck, allot part of your pay to monthly purchase of U. S. savings bonds. As soon as you have a substantial backlog of bonds (at least $1,000, maturity value), begin regular investments in sound common or preferred stocks, mutual funds, or bonds. Do not "play the market" or speculate. Try to build up a portfolio of well-diversified, high-quality stocks or bonds with long records of past dividend payment. Make purchases through a reputable brokerage house—preferably one with an overseas department organized to handle investments for clients who have to be out of the country (your banker can give you a list of reliable brokerage firms). Tell your broker your objectives; capital appreciation or liberal dividends or otherwise, as the case may be. He can then advise you. One investment scheme attractive to most officers is monthly installment-plan purchases of "blue-chip" securities, under which you may allot any given monthly amount in given modest multiples, for installment buying of gilt-edge securities. This allows you to buy high-grade stock without having to save up between purchases, and also gives the advantages of what is known as "dollar averaging," which is the most economical way to accumulate shares of any given issue.

Quinby & Company, Rochester, N.Y. 14603, is an old-line firm which specializes in installment-plan accounts for purchase of top-quality investment-grade securities, and can arrange to handle this via their banking correspondent as a savings allotment (an arrangement few brokerage houses will trouble to effect).

Many economists believe that regular acquisition of high-grade common stocks is more advantageous than investment in bonds or annuities, since stocks increase in value with inflation or with an expanding economy, in contrast to the fixed return of the ultraconservative bonds and annuities.

2312 ■ Borrowing Money and Loans

Since the first few years of a junior officer's career may well be spent paying off debts, it may be wise to underscore Shakespeare's advice, "Neither a borrower nor a lender be . . ." if you can help it. Avoid loansharks, and *equally avoid private loans to brother officers*, however deserving the case may appear; particularly avoid acting as co-signer to any note—this makes you just as liable as the borrower.

Interest Rates. The amount of interest you pay on a loan is a matter of vital concern and often a source of confusion. This is because most lenders charge different rates from those they quote. There are four ways of quoting interest: (1) monthly (½ of 1% per month); (2) add-on rate (6% per year); (3) discount rate (6% per year); (4) simple annual rate (6% per year). Only the last—simple annual interest—is quoted in *true* terms.

To convert quoted rates to simple annual interest, and thus to *true* interest:

Multiply a monthly rate by 12;
Multiply "add-on" or "discount" rate by 2.

Whenever you borrow, ask what kind of interest is being charged—add-on, discount, monthly, or simple. Convert the quoted rate to true annual interest. Then compare interest costs and other charges to determine which lender offers you the terms which are truly best.

If you need credit, investigate the *Navy Federal Credit Union*. Both for borrowing and saving, whether by mail or in person, this nonprofit organization is tailored to the needs of the young officer and its charges are very low, while interest is generous. Navy Federal Credit also covers loans with life insurance at no extra cost, which is of great benefit to a young officer.

Should you find yourself in debt, however, send each creditor something each month, no matter how little. Those who have given you goods on your credit as an officer and a gentleman are entitled to this consideration, which will show that you have not forgotten them and that your obligations will be paid.

2313 ■ Allotments

Upon receipt of orders to sea, expeditionary, or foreign service, allot all pay that you will not actually need. *Be sure that allotments to dependents, or to a bank for dependents, and for insurance, are so described in your pay account.* This is vital, because if a Marine becomes a prisoner or is missing in action, allotments are continued in force under supervision of the Secretary of the Navy, who has authority to start, stop, increase, or decrease allotments for any proper purpose, or as requested by the next of kin.

When a Marine has been missing for 12 months, and no official report of death, or of his being a prisoner or interned, has been received, the Navy Department reviews the case. If the facts do not warrant a finding of death, missing status may be continued. Without allotments, although your right to receive and accumulate pay would not be impaired, your insurance premiums might not be paid and your family might suffer hardship. Section 1338 covers allotments.

2314 ■ Safeguarding Personal Funds

Here are some ways to safeguard personal funds while in combat, overseas, or serving under hazardous conditions:

Allot most of your pay to a dependent or to your bank for support of dependents. Even if an allotment is for your personal use, make it a joint account, naming the dependent upon the allotment request; for example, "Riggs National Bank, Washington, D.C., a/c Sarah Roe." The allotment will continue without interruption if you are missing or become a prisoner.

If a bachelor, allot the bulk of your pay to savings deposit with your bank, or for U. S. savings bonds, depending on which rate of return happens to be more favorable, information your banker can provide. In addition, you may want to register a savings allotment for installment purchase of blue-chip securities (see Section 2311). Draw only enough money to meet immediate expenses.

Do not accumulate cash in the field; pay bills by check, and be sure to save your canceled checks.

Pay can "ride" on the books, but it draws no interest there, and, if field records are lost, it may be difficult to establish claim for more pay than would be normally due after the disbursing officer closes his accounts for a pay period.

In any case, keep a safe-deposit box in your bank for important papers and valuables. Your wife or attorney should have access and keys to the box through joint tenancy or power of attorney.

TAXES

2315 ■ Income Taxes

Federal Income Taxes. The bulk of your U. S. income taxes are collected by withholding, as described in Section 1339. With this information you fill out your return, which is due on 15 April, payable either to the District Director of Internal Revenue in whose area you are stationed, or (if you are out of the U. S.) the Director of International Operations, IRS, Washington, D.C. 20225. Your Director of Internal Revenue can provide the necessary returns and instructions, but it is usual for the paymaster to keep a supply of these, and you can find answers to most of your tax problems in the pay office. Here you should ask for a copy of the pamphlets issued periodically by the Judge Advocate General of the Navy (*Federal Income Tax Information for Service Personnel*), covering tax problems and exemptions applicable to Navy and Marine personnel. In general, members of the Armed Forces are taxed in the same way as civilian wage earners. In time of war or emergency, however, tax exemptions are granted on pay earned in combat zones prescribed by executive order of the President, as well as on pay earned while hospitalized as a result of wounds sustained in action (see Chapter 13). If in a combat zone, you are also ordinarily allowed generous delays without penalty in filing and payment of returns. Up-to-date information on this, too, can be obtained from your disbursing officer.

It goes without saying that you should keep copies of your income tax returns, and the canceled checks with which you paid any taxes above those withheld, as well as checks, receipts, or vouchers covering all deductions claimed.

State Income Taxes. Under the provisions of the Soldiers' and Sailors' Civil Relief Act, you are only liable for income (and other) taxes of the state of which you are a citizen, regardless of whether or not you may be stationed in another state. This law, however, does not exempt retired and retainer pay, the separate income of a spouse or of your family, your income derived from business, investments, and rents, and income earned through outside employment or other sources. Although the provisions and intent of this Act are clear, some states, counties, and municipalities try to impose local taxes on all military personnel on duty within their jurisdiction, even though these may be citizens of other states. Do not pay state or local taxes of this type until you obtain competent verification of your liability from your legal assistance officer (see Section 2319).

2316 ■ Inheritance Taxes

In whatever manner your estate is left to your heirs, certain taxes must be paid to the state and Federal governments—*in cash.*

If you do not provide cash in your estate, the authorities will claim liens against the estate. These "death" taxes are often the largest expenses to an estate, yet many persons give no consideration to their estate tax problem.

After death, nothing can be done but liquidate enough assets to pay the inheritance tax in cash on time. However, you can, by proper planning, arrange your affairs so that substantial tax savings result.

2317 ■ State Inheritance, Estate, and Succession Taxes

Some states impose taxes upon the inheritance of property, upon transfers (without consideration) to take effect after the death of the transferrer, and upon transfers made in contemplation of death. In some states, joint property is taxable upon death of one owner, unless the survivor can prove he was the original owner.

Most states distinguish between separate property and community property; the right to succeed to separate property is taxable, but succession to community property is subject to certain exemptions.

In most states recognizing community property, if either spouse dies without a will, title to the community property vests in the survivor; in any case, half the community property is taxable in most such states.

It is unwise to have insurance payable *to the estate of the insured* as this usually makes the entire proceeds taxable, while in some states up to $50,000 is exempt if payable to named beneficiaries.

Inheritance tax is usually not levied upon the whole estate, but is computed on the share going to each heir. It is a graduated tax, which varies widely according to classification or relationship of the beneficiary.

To prevent escape from inheritance tax by gift before death, many states impose a gift tax, the amount of which depends upon the value of the property transferred and the relationship between receiver and donor.

2318 ■ Federal Gift and Inheritance Taxes

As this edition of *The Guide* goes to press, a wholly new and complex Federal law is likewise taking effect (1977), updating, revising, and substantially altering past laws covering gift and inheritance taxes. It is too early to present the substance or important features of this law in any assured or simplified form. For any estate-planning decisions involving gift or inheritance taxes, consult your lawyer or legal assistance officer.

2319 ■ Soldiers' and Sailors' Civil Relief Act

This legislation, as amended, is designed to relieve officers and men of the Armed Forces from worry over certain civil problems and obligations.

The Civil Relief Act—as it is sometimes short-titled—temporarily suspends enforcement of some civil liabilities of military personnel on active duty, but only if your inability to meet your obligations results from your military status. It does not release you from your obligations, which you still have to meet. However, the law provides machinery for the postponement of legal actions that might be taken against you. The act also provides that in most cases, if you are a legal resident of one state but are stationed in another, you may not be subjected to certain taxes imposed by the latter state. Included are personal property taxes and state income taxes. It applies to auto licenses only if you purchase home state tags. No state tax exemption is applicable to your dependents.

Legal technicalities are frequently involved in any application of the Civil Relief Act, so legal advice is necessary in each case. Remember that the act is designed to provide a shield against hardship, not a device to evade civil liabilities. Information and advice may be obtained from your legal assistance officer.

> NOTE: In the case of debt, invoking the act increases the total amount you owe, since interest charges up to 6 percent are authorized on the postponed payments.

POWERS OF ATTORNEY AND WILLS

2320 ■ Powers of Attorney

A power of attorney authorizes someone else to act in your name in the same manner and extent as you yourself could act. The power permits your representative to do only acts expressly stated therein.

When you are on expeditionary duty "beyond the seas," a power of attorney enables your dependents to carry on your affairs without interruption. Thus, when ordered overseas, you should consider whether to execute a power of attorney covering your affairs generally, and particularly with reference to reimbursements from the United States. If you wish to include authority to transact business in general, consult the legal assistance officer to ensure that the power is legally sufficient, but not needlessly broad.

Whether general or limited, your power of attorney should include phraseology as follows:

> To execute vouchers in my behalf for any and all allowances and reimbursements payable to me by the United States, including, but not restricted to, allowances and reimbursement for transportation of dependents or shipment of household effects as authorized by law or Navy or other regulations; to receive, endorse, and collect the proceeds of checks payable to the order of the undersigned drawn on the Treasurer of the United States for whatever account, and to execute in the name and on behalf of the undersigned, all bonds, indemnities, applications, or other documents, which may be required by law or regulation to secure the issuance of duplicates of such checks, and to give full discharge of same.

Such a power of attorney for Governmental transactions should be executed in the presence of three witnesses and acknowledged before a notary public, or, if outside the United States, by any officer of the United States authorized to administer oaths. For purposes other than those given, each specific power must be mentioned to be effective.

> NOTE: Bear in mind that not everyone needs to execute a power of attorney, as it can be a dangerous tool in the hands of the uninitiated.

2321 ■ Your Will

Considering occupational hazards of our profession, it is important that you have a will. A will simplifies settlement of your estate, reduces expenses, conserves assets, and enables your last wishes to be carried out.

Definitions. A *will* is the legal document by which an individual leaves instructions for disposition of his property after death. A *holographic will* is in the handwriting of the testator.

Settling an estate is a general term used to denote the entire process of collecting assets, filing inventories and accounts, paying claims, distributing assets in accordance with the will or laws of descent and distribution, and filing final accounting with the court.

A *testator* is a person who leaves a will. An *intestate* is a person who dies without a will.

An *executor (executrix)* is appointed by your will to execute its provisions after your death. If you die intestate (see above) the court appoints an *administrator*, who discharges the duties of an executor in settling the estate.

A *codicil* adds to, or qualifies, a will; it revokes the will only to the extent that it is inconsistent therewith. Will and codicil are construed together. A codicil is drawn in the same way as a will. When possible, the best procedure is to make a new will rather than to add a codicil to an old will.

Probating a will is the process of presenting the will for record to the proper authority in the county where the deceased had *legal* residence.

Before Making a Will. Analyze your estate; estimate state and Federal taxes and plan to minimize them. Then, if not sooner, confer with a competent legal adviser. See your legal assistance officer or a member of the local bar for this advice. It is well at this time to discuss, and preferably put in writing, the lawyer's expected charges for settling the estate.

Confer with the trust officer of your bank.

Provide liquid assets in your estate to meet taxes.

List the property that you may not dispose freely—property limited by joint tenancy, community property, your share of trust funds, life insurance already disposed to individuals, and so on—and put down opposite each the amount that is yours to dispose:

Cash
Real Estate
Securities
Life insurance payable to the estate
Business interests
Automobiles in your name
Household furniture and furnishings not community property
Personal effects
Other property

Estimate expenses and debts to be paid from your estate, such as:

Expense of last illness
Funeral expenses
Unpaid household bills
Personal debts
Mortgage or notes payable (just your share, if joint)
Expense of administering estate

Taxes:
    Real Estate
    Estate
    Inheritance

For help with administering expenses and taxes, consult the trust officer of your bank. Federal and state taxes are usually the largest liabilities.

Be careful in making cash bequests; if your estate decreases, you may cut off residual legatees with little or nothing.

For a small estate, it is probably best to make your wife your executrix and for her to make you her executor. *For a large estate or a complicated will*, give consideration to your bank as executor or co-executor. The trust department of a bank has officials trained to handle large or complicated estates; a bank is a continuing institution; a bank is financially responsible; a bank receives no more for its services than an individual. Seek advice from your lawyer and banker.

When your executor has accepted the assignment, discuss your plans, go over your affairs—and take her or him into your confidence.

If your estate is within the amount exempt from a tax, a simple holographic will may be adequate. Your attorney can express your thoughts and wishes in legal language; he will give you advice that will help ensure a well-planned estate.

File the original of your will in your safe-deposit box (or with the Navy Mutual Aid Association, if you belong), and be sure that your wife and your executor know the location and the date of your latest will.

During your lifetime, you will probably make a new will several times. Certain milestones indicate when to reconsider your will and bring it up-to-date:

Change of legal residence
Removal of executor to another state, or his death
Radical change in your estate
Sale of property mentioned in your will
Major changes in tax laws
Marriage, divorce, or remarriage
Birth or death of child
Death of spouse

Check your will on changing legal residence (*not* change of station) to another state, as the provisions of your will may not be legal in that state, and your executor may not be able to function there.

2322 ■ Drawing a Will

A will must be in writing, but no particular form is required if its wording intelligently expresses your intent.

If you dispose of real property, your will must be made in accordance with the law of the state where the real property is located. If you dispose of personal property, your will must be made in accordance with the law of the state in which you are a resident. If both realty and personalty are disposed, your will must conform to the laws of all states among which the property may be distributed.

The number of witnesses required for a will varies from none (for a holographic will) to three; *have three witnesses* to your will, and be safe. Sign in their presence, so that they see you sign and understand that it is your will you are signing. Witnesses, then, in your presence and in the presence of each other, sign the attestation. Be sure that none of the witnesses is mentioned in the will; attestation by an interested witness may invalidate the will, or the witness may lose his legacy. A witness should write opposite his signature his place of residence. Since the authenticity of signatures must be proved in court at the time of probating, care should be exercised to select witnesses who will be available.

A Simple Will. The following short form of a simple, holographic will has been used by many officers:

> All my estate I devise and bequeath to my wife, for her own use and benefit forever, and I hereby appoint her my executrix, without bond, with full power to sell, mortgage, lease, or in any other manner dispose of the whole or any part of my estate.
>
> JOHN WHARTON (Seal)
>
> (Date)
>
> Subscribed, sealed, published, and declared by John Wharton, testator above named, as and for his last will in the presence of each of us, who at his request and in his presence, in the presence of each other, at the same time, have hereto subscribed our names as witnesses this (date) at the City of Washington, in the District of Columbia.
>
> (Signatures and addresses of witnesses, *preferably three in number*. Be sure that the word "Seal" is written in parenthesis after each signature as shown above.)

A Simple Codicil. If possible, write a codicil on the same sheet of paper as the will; if written on a separate sheet, fasten the two together securely. Here is a simple form of codicil:

> I, John Wharton, of Washington, District of Columbia, make this codicil to my last will dated . . . , hereby ratifying said will

in all respects save as changed by this codicil. Whereas, by said will I gave Robert Anderson Wharton, my son, a legacy of $5,000, I now give him a second legacy of $10,000, making $15,000 in all.

(Then follow the testator's signature and seal, the attestative clause, and the witnesses' signature and seals.)

If you decide to modify your will, once it is executed, do not make alterations or interlineations. Consult your legal assistance officer on what to do.

Probate. If executed according to law of your legal domicile, a will made anywhere in the world will be admitted to probate in the jurisdiction of your domicile without question.

Probate establishes the validity of the will and evidences the right of beneficiaries to succeed to title to property in the estate. The place of probate is usually the county and state in which you are domiciled at the time of death; the will must also be probated in any other county and state where you own real property.

DEATH AND BURIAL

2323 ■ Burial Arrangements

When Death Occurs Near a Navy or Marine Activity. When a Marine officer on active duty dies at or near his station, the commanding officer takes charge and arranges for local burial, or for shipment of the body at Government expense. (For burial in Arlington see Section 2325 and *Marine Corps Casualty Procedures Manual.*)

In case of death in a naval hospital, the hospital authorities handle the arrangements.

Death at a Remote Place. When an officer on active duty dies at some distance from a Marine Corps or naval station or hospital, the next of kin should contact the nearest Marine Corps or Navy activity for aid; if unable to contact a local activity, he or she should telegraph the deceased's commanding officer, or the Commandant of the Marine Corps, Washington, D.C., giving the deceased's full name, rank, and service number; the date, place, and cause of his death; and the place where burial is desired. The telegram should request instructions as to burial arrangements, and should give the address to which a reply may be sent. A telephonic report may be made to Marine Corps Headquarters (Code 202 OXford 4-1787 or 4-1788); if neither of the foregoing answers, to OXford 4-2645 (HQMC Duty Officer).

Death in a Naval, Military, or Veterans Hospital (while in Inactive Status). Where death of a veteran or a retired or inactive officer occurs in a naval or military hospital, or in a Veterans Administration facility, the hospital authorities will make necessary arrangements upon request of the next of kin.

> NOTE: The Navy Mutual Aid Association has an outstanding pamphlet *What To Do Immediately in Case of Death*, which is available on request. Obtain a copy of this and keep it with your important papers, as it is a complete checklist of essential information and actions required. Your wife should know of this pamphlet.

2324 ■ Burial Expenses

The expenses of burial or shipment of the remains of Marine officers who die on active duty which are borne by the widow or another individual may be reimbursed within limits allowed by the Government. When the place of death is remote from a Marine Corps or naval station or hospital and it is impossible to obtain instructions from Marine or Navy authorities, the widow may employ a local undertaker, and, if necessary, arrange shipment of the body to the place of burial, but in such cases *she should obtain itemized bills*. The limit of reimbursement in the case of a regular or Reserve officer who dies on active duty in an area where an Armed Forces contract is available and not used, is an amount not to exceed what such authorized services and supplies would have cost the Navy, or where an Armed Forces contract is not available, an amount not to exceed $700; in the case of an honorably discharged, inactive, or retired veteran of any war, the limit of disbursement is $250 (payable by the Veterans Administration). In addition, Social Security will pay a death payment up to $255 if the deceased was covered. This claim must be filed with the nearest Social Security office. A $150 burial-plot allowance will be paid by the Veterans' Administration if burial is in a private cemetery.

Generally, no expenditure is authorized for shipping the remains of an officer who dies on inactive duty. There is no interment expense for burial in a Federal cemetery.

If the remains of an officer who dies on active duty are forwarded to the next of kin for private burial, the expenses of preparation, encasement, and transportation will be borne by the Department of the Navy; and after the body has arrived, the Department will allow a total of not over $700 (for burial in a private cemetery), or $450 (for remains consigned to undertaker prior to burial in a National or Naval Cemetery), or $75 (for remains consigned directly to a National or Naval Cemetery for burial), for one or more of the following items: hearse; transportation of immediate relatives to the cemetery; undertaker; clergyman (fee not to exceed $5); cost of a single grave site in a private cemetery when burial plot is not already owned by relatives of the deceased; digging and closing of grave.

If funeral expenses have been paid, claim for reimbursement should be submitted by letter to the Commandant of the Naval District in which burial was made, stating name and rank of deceased, date and place of burial, and enclosing itemized bills in triplicate, receipted to show by whom payment was made, and the dates when rendered. Each receipted bill should be certified, *"Correct and just; payment not received,"* signed by the person who paid.

If funeral expenses have not been paid, unpaid bills in triplicate are forwarded to the Commandant of the Naval District in which burial was made, as above. Bills should be itemized as described above and have the same certificate, which should be signed by the payee. The bills should bear on their face a statement by the next of kin as follows: *"Services were satisfactorily rendered, and it is requested that payment be made directly to the party rendering the services."* When a bill is signed or receipted in the name of a firm, the signature must appear, thus: *"William Poe Company, by Robert Jones, Secretary (Treasurer)."*

Claims for expenses incurred in an area outside the continental United States may be forwarded to the Bureau of Medicine and Surgery, Washington, D.C. 20390.

(All claims for burial expenses must be submitted within two years after permanent burial or cremation.)

If remains are claimed at the place of death for private burial, and the service of the Government is refused, the next of kin thereby relieves the Government of any obligation for funeral or transportation expenses.

2325 ■ Place of Burial

You may be buried at the place of death, in a private cemetery near your home, or in an open National Cemetery. Leave written instructions as to your choice. If burial is to be in a National Cemetery, your widow or undertaker should telegraph the Superintendent of the National Cemetery selected; if your family lives near the selected National Cemetery, the next of kin may request burial directly from the cemetery superintendent.

Remains are cremated only on written request from the next of kin.

When practicable, should *burial at sea* be desired, arrangements may be initiated either via the Marine Corps Headquarters funeral director, or local naval authorities. However, burial at sea is not a right but a privilege, which it may not be feasible to accord. Expenses incurred for delivery of remains to point of embarkation aboard a naval vessel must be paid by you.

Military Funerals, Arlington National Cemetery. Funeral arrangements for burial in Arlington National Cemetery are made with the Superintendent by the shipping activity, undertaker, or next of kin. Headquarters Marine Corps can make hotel reservations for family and friends; can meet trains or planes; can explain the different types of military funeral; can assist in selection of honorary pallbearers and furnish transportation therefor. Headquarters Marine Corps will also put the widow in touch with the Navy Mutual Aid Association (if deceased was a member) or another organization that can assist in preparing applications for pensions, compensation, or other claims on the Government.

Arlington National Cemetery, with Amphitheater in background, and the mast of USS Maine *in right foreground. Small tombstones are the standard Government marker which will be provided free.*

After the next of kin has received confirmation from Arlington of the request for burial, he or she should telegraph Arlington (with an information copy to Headquarters Marine Corps), stating the number in the funeral party; the means of transportation, and date and hour of arrival; and whether local transportation and hotel reservations are required. The phone number of the Office of the Superintendent, Arlington National Cemetery, is Code 202, OX 5-3175, OX 5-3250, or OX 5-3253.

> NOTE: When you think of burial, remember that Quantico now has a National Cemetery which is, or ought to be, considered a special place for our own in the Corps.

2326 ■ Other Information

Funeral Flag. A United States flag accompanies the remains and may be retained by the family. When death is remote from a naval or Marine activity, the postmaster of the county seat may furnish a flag.

Honors. When practicable, and if requested, full military honors will be provided at the funeral of an officer (see *Marine Corps Manual*). But at cemeteries remote from Marine Corps or naval stations, military honors are not practicable, and relatives must make their own arrangements for funeral services. Veterans organizations usually can assist.

Gravestones. The Government will provide a standard white headstone inscribed with the name, grade, and branch of Service of the deceased. If burial is in a National Cemetery, do not order a private monument until the design, material, and inscription have been approved by the Monument Service, National Cemetery System, Veterans' Administration, Washington, D.C. 20420 (which administers this phase of National Cemeteries). The superintendent of the National Cemetery concerned should be informed of plans for the headstone when you apply for the burial lot, since many National Cemeteries allow private markers only in certain areas. The Veterans' Administration will furnish a Government headstone prepaid to the railroad station nearest to the place of private burial.

Government headstones are provided for officers' dependents buried in National Cemeteries.

Transportation for Family. One person may escort the body of an officer who dies on active duty to the place of burial. The escort may be a relative or friend (not in the Service), with the Government providing transportation in kind. If private burial is desired, a military escort usually accompanies the remains.

Household Effects. The household and personal effects of an officer who dies on active duty may be shipped from his last duty station or place of storage to the place the next of kin selects as home. Arrangements are made in the usual manner with the local supply officer, but shipment must take place within a year of death.

Death Certificates. For a death on inactive duty, the undertaker will obtain as many certificates as may be requested, at a nominal cost ($1 to $2 each, depending on the locality). They are needed for each insurance company, for the will, for each claim, for the Commandant of the Marine Corps, for the pay office carrying your accounts, and for the transfer of each security held in joint ownership. For deaths on active duty, Marine Corps Headquarters furnishes five copies of the official Report of Death, which will serve as a legal death certificate. Additional copies may be obtained on request.

The next of kin should also ask two officers or other friends, who knew the deceased, to identify his remains. These witnesses will then be prepared, if required, to furnish the affidavit of death sometimes demanded by commercial insurance companies.

Burial Privileges for Dependents. If a dependent of any Marine on active duty dies at or while traveling to or from place of active duty, certain Government allowances are payable for transportation of remains. Additionally, the wife, husband, widow, widower, minor child, and (if approved by the Veterans' Administration) unmarried adult child of any Marine or former Marine whose service ended honorably is entitled to burial in a National Cemetery. If this is desired, telegraph the Superintendent of the National Cemetery selected, requesting burial arrangements.

SURVIVOR BENEFITS AND ASSISTANCE

2327 ■ Death Benefits

Back Pay. Pay and allowances to the credit of a deceased are payable to the persons he designated on his Record of Emergency Data to receive them. If he does not make such a designation or if the person designated dies first, this payment is made to his widow, or if he is not survived by a widow, to his children, then his parents, and so on. Headquarters Marine Corps will send the necessary form to the person(s) eligible to receive this payment.

Death Gratuity to Active Personnel. When death occurs on active duty, 6 months' pay (including flight and hazardous duty pay), in an amount not less than $800 nor more than $3,000 is payable to the widow of the deceased, or if he has no widow, to children; if he is not survived by children then to any of his parents or brothers and sisters he designated. Payment to the widow or children is mandatory and is not affected by designations. *This gratuity cannot be checked to liquidate overpayment or any debt to the United States, and is nontaxable.* In most cases, payment by the cognizant commander is proper and authorized, and in fact should be paid within 24 hours after receipt of notification of death. In all other cases, Headquarters Marine Corps will institute claim for death gratuity to the eligible beneficiary upon notification of death of a Marine; therefore, it is not necessary for your next of kin to request this benefit. In case of financial distress, your widow may apply to the nearest Marine command for help. The Navy Relief Society will also help with either a grant or loan (see Section 2330). The death gratuity just described is also paid if death occurs within 120 days after retirement or separation from the Service, provided death is due to disease or injury incurred or aggravated while on active duty.

Compensation for Unused Leave. Your surviving dependent is eligible to claim and receive compensation for any unused leave to your credit, should you die on active duty.

Death Compensations and Pensions. For veterans who die on active duty as a result of disease or injury received in service, the award granted dependents is referred to as "compensation." If an award is based only on the fact of service in a war, it is called a "pension." Pensions may be payable by the Veterans Administration to widows and children of certain retired Marines who die from causes not connected with their active military service. If, for example, you served in World Wars I and II, the Korean War, or Vietnam, your unremarried widow and unmarried children under 18 (21, if attending an approved school), are eligible for pension. So would be any children over 18 if permanently incapable of self-support. If your widow has other income of given categories, it may reduce or terminate her pension during periods when she receives income in disqualifying amounts.

Payment of pension commences as of the day following death of the veteran, if a claim is filed within a year following death; otherwise, it commences from the date of receipt of the application. Claims should be filed with Dependents' Claims Service, Veterans Administration, Washington, D.C. 20420. Considerable investigation is required before a pension claim can be approved, and the following evidence must accompany a claim:

Proof of death (death of a veteran in active service, on the retired list, or in a Government hospital does not need to be proved).

Proof of marriage of the claimant to the veteran (if either had been married previously, proof of death or divorce of the former spouse is required).

Proof of date of birth of children.

Proof of birth of veteran (showing filial relationship, if a parent makes the claim).

If claimant qualifies under more than one rate, the maximum rate is awarded.

> NOTE: If the veteran in question has ever previously filed a VA claim, he has been assigned what is called a " 'C' Number," a number of permanent record which relates to all future VA correspondence or dealings involving the person in question. It is important that a living veteran or next of kin know and record the "C" Number and use it in all VA claims or contacts.

2328 ■ Other Benefits

Hospital and Medical Care. Dependent parents, widows, and children under 21 of deceased Marines are eligible for certain civilian medicare on a cost-sharing basis, and in general for admission to Armed Forces hospitals, and may receive outpatient medical service where such service is available (see Section 805).

Educational Assistance for Children of Marine Corps Personnel. From time to time, the Bureau of Naval Personnel publishes a list of schools, colleges, universities, and other organizations that grant concessions and scholarships to Service children. For this information, write to the Commandant of the Marine Corps (Code MSPA-3).

Navy Relief Educational Loans. The Navy Relief Society (see Section 2330) will lend up to $1,000 per year, *interest-free*, for college or vocational education above high school, or for preparatory work to enter a state or Service academy, to eligible dependents (not over age 23) of active or retired regular Navy or Marine Corps officers or men, and to children of reservists on extended active duty. Information on this valuable privilege may be obtained from the Navy Relief Society, Washington, D.C. 20360.

Dependents' Educational Assistance. Administered by the VA, this program provides up to 36 months' schooling for widows of deceased veterans, wives of living veterans, and children of either (age 18–26) when death or total, permanent disability arose from service. Wives and children of Marines missing in action, POW, or forcibly detained or interned by a foreign power for over 90 days are also eligible.

In certain cases handicapped children may begin special courses as early as age 14. In most cases, a child's eligibility ends with the twenty-sixth birthday.

Generally, eligibility for a wife or widow extends to 30 November 1978, or ten years from the veteran's date of death or total, permanent disability, whichever is later. For wives of MIAs or POWs, eligibility extends to 24 December 1980, or ten years from the date the husband was listed, whichever is later.

Naval Academy Preparatory Scholarships. The Society of Sponsors of the U. S. Navy awards preparatory school scholarships to enable high school seniors who are the children of active, retired, or deceased Marine Corps, Navy, or Coast Guard personnel, to prepare for entrance to the Naval Academy. For information, write Commandant Marine Corps (Code MSPA–3).

Exchange and Commissary Privileges. Armed Forces exchange and commissary privileges are available to the families of Marine Corps personnel upon presentation of a valid Dependents Identification Card. Before going overseas be sure that these ID cards are current for each dependent, and that members of your family over 10 years old have their own cards.

Employment. Important Civil Service preference benefits are granted to Service widows, not remarried, in connection with examinations, ratings, appointments, and reinstatements under Civil Service, and in connection with Government reductions in force ("firings"). Interested widows should apply also to the nearest U. S. Employment Service office for information about jobs administered by that agency.

2329 ■ Marine Corps Casualty Procedures

The next of kin recorded on your Record of Emergency Data is notified in case you are seriously injured, wounded, killed, or missing. Your next of kin is kept advised of your condition while you are on the critical list.

When an officer dies or is missing in action, a Casualty Assistance Officer is appointed from a nearby Marine Corps organization, to provide advice and assistance to the survivors. The Casualty Assistance Officer outlines rights and benefits of survivors, helps prepare claims, and so forth.

The Chaplain. The survivors of a deceased Marine (active or retired) should not fail to seek assistance from the chaplain of the nearest Marine Corps or naval activity. Not only can chaplains minister spiritually at this difficult time, but they are also ready to help with burial arrangements, transportation, and all the problems that arise after the death of the head of the family. Chaplains will likewise arrange for burial in Government or civilian cemeteries, and will conduct the funeral service.

Marine Officials. The nearest Marine commanding officer, recruiting officer, or inspector-instructor is competent and glad to assist families of deceased Marines with their problems.

2330 ■ Aid from Organizations

Dependents of deceased officers can get advice and assistance from several organizations listed below:

American Red Cross. This organization assists dependents with all types of Government claims as well as other problems. Proof of dependency is necessary; dependents should consult the Red Cross field director at the post or station, or the Red Cross chapter in their town.

Navy Mutual Aid Association. In case of death of a member, the Secretary of the Association should be the very first resort and can be depended upon to handle all matters pertaining to pensions and other Government claims; *appeals to attorneys or agents are unnecessary and may delay settlement* (see Section 2305).

Navy Relief Society. This organization provides aid to members of the Naval Services and their dependents. Aid includes financial assistance (loan or gratuity); services of the Navy Relief Nurse; help with transportation and housing; securing information about dependency allowances, pensions, and Government insurance; locating and communicating with naval personnel; and advising about community service available locally. Apply to the Navy Relief Society, Navy Department, Washington, D.C. 20360.

Veterans' Groups. The Retired Officers Association, American Legion, Veterans of Foreign Wars, Disabled American Veterans, Military Order of the World Wars, and other veterans' groups may also render aid to surviving dependents of Marine veterans.

RETIRED OFFICER BENEFITS

When you retire, you rate various Marine Corps benefits and perquisites by virtue of your retired status, and you may also be entitled to various veterans' benefits if you apply and qualify for them. Some of these have already been mentioned, and this section sums up the most important ones remaining. In connection with most veterans' benefits, it is important to know that, although your retired pay is taxable, it is not classed as "other income," and thus does not bar you from receipt, or limit the extent, of benefits for which you are otherwise eligible, except that, if you receive disability compensation from the VA, your retired pay is reduced by the amount of the VA disability payment.

2331 ■ Medical Care and Hospitalization

Retired Marines entitled to retired or retainer pay are entitled to medical and dental care in Uniformed Services medical facilities on the basis of availability of space and capabilities of the medical staff. Their dependents are also eligible on this basis, except as to dental care.

Retired personnel of the Marine Corps, their spouses and children, and the spouses and children of deceased active or retired Marines, are now eligible for civilian hospitalization and out-patient services (routine doctor visits, drugs, etc.) on a cost-sharing basis whereby the Government pays the major part of fees involved. For outpatient services, a single family pays the first $50 per person treated per year (but not over $100 per family) and 25 percent of the remaining cost; the Government pays the remaining 75 percent. For hospitalization, the Government pays 75 percent of the cost (including doctors' bills), and the retired Marine or dependent pays the remaining 25 percent. On reaching age 65, when you become eligible for Social Security Medicare, you no longer rate the civilian-type military medicare just described, but your access to Armed Forces medical facilities remains undisturbed.

> NOTE: At the time of retirement, when your active-duty health record is closed out, ask your sick bay to make (or let you make) a Xerox copy for personal retention in case of future VA claims or whenever you need attention from any military medical facility.

2332 ■ Eligibility for U. S. Naval Home

Aged and infirm retired regular officers are eligible for admission to the U. S. Naval Home, Gulfport, Mississippi, on approval by the Secretary of the Navy (see *Marine Corps Civil Readjustment Manual*). Residence at the Home does not entail forfeiture of retired pay. Retired Navy and Marine officers and their wives, or widows, are also eligible for Vinson Hall, the Navy-Marine Residence Foundation home at Fairfax, Virginia.

2333 ■ Veterans' Privileges

As a veteran of World Wars I and II, Korea, and Vietnam (and probably of any future war), you have certain privileges. Some of these are subject to expiration, while Congress adds others from time to time. The most important now current are:

The Veterans Administration (VA). The VA affords a multitude of special benefits, such as care for the blind; special housing for wheelchair invalids; special hospitalization and domiciliary care; vocational rehabilitation for the handicapped; and admission to VA hospitals for some types of care not always available from Navy medical resources. In addition, it may sometimes be advantageous to receive VA Disability Compensation instead of a portion of your retired pay. Consult the nearest VA field office.

Educational Assistance. Any veteran having served in the Armed Forces since 31 January 1955 is eligible for educational assistance at college or university level in approved courses and institutions for a maximum period of 45 months. Veterans who enlisted prior to 1 January 1977 have ten years after discharge (but not later than 31 December 1989) to complete their education. According to dependency status, assistance for full-time courses ranges from $292 per month to $396, with $24 a month added for each dependent in excess of two.

Marines who enter the Service from 1977 on are not eligible for the above, but may contribute to a voluntary contributory educational program jointly run by the VA and the Defense Department. In this, you contribute at least 12 monthly installments of $50 or $75, up to $2,800, and in return the VA puts in $2 for each dollar from you. You may use the resulting benefit at any time within ten years after your discharge or retirement.

Direct educational loans up to $1,500, at low interest rates, are also available.

Home and Farm Loans. The National Housing Act (P.L. 83-560), as well as P.L. 89-117, establishes an FHA mortgage insurance program to assist members of the Armed Forces and veterans in buying or building homes. This program is a continuing one, and has no date of expiration. See your banker.

U. S. Employment Service. Both the U. S. Employment Service and all state employment offices have counseling and placement facilities for veterans, to which you are entitled.

2334 ■ Travel on Government Aircraft

Retired officers (but not dependents) may travel, space available, in government aircraft within the continental limits of the U. S. To arrange this, check with operations at the air station from which you wish to depart. Retired officers and accompanying dependents may travel, space available, on Military Airlift Command overseas flights. To do so, report in person (with an authenticated copy of retirement orders and/or DD Form 214, and of course, passport, visas as required, and immunization record) to the appropriate MAC "gateway" base. If your destination is the Pacific or the Far East, Travis AFB, California, is the place to report; for the Mediterranean or Latin America, it is Charleston AFB, South Carolina; for Europe, McGuire AFB, New Jersey; and for Alaska, McChord AFB, Washington.

Space-available travelers are subject to being "bumped" by duty travelers at any point en route. Their baggage is limited to 66 pounds, and they must of course be physically present at the MAC terminal to be placed on the space-available waiting list. Finally, anyone traveling in this category should have sufficient ready cash to proceed via commercial means if "bumped."

> NOTE: Because of tighter, Defense Department computerized control of aircraft spaces, space-available travel is becoming increasingly difficult and is fraught with long delays.

2335 ■ Government Employment of Retired Officers

The Dual Compensation Act of 1964 repealed long-standing restrictions that hampered retired regular officers from holding Government jobs. Today, with a few exceptions, retired officers, regular or reserve, may hold any civilian Federal office. One important exception is that, in general, retired military people must wait 180 days before accepting a Defense Department civilian post.

Regular officers retired for any cause but combat disability receive the full pay of the civilian job and (as of 1964, when enacted) the first $2,000 of their military retired pay, plus one-half of the balance of retired pay. Each time retired pay increases by reason of cost of living, the $2,000 increment increases by the same percentage. As of 1977 it is $4,000.

Even with the present greatly simplified, more reasonable legislation on this subject, it remains complicated. Restrictions still in force prohibit retired officers from selling to the Department from which retired, from prosecution of claims against the United States, from acting as an agent or attorney against the United States, from employment by foreign governments without executive sanction, and from participation in certain types of foreign commercial enterprises, to cite the most important examples. If you have any doubt as where you might stand in either contemplated Federal or private employment, you should consult the excellent pamphlet put out by the Judge Advocate General of the Navy, *Reference Guide to Employment Activities of Retired Personnel*, as well as the latest edition of the *Handbook for Retired Marines*. And the Judge Advocate General encourages retired officers in doubt as to the legality of some specific employment to request an advisory opinion.

> NOTE: Civil Service gives a ten-point preference to any Service-disabled veteran seeking U. S. employment, as well as a five-point preference to any veteran (1) who served 180 days between 31 January 1955 and 15 October 1976; or (2) who, after the latter date, has served in a conflict or campaign officially certified by the Defense Department.

2336 ■ **Information for Retired Officers**

Marine Corps Headquarters issues from time to time updated editions of *The Marine Corps Retirement Guide*, a comprehensive guide and compendium on benefits, privileges, and special provisions affecting retired Marines. A copy of this handbook will be furnished you on retirement. In addition, the Bureau of Naval Personnel has several comparable excellent booklets, all of which may be had on application.

> NOTE: When and after you retire, you are required to keep the Commanding Officer, Marine Corps Finance Center, Kansas City, Mo. 64197, advised of your current address of record at all times.

FOR BENEFIT OF YOUR DEPENDENTS

2337 ■ **For Benefit of Your Dependents**

Secure *at once* copies of the following, certified by the seal of the issuing official: marriage certificate, decree of divorce or annulment of former marriage of your spouse or yourself; death certificate of any previous spouse of your spouse or yourself; birth or baptismal certificates for your spouse, children, and yourself; and your retirement orders.

Have photostatic copies made of these certificates, and then have the copies certified by a notary public. Place these certificates, along with your will and other valuable papers, in an envelope and stow them in your safe-deposit box. Discuss these papers, and the plans they represent, with your spouse and anyone else who will have to act upon them.

Record of Emergency Data (Form NAVMC–10526). Should you die while on active duty, your Emergency Data Form, which is filed in your qualification jacket, becomes the most important document in your entire official files. The information you have furnished on this form indicates the person or persons to: (1) be notified of your death; (2) be paid death gratuity; and (3) be paid arrears of pay. If, after retirement, you desire to include a new wife, add new children, change beneficiaries, change insurance companies, it is up to you to request the Commandant for a new Record of Emergency Data.

In addition, your record of emergency data enables you to designate anyone in your immediate family who, because of ill health, *should not* be notified of your death. Moreover, there are spaces in which you indicate the commercial insurance (and policy numbers) that you carry, as well as all Government insurance (see Section 2303). The latter item becomes important to your next of kin when Marine Corps Headquarters prepares an official Marine Corps Report of Death and sends a copy to all such insurance companies you may have listed.

In short, every item of information on your Emergency Data Form is important to your widow to be, your children, and/or your next of kin. It is your responsibility to them to make certain that this form is kept current and in accordance with your desires. *If you die tomorrow, the most important thing you did today* will have been to bring your Emergency Data Form up to date.

> NOTE: If you hold any category of Government insurance, remember that the Emergency Data Form does not of itself effect designation or change of beneficiary, which is a separate transaction between you and the VA. If in doubt, query the Veterans Administration, Insurance Center, P.O. Box 8079, Philadelphia, Pa. 19101.

2338 ■ Foreign Divorces

Medical care, quarters allowances, and other dependents' benefits have been denied in the cases of military personnel who have obtained foreign divorces, usually Mexican, and later attempted to marry some other person. In almost all instances in which both parties to foreign divorce are U. S. citizens, the validity of the divorce and any later marriage is not recognized for allowance or for many other administrative purposes. While it is hardly to be expected that officers will find themselves in such a predicament, you have the responsibility of preventing members of your command from the often tragic difficulties that can result from marital entanglements of this kind.

2339 ■ Instructions Concerning Your Personal Affairs

Death is always unexpected. Even in peacetime one Marine dies every day. Thus it is essential that you gather together all information and instructions concerning your personal affairs, and record this in suitable form for use by your surviving dependents. If you are a member of the Navy Mutual Aid Association (see Section 2330), obtain a copy of their *Personal Log*, a time-tried and comprehensive personal record for the affairs of officers. *Handbook for Retired Marines* also contains an excellent form for recording your personal affairs record. As a minimum, collect and file in one place the following in up-to-date form: burial information and instructions; official letters and correspondence with Government agencies; your will (location and any supplementary instructions); important certificates (see Section 2337); safe-deposit-box information (location, keys, inventories); bank-deposit records and insurance policies; income tax returns; money your beneficiary will receive from the Government (back pay, death gratuity, pension, etc.); inventory of real estate, securities, and personal property.

On the blank pages, you should list: debts or outstanding obligations that will have to be paid by your estate; other assets (debt owed you, investments in partnerships, and any other interest or investment of value); notes or advice that may be useful to your executor or dependents in management of your estate after death.

Date and sign your record of personal affairs. Show it to your wife; tell your lawyer and your executor where this record and your will are.

Your family has the right to expect long-range planning and consideration from you. Don't procrastinate! *Do it now!*

---

*Three things come not back: the arrow that is flown,
the spoken word—and lost opportunities.
—Omar Ibn, 581–644*

24

Marine Corps Social Life

Marine officers occupy a special position in the military establishment of the United States, since they are members of a small, elite, and unique Corps. The Marine officer is well-known for his prowess on the battlefield and his gallantry in action. He is equally known for his gentlemanly demeanor and conduct. In 1775, when the Marine Corps and Navy had barely come into existence, John Paul Jones (a firm supporter and admirer of Marines) is said to have written, "It is by no means enough that an officer of the Navy should be a capable mariner. He must be that of course, but also a great deal more. He should be as well a gentleman of liberal education, refined manners, punctilious courtesy, and the nicest sense of personal honor." Today, as well as in the days of John Paul Jones, the Marine and Navy officer should be as ready in the niceties of the drawing room and the ballroom and in the common courtesies of daily intercourse, as in his courage and ability as a soldier. Being a Marine (and an officer) therefore puts you in a select society which operates on the basis of well-established rights, privileges, customs, and rules. These add up to a code that determines the way most Marine officers and their families behave.

This code, however, should not make you feel that rigid social conformity for its own sake is greatly admired among Marines. Quite the reverse. There is probably no military group in the world where social and professional individuality are more applauded even when not always fully rewarded.

Nevertheless, as you become a Marine and join the corps of officers, you must be prepared to share the time-honored and pleasant social traditions of your Corps. As you do so, remember one thing above all:

the phrase, *"an officer and a gentleman,"* is a current one in the Marine Corps. It means what it says.

To convey a more concrete sense of what life in the Service—especially its social life—means to those who serve, officers and their ladies alike, here is the text of a letter addressed a few years ago to one of our Commandants by the widow of a Marine officer:

>Dear General . . .
>
>Lately I saw an article which set me thinking deeply. It was an article concerning the kind of life which I, myself, lived for many years and somehow the tone of this article gave me a certain sense of complete astonishment—it was an account of some of the experiences which fall to the lot of anyone who marries an officer or enlisted man in the Navy or Marine Corps. As I read, it seemed to me that the writer had, in some strange fashion, missed the whole point of what such a life is meant to be. She had translated it entirely into terms of advantages and disadvantages to herself; it became a recital of what "the Service" had to offer her, personally—or of what it had deprived her.
>
>Now, "service" is a curious word. It may be the proudest word in the world—or the most menial. It depends on the point of view and in how high esteem you hold its importance. But one thing is certain; anyone who enters it, or who is even connected with it, must, perforce, give up the idea that his own comfort and convenience is henceforth to be his paramount object. The very word presupposes that he has undertaken to dedicate his life and his best efforts to some thing which will have first call upon them, and there is no use blinking the fact that it means self-sacrifice for the man and self-abnegation for the woman who marries him.
>
>In our Army, Navy, and Marine Corps there's a phrase which justifies the demand for resignation or discharge of anyone who falls short of the high traditions of his profession. It reads, simply: "for the good of the Service," and that, once proved, is considered sufficient reason to sever his connection with the organization whose standards he was bound to uphold. The Service comes first, and he was supposed to have understood that before he "took on."
>
>And much the same code applies to the wives. It's not that they are such a noble lot, or that they can bear separation, comparative poverty, and discomforts more gladly than others, but if they have any comprehension of what their husband's profession demands of him they do try to help rather than hinder. The Service wife has got to know that she can "take it." If not, she'd better follow *Punch's* advice to the young man about to marry: "Don't." For the Service is not maintained by the citizens of a nation with any altruistic idea of material benefits to wives. There's no question of what good the Service is going to be to them, but there's every question of what good they're going to be to the Service. And there's quite a lot they can do. If they bear in mind that

their husband went into this life of his own free will, considering it, probably, the finest profession in the world, they can help to keep him feeling that way all his life, and thereby contribute no little to his efficiency and happiness in his chosen job.

On the other hand, they can make it a series of grievances and inconveniences for him by their complaints of its disadvantages to themselves, their suspicion that his general is ill-advised or that his admiral is showing favoritism, and that any sudden or unwelcome orders are probably due entirely to lack of consideration or a personal grudge on the part of the authorities. Persistent wifely persuasion of this sort is enough to convince nearly any man that the game's not worth the candle and that the Service is being run all wrong by those at the top.

Nothing which depends on human agency is ever perfectly administered, but if a ship in which one's husband is serving puts to sea just before Christmas, ten to one it's due to some important reason, for, after all, he's not the only man aboard who's being separated from his family—even admirals have families occasionally—and it's not likely that the fellow higher up is just being hardboiled and taking hundreds of men out to sea just in order to cause distress to one wife who'd like to have her family together for the holiday.

It's even possible, when there's a war on, that an officer may have to leave to join his regiment on less than 24 hours' notice—it's one of the things he's trained and paid for, to be ready for such emergencies—and even civilians have had to do as much during the stress of war time. The parting, though, becomes no easier if the wife insists on treating it as a personal grievance.

Certainly she's got no easy road to follow if she's going to be an asset and not a liability. It's a struggle to keep a brave spirit and a high heart in the face of separation, discomfort, and maybe danger to the man she cares for. To keep her letters cheerful and encouraging, with the minimum of complaint all the while her heart fails within her and her spirit sags drearily from loneliness, so that she won't weigh him down the more with matters for which he has no remedy. He is already burdened with the care and welfare of a hundred—or a thousand—or ten thousand men for whose safety he is responsible; she must spare him that extra anxiety—the recital of inconveniences, hardships, and injustices which she is, temporarily, undergoing. That wail of everything being just wrong.

Once I heard the finest officer I ever knew say: "Unless you have constructive criticism to offer, or hopes of developing some in the course of discussion, keep your destructive criticism to yourself—otherwise you're just airing your own private whine."

And that's what the understanding Service wife must learn, in great measure, to do. It's a hard lesson to master, but it can be and is done. There's one great thing that upholds us—a thing which is almost impossible to put into words, but it wells up within us, an instinct rather than a reason—and gives that which enables us to endure the lesser trials and, sometimes, the greater ones. No matter how craven we are, or how little we like being heroic, none of us can look at that column of the London *Times* where the soldier dead are honored—"In sad but proud memory of my dear husband, dead on the field of honor, July, 1942"—without suddenly tasting tears and feeling that there is something stronger than sadness—stronger than death. The pride of Service, which makes it right, and even sweet, to lay down one's life for one's country.

It need not be an actual, physical life; it can be the best years of life's effort, laid as a sacrifice "in line of duty." We can disapprove, or find futile the necessity of such sacrifice—even ridicule it—but we can't get away from the feeling of pride in it. It is bred more strongly and deeply in us than we know. There is some fundamental part of a human being which is stirred by such action and responds to it—something in the spirit which rises to meet gallantry and sacrifice, with a tragic pride, in all humanity for its sake.

That's what makes it all worthwhile. For it takes a lot of living up to—the Service.

2401 ■ Helpful References

This chapter is not intended to be an "Emily Post" for Marine officers, but simply to deal with the military, and more especially, the Marine aspects of Service social life. For more general reference, consult the following publications:

Service Etiquette (Swartz, U. S. Naval Institute), which is a sound general guide; indispensable in certain matters.

Social Usage and Protocol (Foreign Liaison Section, Office of Naval Intelligence)—this handbook, for many years withheld from general circulation by civil servants in ONI, is a comprehensive, useful *vade mecum* within its limited parameters. Any officer going on attaché, MAAG, or Naval Mission duties should obtain a copy.

Social Usage in the Foreign Service (Department of State), which is a highly useful compilation containing many excellent suggestions and much good advice.

SOCIAL OCCASIONS

2402 ■ The Marine Corps Birthday

As every Marine knows, the Corps was founded on 10 November 1775. From that day to this, the Tenth of November has been the climax of the Marine Corps year. November 10th is the top social occasion of the Corps.

The Birthday of the Marine Corps is celebrated officially and socially by all Marines throughout the world. Not only do Marine units carry out the prescribed ceremony, but wherever one or more Marines are stationed—on board ship, at posts of other Services, even in the field—November 10th is habitually celebrated.

For example, take 10 November 1951, in Korea.

At high noon the 1st Marine Division fired every weapon that could bear on enemy territory. Seconds later, every combat airplane in the 1st Marine Air Wing screamed down in ultra-close air strikes, while the riflemen cheered. When these preliminaries ended, units gathered together as best they could without letting the war get out of hand, and in every Marine command, great or small, the traditional words of the *Marine Corps Manual* were published to all hands.

This article, which must be published each 10 November, and was originally composed by General Lejeune in 1920, may be found in Appendix VI.

And so, with a holiday ration to top off, the Marine in Korea again proved what November 10th stands for in the Corps: an all-out day for Marines—professional, social, official, unofficial, officer and enlisted, regular and reserve.

How a Command Observes November 10th. For a Marine command, the Birthday includes prescribed or customary features, which are observed as circumstances permit. For Marines with other Services, many of these items cannot be fulfilled exactly, but this list may serve as a guide:

A troop formation (preferably a parade) for publication of the article from the *Marine Corps Manual*. The uniform should be dress blue A (which includes large medals). If blues cannot be worn, medals should be prescribed on the service uniform for this occasion. On shipboard, hold a special formation of the Marine detachment and get permission from the Captain to pipe the birthday article over the public-address system. If you are with some other Service, and only a few Marines are present, you may defer publishing the article until the evening social function.

Holiday rations and, if the recreation fund can stand it, beer for the troops.

Maximum liberty and minimum work consistent with the missions of the command.

Memorial service at the Post Chapel for Marines who have died in the service of the nation. This service should follow the military ceremony, and should be attended by the Commanding Officer, officers, and—if practicable—by the command in a body.

A birthday ball for officers and one for enlisted Marines. At each, a cake-cutting ceremony takes place.

At any schools or instruction scheduled for 10 November, you should emphasize the traditions and this history of the Corps.

The Birthday Ball. It is up to you to arrange the annual birthday ball with pride, forethought, and loving care. Every Marine command must have one. If on detached service away from the Corps, the senior Marine officer present must arrange a suitable birthday ball, and it is up to you to chip in to support it.

The birthday ball is formal, which means evening dress for officers possessing that uniform, or dress blue (with large medals) as a substi-

Cutting the birthday cake is the high spot in the traditional birthday ball.

tute. If you are not required to possess either evening dress or blues, wear service uniform. For the ladies long dresses, of course. Should November 10th fall on a Sunday, the accepted way to hold the birthday ball without profaning the Sabbath is to have it on the Saturday night preceding, and time the ceremony so that the reading of the Article takes place at one minute after midnight.

The birthday ball is a command performance. Unless duty prevents, you attend. If resources permit, distinguished civilian guests and officers from other Services should be invited, but not too many. *Be sure that retired Marine officers and any Marine officers present from other countries are included.*

The procedure for a birthday ball ceremony is described in Appendix VII. This procedure, of course, is a guide, and details may vary according to facilities, numbers of officers and guests, and local traditions. There is only one ironclad rule for the birthday ball: *Make it a good one.*

2403 ■ Mess Night

A Mess Night (sometimes called a "Guest Night" or a "Dining-In Night") is a formal dinner in mess by all members, or by the officers of a particular post or unit. "Wives," as remarked by Brigadier General R. H. Williams in the December 1955 *Gazette*, "do not attend. This may cause some irritation on the distaff side, but must cheerfully be accepted as one of the many hardships which a soldier's wife must bear."

Mess nights may be held periodically; for example: on special anniversaries (such as that of a battle in which the unit has participated), to "dine out" officers being detached, or to honor a distinguished guest, or guests from another unit, Service, or country.

In the U. S. Armed Forces mess nights date back to the Army's regimental messes of the pre–World War I days and to the days of the wine mess in the wardroom afloat, which ended abruptly in 1914 when Secretary Josephus Daniels imposed prohibition on the Navy. In this early era of a small Marine Corps with only a couple of hundred officers, the only permanent Marine officers' mess was that at 8th and Eye, and here in the Old Center House (torn down in 1908) the officers of Headquarters and the Barracks had their mess nights. Happily, the custom continues in today's successor Center House, as elsewhere.

Preparations. The first step in preparing for a mess night is to designate the officer who will act as *vice president* (see below). In some units, the vice president is traditionally the junior lieutenant present. However, it is good practice to rotate the post among all company officers on board so that all may gain experience. In any case, the function of the vice president—at least beforehand—is to undertake all preliminary arrangements, i.e., guest list (to be approved by the mess president), seating diagram (also to be approved), menu and catering, music, decorations, and so forth. The success of the evening depends on the vice president.

Subject to local or unit customs, and to facilities that are available, here are specific arrangements which should be made for a mess night:

1. After approval of the guest list, invitations should be prepared and mailed or delivered at least two weeks in advance of the mess night. Each guest, regardless of organization or of sponsoring officer in the host unit, is a guest of the Mess and should be so treated.

2. The table is set with complete dinner service—wine glasses, candles, and flowers. Unit or post silver and trophies should be used.

3. Unless the CO desires to preside, a field officer is detailed as President of the Mess for the occasion; a company officer acts as Vice President.

4. Uniform is evening or mess dress, or dress blues or whites. Civilians invited to a mess night should wear full dress with miniature medals if uniform is evening or mess dress; dinner jacket with miniatures, if blues or whites.

5. The National Color and the Marine Corps Color are placed behind the president's chair; guidons and drums may also be used as decorations.

6. The mess president sits at the head of the table, the vice president at the foot. Other guests and members take seat by rank (as in the wardroom on board ship), except that guests of honor are on the right and left of the president. A seating diagram should be posted in advance, and place cards and menu cards prepared. All preliminary arrangements are supervised by the vice president.

7. If available, a three- or four-piece military string orchestra should be detailed to provide dinner music, and should know the national anthems and regimental marches of guest officers. If suitable, "live" music is not available, a good-quality PA system with taped or recorded selections will serve as a substitute. The musical program should be checked and timed by the vice president, and should always include "Semper Fidelis" and the regimental march of each guest.

Procedure. Officers assemble in an anteroom 30 minutes before dinner, for cocktails and to greet guests. This should be the occasion for all officers to speak to each guest and make him feel welcome; and also for each officer to pay his respects informally to the senior officers present, COs especially. Dinner is announced in accordance with local custom. In some messes, "Semper Fidelis" is played; elsewhere, *Officers' Call* is sounded followed by a march (when drum and bugle corps is

Officers and guests gather beforehand in a mess anteroom for cocktails. Uniform is evening or mess dress (or dress blues or whites).

available, "Sea Soldiers" is a suitable march); still another variation is to play "The Roast Beef of Old England" (known and used in the "Old Navy" as "Officers' Mess Gear") on fife and drum. Whatever the signal, officers and guests proceed to their places. Each guest should be escorted by a member of the mess. Grace is said by the chaplain, if present, otherwise by the president. (Note for Chaplains: *Don't make a sermon out of grace.*) Officers then take seats. The ranking guest, seated at the mess president's right, is served first, then the president, and so on counterclockwise without further regard to seniority. Appropriate wines are served with each course. There should be no smoking during dinner, and no officer may leave the table until after the toasts, except by permission from the president. (If for any reason, official or otherwise, you arrive late, you should express your regrets to the mess president before taking your seat.)

After dessert, there is a short concluding Grace, the table is cleared, and port decanters and glasses are placed on the table. The port passes clockwise until all glasses are charged. When the decanter (or both decanters, if two are used) has completed the circuit, the president raps for silence. If a foreign officer is present, the president rises, lifts his glass, and says, "Mr. Vice, His Majesty, King ———————of———————." The vice president then rises, glass in hand, waits until all have risen, and gives the toast. "Gentlemen, His Majesty, King ————— of —————." The orchestra plays the foreign national anthem, following which all say, "King —————of—————," drink, and resume seats. After about a minute, the president again raps for silence, the senior foreign officer rises, and says "Gentlemen, the President of the United States," and the orchestra plays the National Anthem. If no foreign guests are present, the first toast is to the President of the United States, and—in any case—the *concluding* toast is to the Marine Corps, during which, if music is available, "The Marines' Hymn" is played. The wording of this toast should be, *"Mr. Vice, Corps and Country,"* and the custom has grown up (proposed long years ago by Colonel A. M. Fraser) that the vice president reply in words taken from a Revolutionary War recruiting poster of the Continental Marines—*"Long live the United States, and success to the Marines!"* If the guest of

Mess night calls for best linen and silver, table set as shown, Colors behind Mess president, appropriate wines and candles. If music is available it should be included.

honor be a Marine general officer, he may take this occasion to proceed to a few remarks. If the guest of honor is from another Service, a toast to his Service is in order. He may respond and speak. Toasts are not "bottoms up."

Before leaving the subject of toasts, note that toasts may be divided into four classes, and that they are given in the following order:

Toasts of Protocol: Toasts to foreign governments or chiefs of state; toast to the President of the United States.

Official Toasts: Toasts to other Services, military organizations, Government departments, agencies, or institutions.

Traditional Toast: "Corps and Country."

Personal Toasts: Toasts to individuals (distinguished guests, officer being dined out, and so on).

The traditional toast ends the formal part of the evening. Personal toasts and speeches may follow at a suitable interval afterward, as described below.

> NOTE: In some messes and commands, the custom has grown up of emulating the Continental Marines by drinking toasts in rum punch rather than port. Here is the mix for "1775 Rum Punch": four parts dark rum; two parts lime juice; one part pure maple syrup. Add small amount of grenadine syrup to taste. Ice generously and stir well. The maple syrup was originally used during the Revolution because of the British blockade that cut off supplies of West Indies sugar cane.

Following the toasts, coffee is served, the smoking lamp is lighted, and individual drinks or liqueurs may be ordered. At this point, or whenever the orchestra is released, the president may send for the leader, and offer him a drink. If speeches are planned (other than remarks associated with toasts), they are made now. In "dining out" an officer, the commanding officer makes brief, usually humorous remarks, whereupon the officer being honored replies in the same vein. In some messes the orchestra remains and plays the regimental march of each guest, during which the individual stands. When speeches are over, the president announces, "Gentlemen, will you join me in the bar?" and the senior officers rise, following which the remainder of the party adjourn individually to the bar and anteroom, where songs are generally sung and games played. All hands should remain until the ranking guest and the Commanding Officer leave, after which anyone may secure at discretion.

Circumstances will frequently not permit a mess night with all formalities as to uniform, catering, and table service that are outlined herein, or it may not be desired. This should not deter an organization from making the effort. The idea is to do the best you can with what you have, and let the spirit of the occasion take care of the rest. *Do not, in particular, let yourself be overcome or stultified by the apparent formality of mess nights; the object is the pleasure and comradeship of all hands.* Reports that a few commands have actually rehearsed mess nights, if true, make the occasion ridiculous. A mess night is not a minuet.

As to timing, it is better not to schedule mess nights regularly. It is much preferable that officers begin asking when the next one will take place. Thus a mess night will be looked forward to with anticipation and never as a burden.

NOTE: The costs of a mess night, like other "chip-in" Marine Corps functions, should be prorated by rank so that officers who make the most, pay the most. Here is the famous Schatzel formula for prorating by rank, which, even though complicated-looking, is actually quite simple:

Rank	Base Pay	Number Participating
Col	x	δ
Lt Col	y	σ
Maj	z	ν
Capt	α	ϕ
1st Lt	β	Δ
2d Lt	λ	η

K = total cost of the function
$E = x\delta + y\sigma + z\nu + \alpha\phi + \beta\Delta + \lambda\eta$

$$\text{Col share} = \frac{\delta K}{E} \qquad \text{Lt Col share} = \frac{\sigma K}{E}$$

$$\text{Maj share} = \frac{\nu K}{E} \qquad \text{Capt share} = \frac{\phi K}{E}$$

$$\text{1st Lt share} = \frac{\Delta K}{E} \qquad \text{2d Lt share} = \frac{\eta K}{E}$$

2404 ■ Military Weddings

As a Marine officer, you enjoy the privilege of having a military wedding. A military wedding is simply a formal wedding with traditional Service embellishments. The characteristic features and ground rules of a military wedding are as follows:

Uniform. Marine members of the wedding party wear dress blue or white A with sword. Dress A uniforms call for medals, not ribbons. If the weather requires, wear boat cloak rather than overcoat. Even though wearing sword, and thus under arms, the groom should not wear gloves, whereas the ushers should wear gloves throughout the ceremony.

Needless to say, all members of the wedding party wear the same uniform. If officers from other Services are included, they wear their nearest equivalent uniform. For an evening wedding, evening dress or mess dress is worn.

Best Man and Ushers. Since your wedding is to be military, your best man and ushers should be regular or reserve officers. Inactive reserve officers may don uniform for the occasion. It is usual but not necessary for ushers to be the same rank as the groom. The senior usher coordinates the military side of the ceremony and gives commands or signals for movements by the ushers and for the arch of swords.

The best man looks out for the groom. It is a nice compliment to your immediate commanding officer (if you are on those terms) to ask that he be your best man. In any case, however, your CO and brother officers should be invited to the wedding, and all will attend.

If the wedding takes place away from the bride's home, and her parents or near relatives cannot attend—as is sometimes the case in the Service—it is appropriate for the commanding officer, or other senior officer friend of the groom, to give away the bride.

Bachelor Dinner. The night before the wedding, the best man and ushers should give the groom a bachelor dinner. The best man and senior usher attend to the arrangements. On this occasion the groom usually gives his best man and ushers their customary presents (cuff links, cigarette cases, etc.). The toast to the groom is proposed by the best man. Civilian clothes are generally worn.

The Clergyman. You may choose either a chaplain or a civilian clergyman. A chaplain performs the ceremony in uniform or in vestments, according to the customs of the denomination. In some denominations (such as the Episcopal Church) ministers, whether chaplain or civilian, are permitted to wear military ribbons on their vestments and will do so if you request. Your best man should see to this. If your wedding is an evening affair, it is appropriate for the clergyman to wear miniature medals on his civilian coat at the wedding reception.

Do not pay a chaplain for officiating at a military wedding. If you have a civilian clergyman, follow civilian custom regarding fees. Again, this is something your best man should attend to. The same applies to fees for organist and music at the church.

Wedding Under the Colors. If you wish, and if your denomination permits, the National Color and Marine Corps Color of your unit may be crossed above and in rear of the chaplain, or displayed during the ceremony in the chancel of the church. This is known as "A wedding under the Colors." It is an old tradition, signifying your wife's acceptance into the Corps.

Handling the Colors for this ceremony is the responsibility of the senior usher, who, with designated ushers, receives the Colors (cased) from the adjutant, places them before the ceremony, and removes, cases, and returns them immediately afterward.

Wedding Present from the Unit. The officers of the groom's battalion, squadron, or headquarters, or his service school classmates (if he is at school when married), should present him with a piece of plate, such as a silver tray, water pitcher, or cocktail shaker, which is appropriately engraved. Some units have a standard type of wedding gift, which it is the duty of the adjutant to procure, engrave, and collect for. A typical inscription for such a piece of plate might read:

From the Officers of the 1st Battalion, 5th Marines

If the wedding takes place at a school, it is up to the senior Marine officer in the class to see to this wedding present.

Arch of the Swords. The "Arch of Swords" is probably the best known feature of a military wedding. It is carried out in this fashion:

After the ceremony, the senior usher forms the ushers in column of twos, and places them *immediately outside* the exit of the church, facing inboard. As the newly married couple pass through the portal, the senior usher commands: 1. Officers draw; 2. SWORDS. At the command of execution, ushers carry out *only* the first count of the movement leaving their swords raised, with tips touching, to form an arch under which the couple pass. After the newlyweds have passed, swords are returned on command by the senior usher.

Cutting the Wedding Cake. The wedding cake is cut by the bride and groom together, using the groom's sword. If a Marine Corps daughter is being married to a civilian, it is proper to use her father's sword. After the cake has been cut, the best man proposes a toast to the bride and groom, and, as the guests drink, the orchestra plays "Auld Lang Syne."

2405 ■ Christenings

It is customary to celebrate the christening of a Service child with a party after the ceremony. As a minimum, invite the child's sponsors in baptism, the officiating clergyman, and friends of the family. Like weddings, christenings too may be performed "under the Colors," as described in Section 2404. At a christening party, one of the godparents proposes the health of the child. It is also a Service tradition that the officers of the father's immediate unit present the child with a silver "christening cup," engraved with its name, date of christening, and the title of the donors, as:

From the Officers of the 4th Battalion, 11th Marines

2406 ■ **Dances**

In most commands it is usual to have one or two formal dances, either dinner or buffet, at which the officers wear uniform. The appropriate uniform on these occasions is evening or mess dress, with undress blue or white optional for those not required to possess evening dress. During the course of such an evening, you should try to dance with each lady at your table, and make it a point to dance with the wife of your commanding officer. If you feel you should leave before your commanding officer, be sure to bid him and his wife good night, and ask his permission to depart. After the last dance, to indicate the end of the evening, the orchestra should play the National Anthem.

CALLS AND SOCIAL OBLIGATIONS

2407 ■ **Calls**

The exchange of calls among officers and their families receives more emphasis in military social life than "on the outside." At first glance the routine and punctilio of formal calling may seem stilted and boring. The truth is, however, that calls prod us into widening our acquaintance, break down working-hour barriers of seniority, and often disclose mutual interests that might otherwise go undiscovered.

We used to pride ourselves that every Marine officer knew every other Marine officer. Calling is thus essential to the social system. It is the hallmark of a good officer that he pays calls promptly, correctly, and pleasantly—remembering as he does, the wise axiom, "If you want to know someone, it's up to you to go three-quarters of the way."

2408 ■ **Kinds of Calls**

Calls are of two kinds—official and personal. You will find the former covered in Sections 1820–1821. Official calls are rendered only between commanding officers, officials of state, and officers' messes. Personal calls are exchanged between officers and their families.

Although established Marine Corps customs govern personal calls, some commanding officers have special preferences as to when and how calls are paid. Thus, before making any calls, check with the adjutant, and, if necessary, with the general's aide (or, if serving with the Navy, the flag lieutenant) in order to find out the local policies.

The procedures described below apply to small and medium-sized posts or organizations (individual ships and units no larger than a regiment). At large posts, at schools, and in Washington, the large officer population and the constant turnover prevent adherence to the protocol (but not the spirit) of formal calling.

Initial Calls. Whether married or single, you must pay a "visit of courtesy" on your commanding officer or reporting senior within 48 hours after you report. This call is required by *Navy Regulations*, and is in addition to the occasion when you report for duty. If your commanding officer or reporting senior is married, your wife (if you are married) should accompany you. Unless your CO indicates otherwise, *the uniform for this call is undress blue or white.*

> NOTE: Some commanding officers will suggest a convenient time for this "visit of courtesy," even though it may not be strictly within the 48-hour time limit. The important thing is that the call be made promptly and at a time when the CO and his wife will be in, the object being for you to meet them socially, and vice versa.

If you are a bachelor, after you have completed the foregoing call on the commanding officer, you should call in quarters on all field officers, and, if possible, on all married officers of the command. Unless local rules decree otherwise, wear civilian clothes for these calls.

If you are married and have your wife with you, as soon as you are settled in quarters, on or off post, you may expect calls from other officers. Return each call within ten days. Wear civilian clothes for such return calls.

If you are married but not accompanied by your wife, pay the same calls as a bachelor, provided your wife is not expected to join within a month. Once your wife arrives and the family is settled, married calls will be paid as if you both had just arrived.

Calling Hours. The hours for formal calling ("calling hours") are from 1700 to 1900 in the Marine Corps, Navy, Coast Guard, and vireually all foreign forces. In the Army and Air Force, however, you call between 1930 and 2100, after dinner.

Twenty minutes (or "one drink") is the accepted duration of a formal call.

"At Homes" and "Calling Parties." On large posts, at schools, and in Washington, calling obligations are usually discharged through "At Homes" or "Calling Parties."

An *"At Home"* is a specified day and time (once a month or once a quarter) when a senior officer, such as a head of division at Marine Corps Headquarters, or the director of a school, desires that callers present themselves at his home. Refreshments are served, and, as on all formal calls, cards are "dropped." Attending an "At Home" is equivalent to a formal call made and returned between the host and those who leave cards. "At Homes" are rarely held by officers below general or flag rank, unless the host has an independent command.

A *"Calling Party"* is like an "At Home," but is given by the command as a whole rather than by an individual officer. In large commands, where general calling is not practicable, it is customary, once every six months, or on the reporting of a new commanding general, to hold a "Calling Party" at the officers' mess. At a "Calling Party," the commanding general, other general officers, and their wives, form the receiving line; all other officers and their ladies attend *and go through the line*, although cards are not dropped. The mess provides refreshments. Attending a "Calling Party" is equivalent to required formal calls paid and returned—not only between the receiving officers and those who attend, but among all families at the party. Dress blue or white B is worn at calling parties.

"Calling Book." In some commands, the CO or commanding general may have a "calling book." The "calling book" is kept in the CO's or CG's office (usually by the adjutant or aide). Officers who would otherwise be required to pay a personal "visit of courtesy" as outlined above, may meet this obligation by inscribing their names in the "calling book."

2409 ■ Calling Cards

Correct selection and use of calling cards is of some social importance to an officer. For this reason the following paragraphs cover a few essentials of "card etiquette."

Selecting Cards. You can make few social mistakes as avoidable, as conspicuous, or as lasting as selecting the wrong kind of personal cards. The best way to avoid such bobbles is to have your cards engraved by a good stationer. In large cities, the better department stores usually have engraving services well-qualified to advise you on the layout for a correct card.

Marine Corps Exchanges can order all types of calling cards, and have brochures to illustrate the various types and styles of approved cards. Should you wish to purchase calling cards from commercial sources, any reputable engraver or stationer familiar with the correct use of letters and social forms, and the proper type and paper, can advise you competently. Webb's, 2600 28th Street, N.E., Washington, D.C. 20018, is an engraving firm which does excellent, correct work and, at the time of writing, gives a military discount. No matter with whom you trade, however, the general rules on cards are:

1. Give your full name.
2. Show full rank in the lower right corner. Avoid abbreviations: use "United States Marine Corps" or "United States Marines," *not* "USMC"

or "U. S. Marine Corps." *Never* use "U. S. Marine Air Corps" or suchlike incorrect terms.

3. Be sure the card is engraved, not printed.

4. Avoid fancy type-faces. Preferred types are: shaded antique Roman, solid Roman, or Ideal Script.

The following specifications are appropriate for Marine officers and their families, in the purchase of personal calling cards:

Dimensions (approx.)	*Engraver's card size*
Male officer: 3⅜″ × 1¾″	123
Woman officer or Miss: 2⅞″ × 2″	128
Officer's wife: 3⅛″ × 2¼″	129
Joint family card: 3½″ × 2½″	130

Name and rank are engraved no larger than 9/64″ for the capital letters, with lower case letters of appropriate size in keeping with the size of the capital letters across the center, where the name is printed. Rank and Service, in the lower right-hand corner of the card are printed in letters 7/64″ high.

The rank of general and field grade officers precedes the name, whereas the rank of company officers must be engraved one line above the Service designation in the lower right-hand corner of the card. First and Second lieutenants may use the rank designation of *Lieutenant*. Officers retired from the Marine Corps should have the word *Retired* engraved and centered beneath the Service designation in the lower right-hand corner.

If married, you should also have a *joint family card*, as mentioned above, which lists your rank (but not the Corps), with a legend such as this:

Lieutenant and Mrs. Fuller Barnett Henderson.

In addition to the general rules summarized here, much detailed information on card etiquette and customs is to be found in *Naval Customs, Traditions, and Usage*, by Vice Admiral Leland P. Lovette (published by U. S. Naval Institute).

It is convenient, if you are married, to have what are known as "informals," or folded cards inside which you or your wife may pen short messages, invitations, or acknowledgments. The legend on the face of the "informal" is the same as that shown above for a joint family card.

After preparing your first batch of cards, most engravers will retain your plate. This makes it easy for you to reorder. It also permits you to modify the basic plate each time you are promoted. This is considerably less expensive than having a new plate engraved each time you wet down a commission.

Leaving Cards. Aside from obvious miscellaneous use (exchange between new acquaintances, enclosure with gifts, name-plates on doors, and so on), you should leave cards on the occasions discussed below:

Formal Calls. When you pay or return a formal call (discussed in Section 2408), always leave cards.

As a bachelor paying or returning a call, leave one card for your host, one for your host's wife, and not more than one extra for other adult ladies in the household—a ceiling, that is, of three cards.

If you are calling as a married couple, you (the husband) leave cards on the basis just described for a bachelor; for your wife, add one of the family's joint cards. Alternatively, your wife may leave one of her cards for each adult lady in the household visited.

Example: A married couple call on another married couple whose sister is visiting. *Cards left:* (husband's card) one for the host, one for the hostess, one for the visiting sister; (joint family card) one only; or, (wife's card) one for the hostess, and one for the sister.

Leavetaking. When you are about to leave a post, you should leave a "P.P.C." card. "P.P.C." stands for *"pour prendre congé,"* which is French for "to say goodbye." Write the initials "P.P.C." on the face of your card, and affix it to the bulletin board of the station mess. If you are a bachelor, use your personal card. If married, use the family's joint card.

If, as a visitor, you have been accorded the privileges of a club, mess, or wardroom, ashore or afloat, always leave a "P.P.C." card. Some clubs and messes have a section of bulletin board reserved for such cards. Other messes may have a mess "calling book," like the CO's calling book described in Section 2408.

If you desire to leave "P.P.C." cards for individual friends, you may mail them, have them delivered by a chauffeur, or drop them in person at the door.

"To Inquire." When a brother officer or a lady of his household is seriously ill, or hospitalized and unable to receive visitors, it is an appreciated courtesy to leave your card, inscribed "To inquire." This signifies that you have the invalid in mind, and have asked after his or her health.

Some general rules:

Never leave more than three of any type of card.

Gentlemen leave cards for ladies *and* gentlemen in the household receiving the call, whereas ladies never leave cards (including joint cards) except for a household that includes one or more ladies.

When you call in person (which means in most cases), follow custom and turn down the top left-hand corner of your card(s).

While all women officers are of course ladies, they pay and return calls in their status as officers.

CLUBS AND MESSES

2410 ■ Commissioned Officers' Messes

Every post or station has a commissioned officers' mess. The mess acts as a social focus for the officers and ladies of the post, it serves meals, and it sometimes provides accommodations for visiting officers. A few large messes also have limited guest accommodations for officers' wives and children.

The commissioned officers' mess is often referred to as "The Club" or "The Officers' Club." Like any club, the mess is a private association operated for the convenience of its members, who share its expenses. Although you are not automatically a member of any commissioned officers' mess just because of your rank and status, it is nonetheless habitual to extend privileges of a station mess to any visiting officer and his family. At your home station, you are entitled to join the commissioned officers' mess upon payment of the required fees, if any, but you may be denied the privileges of the mess if you abuse them.

2411 ■ Officers Clubs and Bachelor Quarters

Clubs and bachelor quarters (BOQ) have three main functions: social recreation, meal service, and housing for bachelor, temporary bachelor, and visiting officer. Any officer assigned temporarily or permanently to a BOQ pays a fixed charge to take care of cleaning services, linen, and so forth. In certain BOQs which have messing facilities, officers living therein pay for and take meals at given rates on a menu catered by the base officers' club. In a few cases (e.g., Camp Barrett at Quantico), the post central mess operates the officers' dining facility and serves (and charges for) the basic enlisted ration, which is habitually a good one. Where a BOQ does not have a messing capability, you take your meals at the officers' club (or possibly in an officers' section of a troop mess).

Closely related to the BOQ are its derivatives, "MOQ," "WOQ," and "TOQ." An MOQ is a Married Officer(s)' Quarters, usually an apartment-type building. WOQ are "Women Officers' Quarters," and TOQ stands for "Transient Officers' Quarters."

2412 ■ The Wardroom Mess

A wardroom mess is a commissioned officers' mess on board ship. The wardroom is the common room, recreational space, and dining room for the officers of a man-of-war. The wardroom mess is the organization through which the ship's officers cater their meals and meet most of their social and recreational needs while on board. Like any closed mess, the wardroom mess is a private association whose operation is paid for by members. Because wardroom messes fulfill essential functions of feeding and accommodation, they receive some Government support. You will find notes about wardroom mess etiquette in Chapter 22. If you are going to sea duty, be sure to look these rules up—and observe them.

2413 ■ "Cigar Mess"

The "Cigar Mess," although called a mess, is a shipboard purchasing, bookkeeping, and retailing agency rather than an actual mess. It is descended from the wardroom wine mess that each ship had before 1914. Today the cigar mess sells tobacco, soft drinks, candy, etc., to wardroom officers.

2414 ■ Mess Etiquette

Before we leave the subject of clubs and messes, here are some general rules of conduct and etiquette which have always maintained the tone and correctness of Marine officers' messes.

Remember that the mess belongs to the members who support it. As a guest, defer to their ways and rules; as a member, assume responsibility for it and support it as *your* mess.

Attend mess meetings whenever they are held. You have no right to complain about the way a club is run if you are unwilling to attend meetings and voice your ideas at the proper time.

Dress conservatively and correctly at "the club." You can't go wrong, ordinarily, if you wear full uniform of the day or complete civilian clothes, including coat and suitable neckwear. Most messes publish and post their uniform rules. You, your guests, and your dependents must abide by them if you expect to use the club.

Pay club bills promptly, sign chits legibly and accurately, and always be sure your checking account is in shape to meet any checks you write. The officers' mess is founded on the proven concept that a Marine officer's word or signature is his bond. Dishonorable disregard of your obligations as an officer and a gentleman will destroy your personal standing, weaken your mess, force irksome restrictions on other members, and bring swift retribution, which will mar your record.

Do not tip mess servants or employees unless club rules expressly so authorize and encourage (and be very leery, as a club member, of giving approval to any such relaxation of rules; individual tipping at a club is a sure way to tarnish both service and attitude of servants).

When you are a *guest* in a club or mess—unless the place is on a cash basis—do not attempt to stand drinks. If you are on temporary additional duty and thus become a member of the mess, however, you pay for your fair share. If, as a member, you see a strange officer alone in your mess, introduce yourself and extend him all hospitality, remembering that a guest of any officer is a guest of the mess.

When you bring guests to the mess, be sure they are those you would entertain in your own home or introduce, as your friends, to the commanding general and his wife.

Whether in a private club or a service mess, remember that an officer of Marines is a gentleman. If in doubt as to some nicety or ground rule, do the gentlemanly thing. You will never go far wrong.

WASHINGTON DUTY

2415 ■ White House and Diplomatic Functions

The White House is the focus of social and official Washington. Some officers on duty in Washington may expect to be entertained at the White House, and a few are detailed to additional duty as Marine aides-de-camp at the Executive Mansion.

Since a White House invitation constitutes a Presidential command, it takes precedence over any other social commitment, previous or not. If you receive a White House invitation, consult one of the aides to the Commandant, at Marine Corps Headquarters. He will tell you the uniform, and give you whatever briefing may be in order.

Second only to White House functions in their requirement for fine attention to dress and etiquette are those conducted by the diplomatic corps.

Uniform is ordinarily worn for official parties at embassies or legations, or given by a military or naval attaché. The general rule is: undress blue or white for afternoon receptions and cocktail parties; evening or mess dress for formal evening parties, black or white tie.

When at a foreign diplomatic party, be alert for and familiar with foreign badges and insignia of rank, and with their national anthems. On occasions of this kind, your dignity, courtesy, and smartness set you apart not only as a Marine but as a representative of the United States.

2416 ■ Washington Calling Etiquette

Call on your reporting senior as if he were your commanding officer on a post, but don't wear uniform for this call as you might elsewhere. Because of the relatively large number of Marine officers in the Washington area, the Commandant does not require calls. Field officers (and selected company officers) on duty in Washington may be invited to a reception or some other function at the Commandant's House, and attendance constitutes a call made and returned.

It is permissible for officers on duty in Washington to drop cards at the White House and at any or all foreign embassies and legations, but this practice seems to be going out of fashion.

Subject to the foregoing ground rules, when in Washington you call on particular friends and acquaintances only, following the calling procedures that obtain elsewhere.

2417 ■ Recreation in Washington

Navy and Marine Messes. The two Navy commissioned officers' messes are at Washington Navy Yard and Naval Hospital, Bethesda. Officers at MB, Eighth and Eye, have a closed mess for members only at Center House.

Other Services. The Army has three excellent (dues-charging) clubs: Fort Myer, Ft. McNair (National War College), and Cameron Station (outside Alexandria, Va.). The Air Force has an outstanding club at Bolling AFB, and one at Andrews AFB. Marine officers, if they choose to frequent clubs and messes of other Services, are eligible for membership in all of the above.

Private Clubs. The Washington area boasts two of the foremost military and naval clubs in the country: the Army and Navy Club (the "Town Club"), and the Army and Navy Country Club (the "Country Club").

The Army and Navy Club, located on historic Farragut Square, is one of the senior private clubs in the United States, and provides all amenities (including rooms for members and wives). The "Town Club"—whose cuisine is renowned—is a traditional meeting place for officers and their friends. The Army and Navy Country Club, in Arlington, Virginia, overlooking the city, is one of the coolest summer spots in the metropolitan area, and a country club of first rank (with a first-rank golf course).

> NOTE FOR NEW REGULARS: Both the Town and Country clubs allow newly commissioned Regular officers to join, as nonresidents, with greatly reduced entrance fees well within your pocketbook. If you fail to take advantage of this privilege at the outset of your career, you must later buck long waiting-lists and pay relatively large initiation fees, which may make it impossible for you to be a member of these fine clubs. *Membership on these terms is one of the best bargains open to a new officer; lose no time in taking advantage of it.*

Another important social aspect of Washington (and in fact of East Coast) duty is the Army and Navy football game at Philadelphia, for which seats are always at a premium. The way in which Marine Corps and Navy officers obtain their tickets to this game (and to other Naval Academy athletic events) is by joining the Naval Academy Athletic Association, through which tickets are distributed; it is open to regular-officer non-Academy graduates on the same basis as to alumni. Members may also obtain game tickets from the Army and Navy Club, which runs a private railroad-car to the game.

654   MARINE CORPS SOCIAL CUSTOMS

2418 ■ **Marine Corps Social Customs**
Certain social customs are observed throughout the Corps and deserve mention here.

Wetting Down Your Commission. Whenever you are promoted, you are obligated to hold a "wetting-down party." At this affair your new commission (which is usually displayed at some conspicuous but safe vantage point) is said to be "wet down." When several officers are promoted together, you may join in a single wetting-down party.

Cigars. If you are either newly promoted or a new father, you distribute cigars to all officers and staff NCOs of your unit. If you are a newly promoted woman officer, candy (unless you happen to smoke cigars) is an acceptable substitute.

Five Aces. Any officer who rolls five aces when throwing dice for refreshments in a mess is obliged by tradition to buy a complete round of drinks for all the mess-mates present. In large messes this custom is eased to the extent that you have to buy drinks only for your own party.

Entering a Mess Covered. Unless you are on duty and under arms, if you enter a mess covered, you are liable to buy a round of drinks. Most messes adhere to and post the old rule: "He who enters covered here buys the house a round of cheer." In fact, some even have a bell and lanyard that may be rung by anyone present who spots an offender against this rule, thus signaling a free round.

Drawing Your Sworn in a Mess. The seagoing rule that any officer who unsheathes his sword in the wardroom must buy a round also applies on shore, if you are so unwary as to draw sword in any public room of an officers' mess. The custom goes back to the days of duelling, when this was one method of cooling off hotheads and restricting indiscreet sword-play.

Welcome on Board. Whenever a new unit arrives at a post, or a transport brings in an appreciable number of Marines or Marine dependents, the local Marine commanding officer or his representative, together with the post band, greets the newcomers.

Departure from a Post. When a unit or draft leaves, the commanding officer, band, and friends see them off. If the move is routine, the band plays "Auld Lang Syne" as aircraft embarkation is completed, or as the transport casts off her last line or the train gets under way. If the unit is on war or expeditionary service, "The Marines' Hymn" is the send-off. In either case, the departing unit should be played down to the airfield, dock or loading platform by "Semper Fidelis."

Christmas Cards. At Christmas time, it is customary for the officers' mess to reserve special display space for individual Christmas cards. All families affix personal greeting cards on this board. This not only adds Christmas color to the mess but constitutes an exchange of greetings among all who participate.

Special Courtesy to COs and Senior Guests. At any social functions—cocktail parties and receptions especially—you have certain special obligations to your commanding officer and his wife, and to the guest of honor, if any. On your arrival (or his, if he arrives after you do) both you and your wife should make it an immediate point to approach and speak to your CO and his wife. This is known as "making your number." Except when absolutely necessary, you should not depart before your CO and the guest of honor do so. If you must leave early, however, express your regret to your CO and ask his permission. It is a mark of the worst military manners and social upbringing if either you or your wife fail to observe these courtesies.

SOCIAL DOS AND DON'TS

2419 ■ Social Dos and Don'ts

Common sense, tact, and ordinary courtesy are the fundamentals of social success in the Marine Corps. For fine points you may wish to refer to the tested references listed in Section 2401, or to Emily Post, or Jean Ebbert's *Welcome Aboard*, or *The Marine Corps Wife*, by Sally Jerome and Nancy Shea, or some other recognized social guide. The following pointers are supplementary, therefore, but worth your perusal.

When you are on a post, on board ship, in uniform ashore, or otherwise recognizable as a Marine officer, your conduct must be impeccable. "If you must raise hell," runs an old Marine proverb, "do it at least a mile away from the flagpole."

It was once written, "The ideal income is a thousand dollars a day—*and expenses.*" Obviously you don't stand much chance of attaining this on Service pay, although a few inexperienced or improvident officers try to live as if they had it. You cannot fool anybody as to how much you make, so *live within your income.*

"Good clothes open all doors"—be sure yours are correct both for style and occasion. *And always check to see which uniform is prescribed, before you attend a social function.*

"Whoever gossips to you will gossip of you"; "It is easier to be critical than correct"—avoid personalities about other officers, and never vent destructive criticism of your Service, your unit, or your superiors.

Never serve bad liquor—"Use hospitality to one another without grudging."

Be punctual. It is never wrong to arrive exactly on time. For large cocktail parties, dances, receptions, and debuts, you may arrive *not later* than a half-hour after the announced time. For meals, be exactly on time. "Punctuality is the politeness of kings."

Don't load down social conversation with technical language or with labored application of Marine Corps terms to civilian matters. On the other hand, as a professional, learn the talk and nomenclature of the Corps. Use precise terms to convey precise meanings. Avoid undue shop talk and thus avoid the character whom Addison so well described:

> The military pedant always talks in a camp, and is storming towns, making lodgements and fighting battles from one end of the year to the other. Everything he speaks smells of gunpowder; if you take away his artillery from him, he has not a word to say for himself.

At a mess or at any official function, politics, religion, and ladies are discussed (if at all) only with the greatest discretion. Whatever you do, never speak ill of your Corps or of any fellow officer in the presence of outsiders, civilians, or members of any other Service. And remember always, insofar as public utterances are concerned, an American soldier has no politics and espouses no political party or cause.

Polite society is no place to play "the tough Marine." Courtesy and personal modesty are never more becoming than in an officer. Rudeness, abruptness, gory tales of blood and thunder, and coarse language usually show up the greenhorn or counterfeit, and certainly the ill-bred. "The bravest are the tenderest; the gentlest are the daring."

Remember that your wife does not and cannot wear your rank. Be certain that she understands this quite clearly and does not exhibit a tendency to put herself ahead of the wives of juniors or subordinates.

This will only belittle your rank in the eyes of others. Insist, however, that your wife, as an officer's lady, receive the courtesy due her from all.

"Be prepared" is just as good a social motto for Marine officers as for Boy Scouts. Before you attend any social function, ascertain the dress, whether there will be a receiving line, who will receive, when the line closes, who of importance to the Marine Corps may attend. All these are the "EEIs"—essential elements of information—which help place you at ease and prepare you for any social eventuality.

Teach me to be obedient to the Rules of the Game.
Teach me to distinguish between sentiment and sentimentality, admiring the one and despising the other.
Teach me neither to proffer nor to receive cheap praise.
If I am called on to suffer, let me suffer in silence.
Teach me to win if I may: teach me to be a good loser.
Teach me neither to cry for the moon nor to cry over spilt milk.
—Lines framed in his cabin by King George V of England, while serving as a naval officer.

We are all members of the same great family. . . . On social occasions the formality of strictly military occasions should be relaxed, and a spirit of friendliness and good will should prevail.
—John A. Lejeune

Appendices

- I: THE MARINES' HYMN
- II: COMMANDANTS OF THE MARINE CORPS
- III: THE IMPORTANCE OF BEING INSPECTED
- IV: READING FOR MARINES
- V: BROTHER MARINES
- VI: ART. 38, MARINE CORPS MANUAL, 1921
- VII: BIRTHDAY BALL CEREMONY

I: THE MARINES' HYMN

From the Halls of Montezuma
To the shores of Tripoli,
We fight our country's battles
In the air, on land, and sea.
First to fight for right and freedom,
And to keep our honor clean,
We are proud to claim the title
Of United States Marine.

Our flag's unfurl'd to every breeze
From dawn to setting sun;
We have fought in every clime and place
Where we could take a gun.
In the snow of far-off northern lands
And in sunny tropic scenes,
You will find us always on the job—
The United States Marines.

Here's health to you and to our Corps
Which we are proud to serve;
In many a strife we've fought for life
And never lost our nerve.
If the Army and the Navy
Ever look on Heaven's scenes,
They will find the streets are guarded
By United States Marines.

II: COMMANDANTS OF THE MARINE CORPS

Major Samuel Nicholas, 1775–1781
Lieutenant Colonel William Ward Burrows, 1798–1804
Lieutenant Colonel Franklin Wharton, 1804–1818
Lieutenant Colonel Anthony Gale, 1819–1820
Brigadier General Archibald Henderson, 1820–1859
Colonel John Harris, 1859–1864
Brigadier General Jacob Zeilin, 1864–1876
Colonel Charles G. McCawley, 1876–1891
Major General Charles Heywood, 1891–1903
Major General George F. Elliott, 1903–1910
Major General William P. Biddle, 1910–1914
Major General George Barnett, 1914–1920
Major General John A. Lejeune, 1920–1929
Major General Wendell C. Neville, 1929–1930
Major General Ben H. Fuller, 1930–1934
Major General John H. Russell, Jr., 1934–1936
Lieutenant General Thomas Holcomb, 1936–1944
General Alexander Archer Vandegrift, 1944–1948
General Clifton B. Cates, 1948–1952
General Lemuel C. Shepherd, Jr., 1952–1956
General Randolph McC. Pate, 1956–1960
General David M. Shoup, 1960–1964
General Wallace M. Greene, Jr., 1964–1968
General Leonard F. Chapman, Jr., 1968–1972
General Robert E. Cushman, 1972–1976
General Louis H. Wilson, Jr., 1976–

III: THE IMPORTANCE OF BEING INSPECTED

(This appendix consists of a slightly abridged and updated version of an article of the same title, by Lieutenant Colonel W. C. Stoll, which appeared in the May 1955 *Marine Corps Gazette*, and is reprinted by kind permission of the *Gazette*. As a compilation of highly practical advice for the junior troop leader, this article is most valuable.)

Inspections have a very important significance in every Marine officers' career. It is by inspections that the efficiency, morale, discipline, training, and leadership of a unit are often determined. Quite often a superior has little opportunity to observe certain subordinates. Thus, he often must rely on inspections to evaluate that officer's efficiency for reporting purposes, causing an inspection to be a very serious and important event.

Through numerous Inspector General, Fleet Marine Force, Division, Force Troops and Commanding Officer's inspections, it has become obvious that most junior officers and staff NCOs do not prepare their units properly for these important events.

Why is this? In search of an answer to this $64 question I have observed many inspections and made many queries with an attempt to correlate results of inspections with pre-inspection preparation, procedures, and planning.

As a result, I am of the opinion that many competent leaders do not have a follow-through program of checks and double checks to insure that all discrepancies are corrected. The following 5-step program for improving inspection results is offered, and excellent results are guaranteed if these steps are meticulously followed. Perhaps outstanding inspection results may be obtained without actually performing the below-listed steps, but why gamble when you can take a guaranteed course of action?

Step 1: *Check-Off List.* A detailed check-off list is a must. This list must not only be available to officers and NCOs but must be disseminated to all hands. If the paper supply permits, issue one copy to every man in your unit so that he can actually inspect himself as well as others.

Step 2: *Proper Inspection.* Instruction on how to prepare for inspections, including what and how to inspect, should be held periodically and included in the regular training schedule. Each Marine should be trained to inspect himself first and then others. Practice inspections should be held with Marines in ranks and certain discrepancies noted. Every other Marine is then given the opportunity to inspect and jot down the discrepancies. This not only gives each man valuable experience and sharpens his observation, but it also gives him interest in an otherwise dull subject.

Step 3: *Routine Weekly or Monthly Inspection.* The routine inspection must include all hands. Those persons on leave, in the hospital, on working parties, or otherwise absent from an inspection, must be inspected either in a special group or individually as appropriate after their return. Often a man will miss a routine inspection and therefore has discrepancies which are allowed to continue without correction. Measures must be taken to insure that inspections are thorough and that all men with descrepancies are re-inspected and discrepancies are corrected.

Step 4: *The Preliminary Inspection.* When an inspection is announced beforehand, as most Inspector General, Commanding General or Commanding Officer's inspections are, a detailed preliminary inspection by subordinate commanders should be held immediately after the announcement is made so that maximum time will be available to correct all discrepancies beforehand.

Step 5: *Last Minute Inspection.* It is always irritating and infuriating to a commanding officer to make an inspection and find troops with buttons unbuttoned, shoes not shined, fingernails dirty, cartridge belts not properly adjusted, hats not properly placed on heads, field scarves not properly tied and other numerous discrepancies that take but a moment to remedy on the spot. A thorough inspection administered a half an hour or so before the superior's inspection will correct these discrepancies.

If the program outlined below is followed, step-by-step, by all officers and NCOs in charge of units, the results of inspections will improve considerably and the efficiency, morale, discipline, and leadership of the units will improve proportionally.

SAMPLE INSPECTION CHECK-OFF LIST
UNIFORM
1. Is each item of regulation material?
2. Is each item tailored correctly?
3. Does each item fit?
4. Is item neat, clean, and pressed (if applicable) and properly marked?
5. Is each item in good repair, free from frayed edges, holes, and tears?
6. Are shoes shined and in good repair? Soles and heels not worn down?
7. Do shoe laces match shoes?

8. Are socks regulation color and material?
9. Are field boots clean, in good repair, and properly shined?
10. Headgear worn properly?
11. Garrison cap sewed properly?
12. Are chevrons and hashmarks worn properly?
13. Visibly bulging pockets?
14. Field scarves knotted properly?
15. Collar neat in appearance and unfrayed?
16. Proper insignia worn, and worn properly?
17. Emblems proper color, and neither shiny nor corroded?
18. Buckle and tip of belt shined?
19. Current authorized marksmanship badge worn in proper place?
20. Authorized ribbons worn in correct sequence, clean, unfrayed, positioned in accordance with *Uniform Regulations*, battle stars pointed correctly?
21. All insignia worn properly, anchors on collar emblems pointed inboard, cap emblem pointed correctly?
22. Is belt proper length of overlap (2⅔"–3¾")?
23. Are nonregulation items showing?
24. Are nonregulation markings showing?
25. Trousers proper length?

WEAPON
1. Is weapon clean and does it function properly?
2. Are magazines clean and functioning properly?
3. Bayonet clean and bayonet ring turned properly?
4. Does each man know the number of his weapon, name of weapon, safety device, nomenclature of principal parts, and field-stripping procedures?
5. Does each man know the zero of his weapon and his windage and elevation rules?
6. Rifle: Is sling properly placed and cared for, is stock properly cared for?
7. Is inspection arms done correctly and uniformly throughout unit?
8. Has man fired weapon he is armed with, at least for familiarization?

MISCELLANEOUS
1. Satisfactory posture?
2. Hair cut neatly, face shaven and mustache, if authorized and worn, neatly trimmed?
3. Hands, face, and fingernails clean?

4. Web equipment clean, serviceable, in good repair, and uniformly placed?
5. All items on cartridge belts worn uniformly?
6. Leather holsters shined?
7. Identification tags and identification card present?
8. ID card up to date and not cracked?
9. Does man hold up-to-date immunization card?
10. Eyeglasses (if required)? Two pairs?
11. (When helmet worn) Helmet cover skin-tight, right color out, chinstrap properly adjusted and on exact point of chin?

CLOTHING AND EQUIPMENT ON BUNK
1. Display neat, regulation and uniform with other displays throughout unit?
2. Is all clothing displayed, regardless of required amount?
3. Clothing clean, in good repair and marked in accordance with regulations?
4. Missing clothing accounted for by a laundry, cleaning, or cobbler's chit and certified by platoon leader?
5. Equipment all present and uniformly displayed?
6. Canteen corks serviceable?
7. Extra shoe laces present?
8. Are all improper markings on clothes blocked out neatly?
9. Has shelter half all buttons, tie-down loops, and no holes?
10. Tent pins, poles, and line, in serviceable condition?
11. All snaps (e.g., first-aid pouch and canteen cover) working easily, but holding firmly?
12. All metal ends on web straps present and not deformed?
13. Do hooks for D-rings have spring catch?
14. Does mess gear close tightly and is it free of dirt or corrosion?
15. Canteen cup free of rust and dirt?
16. Is entrenching tool clean and does locking nut move freely?
17. First-aid packet present and unopened?

FIELD DISPLAY
1. Does each individual know proper method of laying out equipment?
2. Is display uniform thoughout unit?
3. Are all items present?
4. Does first-aid packet have necessary component items?
5. Canteen cork serviceable?
6. All equipment serviceable?
7. Identification tags present?

OTHER ITEMS
1. Locker boxes, locker, and other personal gear neatly stowed in accordance with local requirements?
2. Name tag on bunk?
3. Sheets, bedding, and blankets clean and in serviceable condition?
4. Bunk in good repair, all parts present and double-deck bunks secured?

CHECK-OFF LIST FOR OFFICERS
1. Does each unit commander in a company know each and every man by name?
2. Does each platoon leader or section leader keep a record with pertinent data on each of his men?
3. Is each subordinate leader thoroughly familiar with messing of troops and do they visit troop mess halls frequently?
4. Do subordinate leaders know location and time of operation of all recreation and post exchange facilities available to the troops?
5. Are subordinates familiar with liberty transportation schedules?
6. Does each leader know how many men are enrolled in MCS/MCI correspondence courses and which; and what recent disenrollments and why?
7. Article 137, UCMJ, being fully complied with and carried out?
8. Do all of your officers and NCOs know capabilities, use and characteristics of all equipment in their charge?
9. Do they keep a record of people absent from formations, where orders are promulgated or important information is put out? If a record is kept, are plans made to get this information to absentees as soon as possible?
10. Does each officer and NCO have a personal interest in basic training of his troops? Does he keep a personal record of troop training other than the record kept by the companies?
11. Do the subordinates give personal attention to their troops and make themselves available to troops for instructions after hours when special instruction is needed?
12. Do subordinates comply with spirit of request-mast rights of troops and make it easy for a man to see his commanding officer without a lot of red tape?
13. Does each subordinate leader know how many troops he has, exactly where they are and what they are supposed to be doing?
14. Does each leader know what equipment or supplies his unit should have but is unserviceable or missing, and has he initiated action to remedy the situation?

15. Are all needs for repairs, supplies, or shortages of personnel covered in writing by appropriate work requests, requisitions, or personnel requests?
16. Are any men on mess duty in violation of current regulations?
17. Are all pertinent regulations and orders affecting troops repromulgated regularly to individuals just joining the command?
18. Are all men joining the organization properly indoctrinated by their company officers and NCOs in a planned program?
19. Are people in brig being paid health and comfort allowances; are hospital patients and prisoners visited by their company officers regularly?
20. Are all hands aware of various benefits of a Marine Corps career?
21. Do troops have access to all pertinent regulations, chart instructions for marking clothes, and uniform regulations?
22. Do all troops and leaders know meaning of various security classifications?

IV: READING FOR MARINES

> *Reading maketh a full man, conference a ready, and writing an exact man.*
>
> —Francis Bacon

This appendix suggests a few books that it will profit you as a Marine and as a professional soldier to know. *See if you can finish these by the time you finish Basic School.* The method of choice has been arbitrary, and there are many more which you will enjoy and which will benefit you greatly that are not listed here. Certain professional journals are listed. To be an up-to-date, well-rounded officer abreast of the times, you should subscribe to those mentioned in this appendix. In addition to the following listings, you should seek out and read the excellent official historical publications of Marine Corps Headquarters—not only the operational histories of World War II and Korea, but the comprehensive monographs on each of the World War II operations—and also the entertaining and vivid biographies of such Marines as Lejeune, Butler, Wise, Holland Smith, Vandegrift, and Puller. For the feel of the "Old Corps," nothing can surpass the several collections of stories by Colonel John Thomason (his own selection of which is embodied in an anthology, *... And a Few Marines*). When you branch out, remember that, as a Marine officer, you are also in the broadest sense a naval officer, and learn your naval history and the lore of the sea.

Most of the works recommended can be obtained through the U. S. Naval Institute, at a discount to members; the *Marine Corps Gazette* Bookshop gives a comparable discount. For professional books out of print, try Goodspeed's, in Boston; or Francis Edwards, in London (whose military collection is perhaps the finest in the world).

BASIC BOOKS FOR MARINE OFFICERS:

Earle, Craig, and Gilbert, *Makers of Modern Strategy*
Fehrenbach, T. R., *This Kind of War*
Fuller, J. F. C., *A Military History of the Western World*
Griffith, S. B., *The Battle for Guadalcanal*
Heinl, R. D., *Soldiers of the Sea*
Heinl, R. D., *Victory at High Tide*
Liddell Hart, Sir B. H., *Strategy*
Lovette, L. P., *Naval Customs, Traditions, and Usage*
Masters, John, *Bugles and a Tiger*
McLeave, Hugh, *The Damned Die Hard*
Millis, Walter, *Arms and Men*
Moorehead, Alan, *Gallipoli*
Morison, S. E., *Two-Ocean War*
Sherrod, R., *Tarawa*
Stallings, Laurence, *The Doughboys*
Thomason, J. W., *Fix Bayonets!*
West, F. J., *The Village*

PROFESSIONAL JOURNALS:

Marine Corps Gazette (PO Box 1775, MCS, Quantico, Virginia 22134)
U. S. Naval Institute Proceedings (Annapolis, Maryland 21402)
Navy Times (475 School St., S.W., Washington, D.C. 20024)
Military Affairs (Dept. of History, Kansas State University, Manhattan, Kansas 66506)

> NOTE: The cost of all books, and subscriptions to all journals recommended in this appendix, are tax-deductible on your income tax as professional books and periodicals.

V: BROTHER MARINES

The bonds of professional comradeship that knit most of the Marine Corps in existence today are unusually strong. As a member of the world's largest (though not oldest) Marine Corps, you should know of the distinguished bodies of Marines under other flags, with which our Corps has associations.

The Royal Marines

I never knew an appeal to their courage or loyalty that they did not more than realize my expectations. If ever the hour of real danger should come to England, the Marines will be found the country's sheet-anchor.
—Lord St. Vincent

Britain's Royal Marines, elder brothers of the U. S. Marine Corps, were 111 years old in 1775 when our own Corps was founded. From inception, the infant American corps was modeled after its illustrious British prototype, and many of the traditions of our Corps today can be traced to the Royal Marines.

As a result, despite early fallings-out (as in 1775, at Bunker Hill, and in 1814, when Royal Marines burned Washington after the Bladensburg fight), the camaraderie between U. S. and British Marines is a tradition of both Corps, and knowledge of the Royal Marines is part of every U. S. Marine's fund of information.

The Royal Marines perform much the same duties as U. S. Marines, with certain variations. They do not, for example, provide security personnel for naval stations, nor do they have a Fleet Marine Force, although the Commando Brigade, Royal Marines, units of which are always embarked in "commando ships" (LPH), performs many similar functions. Royal Marines provide all bands for the British Navy, and the Corps has the combat mission of maintaining commando, or amphibious raiding, troops for the British armed forces. Although amphibious warfare is not a primary mission or responsibility for the Royal Marines, but rather (in the United Kingdom) a joint, inter-Service affair, the British Marines have, through natural inclination, tradition, and background, always played a major role in England's long amphibious history.

The Royal Marines' badge, like our own, is the globe, though in this case the half shown is the eastern rather than the western hemisphere. Their right to wear "the Great Globe itself" as their emblem was conferred in 1827 by King George IV. Surrounding the globe is a laurel wreath, which was won in 1761, in recognition of the Royals' storming of Belle Isle. The motto of the Royal Marines is truly descriptive: *Per Mare, Per Terram* ("By Land and by Sea"). Like all Royal regiments, they are entitled to display the Lion and Crown as part of their Corps device and on their Colors.

The uniforms of the Royal Marines are much like our own. Aboard ship and on certain shore duties, they wear blues. Their ranks, rank insignia, and field undress unforms are those of the British Army, but the color of the service uniform is forest green. All Royal Marines wear a blue beret as one type of headgear, but members of commando units wear green berets, since green has always been the traditional commando color.

The official colors of the Corps are scarlet, yellow, green, and blue. These colors appear on the Royal Marines necktie, which is worn with civilian clothing by all members of the Corps, in the same way as our own Corps necktie.

The "Birth of the Corps Day," which corresponds to our November 10th, is 28 October of each year. The Royal Marines were organized in 1664.

The Sovereign, or a member of the royal family, is Captain-General of the Corps. At present this post is filled by the Duke of Edinburgh, husband of Queen Elizabeth.

The principal stations of the Royal Marines are the major barracks (Stonehouse) at Plymouth; Commando School, Bickleigh; Infantry Training Center, Lympstone; Amphibious School, Poole; and a recruit depot and school of music at Deal.

Although U. S. and British Marines have served side by side on many occasions, both Corps particularly cherish associations stemming from the Boxer Uprising and from the Korean War. In the Boxer Uprising, U. S. and Royal Marines formed the backbone of the band of Western troops who defended the Legation Quarter in Peking throughout a long and bloody siege in 1900. In addition, in the International Brigade which finally relieved both Peking and Tientsin, U. S. and British Marines were formed side by side. Fifty years later, a Royal Marine commando was attached to the 1st Marine Division in Korea, and served with the Division throughout the Chosin Reservoir campaign. And it was at Suez, in 1956, that the Royal Marines conducted the first carrier-based helicopter assault landing ever executed in combat.

Royal Netherlands Marines (Korps Mariniers)

The *Korps Mariniers*—as the Dutch Marines are officially entitled—were founded on 10 December 1665 in the Dutch Wars which caused the British to form the Royal Marines. One of the most important early operations of the Netherlands Marines was the amphibious raid up the Thames in 1666, one of the few occasions when foreign troops have landed in Great Britain since the Norman Conquest. Subsequently the *Korps Mariniers* performed normal sea duty and garrison duty throughout the Dutch empire. During World War II, when Holland was overrun by the Germans, several thousand Dutch Marines were trained at Camp Lejeune as the basis for reconstitution of the Corps, and the relationship between our two Corps has since been close. *Qua Patet Orbis* ("To the Ends of the World") is the Dutch Marines' motto; their uniforms, both service and dress, are similar to those of the Royal Marines and of our own Corps.

Today, the *Korps Mariniers* serves in ships' detachments, in colonial garrison duty, and at the Corps depot, Marine Barracks, Doorn. In addition to Doorn, the Dutch Marines have an amphibious training center at Texel and another marine barracks at Rotterdam, where Netherlands Marine Corps Headquarters is also located.

The basic tactical formation of the *Korps Mariniers* is the QPO Company (standing for the words in the Corps motto, *Qua Patet Orbis*). A QPO Company is a heavily reinforced rifle company with added crew-served weapons, an amphibious reconnaissance element, and headquarters and service personnel for independent operations. These units, which are stationed in Holland and abroad, comprise the force-in-readiness of the Netherlands Marines.

The ceremonial functions of the Corps are to provide the Netherlands Marine Band and to act as ceremonial troops for the Dutch government. Marines are the only Dutch troops authorized to have fifes and drums, and this music is the Marine trademark in Holland.

Spanish Marines (Infanteria de Marina Española)

Dating from the *Tercios de la Armada Naval* of Spanish Armada days and earlier, Spain's *Infanteria de Marina* can claim four centuries of service. Its men fought at Lepanto (1571), and with the Armada in

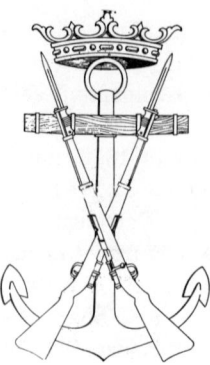

1588; defended Cartagena in 1741; took Sardinia in 1748; and served gallantly in the Peninsular War, Cuba, the Philippines, Guam, Morocco, Cochin China, and the Spanish Civil War.

The missions of the Spanish Marines are to provide ships' detachments and garrison and defense forces for the naval districts and bases of Spain and to maintain a BLT-scale expeditionary landing force.

The *Infanteria de Marina* today maintains *Tercios* (light infantry regiments) at El Ferrol, Cartagena, and Cadiz, together with separate battalions in Madrid and the Canary Islands. The Cadiz *Tercio* is part of the so-called Special Group, including the Marine Corps School, school troops, and research development activities. The major general commandant of the Corps, officially titled "Inspector General of the Marine Corps," has his headquarters at Madrid.

Spanish Marine officers' uniforms are those of the Navy, with distinguishing badges; enlisted Marines wear dress blue similar to those of the Royal Marines, and combat/utility uniforms resembling those of our own Corps. The emblem of the *Infanteria de Marina* is an anchor (up-and-down) with crossed rifles, surmounted by the crown of Spain.

The Spanish Marines' motto is "Valiant on Land and Sea." Since 1701 the traditional colors of the Corps have been red and blue. As in the case of our own Corps, Horse Marines are both a tradition and a joke with the Spanish, dating from the fact that, during the nineteenth century guerrilla operations in Cuba, "Navy Cavalry" mounted units were formed of Marines.

Argentine Marine Corps

The origin of the Argentine Marines goes back to 1807 when a naval battalion was organized to defend Buenos Aires against British attack. Subsequently, during Argentina's War of Independence, Marines served

on board warships and conducted landing operations. In 1879, a Marine artillery battalion was formed, to man coast defenses at Argentina's seaports and naval bases. In 1947, following World War II, the Corps was reorganized along modern amphibious lines, and a U. S. Marine advisor was provided.

The major operating units of the Argentine Marine Corps includes a cold-weather center garrisoned by a battalion in Patagonia, a riverine battalion on the River

Plate delta, and another battalion at Rio Santiago naval base. The 2d Marine Force includes the Marine Brigade, an amphibious RLT, based at Puerto Belgrano.

The uniforms and ranks of the Corps are similar to those of our own. The annual birthday ceremonies are held on "Day of the Marine Corps," 19 November.

Brazilian Marine Corps (Corpo do Fusileiros Navais)

The *"Fusileiros,"* as the Marines are known throughout Brazil, date their lineage back to the Portuguese Marines, which were founded in 1797. Units of this organization first came to Brazil in 1808, and 7 March, the date of their landing, is the birthday of the Corps in Brazil—which was then an overseas dominion of Portugal and subsequently separated amicably from the mother country.

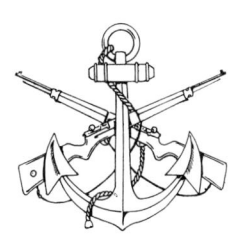

Brazilian Marines fought in their country's wars throughout the nineteenth century, including major riverine operations along the River Paraguay. The most recent expeditionary service of the *Fusileiros* was as part of the Inter-American Peace Force, which kept order in the Dominican Republic for 15 months in 1965 and 1966, side by side with U. S. Marines during part of that time.

The Brazilian Marine Corps is divided into operating forces (which include a Fleet Marine Force and security forces and ships' detachments) and a supporting establishment, which functions in the same way as our own. The *Fusileiros'* headquarters and FMF (shaped around a divisional nucleus) are located at Rio de Janeiro.

Colombian Marine Corps

The first combat landing by Colombian Marines took place on 11 November 1811, less than a year after their organization during their country's War of Independence. Throughout the nineteenth century the

Corps had its ups and downs, but it was permanently constituted as amphibious and expeditionary troops in 1936. Ever since 1948, during Colombia's prolonged struggle to win over communist banditry, the Corps has been continually engaged in riverine, amphibious, and pacification duties. Like our own Corps, the Colombian Marines carry out operations in both the Atlantic and the Pacific. The corps includes one tactical battalion. Virtually all its officers today are graduates of Basic or Amphibious Warfare Schools at Quantico, and many of its NCOs are also graduates of U. S. schools.

Venezuelan Marine Corps (Infanteria de Marina Venezolana)

The Venezuelan Marine Corps was formed on 22 July 1822—as in the case of most of the other South American Marines—during their country's War of Independence. During the nineteenth century, however, it became inactive, and it was not officially reconstituted until 1938.

The missions of the Venezuelan Marines includes amphibious operations, counterguerrilla and pacification duties, and naval base security. The Corps regularly conducts BLT-level landing exercises, and over a third of its officers are graduates of Marine Corps Schools.

The Venezuelan Marine Corps is made up of three battalions, based at Puerto Cabella and Maiquetia, with headquarters at Caracas. All these units have been active in Venezuela's defense against Castro-communist guerrillas and seaborne infiltrations from Cuba.

Republic of Korea Marine Corps (ROKMC)

The Korean Marine Corps, which has fought side by side with U. S. Marines in two wars—Korea and Vietnam—was founded on 15 April 1949, at Chinhae, destined to become the Quantico of Korea. Within less than two years, the 1st Korean Marine Regiment had become an integral part of the 1st U. S. Marine Division and played an outstanding part in the three years of hard fighting.

The primary mission of the ROKMC is to conduct amphibious landings as part of the national mobile striking force and to serve as a portion of the natioanl force in readiness. In addition, like our Corps, they perform security duty for the naval shore establishment and are responsible for the development of amphibious warfare doctrine, tactics, techniques, and materiel.

In addition to maintaining a brigade in the main line of resistance at Kimpo, the ROKMC has two Marine Corps Bases, Chinhae and Pohang. Marine garrisons or security units are found at Seoul, Paeng Yong Do, Cheju-Do, Pusan, Muk-Ho, Inchon, and Mok Po.

The uniforms of the ROKMC are similar to those of the U. S. Marine Corps. The official color of the Korean Marine Corps is scarlet. The creed of the Corps, which serves as its motto, is as follows:

Loyal to the nation
Be ever victorious
Unite as a family
Honor is worth more than life
Love your fellow countrymen.

Chinese Marine Corps

The Marine Corps of the Republic of China—at the time of writing, the second largest Corps of Marines in the free world—dates from 1917. It has a combat record extending throughout China's anti-Japanese war up to the defense and resupply of the Kinmen Islands in 1958, for which units of the Chinese Marine Corps were awarded the Banner of Honor.

The missions of the Corps closely resemble those of our own Corps, and are threefold:

1. To provide a Fleet Marine Force for the conduct of amphibious landing operations.
2. To provide security forces for naval shore bases.
3. To develop doctrine, tactics, technique, and equipment for amphibious operations.

Chinese Marine Corps Headquarters, FMF Headquarters, Schools, Reserve Training Center, and a Marine brigade are grouped at Tsoying on Taiwan. The 1st Chinese Marine Division and the Recruit Depot are located in the vicinity of Fang-Shan.

The colors of the Chinese Marines are scarlet and gold. The Corps has a Chinese motto which is translated *Semper Fidelis.*

Royal Thai Marine Corps

The Royal Thai Marine Corps was founded in 1932. Its missions are amphibious operations, naval base security, counterinsurgency, and support, when required, of the Thai Army. Because it is part of the Navy, the Corps uses naval rank and rating titles.

The principal operating unit of the Thai Marines is the Marine brigade, composed of three rifle battalions, an artillery battalion, and supporting headquarters and service units.

Sattahip, on the Gulf of Thailand, is the main base of the Thai Marines, but Chantaburi, near the Cambodian frontier, is the secondary base. In addition to being headquarters for the Corps, Bangkok is also the home station for a Marine garrison.

VI: ART. 38, MARINE CORPS MANUAL, 1921

Every November Tenth, the central part of the ceremony is the publication to all hands of Article 38, *Marine Corps Manual*, 1921, which was written especially for this purpose by John A. Lejeune, the 13th Commandant. While this text and its introduction are found in the *Marine Corps Manual* today, it is reproduced here as a matter of convenience for those who may not have a *Manual* within easy reach.

On November 1st, 1921, John A. Lejeune, 13th Commandant of the Marine Corps, directed that a reminder of the honorable service of the Corps be published by every command, to all Marines throughout the globe, on the birthday of the Corps. Since that day, Marines have continued to distinguish themselves on many battlefields and foreign shores, in war and peace. On this birthday of the Corps, therefore, in compliance with the will of the 13th Commandant, Article 38, United States Marine Corps Manual, Edition of 1921, is republished as follows:

"(1) On November 10, 1775, a Corps of Marines was created by a resolution of the Continental Congress. Since that date many thousand men have borne the name Marine. In memory of them it is fitting that we who are Marines should commemorate the birthday of our Corps by calling to mind the glories of its long and illustrious history.

"(2) The record of our Corps is one which will bear comparison with that of the most famous military organizations in the world's history. During 90 of the 146 years of its existence the Marine Corps has been in action against the Nation's foes. From the Battle of Trenton to the Argonne, Marines have won foremost honors in war and in the long era of tranquility at home generation after generation of Marines have grown gray in war in both hemispheres, and in every corner of the seven seas that our country and its citizens might enjoy peace and security.

"(3) In every battle and skirmish since the birth of our Corps, Marines have acquitted themselves with the greatest distinction, winning new honors on each occasion until the term "Marine" has come to signify all that is highest in military efficiency and soldierly virtue.

"(4) This high name of distinction and soldierly repute we who are Marines today have received from those who preceded us in the Corps. With it we also received from them the eternal spirit which has animated our Corps from generation to generation and has been the distinguishing mark of the Marines in every age. So long as that spirit continues to flourish Marines will be found equal to every emergency in the future as they have been in the past, and the men of our Nation will regard us as worthy successors to the long line of illustrious men who have served as 'Soldiers of the Sea' since the founding of the Corps."

The inspiring message of our 13th Commandant has left its mark in the hearts and minds of all Marines. By deed and act from Guadalcanal to Iwo Jima, from Inchon to the Korean Armistice, from Lebanon to Taiwan, the Marines have continued to epitomize those qualities which are their legacy. The success which they have achieved in combat and the faith they have borne in peace will continue. The Commandant and our many friends have added their hearty praise and congratulations on this, our . . . . . . birthday.

VII: BIRTHDAY BALL CEREMONY

The following is an outline for conducting the Marine Corps birthday-ball ceremony in a medium-sized command with drum and bugle corps (or at least a field music) and an orchestra available. Bear in mind that this is a guide, and may be modified or improvised according to local resources and traditions.

- At H-15 minutes, drum and bugle corps sounds *Officers' Call*.
- Adjutant (who acts as announcer) requests that officers, guests, and ladies clear the floor for the ceremony. Floor Committee place line and stanchions (if used) to define ceremonial aisle and area.
- At H-5, D&B, color guard, and honor guard (see below), form at exit, prepared to march on.
- At H-1, adjutant takes post on floor, adjacent to exit, and, at H-hour, when all hands are posted, commands, "Sound *Adjutant's Call*."
- D&B sounds *Adjutant's Call*, then marches up the aisle to designated post, playing "Foreign Legion March," or "Sea Soldiers."
- (When D&B halts, historical pageant, if any, commences. At conclusion of pageant—or next event, if no pageant—orchestra plays "Semper Fidelis.")
- On first note of "Semper Fidelis," honor guard steps off.

   NOTE: For an officers' birthday ball, honor guard consists of two officers of each grade; at small posts, where the ball is an all-hands party for the whole command, honor guard consists of two lieutenants, two staff NCOs, two sergeants, and two corporals. All honor guard members are covered and wear Mameluke or NCO Sword as appropriate.

- Honor guard, junior rank in lead, proceeds up the aisle two abreast, each pair at six-pace intervals. At six paces inside hall, senior man in leading pair commands, 1. Officers 2. HALT. Without further

command, pair face outboard, take three paces, halt, and face about. Six paces farther, the next junior pair repeats this evolution, etc. In each case the only spoken command is 1. Officers 2. HALT, the remaining movements being executed simultaneously in cadence without command. When the honor guard is posted, the orchestra stops playing.
- D&B sounds *Attention*.
- Senior Marine commander and honored guests (the official party) enter and march up aisle, face about, take post at head of aisle abreast of senior pair of honor guards, and receive honors (if a flag or general officer is present) from D&B.
- Orchestra commences "Stars and Stripes Forever." Color guard enters from exit and marches up aisle, halting abreast of next senior pair of honor guards. Music ceases when color guard halts.
- Adjutant, from original post at rear, proclaims, *"Long live the United States, and success to the Marines!"*
- D&B plays *To the Color*. All covered officers come to hand salute. Colors then take designated post.
- Fanfare by D&B.
- Orchestra commences *The Marines' Hymn*. Birthday cake is wheeled in from exit by four-person cake escort, followed by the adjutant. Cake is posted abreast of second senior pair of honor guards. Cake escort takes post in rear of cake.
- Adjutant steps front and center between cake and official party.
- Senior Marine commands, "Publish the Article, Sir."
- The adjutant then publishes Art. 38, *Marine Corps Manual 1921*, and resumes his post.
- Senior Marine steps forward to make remarks, followed by remarks, if any, by honored guest.
- At conclusion of remarks, adjutant steps forward and hands senior Marine an unsheathed Mameluke Sword (previously placed on cake table), with which senior Marine cuts cake while orchestra plays "Auld Lang Syne."
- Senior Marine then introduces and presents cake slice to youngest and oldest Marines present.
- Cake escort then retires cake to a flank where it is received by waiters.
- D&B commences "Semper Fidelis." Senior Marine and official party retire from post and proceed to head table or box.
- Color guard marches off, followed by the honor guard in reverse sequence (senior pair leading). As the rear rank of honor guard comes abreast of next pair, the senior of that pair commands 1. Forward. 2. MARCH, and the pair marches three paces inboard,

face right and left respectively, and step off without further command. The D&B marches off at six paces behind final pair of honor guards. On passing through exit, each D&B player mutes his instrument so the music will seem to fade away in the distance.
- Floor Committee removes line and stanchions. D&B ceases playing, and ceremony is ended.

> NOTE: When senior to senior Marine (for example, an ambassador, Secretary of the Navy, etc.), honored guest is asked to cut cake by senior Marine, who then introduces youngest and oldest Marines, who in turn receive slices from honored guest.

Glossary

The glossary which follows contains a compilation of terms currently peculiar to the Marine Corps. Certain of these may be recognized as belonging also to one of the other Services; in such cases, however, the term has been incorporated here only by virtue of long inclusion as part of the Marines' distinctive vocabulary.

Airedale: Aviator.
All hands: All members of a command; everybody.
Ashore: (1) On the beach, as differentiated from on board ship; (2) any place off a Marine Corps or Government reservation. **Go ashore:** go on liberty, or leave the reservation.
Asiatic (adj): Mildly deranged or eccentric as a result of too much foreign duty; **(n)** one who has "missed too many boats."
Aye, Aye, Sir: Required official acknowledgement of an order, meaning, "I have received, understand, and will carry out the order or instructions."
Barracks cap: Frame type, visored cap, so-called because this type of headgear was traditionally prescribed for non-FMF organizations.
B & W (n): Solitary confinement on bread and water, now only authorized on board ship; sometimes spoken of as "cake and wine."
BCD (n): Bad-conduct discharge.
Binnacle list: List of men placed on light duty by the surgeon; in old days it was posted on or near the binnacle.
Blue Book: *Combined Lineal List of Officers of the Marine Corps on Active Duty*; also the *Register of Commissioned and Warrant Officers of the U. S. Navy and Marine Corps*.

Blues: Dress or undress blue uniform.

Boondockers: Field shoes or boots.

Boondocks (n): Woods, jungles, faraway spaces; semifacetiously defined as "that portion of the country which is fit only for the training of Marines."

Boot: A recruit.

Boot camp: Recruit depot.

Break out (v): (1) To unfurl; (2) to remove from storage; (3) to arouse.

Brig: Place of confinement aboard ship or ashore at a Marine Corps or naval station; the post prison. **Brig time:** confinement.

Brig rat: One who has served much brig time, a habitual offender.

Bulkhead: (1) **(n)** A wall; (2) **(v)** to complain against or asperse a superior while superficially pretending not to.

Calk off (v): (1) Take a nap; (2) loaf on the job.

Cannon-cocker: Artilleryman.

C & S (adj): "Clean and Sober," notation formerly entered on the liberty list beside the names of men returning from liberty in that condition.

Charge-of-Quarters: Duty noncommissioned officer responsible for safety and good order in a barrack or billet. Abbreviated as "CQ."

Charger: Highly motivated, aggressive Marine (contraction of "hard-charger").

Chaser: Contraction of "prisoner-chaser," an escort for a prisoner or detail of prisoners.

Chew out (or on): Reprimand severely.

Chief messman: Permanently detailed assistant to the mess sergeant, in charge of all messmen and responsible for the police and good order of the messhall.

Chit: Acknowledgment of indebtedness to a mess; a receipt or authorization; in general, a small piece of paper.

Chopper: Helicopter.

Chow: Food, rations.

Chow bumps: Two short blasts sounded by the field music five minutes before Mess Call.

Chow-down: Phrase meaning, "a meal (or food) is ready."

Chow hound: One who appreciates his food.

CG: The Commanding General.

Class VI: Alcoholic beverages of any kind.

CMC: Commandant of the Marine Corps.

CO: The Commanding Officer.

Clutch (n): A serious, sudden emergency.

Clutched-up: Nervous, panicky.
Color sergeant: A distinguished noncommissioned officer given the privilege of carrying the National Color and of commanding the color guard.
Communicator: Officer or enlisted man assigned to or specializing in communication duties.
Corpsman: Enlisted man of the Navy Hospital Corps.
Cover (n): A Marine's cap or hat; headgear.
Cruise (n): An enlistment, sometimes erroneously spoken of as a "hitch" (Army term).
Crumb: Untidy or uncleanly individual.
Crummy: Untidy or uncleanly in person or uniform.
Crum up (v): To neaten one's person and uniform, or articles thereof.
Cruncher: Aviation term for Marine assigned to a ground unit; contraction of "gravel cruncher."
Crying towel: A towel said to be employed by those with many troubles or complaints, to wipe away their tears; a crying towel is said to hang in every chaplain's office.
Cumshaw: (1) (n) Something free, gratis, obtained at no cost; (2) (v) To obtain something at no cost or with no accountability in the supply system.
Cut it: See "hack it."
D & D (adj): Drunk and disorderly, an entry formerly made on the liberty list beside the name of any man returning from liberty in that condition.
DD (n): Dishonorable discharge.
Deck: (1) (n) The floor, the surface of the earth; (2) (v) to knock down with one blow.
D.I. (n): Recruit-depot drill instructor, ordinarily an experienced drillmaster.
Dinged (adj): Hit, as by a bullet; *to be* . . . , to be hit by enemy fire.
Doc: Navy hospital corpsman.
Doggie (n): Diminutive for "dog-face," an Army enlisted man.
Dope: (1) Information; (2) sighting and/or wind correction for a rifle under given conditions; *bad* . . . , misinformation.
Dungarees: (1) Navy denim working uniform; (2) Marine Corps green utility clothing.
Eight-ball (n): Worthless, troublesome individual; one who deservedly remains "behind the eight-ball."
Emblem: United States Marine Corps Emblem, or Corps badge, adopted in 1868, frequently referred to as the Globe-and-Anchor.
EPD (n): Extra police duties.

Extend: To lengthen a current enlistment by contracting to remain in the service one or more years after the enlistment would ordinarily expire.

Fall out (v): To assemble outside barracks, immediately prior to a formation.

Field boots: Heavy half-boots designed and issued for field service; boondockers.

Field Day: Day or portion of a day set aside for general cleanup or police of an organization or area.

Field hat: Broad-brimmed felt hat with four-dent crown, formerly worn on expeditionary service by the Marine Corps, but now only worn at rifle ranges and recruit depots; often erroneously called "campaign hat" (Army term for same type headgear).

Field music: (1) Drummer or trumpeter; (2) a small drum and bugle corps organized by grouping together all the field musics within a command.

Field scarf: Regulation Marine Corps khaki necktie (obs).

First Soldier: First sergeant.

Flag allowance: Marines assigned to duty in an admiral's headquarters.

Flatfoot (n): Bluejacket, sailor.

FMF (n): Fleet Marine Force.

Fore-and-aft cap: Garrison cap, also referred to as a "p-ss cutter."

Foul up: (1) (n) A mistake, botch, bungle, or confused situation; (2) (v) to confuse or bungle. **Fouled up:** badly confused.

Frock (v): To grant official permission for an officer who has been selected but not yet made his number, to assume the style, title, uniform, and authority of the next higher grade.

Frost-call (n): A procedure within a command whereby all officers and other key personnel may be alerted by sequential telephone calls or other notification.

Furlough: Period of authorized leave for an enlisted man, not to be confused with a "48" or "72."

Galley: (1) Kitchen of a mess hall; (2) mobile field kitchen; (3) ship's kitchen.

Gear: Equipment; ... *pack the gear*: measure up to Marine standards.

General mess: The enlisted men's mess.

Giz (n): Diminutive for "gizmo," any miscellaneous, nondescript, unidentified thing or gadget.

Gizmo: See "giz," above.

Globe-and-Anchor: Marine Corps Emblem.

Gravel cruncher: See "cruncher."

Greens: Marine Corps green service uniform.

Grinder: Drill field.

Ground-pounder: See "cruncher."

Grunt (n): Aviation term for a rifleman. See also, "cruncher," above.

Gung-ho: (1) **(n)** Aggressive *esprit de corps*; (2) **(adj)** Hard-charging.

Gunner: Contraction of "Marine Gunner," the title for line warrant officers.

"Gunny": Contraction for gunnery sergeant.

Gunship: Armed helicopter.

Hack (n): Arrest, officer's; *to be in or to be under* . . . : to be under arrest.

Hack it: To be competent or successful in a job or assignment, as, "Do you think Corporal Calkoff can hack it as a squad-leader?"

Hands, all: All members of a command.

Happy Hour: (1) An evening smoker of athletic and recreational events for all hands (obs.); (2) late afternoon period during which the price of drinks at an officers' or NCOs' mess is sharply reduced.

Hard-charger: Aggressive, dynamic, zealous, indefatigable officer or enlisted Marine; one who is professionally keen.

Hashmark (n): Service stripe worn on the uniform sleeve by enlisted men for completion of an honorable four-year enlistment in any of the U. S. Armed Services.

Head (n): Toilet facility; latrine.

Heel-and-toe watch: A condition during which watch-standers alternate tours, one individual relieving the other, and vice versa, for an indefinite period.

Hill, to go over the (v): To desert.

Hill, to run over (v): To force an individual to desert, apply for a transfer, or for retirement, as, "Captain Hardnose certainly ran that brig-rat over the hill."

Holiday routine: Condition during which routine drills, instruction, training, and work are knocked off (q.v.) throughout a command; routine followed on authorized holidays and Sundays.

I & I (n): Inspector-instructor, a regular officer assigned to supervise the training of a Reserve unit.

ID card: Armed Forces identification card, issued to every member of the U. S. Armed Forces.

IG (n): The Inspector General.

IG Inspection: An official inspection of a command or unit (usually annually) by the Inspector General or his representatives.

Iron Mike: Nickname bestowed on statue of World War I Marine in front of old Post Headquarters, Quantico (now the Marine Corps Association offices).

JO (n): Junior officer.
Joe: Coffee.
Joe-pot: Coffee pot, percolator.
Junk on the bunk: Periodic inspection of equipment or, more loosely, of clothing and equipment, displayed on the bunk.
Khakis: Summer service uniform (obs).
Knock off: To cease forthwith.
Lad: Generic term of address for any enlisted Marine, regardless of rank.
Ladder: (1) **(n)** Stairs or stairway; (2) **(v)** to adjust gunfire by a series of graduated spots in range.
Liberty: Authorized free time ashore or off station, not counted as leave.
Liberty list: Periodic (usually daily) list prepared by the first sergeant, containing the names of enlisted men entitled to liberty, employed by the guard in checking enlisted personnel on and off the ship or station.
Line company: Originally, a separate, numbered Marine company performing infantry duties (obs); now, the aviation term for ground units or organization.
Line duty: General duty in a ground organization of the Marine Corps.
Lock up (v): To confine in a brig (enlisted); to place under arrest in quarters (officer).
Locked up (adj): Confined or under arrest.
Main gate: Main entrance to a post, station, reservation, camp, or compound, at which a guard post is maintained.
Manual, the: *Marine Corps Manual.*
MarCorps: Abbreviated title for U. S. Marine Corps Headquarters, Washington, D.C.
Mast: Navy equivalent of Office Hours (q.v.); upright spar supporting signal yard and antennas in a naval ship.
MCM (n): *Marine Corps Manual* (not to be confused with *Manual for Courts-Martial*).
Messman: Nonrated enlisted man assigned to duty in the mess hall for a period of one month; aboard ship, called "mess cook."
Mess sergeant: Noncommissioned officer in charge of an enlisted mess.
MGC (n): Abbreviation for "Major General Commandant," official title of the Commandant of the Marine Corps from 1901 through 1942.
Mount-out (v): To load and embark for expeditionary service in amphibious shipping or transport aircraft.

Music: Contraction of "field music"; specifically, the field music assigned to the Guard of the Day.

NCO: Noncommissioned officer.

Nervous in the Service: Jittery, fearful, apprehensive, especially when in forward areas.

Nonrated (adj): Not of noncommissioned or petty officer rank; . . . *man*, private or seaman.

Number, to make: (1) To be promoted, when a vacancy occurs, to a higher grade for which previously selected; (2) (colloq) to pay one's respects to a senior.

OD (n): Officer of the day.

Office hours: Periodic, usually daily, occasion when the Commanding Officer receives requests, investigates offenses, reenlists and discharges enlisted men, and awards commendations.

Officers' Country: (1) Officers' living spaces on board ship; (2) any portion of a post or station allocated for the exclusive use of officers.

Old man: The commanding officer.

Old salt: (1) Old-timer, experienced Marine; (2) sardonically, person who thinks he knows all the answers.

Out-of-bounds: An area or space restricted from use by normal traffic, or prohibited to enlisted men, sometimes called "restricted area." Avoid "Off Limits," the equivalent Army/Air Force term.

Outside: Civilian life, sometimes colloquialized as, "Sergeant Boatspace is now serving on the USS *Outside*."

Overhead: Ceiling of a room (ashore) or compartment (aboard ship).

Paid off (adj): Discharged at the end of an enlistment.

PAL: Prisoner-at-large, formerly the legal term to define an enlisted man in disciplinary status confined to the limits of ship or station; slang term for individual undergoing punishment by restriction to limits.

Pass over (v): To omit an officer or staff NCO from a promotion list by promoting one junior to him in rank.

Passed over (adj): In the status of having failed of selection for next higher commissioned or staff NCO rank.

Pay office: Disbursing office.

Paymaster: Disbursing officer.

People: (1) Enlisted seamen or Marines; (2) one's subordinates, regardless of rank.

Pick up (v): To promote an officer who has previously been passed over (q.v.).

Picked-up (adj): In the status of having been selected for next higher rank after having been passed over one or more times.

Piece: (1) A Marine's rifle; (2) artillery piece.

Pipe (v): To notice, as, "Hey! Pipe that babe at the next table."
Pipe up: Speak up.
Platoon sergeant: Senior noncommissioned officer in a platoon, executive to the platoon leader.
Pogey-bait: Candy, snacks.
Pogey-rope: Fourragère.
Police: (1) (v) To straighten or tidy up an individual, area, or structure; (2) (n) condition of neatness or cleanliness.
Police gang: Permanent working force assigned to the police sergeant.
Police shed: Structure or space assigned to the police sergeant for stowage of tools, gear, and supplies; the police sergeant's workshop.
Police up (v): See "crum up."
Prisoner-chaser: See "chaser."
Property room: Storeroom for unit property, sometimes called "property shed."
PX (n:) Marine Corps Exchange, a store maintained within the organization for sale of articles necessary for the health, comfort, and morale of the command.
Qualify: To attain the minimum qualifying score in weapons proficiency, to attain the rating of marksman.
Quarters: (1) Government housing at a post or shore station, for officers and NCOs with authorized dependents; (2) periodic, usually daily semimilitary muster of a ship's company (Navy).
Rack: Bed, bunk; sometimes referred to as "sack."
Raider: Former member of one of the World War II raider battalions.
Rated man: Noncommissioned or petty officer.
Rating: Noncommissioned rank, sometimes shortened to "rate."
Read off: (1) To reprimand severely; (2) to publish the findings and sentence of a court-martial.
Reading, take a (v): To sound out.
Record day: The day on which a Marine fires an individual weapon for record of qualification.
Recruiter: Marine assigned to recruiting duty.
Regulation (adj): (1) Strictly in accordance with regulations or adopted specifications; (2) issued from Government sources (equivalent Army term, "GI").
RHIP: Colloquial abbreviation for the Service phrase "Rank hath its privileges."
Rock-happy (adj): Eccentric or mildly deranged as the result of long overseas duty at a remote station, usually an island; akin to "Asiatic" (q.v.), but without cosmopolitan connotations.

Rocks and Shoals: Punitive articles of the Uniform Code of Military Justice.
Ropeyarn Sunday: Weekday afternoon, usually Wednesday, when routine drills, instruction, training, and work are knocked off for organized or individual recreation.
Runner: Messenger, usually the field music.
Running Guard: Guard duty in which individuals have one tour on duty, one off, and then back on again with no intervening free period.
Rustbucket: Old, worn-out ship; Navy transport.
Saddle up (v): To put on packs and prepare to move out.
Salty: A seasoned Marine of any rank.
Sack: See "rack."
Scoop, the: Late news, information.
Scope out: To ascertain or verify a piece of information, as, "I'm not sure whether that's good or bad dope—you'd better scope it out."
Scuttlebutt: (1) Drinking fountain, or a container of drinking water; (2) unconfirmed rumor.
SDO (n): Officer detailed to supply duties only, formerly described as a "QM" (*obs*).
Seabag: Canvas duffle bag issued to each enlisted Marine for storage and transportation of uniforms and personal gear.
Seagoing: (1) **(n)** Sea duty; (2) **(adj)** pertaining to or assigned to sea duty; (3) **(n)** The uniform combination of blue trousers and khaki shirt.
Sea soldier: Marine.
Sea story: Yarn calculated to impress recruits or other gullible individuals.
Secure: (1) **(v)** To anchor firmly in place; (2) **(v)** to cease or terminate an activity or exercise; (3) **(n)** an outdated movement in the manual of arms.
782 Equipment: Individual combat equipment issued on memorandum receipt to Marine officers and enlisted men, so called because of the designation of the receipt-form employed.
Shanghai (v): To get rid of an individual by involuntary or surprise transfer.
Shift (v): To change uniforms, or from uniform into civilian clothing and vice versa.
Ship over (v): To reenlist.
Shipping-over music: Martial music, supposedly calculated to inspire Marines to ship over; loosely speaking, any inducement calculated to make a Marine desire to ship over.

Shook (adj): Dazed, groggy.

Shoot the breeze (v): To chat or conduct casual conversation.

Shooter: Marine whose avocation is marksmanship with the rifle or pistol; loosely, a Marine who has displayed special prowess with rifle or pistol, or who has served with distinction on a Marine Corps rifle or pistol team.

Short-fused: Very quick-tempered. Sometimes derivatively used in the nominative sense, as, "Gunnery Sergeant Piledriver sure has a short fuse."

Short-timer: One whose enlistment or current tour of duty is about to expire.

Shove off: To depart or leave, to get under way; an order to a boat to leave a landing or a ship's side.

Sick bay: Ship or unit aid station, dispensary, or infirmary.

Sick-bay soldier: (1) Individual who spends undue time in hospital or at sick call: (2) malingerer.

Sick call: Daily period when routine ailments are treated at the sick bay.

Sight in (v): In general, to aim a weapon at a target; loosely used as synonym for "zero."

Skipper: Commanding Officer.

Skivvies: Underwear.

Slopchute: Post exchange restaurant or beer garden (equivalent of "Geedunk" on board ship).

Slop down (v): To drink in quantity and rapidly, beer especially.

Slop up (v): To eat in quantity and rapidly, without regard to table manners; to gourmandize.

Small chow: Hors d'oeuvres.

Smoking lamp is lighted (out): Smoking is (is not) permitted (originally, a lamp on board old-time ships used by men to light their pipes).

Snap in (v): (1) To conduct sighting and aiming exercises with an unloaded weapon; (2) to try out for, or break in for, a new job.

Snow (v): To fool, bewilder, mislead, or exaggerate.

Snow job: Misleading or grossly exaggerated report or sales talk.

Spit and polish (n): (1) Extreme individual or collective military neatness; (2) extreme devotion to the minutiae of traditional military procedures ceremonies.

Spit-shine: (1) **(v)** To shine leather, employing spittle or tap water to remove excess grease and produce a high polish; (2) **(n)** an extremely high polish on a piece of leather.

Squadbay: Barrack room occupied by privates and junior NCOs.

Square away (v): To align, set in place, or correctly arrange an article, articles, or living space; when applied to individuals, to take in hand and direct.
Staff NCO: Noncommissioned officer above rank of sergeant.
Staff returns: Individual administrative records of a Marine; *Transfer by* . . . (1) **(v)** to effect a paper transfer of an enlisted man without changing his physical whereabouts; (2) **(n)** paper transfer.
Striker: (1) Apprentice or aspirant, attempting to learn a military specialty; (2) aboard ship, the Marine entrusted with the ordnance maintenance of a single gun, sometimes designated "gun-striker."
Swabbie: See "swab-jockey."
Swab-jockey: Sailor.
Survey: (1) **(n)** Medical discharge; examination by authorized competent personnel to determine whether a piece of gear, equipment, stores, or supplies should be discarded or retained; (2) **(v)** to effect discharge or retirement of an individual for medical reasons; to dispose of an item of Government property by reason of unserviceability; to obtain a second, third, or fourth helping of food.
Sympathy chit: Chit supposedly issued by those in authority, or by chaplains, authorizing an individual with many woes to obtain a prescribed amount of sympathy; expression used derisively to indicate lack of sympathy or concern over the plight of another.
Take off your pack: relax.
Thirty-year man: Marine who intends to make the Corps his career.
Top: First sergeant. Avoid the Army term, "Top-kick."
Train in: (1) To traverse a gun fore-and-aft; (2) to terminate a drill or exercise.
Troop and stomp: Morning troop inspection, followed by close-order drill.
Two-block (v): (1) To hoist a flag or pennant to the peak, truck, or yardarm; (2) to tighten and center a field scarf.
Under way, to get: To depart, or to start out for an objective.
Up the pole (adj): Abstaining from alcohol in any form.
Utilities: Green or camouflaged field and work uniform.
Walking John (n): Traditional nickname for the Marine Corps sergeant in blues (marching) who sometimes appears on recruiting posters (obs).
Watch (n): Official tour of duty of prescribed length, such as guard or officer of the day.
Wet down (v): To serve drinks in honor of one's promotion.
Wetting-down (n): Party in honor of a promotion.
Whirlybird: Helicopter.

White-blue-whites: Dress blue uniform C.
Whites: Marine Corps or Navy white uniforms; in the Marine Corps, worn only by officers.
Wing-wiper: Enlisted aviation Marine.
WMs (n): Women Marines.
Word, the: Late news, usually well verified and reliable.
Work one's bolt: To resort to special measures, either by energy or guile, to attain a particular end.
Work over (v): To reprimand severely; (2) to place heavy fire on a target or area.
Working over (n): (1) Severe reprimand; (2) heavy attack by fire.
Zapped (adj): Killed in action. Occasionally used as a verb, as, "Corporal Buttplate sure zapped that sniper."
Zero: (1) **(v)** To determine by trial and error the sightsetting required to obtain a hit with an individual weapon at a given range; synonymous with "zero in"; (2) **(n)** the sight-setting required to obtain a hit with a rifle at a given range.

TERMS AND USAGES TO BE AVOIDED

In recent years, as a side effect of unification, certain undesirable terms or expressions from outside the Naval Services have been picked up by a few individuals and used to the detriment of the authentic Marine Corps way of talking. Avoid especially the unfortunate usages which follow:

ZI: Use "Conus," or just "the United States."
GI: Use "squared-away" or "regulation." *Never* speak of an enlisted Marine as "a GI."
EM: Just say "enlisted man." Even better, say, "Marine."
O-Club: Speak of it as the "Officers' Club" or "Officers' Mess."
TDY: Army/Air Force term which now appears on many joint forms. Always use the Navy/Marine "TAD."
Trooper: Of Army airborne origin. Refer to an individual Marine as a Marine, never a "trooper." "Troops" as a plural is acceptable, but not "troopers." "People" is best.
Hitch: Use "cruise" or "enlistment." "Hitch" is an Army term dating from the horse cavalry.
Insignia (when you mean Emblem): Even though unified clothing procedures have designated the Marine Corps Emblem as "insignia, branch of service," this terminology should be absolutely shunned. The only acceptable word is "Emblem."

E-4 (and other similar ways of speaking of enlisted rank): Under no circumstances, refer to an enlisted man as "an E-3" or an E-6," etc. This is as bad as calling him a "member." Give people their correct ranks.

"Career": (as in "career officer"): Say "Regular."

Medic: Army/Air Force term for a hospital corpsman or "aid man" (also an Army/Air Force term). Always say, "Corpsman."

> NOTE: The Commandant has expressly forbidden the practice of suffixing the unnecessary word, "hours," after each indication of time of the day. This is another Army usage. Say or write, "1200," never "1200 *hours*."

Index

Accompanied tour, 349
Accountable officer, duties of, 502, 504
Act of 11 July 1708, "Establishing and Organizing a Marine Corps," 73–74, 78–79
Act of 39 June 1834, "For the Better Organization of the Marine Corps," 73, 79
Administrative Discharge Boards, 491
Advanced Base School, 129
Advanced bases, 127; defense of, 129
Advanced-Degree Program (ADP), 359–360
Aides-de-camp, 453–458; characteristics of, 454; duties of, field, 457–458; duties of, garrison, 456–457; White House, 652
Aiea Naval Hospital, 215
Air and Naval Gunfire Liaison Company (ANGLICO), 227
Air liaison parties, 139
Allotments, 325–327, 606
Allowances
  clothing, 311
  dislocation, 323
  per diem, 318
  personal money, 309
  quarters, 307–308
  trailer, 323
  uniform, 309–310
  weight, 322
American Red Cross, 199, 623

Amphibian tractor (LVT), 135
Amphibious development, 130, 134–136
Amphibious doctrines, 146
Amphibious operations, 149
Amphibious Warfare Presentation Branch, 101
Amphibious Warfare School, Quantico, Va. (*See also* Junior School), 100, 356
Anderson, Secretary of the Navy Robert B. (quoted), 83
"And St. David!" 165
Annapolis, Md. (*See also* Naval Academy), 178
Anthony, Pvt. William, 126
Appropriations Committees, House and Senate, 13
Argentine Marine Corps, 672–673
Arlington Annex, 85
Arlington National Cemetery, 617
Armed Forces Expeditionary Medal (*See also* Decorations and Medals), 180
Armed Forces Industrial College, Washington, D.C., 356
Armed Forces Police Detachment, 403
Armed Forces Reserve Act of 1952, 226
Armed Forces Staff College, Norfolk, Va., 19, 356
Armed Services Committees, House and Senate, 13, 74

Armored School, Fort Knox, Ky., 356
Army and Air Force Augmentation Act of 1956, 280
Army and Navy Club ("Town"), 267, 653
Army and Navy Country Club, 653
Army of the Republic of Vietnam (ARVN), 150
Army War College, Carlisle Barracks, Pa., 356
Arnold, Lt. Gen. H. H., 17
Arrest, 488–489
Articles for the Government of the Navy (See also Rocks and Shoals), 368
Asiatic Fleet, 126, 242
Assault helicopters, 146
Assignment and detail, 348
Assignment, Classification, and Travel Manual, 314
Assistant Commandant of the Marine Corps (ACMC), 87, 282
"At homes," 646
Atlantic Fleet, 127
Attributes of a Marine leader, 368–371
Authorized commissioned officer strength, 280
Automobile insurance, 601
Aviation Career Incentive Pay (*See also* Pay), 310
Awards, purpose of (*See also* Decorations and Medals), 378
Azores, 131

Bachelor quarters, bachelor officers' quarters, 650
Baldwin, Hanson (quoted), 141
Barbary pirates, 118
Barnett, Maj. Gen. George, 223
Barney, Commo Joshua, 120
Barrier Forts, 124
Basic School, The, Quantico, Va., 101–102, 257, 261–264, 289, 299, 329, 351–352, 356, 667; objective of, 264
Battle Color of the Marine Corps, 172
Battle colors, 171, 430–431
Battles:
    Aisne-Marne, 75
    Belleau Wood, 2, 75, 130, 162
    Blanc Mont Ridge, 75, 130
    Hatchee-Lustee, 122
    St. Mihiel, 75, 130
    Soissons, 130
    Trenton, 2
Bearss, Capt. Hiram I., 127
Belt, Sam Browne, trousers, 186, 188
Biddle, Maj. W. P., 128
Billet preference form, 350
Bladensburg, Md., 120
"Blue Book," (*See also Combined List of Officers on Active Duty in the Marine Corps*), 352, 455
Bluefields, Nicaragua, 148
Bluejacket's Manual, 526
Boards and councils, 193
Board of Awards, 180
Board of Investigation, 490
Board of Survey, 491
Boat cloak, 183
Book of Remembrance, 169
Boot camp (*See also* Marine Corps Recruit Depots, Parris Island and San Diego), 102–103
Borrowing money and loans, 605, 606
Bougainville, 138–139
Boxer, 148
Boxer Uprising, 127, 165
Bradley, Gen. Omar, 142
Brazilian Marine Corps (Corpo de Fusileiros Navais), 673
Breaking camp, 520–521
Bridgeport, Calif., 105
Brig guards (chasers), 393
Brig Manual, 401
British Imperial Defense College, London, England, 356
British Joint Services Staff College, Latimer, England, 356
Brown, John, 124, 160
Brown, Assistant Secretary of the Navy John Nicholas (quoted), 78, 83
Brown, Maj. Gen. Wilbur S. (Bigfoot) (quoted), 85
"Bunker Hill," 146
Burial arrangements, 614–615
Burial expenses, 615–616

Burke, Adm. Arleigh A., 260
Burrows, Maj. William Ward, 3, 118
Butler, Maj. Smedley D., 130; Maj. Gen., 153–154

Calling book, 647
Calling cards, 647–650; leaving cards, 649
Calling party, 647
"Call out the Marines!" 2
Calls (*See also* Social Obligations), 443–453, 535, 645; exchange of, 535; formal, 649; initial, 646; official, 336, 645; personal, 645
Campaign medals (*See also* Decorations and Medals), 180
Camp Barrett, Quantico, Va., 262–263
Camp Courtney, Okinawa, 217
Camp David, Md., 77
Camp Elmore, Norfolk, Va., 105
Camp Garcia, Vieques, Puerto Rico, 105
Camp Hansen, Okinawa, 217
Camp H. M. Smith, Oahu, Hawaii, 105, 215–216
Camp Lejeune, N.C., 103, 142–143, 329, 332
Camp McTureous, Okinawa, 217
Camp Pendleton, Calif., 142–143, 329, 525–526
Camp Rapidan, Va., 77
Camp Schwab, 217
Camp S. D. Butler, Okinawa, 105
Canton Bell, 162–163
Canton, China, 124
Canton Forts, 162
Captain's mast, 476
Career management, 347–365
Carmick, Maj. Daniel, 121
Casualty procedures, 622
Cates, Gen. Clifton B., 162, 244; quoted, 80
Cavite Navy Yard, 127
Central Intelligence Agency, 9, 11, 13
Central Marine Corps Recreation Fund, 505
Central Pacific campaigns, 137–138
Ceremonies, 458–465; pointers on, 462–465; types of, 458–460
Challenging and countersign, 397

Chapultepec Castle, 122
Checkage of individual pay (*See also* Pay), 503
Chesapeake, USS, 159
Chief of Staff, HQMC, 88, 282
China, 133
Chinese Communist Forces, 144
Chinese Marine Corps, 675
Choiseul, 138
Chosin Reservoir, 2, 144
Christenings, 644
Chulai, 149
Cigar Mess, 651
Ciudad Santo Domingo, 148
Civilian clothes, 187
Civilian Health and Medical Program, Uniformed Services (CHAMPUS), 195, 197
Civil law, 468–469
Civil war, 124–125, 160
Class E messages, 296
Classified matter, handling of, 297
Close air support, 139
Clothing inspections (*See also* Inspecting), 498–499
Clothing issues, 498
Clubs and messes (See also Commissioned Officers Mess), 236, 650–652
Coast Guard, 34; functions of, 35; mission of, 35
Coast Guard Academy, New London, Conn., 35
Cochrane, 2d Lt. H. C., 163
Cockburn, Adm. George, 119
Colclough, V.Adm. O. W. (quoted), 83
Cold weather hints, 519–520
College Degree Program (Operation Bootstrap), 359
Colombian Marine Corps, 673–674
Color sentinel, 396
Colors, flags, and standards, 170–174
Combined Action Platoons, 150
Combined Chiefs of Staff, 17
Combined List of Officers on Active Duty in the Marine Corps (See also "Blue Book"), 286
Command and General Staff College, Fort Leavenworth, Kan., 356
Command and Staff Action (FMFM 3–1), 111

Command and Staff College, Quantico, Va. (*See also* Senior School), 99–100
Commandant of the Marine Corps (CMC), 17, 43, 76, 86, 183, 272, 282; duties of, 87; license plate of, 165; member of the JCS, 74; responsibilities of, 45, 82, 87; status of, 81–82
Command in the field, taking, 339–340
Command presence, 368
Command responsibilities, 500–501
Commander of the guard, duties of, 400–401
Commissaries, 197–198
Commissioned Officers' Mess, 336
Commissions, commissioning, 259–260; from the ranks, 258; roads to commission, 251–259; table, 251–254
Committees on Government Operations, House and Senate, 13
Communication Officers School, Quantico, Va., 100, 356
Completed staffwork, 114
Computer Sciences School, Quantico, Va., 101
Concurrent reports (*See also* fitness reports), 275
Confidential File, 273
Consolidated mess system, 495
Con Thien, 150
Continental Congress, 117
Continental Navy, 118
Contingency Option Act (COA), 600
Control of property, 603–604
Corrections Manual, The, 488, 541
Corregidor, 137
Correspondence and messages, 291–296
Correspondence courses, 357
Correspondence jacket, 272
Counsel, duties of (*See also* Defense Counsel, Trial Counsel), 487–488
Court-martial punishments (Table), 486–487
Court members, duties of, 487
Court of Military Review, 468, 471
Court of Military Appeals, U.S., 468, 471

Courtesy to the flag, 429–431
Courts of inquiry (*See also* Investigations), 489–491
Creeks (Indian tribe), 122
Croix de Guerre (*See also* Decorations and Medals), 179
Cuba, 127–128, 130, 132
Cuban Pacifications, 127
Cunningham, Maj. A. A., 131

"Dagger Thrust" Operations, 149
Da Krong Valley, 150
Daly, GySgt. Daniel (quoted), 269
DaNang, 148, 150
Dances, 645
Daniels, Secretary of the Navy Josephus, 637
"Dead Horse" (*See also* Advance Pay), 309
Death and burial, 614–619
Death benefits, 619–621
Decorations, Medals, and Unit Citations, 153, 174, 237
  Miniature medals, 180
  Presentation of decorations, 377
  Wearing of, 180
Defense counsel (*See also* Counsel, duties of), 488
Delay in reporting, 299
Demilitarized Zone (DMZ), 149–150
Denby, Secretary of the Navy Edwin, 389
Dental care, 194
Department of Defense, 9, 14
  Armed Forces Policy Council, 16
  Assistant Secretaries of Defense, 16
  Defense Advanced Research Projects Agency, 19
  Defense Clothing and Textile Supply Agency, 104
  Defense Communications Agency, 19
  Defense Intelligence Agency, 19
  Defense Logistic Agency, 19, 104
  Defense Nuclear Agency, 19
  Defense Research and Engineering, 16
  Deputy Secretary of Defense, 16

Federal Civil Defense Program, 22
Joint Metereological Group, 18
Joint Strategic Target Planning Staff, 19
Military Staff Committee, 18
Department of State, 9
Department of the Air Force, 14, 27
   Air Command and Staff College, Maxwell AFB, Ala., 356
   Air Force Academy, Colorado Springs, Colo., 33, 179, 256
   Air Staff, 30; organization of, 30–31
   Air University, Maxwell AFB, Ala., 32
   Air War College, Maxwell AFB, Ala., 356
   Components of, 33
   Major air commands, 32
      Aerospace Defense Command, 32
      Air Force Logistics Command, 32
      Air Force Systems Command, 32
      Air Training Command, 32
      Fifth Air Force, 145
   Mission of, 27
   Officer candidates, education of, 33–34
   Organization of (figure), 31
   Primary responsibilities, 28
   Separate operating agencies, 32
   Structure of, 28–29
Department of the Army, 14, 20–27
   Basic branches, 21
   General staff, 22–24
   Major commands, 25
      Army Forces Command, 25
      Army Materiel Development and Readiness Command, 25
      Army Training and Doctrine Command, 25
      Eighth Army, 144
      IX Corps, 144–145
      2d Infantry Division, 75
   Mission of, 20–21
   Officer candidates, education of, 26
   Organization of (figure), 23
   Special branches, 21
   Special staff, 24
   Structure of, 21

Department of the Navy, 14, 34, 37–71
   And the Executive Branch, 39
   And the Judicial Branch, 40
   And the Legislative Branch, 39
   Assistant Chief of Naval Operations, Marine Aviation (OP-O5M), 89
   Assistant Secretaries of the Navy, 44
   Basic policy of, 37
   Bureau of Medicine and Surgery, 52
   Bureau of Naval Personnel, 51
   Chief of Naval Material, 46
   Chief of Naval Operations (CNO), 45, 47–50, 87, 109
   Civilian Executive Assistants, 43
   Components of, 60
   Composition of, 60
   Deputy Chiefs of Naval Operations (DNCO), 49–50
   Financial management, 44
   Functions of, 39–40 (figure)
   Installations and Logistics, 44
   Judge Advocate General of the Navy, 52, 272, 291, 468, 471
   Manpower and Reserve, 44
   Marine Corps Liaison Officer (Op-O9M), 48
   Mission of, 38
   Naval air bases, 55
   Naval Air Systems Command, 50, 104
   Naval districts, 53
   Naval Electronics Systems Command, 51
   Naval Establishment, 34, 78–79
   Naval Facilities Engineering Command, 51, 104
   Naval Material Command, 41, 50
   Naval operating bases, 55
   Naval Reserve, 67
   Naval Sea Systems Command, 50
   Naval staff, functions, and organization of, 68; Marine duties on, 70; organization, 67–71
   Naval Supply Systems Command, 51
   Office of Naval Research, 52

Operating forces, 56–60, 87, 91, 97
Organization of, 34, 40–41
Policy Council, 43
Professional assistants, 46
Research and Development, 44
Responsibilities of, 38
Shore establishment, 52–55; status of the Marine Corps in, 52
Under Secretary of the Navy, 43
Vice Chief of Naval Operations (VCNO), 48–49
Dependents (definitions), 307
Identification cards, 197, 302
Medical care, 195–197
Moving dependents overseas, 341–342
Transportation of, 320–325
Deputy Chiefs of Staff, HQMC, 88
Aviation, 109, 282
Installations and Logistics, 89
Manpower, 282
Plans and Operations, 88, 282
Requirements and Programs, 89
Research Development, and Studies, 89
Derna, Tripoli, 119, 158–159
Detachment commander, duties of, 536–537
Dewey, Commo. George, 127
Dining-In Night (*See also* Mess Night), 637
Diplomatic functions, 652
Director of Command, Control, Communications, and Computer Systems (C4), HQMC, 90
Director of Headquarters Support, HQMC, 89
Director of Information, HQMC, 90, 362
Director of Intelligence, HQMC, 89
Director, Judge Advocate Division, HQMC, 90
Director, Marine Corps Reserve, 90, 239
Director of Marine Corps History and Museums, 90
Director of Special Projects, HQMC, 89
Discipline, bases of, 376; characteristics of effective discipline, 376; defined, 375; military discipline, 375–377; object and nature of, 375
Distinguished Unit Emblem (*See also* Decorations and Medals), 179
Division of Reserve, HQMC, 227
Divine services, 168
Dominican Republic, 130, 147–148
Dominican Revolt, 244
Douglas-Mansfield Bill (*See also* Public Law 416), 73, 141
Drawing a will, 613–614
Drill and Ceremonies Manual, 268, 377, 432, 436, 449, 458–460
Drill instructors (DI's), 102
Dual Compensation Act of 1964, 626
Duty, preference for, 350
Duty, types of
Active duty for training, 234
Administrative duty, 234
Annual training duty, 234
Appropriate duty, 234
Duvalier, Francois, 147

Eberstadt, Ferdinand (quoted), 115
Edson Range, 213
Educational facilities, 199
Eisenhower, Gen. Dwight D., 17
Elliott, Maj. Gen. G. F., 128–129
Ellis, Lt. Col. Earl, 135
Embassy guards (*See also* State Department Guards), 134, 221
Emblem, Marine Corps, 155
Energy Research and Development Agency, 12–13
Engineer School, Fort Belvoir, Va., 356
Enlisted personnel, paying, 310–311
Epibatae, 97
Equivalent instruction, duty, 233–234
Erskine, Gen. Graves B. (quoted), 370
Esprit de corps, 3
Essex, USS, 119
Evening parade, Marine Barracks, 8th and Eye, 164–165
Ewa, Hawaii, 161
Exchange and commissary privileges, 237
Exchange council, 505

Exchange enlisted committee, 505
Exchange officer, duties of, 504
Executive Order 9635, 81
Extension School, Quantico, Va., 100–101, 234, 357

Farragut, Adm. David G. (quoted), 589
Feeding in the field, 515–517
"Fideli certa merces," 3
Field Artillery School, Fort Sill, Okla., 356
Field gear, care of, 521
Field musics, 395
Field sanitation, 513
Field service, useful articles for, 521–523
Fire insurance, 601
First Battle of Manassas, 125
"First to Fight," 159–160
Fiscal Director, HQMC, 90
Fitness reports, 273–278, 350; marginal reports, 276; marking of, 271–278; types of, 274; unsatisfactory reports, 276
Flag Marine detachment, 70
Flags, 153; appurtenances of, 171–172; display of, 438–443; foreign, 440
Fleet Marine Corps Reserve, 229–230, 238, 378
Fleet Marine Force, 82–83, 87, 91–95, 98, 127, 129, 135–137, 139, 151, 225; organization of, 92
Fleet Marine Force, Atlantic, 141, 202
Fleet Marine Force aviation, 95
Fleet Marine Force, Pacific, 141, 210, 215
Fleet Post Office (FPO), 295
Flight training, 357
Food services officer, duties of, 495
Foreign Marine Corps (*See also* Appendix V), 1
Foreign Service Act of 1946, 77
Foreign stations, 340–345
Fort Fisher, 125
"Fortitudine," 157
Fort Riviere, Haiti, 130
Fort Sumter, S.C., 124

France, 130
Fraser, Col. Angus M., 640
Frederick, Md., 120
French fourragere, 179
FTP-167, Landing Operations Doctrine, U.S. Navy, 135
Functions of the Department of Defense and its Major Components ("The Functions Paper"), 21, 28, 76–77, 83
Funeral escorts, 434
Funerals, general information on, 432

Gale, Lt. Col. Anthony, 122
Gamble, Capt. John M., 119
Garrison, reporting in, 330–337
Gendarmerie d'Haiti, 133
General courts-martial, 485–486
General mess, 494
General officers, 282
General (executive) staff, 111–112
Geneva Convention Card, 302
GI Bill of Rights, 359
Gillespie, 1st Lt. Archibald, 123
Globe and Anchor, 153
Glossary of Marine Corps terms, 681–693
Government Life Insurance, 593–595
Green, 1st Lt. Israel, 124
Greenwich, 119
Gridiron Club, 106
Grooming, 187–189
Guadalcanal, 137–139, 161
Guam, 138, 161
Guantanamo Bay, Cuba, 127, 147
Guardia Nacional, 133–134
Guards, guard duty
  Afloat, 538–542
  Commander of the guard, 394
  Duties of, 393–396
  Embarked units, 541–542
  Exterior guard, 392
  Guard mounting, 460
  Guard routine, 396
  Honor guards, 447
  Importance of, 391
  Interior guard, 392, 398
  Special guards, 392
  Supernumerary of the guard, 395
  Types of, 301

Guest Night (*See also* Mess Night), 637
Guidons, 171
Gun-salutes, 447

Hack, under hack, 488
Hagood, Maj. Gen. Johnson (quoted), 151
Haiti, 127, 130, 132–133
Halls of the Montezumas, 122
Hamblet, Maj. Julia E., 243
Hamhung, 144
Handbook For Retired Marines, 595, 626, 628
Harbord, Lt. Gen. James G., 162
Harper's Ferry, W. Va., 124, 160
Harris, John, 124
Headquarters, U.S. Marine Corps (HQMC), 85–91, 98, 100, 118, 227, 232, 236, 239, 242–243, 271, 276, 290, 342–343, 348–349, 351, 357, 617–620, 652, 667
Health records, 273–274, 335
Helicopter assault ship (LPH), 58
Henderson, Col. Archibald, 28, 87, 121–122; Brevet B. Gen, 124
Henderson Field, 137
Henderson Hall, Arlington, Va., 105
Heywood, Col. Charles, 126, 262; B. Gen, 198
Heywood Hall, 263
History and Museums Division, 274, 361
Holcomb, Lt. Gen. Thomas, 17, 136, 139, 225, 242
Honors, 443–453; in the field, 453; rendering of, 446; table of, 450–452
"Hook, The," 146
Household goods, 321; storage of, 322
Housekeeping, 493–507; afloat, 554–556
Hue, 150
Hughes, Chief Justice Charles Evans (quoted), 38
Huntington, Lt. Col. R. W., 127
Huntington's Fleet Marines, 160
Hwachon Reservoir, 145

Iceland, 75, 136, 160
Identification (ID) card, 197, 301
Identification tags (dog tags), 302
Identity devices, 301–303
Imjin River, 145–146
Inchon, 143
Income tax, 326
Individual administration, 271
Individual Clothing Regulations, 497
Individual equipment (*See also* 782 gear), 500, 502
Individual hygiene, in the field, 513–514
Individual Ready Reserve (*See also* Marine Corps Reserve), 230–231
Industrial College of the Armed Forces, 19
Infantry School, Fort Benning, Ga., 356
Infantry Training School (ITS), 525
Initial detail, 351
Inspecting, inspections, 380–384, 459, 662–667; follow-up of, 383; preparations for, 380; purposes of, 380; types of, 380
Inspector General of the Marine Corps (IG), 90, 384; IG inspections, 384
Inspector-Instructor (I&I), 232
Instructor Training School, 100
Insurance Death Claims, 598–600
Inter-American Defense College, 19
Interior Guard Manual, 392
Investigations, 490
Investments, 604–605
Involuntary retirement, 287
Involuntary separation, 287
Issuing orders, 373–374
Iwo Jima, 3, 138–139, 169

Jackson, Andrew, 121–122
James C. Breckinridge Memorial Library, 202
Japan, 124, 146
Jefferson, Thomas, 163
Johnson, Secretary of Defense Louis, 142
Johnson, President Lyndon B., 147
Johnson, Samuel (quoted), 269
Johnston Island, 161
Joint Chiefs of Staff (JCS), 13–14, 16–20, 45, 74, 76, 141, 282; func-

tions of, 18; Marine Corps representation in, 17; organization of, 18
Joint Secretariat, 18
Joint Staff, 18
Joint Travel Regulations, 313, 318–319
Joint Uniform Military Pay System (JUMPS), 306
Jones, John Paul (quoted), 631
Junior Marines, 537–538
Junior NROTC units, 66
Junior Officer of the Deck (JOOD), 543
Junior officers' mess, 528
Junior School (See also Amphibious Warfare School), 100
Jurisdiction, definition of, 469

Kennedy, President John F., 163
"Key West Agreement, The," 21, 39, 76
Khe Sanh, 2, 150
King, Flt. Adm. Ernest, 17
Korea, 74, 141, 146, 244, 635, 667
Korean War, 141–146

Leadership, 367–387; aspects of, 384; maxims of, 385–387; principles of, 367–368
Leahy, Adm. William D., 17
Leathernecks, 161
"Leatherneck Square," 150
Leave and Earnings Statement (LES), 300, 306
Leave and liberty, 297–301
  Accrued leave, 298
  Advanced leave, 298
  Annual leave, 298
  Computing leave, 299
  Earned leave, 298
  Emergency leave, 298
  Excess leave, 298
  Foreign leave, 300–301
  Graduation leave, 299
  Leave of absence, 297–299
  Leave requests and records, 300
  Sick leave, 298
  Unused leave, settlement for, 312
Lebanon, 161

Lee, Col. Robert E., 124
Legal Assistance Officer, 473
Legal Officer, 473
Legion of Merit (See also Decorations and Medals), 179
Legislative Assistant to the Commandant, 89
Lejeune, Maj. Gen. John A., 4–5, 7, 75, 101, 135, 368, 424, 635, 676; quoted, 364, 567
Length of service, for pay purposes, 306
Letter reports, 275
Liberty, (See also Leave), 298, 301
Life insurance, 592–596
Life in the field, 509–523
Limited assignment (overage) category, 231
Limited Duty Officers (LDOs), 258–259, 279, 281, 287; separate promotion zone for, 284
Living under canvas, 511–512
Longevity increases (fogies), 306

Mail guard, 134, 389
Maine, USS, 126
Maintenance officer, duties of, 193
Making camp, 509–511
Mameluke sword (See also Marine officer's sword), 119, 159
Manila Bay, 127
Manpower Department, HQMC, 88, 91
Manual for Courts-Martial, 1969 (MCM), 467–468
Manual of the Judge Advocate General of the Navy (JAG Manual), 467–469
Manual of the sword, 436–437
Marble Mountain, 107
Marine activities in the Hawaiian area, 215–217
Marine air-ground task forces, 95
Marine air-ground team, 106
Marine Air Reserve Training Command (MARTCOM), 110, 227
Marine Air Reserve Training Detachment (MARTD), 229
Marine amphibious brigade (MAB), 95

703

Marine amphibious force (MAF), 95–96
Marine amphibious unit (MAU), 56, 95
Marine aviation, 89, 106–115, 131, 139; aviation staff functions, 113–114; aviation supply, 104; aviation supporting establishment, 110; organization of, 108–110; role of, 106
Marine Band, The, 77, 105, 118, 163, 165, 200
Marine barracks, 97
    Eighth and Eye Streets, Washington, D.C., 103, 105, 120, 166, 172, 199–201, 262, 637
    Guantanamo Bay, Cuba, 220
    Naval Base, Hampton Roads, 202
    Norfolk Naval Shipyard (Portsmouth), 202, 526
    Pearl Harbor, 215
Marine battalion landing teams (BLTs), 56
Marine Corps, 5, 41
    And the Department of the Navy, 80–83
    Attitude, 2
    Force in readiness, 78
    Functions of, 39, 76
    History of, 117–161
    Operating forces, 85, 91–98
    Organization of, 38, 85–115
    Professionalism, 4
    Roles and missions, 78–83
    Status of, 78–80, 83, 141
    Strength of, 136, 139
    Supporting establishment, 85, 98
    What it is, 1
    What it stands for, 2, 5
Marine Corps Air Bases Command, Cherry Point, N.C., 55
Marine Corps Air Bases, Eastern Area, 110
Marine Corps Air Bases, Western Area, 110
Marine Corps Air Facility, Quantico, Va., 55
Marine Corps Air/Ground Combat Training Center, Twentynine Palms, Calif., 210
Marine Corps Air Stations, 104
    Beaufort, S.C., 105, 207–208
    Cherry Point, N.C., 105, 110, 142, 203–204
    El Toro, Calif., 105, 110, 142, 214–215
    Futenma, Okinawa, 105, 217
    Iwakuni, Japan, 105, 218
    Kaneohe Bay, Hawaii, 105, 215
    New River, N.C., 105
    Quantico, Va., 105
    Santa Ana, Calif., 105
    Yuma, Ariz., 105, 209
Marine Corps Association, 237, 267, 593
Marine Corps Association Group Benefit Program, 595
Marine Corps bases, 104
    Camp Lejeune, N.C., 105, 204–206
    Camp Pendleton, Calif., 105, 213–214
    Camp Smedley D. Butler, Okinawa, 217
    Twentynine Palms, Calif., 105, 210–211
Marine Corps Birthday, 158–159, 460, 634–637
Marine Corps Birthday Ball, 183, 636–637; ceremony, 677–679
Marine Corps bulldog, 166
Marine Corps Casualty Procedures Manual, 614
Marine Corps Civil Readjustment Manual, 624
Marine Corps colors, 157, 171
Marine Corps Command and Staff College (*See also* Senior School), 356
Marine Corps Development and Education Command (MCDEC), 99, 105, 201–202, 229, 355, 378; mission of, 99
Marine Corps Development Center, 101
Marine Corps Districts, 105, 212
Marine Corps Drum and Bugle Corps, 200
Marine Corps emblem, 155
Marine Corps Equipment Board, 135–136
Marine Corps Exchange Fund, 198

Marine Corps Exchange Manual, 199, 505
Marine Corps Exchanges (PXs), 198–199, 504–505; missions of, 198
Marine Corps Expeditionary Medal (*See also* Decorations and Medals), 180
Marine Corps Finance Center, Kansas City, Mo., 306, 312
Marine Corps Flag Manual, 171
Marine Corps Gazette, 237, 251, 361–362
Marine Corps Historical Center, 169
Marine Corps Institute (MCI), 200, 232, 234, 357–358, 378
Marine Corps Logistic Support Base, Atlantic, Albany, Ga., 103–114
Marine Corps Logistic Support Base, Pacific, Barstow, Calif., 104, 211–212
Marine Corps Manual, 103, 249, 268, 291, 337, 468, 494, 496, 635, 676
Marine Corps Memorial Chapel, Quantico, Va., 169
Marine Corps Motto, 157–158
Marine Corps Mountain Warfare Training Center (Pickel Meadows), Bridgeport, Calif., 105, 520
Marine Corps Museums (*See also* Marine Corps Historical Center), 169–170
Marine Corps Personnel Manual, 676
Marine Corps posts and stations, 199, 218–221
Marine Corps promotion system (*See also* Selection Boards), 278–279
Marine Corps Recruit Depots, 102
Parris Island, S.C., 206–207
San Diego, Calif., 212–213
Marine Corps Reserve (*See also* Organized Reserve, Standby Reserve), 146, 223–239, 257, 259
Composition of, 229–231
History, 223–226
Organization, 226–231
Reserve retirement, 238–239
Units, 227–229
Marine Corps Reserve Ribbon (*See also* Decorations and Medals), 237
Marine Corps Retirement Guide, 627

Marine Corps Schools (*See also* Marine Corps Development and Education Center), 135–136
Extension courses, 232
Marine Corps Seal, 156–157
Marine Corps Special Services Manual, 505
Marine Corps Staff Noncommissioned Officers Academy, Quantico, Va., 101
Marine Corps stock lists, 103
Marine Corps Subsistence Manual, 494, 496
Marine Corps Supply Manual, 494, 496, 501, 503
Marine Corps supply system, 103
Marine Corps Uniform Regulations, 180–182, 189, 237, 249, 267–268, 445, 497, 499
Marine detachments, 70, 535–538
Marine division, composition of, 93
Marine Gunner, 279
Marine Junior ROTC program, 248
Marine Museum, Treasure Island, San Francisco, Calif., 169
Marine officer's sword (*See also* Mameluke sword), 435–436
Marine regimental landing team (RLT), 95
Marine regiments, composition of, 96–97
Marine reservists (*See also* Marine Corps Reserve)
Perquisites, privileges of, 235–238
Marine security forces, 98
Marine shore activities, 97
Marines' Hymn, 158, 660
Marines in the Revolution, 117
Marinettes (*See also* Women Marines), 241
Marine units
Air
Aircraft, Fleet Marine Forces, composition of, 109
1st Marine Aircraft Wing, 74, 144, 148, 150, 217, 635
4th Marine Aircraft Wing, 226–227, 232
Marine Aircraft Group 16 (MAG-16), 107

Marine units (*cont.*)
- Marine Aircraft Group 36 (MAG-36), 149
- Marine Fighting Squadron 211 (VMF-211), 107, 179
- 1st Aeronautical Company, 131
- Ground
  - III Marine Amphibious Force (III MAF), 149, 217
  - IV Marine Amphibious Force (IV MAF), 227
  - 1st Marine Division, 74, 137, 142, 144–146, 149, 635, 674
  - 2nd Marine Division, 96, 143
  - 3d Marine Division, 96, 148, 150, 161, 217
  - 4th Marine Division, 96, 226–227, 232
  - 5th Marine Division, 149–150
  - 1st Marine Brigade, 132, 143, 215
  - 1st Provisional Marine Brigade, 75, 136, 142, 161
  - 4th Marine Brigade, 75, 130–132, 162, 179
  - 5th Marine Brigade, 132
  - 1st Marines, 96
  - 2nd Marines, 96
  - 3d Marines, 96
  - 4th Marines, 96, 134
  - 5th Marines, 96, 130, 150, 160, 179
  - 6th Marines, 96, 130, 179
  - 7th Marines, 96
  - 8th Marines, 96, 132
  - 9th Marines, 96, 132, 150
  - 10th Marines, 96
  - 11th Marines, 96
  - 12th Marines, 96
  - 14th Marines, 96
  - 23d Marines, 96
  - 24th Marines, 96
  - 25th Marines, 96
  - 28th Marines, 169
  - 1st Defense Battalion, 179
  - 3d Battalion, 6th Marines, 148
  - 6th Machine Gun Battalion, 4th Brigade, 130
- Miscellaneous

Advanced Base Force, 129–132, 136
East Coast Expeditionary Force, 136
Marksmanship badges, 181
Marquesas Islands, 119
Marshall, Gen. George C., 17
Marshall Islands, 138–139
Medal of Honor (*See also* Decorations and Medals), 126–127, 130, 133, 178
Medical care, 194
Meritorious Enlisted Marines, 258
Mess entrance fee (mess share), 528
Mess etiquette, 651–652
Mess management, 494–496, 515; afloat, 554–555
Mess Night, 378, 533, 637–642
Mess officer, duties of, 495
Mess sergeant, 495
Mexico City, 122
Midway Island, 137, 159, 161
Military Academy (*See also* West Point), 25–26, 256
Military Affairs, 361
Military courtesy, 407–417; definition of, 407
Military discipline (*See also* discipline), 375–377
Military etiquette, 413–417
Military funerals (*See also* Funerals), 432–435, 459
Military judge, duties of, 471
Military justice, described, 467–491
Military law, 468–469; sources, of, 468
Military magistrate, duties of, 471
Military Occupational Specialty (MOS), 352–354
Military Pay and Allowances Entitlements Manual, 310
Military police (MPs), 402; and shore patrol, 402–405
Military Secretary, HQMC, 89
Military titles, 408–413
Military weddings, 642–644
Milites classiarii, 97, 525
Monitors, 350
Monterey, Calif., 123

Morning and evening colors, 440–442
Motor transport officer, duties of, 506
Munsan-ni, Korea, 145

MacArthur, General of the Army Douglas, 142; quoted, 7, 143
McCawley, B. Gen. Charles G., 126

Napoleon Bonaparte (quoted), 360
NATO, 19
NATO Defense College, 356
National Aeronautics and Space Administration (NASA), 12
National Anthem, 428–429
National Colors (standard), 170
National Defense University, 19, 355–356
National Ensign, 170
National Guard, 25
"National Intelligence Estimates," 11
National objectives, 13
National policy, 13
National Security Act of 1947, 15, 17, 20, 27, 38–39, 73, 76–77, 79, 83, 140; as amended, 17, 73
National Security Agency (NSA), 19
National Security Council, 9, 11, 13, 16, 18, 36; duties of, 11; functions of, 11
National security, organization for (1977), 10
Naval Academy (See also Annapolis), appointment to, 65; Marine commission from, 255
Naval Academy Athletic Association, 654
Naval Air Station, Barber's Point, Wahiawa, Hawaii, 215
Naval Air Station, New Orleans, La., 227
Naval Air Training Center, Pensacola, Fla., 357
Naval Ammunition Depot, Lualualei, Hawaii, 216
Naval Aviation Cadet Act of 1935, 224
Naval courts-martial, 483–489
Naval Medical Center, Bethesda, Md., 201

Naval Officer's Guide, The, 554
Naval Regional Medical Centers, 194
Naval Reserve Act of 1925, 223
Naval Reserve Act of 1938, 226
Naval Reserve Officers' Training Corps (NROTC), 65, 101, 256–257, 264
Naval service, 81
Naval War College, Newport, R.I., 356–358, 361
Naval War with France, 118
Navy and Marine Corps Awards Manual, 174, 180
Navy Corrections Manual, 399, 526
Navy Correspondence Manual, 291, 293
Navy Cross (See also Decorations and Medals), 179
Navy Enlisted Scientific Program (NESEP), 258, 264
Navy Federal Credit Union, 605
Navy Mutual Aid Association, 199, 593, 595–596, 599, 617, 623, 628
Navy property accounting procedures, 556–557
Navy ratings, abbreviations, 586–589
Navy Regulations, 79–80, 82–83, 171, 183, 268, 276, 291, 297, 368, 405, 418, 432, 443–445, 449, 468, 539–540, 542, 546, 550, 553–555
Navy Relief Society, 199, 620, 623
Navy Security Manual for Classified Matter, 297
Navy Shore Patrol Manual, 403
Navy Times, 361
Navy Travel Instructions, 313
Navy Unit Commendation (See also Decorations and Medals), 179
Nelson, Lord Horatio (quoted), 265, 405
New Britain, 138
New Georgia, 138
New Hall, Philadelphia, Pa., 169
New officers, hints for, 264
New Orleans, 119–120
New Providence Island, 117
Nicaragua, 124, 127, 130, 133–134, 148
Nicholas, Capt. Samuel, 117

Nimitz, Fleet Admiral Chester, 3, 140
Nonjudicial punishment, 476–482; appeal from, 480–481; limits of, 481
Norfolk, Va., 262
North China, 134
North Korean Army, 141, 143
North Vietnamese Army (NVA), 149

O'Bannon, 1st Lt. Presley Neville, 119, 158–159
Occupation duty in Germany, 132
Off-Duty Education Program, 359
Offenses, 473–474
Office hours, 377–380, 476, 488; procedures of, 477–480
Office of Management and Budget (OMB), 9, 12, 15; functions of, 12
Officer Assignment Branch (Detail Branch), HQMC, 348, 350–351
Officer Basic Course (*See also* Basic School), 262
Officer Candidate Course (OCC), 258
Officer candidates, education of, 631
Officer Candidates School, Quantico, Va., 66, 101, 247, 264
Officer categories, 279–282
Officer distribution, 279–280
Officer Grade Limitation Act of 1954, 281
Officer of the Day (OD), 394, 397–398; duties of, 398; relief of, 399
Officer of the Deck (OOD), 542–549; hints for, 543–549; responsibilities of, 542; underway, 549
Officer Personnel Act of 1947, 279–280, 286
Officers' clubs and messes (*See also* Clubs and Messes), 533, 650
Officer's Qualification Record, 273
Officers' records, 271–278
Official calls and visits (*See also* Calls), 443–446, 553–554
Official correspondence, 291–292; letters, 293–295
Official photograph, 302
Official record, 271
"Oil stain" tactics, Vietnam, 149
Okinawa, 138–139, 146, 149, 329

Operation "Dewey Canyon," 150
Orderlies, duties of, 395
Order of the Carabao, 106
Orders, types of, 314
  Blanket or repeat travel, 314
  Original, 333
Organization and Functions of the Joint Chiefs of Staff (JCS Publication 4), 20
Organization Color, 171, 430–431
Organized Marine Corps Reserve Medal (*See also* Decorations and Medals), 237
Organized Reserve (*See also* Marine Corps Reserve), 143, 225, 227, 230, 232–234, 244, 248; aviation units, 228–229; ground units, 228, 280; training of, 232–235
Outposts "Carson," "Reno," and "Vegas," 146
Overseas training missions, 221
Overseas travel, 340–345

Pacific Squadron, 123
Panama, 124
Panmunjom, 145
Parris Island, S.C. (*See also* Marine Corps Recruit Depots), 102, 169, 235, 262
Pass-overs for promotion, 284
Pate, Gen. Randolph McC., 162
Patton, Gen. George S. (quoted), 373
Pay (*See also* Allowances), 236, 500
  Advance pay (*See also* "Dead Horse"), 309
  Combat pay, 327
  Flight pay, 310
  Hazardous duty pay, 308
  Incentive pay, 308
  On promotion, 308
  Sea or foreign duty pay, 311
  Severance pay, 311–312
  System, pay, 306
Pearl Harbor, Hawaii, 161
Peking, 127–128, 134, 154, 165
Peleliu, 138–139
Penobscot Bay Expedition, 118
Permanent change of station, 314, 317

Personal affairs, handling of, 591–629
Personal appearance, grooming, 181–189, 266
Personal correspondence (*See also* Correspondence and Messages), 295–296
Personal decorations (*See also* Decorations and Medals), 174–177
Personal effects, 340
Personal file, 274
Personal Finance Record (PFR), 273, 306
Personal flags and pennants, 442–443; afloat, 552–553
Personal liability insurance, 601
Personal property insurance, 601
Personal radio traffic, 296
Personal security, 296
Philadelphia Navy Yard, 262
Philippine Insurrection, 127
Phnom Penh, 151
Physical Evaluation Board, 290–291
Physical retirement, 290
Place of burial (*See also* Funerals, Military Funerals), 616–618
Platoon Leaders Class (PLC), 101, 224, 257, 264; PLC (Law) Program, 258
Platt Amendment, 128
Police and maintenance, 493–494
Police Sergeant, 493
Port-au-Prince, Haiti, 132
Porter, Capt. David, 119, 127
Port Royal, S.C., 262
Posts and stations, 191–221; organization of, 192–199
Power of attorney, 324, 609–610
Precedence (*See also* Promotion), 286
Precedence of forces, 460
Presidential Unit Citation (*See also* Decorations and Medals), 179
"President's Own, The," (*See also* Marine Band), 77, 163–164
"Proceed" time, 314
Professional competence, 370
Professional examinations, 273
Professional reading, 360–361, 667–678

Professional schooling, 354–360
Profession of Arms, 264–265
Promotion and precedence (*See also* Precedence), 278–286
Promotions, 377; effecting, 285; opportunities, 237; procedure, 282–285; zone, 283
Property, 501; expenditure of, 503
Provost guard, 403
Provost marshal, 193
Public information, 364
Public Law 416 (*See also* Douglas-Mansfield Bill), 73, 75–76, 79–83
Public Law 432, 79–80, 83
Public Law 729, 287
Public Law 810, 238
Public relations, 364
Public speaking, 362–364
Public Works Officer, duties of, 193
Puller, Lt. Gen. Lewis B. (Chesty), 166; quoted, 189
Punchbowl, The, 145
Punitive investigations (*See also* Investigations), 474
Pusan Perimeter, 141–143

Quantico, Va., 99, 103, 107, 135, 139, 146, 169, 247, 257, 261–262, 329, 355–356
Quarterdeck, the, 532
Quatrefoil, 161
Que Son, 150
Quick, Sgt. John H., 127
Quilali, 133

Ready Reserve, (*See also* Marine Corps Reserve, Organized Reserve), 231, 235
Real estate, 602–603
Record of Emergency Data, 627
Records, access to, 273
Recreation, 199
Recreation Councils, 506
Recreation Fund, 505
Reenlistment bonus, 311
Reference Guide to Employment Activities of Retired Personnel, 626
Registered Publications Manual, 297
Regular Military Compensation (RMC) (*See also* Pay), 305–306

Reimbursement, 317–318; reimbursable expenses, 318–320
Relations between officers and men, 4
Remote storage activities (RSA), 103
Reporting in the field, 337–340
Reporting seniors, 275
Reprimand, 378
Republic of Korea, 142
Republic of Korea Marine Corps (ROKMC), 674–675
  1st KMC Regiment, 674
Request mast, 380
Requirements to become an officer, 251
Reserve Forces Act of 1955, 225, 235
Reserve officers (*See also* Marine Corps Reserve), 279, 282
Reserve Officers' Training Corps (ROTC), 27
Resident schools, 101, 356
Resignation, 286
Responsible Officer, 502; duties of, 502–503
Retired officers' benefits, 623–627
Retired pay (*See also* Pay), 311–312; accounts, 312; computation of, 312
Retired Reserve (*See also* Marine Corps Reserve), 231
Retired Servicemen's Family Protection Plan, 600
Retirement and separation, 286–291; for age, 287; requirements for, 287–289
Review and appeals, 471
Revocation of commission, 287, 289
R.H.I.P. (Rank Has Its Privileges), 375
Rhodes Scholars, 355
Rights of the accused, 469
"Rocket Belt," 150
"Rockpile, The," 150
Rocks and Shoals (*See also* Articles for the Government of the Navy), 469
Rommel, Field Marshal Erwin (quoted), 371
Roosevelt, President Franklin D., 17, 75, 162, 179
Roosevelt, Theodore, 128
Ross, Maj. Gen. Robert, 119
Royal Marines, 145, 156–157, 162, 669–670

Royal Netherlands Marines (Korps Mariniers), 671
Royal Thai Marine Corps, 675
Royal Welch Fusiliers (23rd Foot), 165
Ruffles and flourishes, 427
Running mate (*See also* Promotion), 287
Russell, Maj. Gen. John H., 92, 101, 132, 136

Saigon, 149, 151
Saipan, 138
Salee River, forts, 125–126
Salutes, saluting, 417–432
  Group saluting etiquette, 426–428
  Hand salutes, 419–421
  Individual saluting etiquette, 423–426
  Military funerals, saluting at, 427
  Origins of, 418
  Pointers on, 431–432
  Rifle salutes, 421–422
  Saluting distance, 419
  Sword salutes, 422–423
Saluting battery, 447
Samar, 127
San Diego, Calif. (*See also* Marine Corps Recruit Depots), 102, 235, 526
Sanitation, camp and unit, 514–515
Santo Domingo, 127, 130, 132, 147
Schilt, 1st Lt. C. F., 133
School of Application, 262
Schools, types and levels of, 354–355
Scott, Gen. Winfield, 122
Sea duty, indoctrination, 525–527; reporting aboard, 525–529
Sea Duty Supply Officers, 555
Seagoing Marine officers, hints for, 557–559
Seagoing Marines, 97; glossary for, 559–586
Sea School, 526
Secretary of Defense, 13–18, 21
Secretary of State, 13
Secretary of the Air Force, 30
Secretary of the Army, 21
Secretary of the General Staff, HQMC, 89
Secretary of the Navy (SecNav), 41, 283

Security classifications, 297
Security of classified material, 297
Security of information, 296–297
Selected Reserve (*See also* Marine Corps Reserve), 230–231
Selection, eligibility for, 283; mechanics of, 283–285
Selection Board Jacket, 272
Selection boards, 278, 284–285, 289; duties and obligations of, 285
Selective Service System, 12
Seminoles, 122
"Semper Fidelis," 2, 157–158
Seniors, addressing, 408
Senior School (*See also* Command and Staff College), 99
Sentinels, duties of, 102, 394; general orders for, 394; guardhouse, 395; main gate, 395
Seoul, 143, 145
Separation or discharge, definition of, 286
Sergeant Major of the Marine Corps, 89
Sergeant of the Guard, 394
Seringapatam, 119
Service afloat (*See also* Sea Duty), 525–589
Service author, the, 361–362
Service component commanders, 19
Servicemen's Group Life Insurance (SGLI), 237, 594
782 equipment (*See also* Individual Equipment), 339, 500
1775 Rum Punch, recipe for, 641
Seventh Fleet, 149
Shaking down, 336
Shanghai, 134
Shepherd, Capt. Lemuel C., Jr., 526; Gen., 76, 156, 168; quoted, 38, 239
Shipboard courtesy and etiquette, 549–554
Shipboard guard, 538–539
Shipboard life, 527–535
Shipboard posts maintained by Marines, 540–541
Ship's bell, 166
Ships' Detachment Supply Officers (SDSOs), 526
Ship's guards, 392
Ship's organization, 529–531
Ship's regulations, 531–532

Shore patrol, permanent, 403
Sicily, 142
Signal School, Fort Monmouth, N.J., 356
Silver Star Medal, (*See also* Decorations and Medals), 179
"Situation is well in hand, The," 2
Slates, officers', 350
Social customs, life, Marine Corps, 631–651, 654–655
Social obligations (*See also* Calls), 645–650
Social Security, 597–598
Social Security Tax, 326–327
Soissons, 130
Soldiers and Sailors Civil Relief Act, 604, 607, 609
Sources of supply, 104
Sousa, John Philip, 158, 164
South Pacific campaigns, 137
Space available travel, 625–626
Spanish Marines (Infanteria de Marine Espanola), 671–672
Special Courts-Martial, 484–485
Special Education Program (SEP), 359
Special Enlisted Reserve Program (*See also* Marine Corps Reserve), 235
Special fitness reports (*See also* Fitness reports), 275
Special services, 505–506
Special Services Officer, 505
"Special Trust and Confidence," 261
Specified Commands, 19–20
   Aerospace Defense Command (ADCOM), 20
   Military Airlift Command, 20
   Strategic Air Command, 20, 32
Staff groups, 229
Staff Judge Advocate, 471
Staff Organization and Procedure (Field Manual 101-5), 111
Staff organization, procedures, and relationships, 110–114; personal staff, 113; special staff, 112–113; staff supervision, 114
Staff returns, 273, 333
Standard vehicle stickers, 303
Standard Written Agreement, 234
Standby Reserve (*See also* Marine Corps Reserve), 231

State Department guards (*See also* Embassy Guards), 77–78
Status-of-forces agreements, 469
Streamers (*See also* Decorations and Medals, Flags), 172–174
Streeter, Col. Ruth Cheney, 243
Subic Bay, 127
Subordinates, dealing with, 371–373
Subsistence (*See also* Pay), 307, 494–496
Suez Incident 1956, 147
Sullivan, Secretary of the Navy John L., 80
Summary Courts-Martial, 483–484
Suribachi Yama, 169
Survivor Benefit Plan (SBP), 600–601
Survivor Benefits and Assistance, 592, 619–623
Swagger sticks, 166

Tables of Allowance (T/A), 93, 502
Tables of Equipment (T/E), 93, 502
Tables of Organization (T/O), 93
Tactical air control parties (TACP), 110
Tarawa, 138–139
Tarawa, USS, 108
Tartar City, Peking, 128
Tartar Wall, 154
"Task Force Principle, The," 57
Taxes, 607–609
Tax exemptions, 327
"Tell it to the Marines!" 2, 167
Temporary additional duty (TAD), 308, 314, 317
Temporary Disability Retired List, 290–291
Temporary duty (TD), 314, 317
Temporary officer, 259
Tentative Landing Operations Manual, 135
Tet Offensive 1968, 150
38th Parallel, 143
Thomason, Capt. John W., Jr., 167
Tientsin, 127–128, 134, 165
Tinian, 138
Title 10 (Armed Forces), U.S. Code, 79, 81
Total commissioned service, 287
Tour of duty, 348–349

Towle, Col. Katherine A., 244; quoted, 249
Traditions and customs, Marine Corps, 4, 153–169
Transit insurance, 323–324
Transportation, 506–507
Travel advance (See also "Dead Horse"), 319
Travel expense and mileage, 312–313
Travel orders, 313–314
Travel reimbursement, 319
Travel time, figuring, 316–317
Trial counsel (*See also* Counsel, duties of), 487
Tropical service, hints for, 517–519
Truman, President Harry S., 17
Tulagi, 137
Tun Tavern, 117
Twiggs, Maj. Levi, 123

Uijongbu, 143
Unification, 15, 79
Unification Act (*See also* National Security Act of 1947), 74
Unified Action Armed Forces (UNAAF) (JCS Publication 2), 20, 77, 83
Unified Commands, Atlantic; Continental Air Defense; Pacific; Readiness; U.S. European; U.S. Southern, 19
Uniform Code of Military Justice (UCMJ), 35, 465, 468–473; punitive articles, 473–474
Uniforms, 153, 161–162, 182–186, 236, 262
  Accessories, 186
  Belts, Sam Browne, trousers, 186, 188
  Boat cloak, 183
  Collar emblems, 162
  Dress uniform, 183
  Evening dress, 183
  Full service uniform, 182
  Hat, field, 162
  Insignia, 181–189
  Inspections, uniform for, 465
  Mess dress, 183
  Mixed uniform, 182
  Scarlet trouser stripes, 161

Supervision of, 499–500
Undress uniform, 182
Unit decorations (*See also* Decorations and Medals), 174–177, 179
Unit mess officer, duties of, 496
Unit Punishment Book (UPB), 477
United Nations, 18–19
U.S. Naval Home, 624
U.S. Naval Institute, 237
United States Naval Institute *Proceedings*, 237, 361–362
Universal Military Training and Service Act, 237
Unlimited duty, 234

Vandegrift, Gen. Alexander A., 80, 139
Van Tuong Peninsula, 149
Venezuelan Marine Corps (Infanteria de Marina Venezolana), 674
Veracruz, 122, 127, 129–130, 160
Veterans Administration (VA), 12, 624
Veterans' Group Life Insurance (VGLI), 594
Veterans' privileges, 624–625
Vietcong, 149
Viet Minh, 149
Vietnam, 107, 146, 161
Vietnamese Marine Corps, 148
Vietnam War, 4, 148–151
Vinson, Representative Carl (quoted), 78
Voluntary retirement, 289–290
Volunteer Marine Corps Reserve (Class III) (*See also* Marine Corps Reserve), 230
Volunteer Reserve training (*See also* Marine Corps Reserve), 233
Volunteer Training Units (*See also* Marine Corps Reserve), 227, 229, 248

Wake Island, defense of, 107, 137, 139, 161, 179
Waller, Maj. L.W.T., 127–128; Col., 130
Wardroom, the, 532–533, 650; etiquette, 533–534
Warm Springs, Ga., 77
Warnings, 475
War of 1812, 119–121, 159
Warrant officers, 259
Warrant Officers' Basic Course, 262
War with Mexico, 122
War with Spain, 126–127
Washington, D.C., duty in, 652–654
Washington, George (quoted), 71
Watch Officer's Guide, The, 443, 526, 549, 554
Watchstanding, 389–405
Wavell, Field Marshal (quoted), 189
WAVES, 242
Ways and Means Committees, House and Senate, 13
Weapons proficiency, 384
West Point, N.Y. (*See also* Military Academy), 26, 179
Wetting-down party, 654
Wharton, Lt. Col. Franklin, 118
White House duties, Marine Corps, 77
White House Fellows Program, 359
White House functions, 652
White Papers, 13
Wills, 610–614
Women in the Air Force, 33
Women in the Army, 24, 26
Women Marines, 90, 206, 241–249
  Assignments of, 245–246
  Duties of, 245–246
  History of, 241–245
  Training, 246–247
Women Officer Candidate Course (WOCC), 247, 259
Women officer promotion, 285–286
Women Reserve Schools, Camp Lejeune, N.C., 242
Women Reservists (*See also* Women Marines), 231, 243, 247–248
Women's Armed Services Integration Act of 1948, 241, 243, 286
Wonsan, Wonsan Harbor, 143, 145
World War I, 130–131, 160
World War II, 136–140

Yalu River, 144

Zeilin, Maj. Jacob, 124, 126; B. Gen., 155–156
Zukeran, 217